Springer Texts in Statistics

Advisors:

Stephen Fienberg   Ingram Olkin

Springer Texts in Statistics

Continued at end of book

J.D. Jobson

Applied Multivariate Data Analysis

Volume I: Regression and Experimental Design

With 93 Illustrations in 116 parts

With a diskette

Springer-Verlag
New York Berlin Heidelberg London Paris
Tokyo Hong Kong Barcelona Budapest

J.D. Jobson
Faculty of Business
University of Alberta
Edmonton, Alberta T6G 2R6
Canada

Mathematics Subject Classification: 62-07, 62J05, 62J10, 62K10, 62K15

Printed on acid-free paper.

Photocomposed copy prepared from the author's TEX file.
Printed and bound by R.R. Donnelley & Sons, Inc., Harrisonburg, Virginia.
Printed in the United States of America.

9 8 7 6 5 4 3 2 (corrected second printing)

ISBN 0-387-97660-4 Springer-Verlag New York Berlin Heidelberg
ISBN 3-540-97660-4 Springer-Verlag Berlin Heidelberg New York

Some Quotations From Early Statisticians

All sciences of observation follow the same course. One begins by observing a phenomenon, then studies all associated circumstances, and finally, if the results of observation *can be expressed numerically* [Quetelet's italics], estimates the intensity of the causes that have concurred in its formation. This course has been followed in studying purely material phenomena in physics and astronomy; it will likely also be the course followed in the study of phenomena dealing with moral behavior and the intelligence of man. Statistics begins with the gathering of numbers; these numbers, collected on a large scale with care and prudence, have revealed interesting facts and have led to the conjecture of laws ruling the moral and intellectual world, much like those that govern the material world. It is the whole of these laws that appears to me to constitute *social physics*, a science which, while still in its infancy, becomes incontestably more important each day and will eventually rank among those sciences most beneficial to man. (Quetelet, 1837)

The investigation of causal relations between economic phenomena presents many problems of peculiar difficulty, and offers many opportunities for fallacious conclusions. Since the statistician can seldom or never make experiments for himself, he has to accept the data of daily experience, and discuss as best he can the relations of a whole group of changes; he cannot, like the physicist, narrow down the issue to the effect of one variation at a time. The problems of statistics are in this sense far more complex than the problems of physics. (Yule, 1897)

Some people hate the very name of statistics, but I find them full of beauty and interest. Whenever they are not brutalized, but delicately handled by the higher methods, and are warily interpreted, their power of dealing with complicated phenomena is extraordinary. They are the only tools by which an opening can be cut through the formidable thicket of difficulties that bars the path of those who pursue the Science of man. (Galton, 1908)

To Leone

Preface

A Second Course in Statistics

The past decade has seen a tremendous increase in the use of statistical data analysis and in the availability of both computers and statistical software. Business and government professionals, as well as academic researchers, are now regularly employing techniques which go far beyond the standard two-semester, introductory course in statistics. While for this group of users short courses in various specialized topics are often available, there is a need to improve the statistics training of future users of statistics while they are still at colleges and universities. In addition, there is a need for a survey reference text for the many practitioners who cannot obtain specialized courses.

With the exception of the statistics major, most university students do not have sufficient time in their programs to enroll in a variety of specialized one-semester courses, such as data analysis, linear models, experimental design, multivariate methods, contingency tables, logistic regression, etc. There is a need for a second survey course that covers a wide variety of these techniques in an integrated fashion. It is also important that this second course combine an overview of theory with an opportunity to practice, including the use of statistical software and the interpretation of results obtained from 'real' data.

Topics

This two-volume survey is designed to provide a second two-semester course in statistics. The first volume outlines univariate data analysis and provides an extensive overview of regression models. The first volume also surveys the methods of analysis of variance and experimental design including their relationship to the regression model. The second volume begins with a survey of techniques for analyzing multidimensional contingency tables and then outlines the traditional topics of multivariate methods. The second volume also includes a discussion of logistic regression, cluster analysis, multidimensional scaling and correspondence analysis, which are not always included in surveys of multivariate methods. In each volume an appendix is provided to review the basic concepts of linear and matrix algebra. The appendix also includes a series of exercises in linear algebra for student practice.

Mathematics Background

The text assumes a background equivalent to one semester of each of linear algebra and calculus, as well as the standard two-semester introductory course in statistics. Calculus is almost never used in the text other than in the theoretical questions at the end of each chapter. The one semester of calculus is an indication of the ideal mathematics comfort level. The linear algebra background is needed primarily to understand the presentation of formulae. Competence with linear algebra however is required to complete many of the theoretical questions at the end of each chapter. These background prerequisites would seem to be a practical compromise given the wide variety of potential users.

Examples and Exercises

In addition to an overview of theory, the text also includes a large number of examples based on actual research data. Not only are numerical results given for the examples but also interpretations for the results are also discussed. The text also provides data analysis exercises and theoretical questions for student practice. The data analysis exercises are based on real data which is also provided with the text. The student is therefore able to improve by "working out" on the favorite local software. The theoretical questions can be used to raise the theoretical level of the course or can be omitted without any loss of the applied aspects of the course. The theoretical questions provide useful training for those who plan to take additional courses in statistics.

Use as a Text

The two volumes can be used independently for two separate courses. Volume I can be used for a course in regression and design, while Volume II can be used for a course in categorical and multivariate methods. A quick review of multiple regression and analysis of variance may be required if the second volume is to be used without the first. If the entire text is to be used in two semesters some material in each chapter can be omitted. A number of sections can be left for the student to read or for the student's future reference. Large portions of most chapters and/or entire topics can be omitted without affecting the understanding of other topics discussed later in the text. A course in applied multivariate data analysis for graduate students in a particular field of specialization can be derived from the text by concentrating on a particular selection of topics.

This two-volume survey should be useful for a second course in statistics for most college juniors or seniors. Also, for the undergraduate statistics major, this text provides a useful second course which can be combined with other specialized courses in time series, stochastic processes, sampling theory, nonparametric statistics and mathematical statistics. Because the text includes the topics normally found in traditional second courses, such as regression analysis or multivariate methods, this course provides a broader substitute which also includes other topics such as data analysis, multidimensional contingency tables, logistic regression, correspondence analysis and multidimensional scaling. The set of theoretical questions in the book can provide useful practice for statistics majors who have already been exposed to mathematical statistics.

For graduate students in business and the social and biological sciences, this survey of applied multivariate data analysis is a useful first year graduate course which could then be followed by other more specialized courses, such as econometrics, structural equation models, time series analysis or stochastic processes. By obtaining this background early in the graduate program the student is then well prepared to read the research literature in the chosen discipline and at a later stage to analyse research data. This course is also useful if taken concurrently with a course in the research methodology of the chosen discipline. I have found the first year of the Ph.D. program to be the ideal time for this course, since later in their programs Ph.D. students are too often preoccupied with their own area of specialization and research tasks.

Author's Motivation and Use of Text

The author's motivation for writing this text was to provide a two-semester overview of applied multivariate data analysis for beginning Ph.D. students in the Faculty of Business at the University of Alberta. The quantitative background assumed for the business Ph.D. student using this text is equivalent to what is required in most undergraduate business programs in North America — one semester of each of linear algebra and calculus and a two-semester introduction to statistics. Many entering Ph.D. students have more mathematics background but do not usually have more statistics background. A selection of topics from the text has also been used for an elective course in applied multivariate data analysis for second year MBA students. For the MBA elective course much less emphasis is placed on the underlying theory.

Because of the many different fields of interest within business Ph.D. programs — Accounting, Finance, Marketing, Organization Analysis and Industrial Relations — the topical needs, interests and level of mathematical sophistication of the graduate students differ greatly. Some will pursue a strong statistics minor, while others will take very little statistics training beyond this course.

In my Ph.D. class the wide variety of needs are handled simultaneously by assigning portfolios of theoretical questions to the statistics minor student, while the less theoretically oriented students are assigned a paper. The paper topic may involve a discussion of the application of one or more of the statistical techniques to a particular field or an overview of techniques not discussed in the text. A small number of classes are devoted exclusively to the discussion of the theory questions. For the theory classes only the 'theory folk' need attend. All students are required to complete data analysis exercises and to provide written discussions of the results. For the data analysis exercises great emphasis is placed on the quality of the interpretation of the results. Graduate students often have greater difficulty with the interpretation of results than with the understanding of the principles.

Quotations

The quotations by Quetelet (1837) and Yule (1897) were obtained from pages 193 and 348 respectively of *The History of Statistics: The Measurement of Uncertainty Before: 1900*, by Stephen Stigler, published by Harvard University Press, Cambridge, Mass., 1986.

The quotation by Galton (1908) was obtained from *An Introduction to Mathematical Statistics and its Applications*, Second Edition, by Richard J. Larcen and Morris L. Marx, published by Prentice–Hall, 1986.

Acknowledgments

The production of this text has benefited greatly from the input and assistance of many individuals. The Faculty of Business at the University of Alberta has born most of the cost of production. Steve Beveridge and John Brown were helpful in making funds available. John Brown and John Waterhouse also provided much encouragement during the development and early implementation of this text in our Ph.D. program.

The bulk of the typing has been done by two very able typists, Anna Fujita and Shelley Hey. Both individuals have been tireless in providing error-free typing through the uncountable number of drafts. The numerous graphs and figures could not have been carried out without the capable assistance of Berry Hsu and Anna Fujita.

The examples and data analysis exercises have been generated from data sets provided to me by my colleagues and former students. Colleagues who have allowed me to use their research data or have suggested examples include Alice and Masao Nakamura, Rodney Schneck, Ken Lemke, Chris Vaughn, John Waterhouse, Chris Janssen, Bernard Yeung and Jordan Louviére. Graduate students who have gifted me their data include Nancy Button, Nancy Keown, Pamela Norton, Diane Ewanishan, Clarke Carson, Caroline Pinkert-Rust, Frank Kobe and Cam Morrell. I am also grateful to G.C. McDonald for the use of the air pollution data and to SPSS for the bank employee salary data.

Subash Bagui, Sheila Mozejko, Alice Nakamura and Tom Johnson read parts of the manuscript and were extremely helpful in improving the overall presentation. Many graduate students have provided editorial assistance by pointing out errors in various drafts of the text. These students have also been very patient in tolerating the inconveniences resulting from the many errors. Three graduate students who provided editorial assistance by checking grammar, and table and figure references were Caroline Pinkert-Rust, Ellen Nygaard and Mary Allen. Any errors that remain are solely my responsibility.

I am also grateful to the University of Alberta who provided my undergraduate training in mathematics, 1959–1963; to Iowa State University who provided my graduate training in statistics, 1968–1971; and to the University of North Carolina who provided a statistics theory update during my study leave in 1980–81. I am extremely fortunate to have been exposed to so many great teachers.

Last and most importantly I am indebted to my wife, Leone, who cheerfully attended to most of my domestic chores while I was preoccupied with this task. At times I think she was the only one who believed that the project would ever be completed. Perhaps that was wishful thinking, as the book was often spread out all over the house. A vote of thanks is also due my two daughters, Leslie and Heather, who cheerfully accepted DAD's excessive devotion to "the book."

J.D. Jobson

Contents

Contents of Volume II

1
Introduction

This chapter introduces the concept of the multivariate data matrix and discusses scales of measurement. An outline of the techniques to be presented in Volume I is also provided. The chapter ends with a review of statistical inference for univariate random variables.

1.1 Multivariate Data Analysis, Data Matrices and Measurement Scales

The past decade has seen tremendous growth in the availability of both computer hardware and statistical software. As a result the use of multivariate statistical techniques has increased to include most fields of scientific research, and many areas of business and public management. In both research and management domains there is increasing recognition of the need to analyze data in a manner which takes into account the interrelationships among variables. Multivariate data analysis refers to a wide assortment of such descriptive and inferential techniques. In contrast to univariate statistics we are concerned with the jointness of the measurements. Multivariate analysis is concerned with the relationships among the measurements across a sample of individuals, items or objects.

1.1.1 Data Matrices

The raw input to multivariate statistics procedures is usually an $n \times p$ (n rows by p columns) rectangular array of real numbers called a *data matrix*. The data matrix summarizes observations made on n objects. Each of the n objects is characterized with respect to p variables. The values attained by the variables may represent the measurement of a quantity, or a numerical code for a classification scheme. The term object may mean an individual or a unit, while the term variable is synonomous with attribute, characteristic, response or item. The data matrix will be denoted by the $n \times p$ matrix $\mathbf{X}$, and the column vectors of the matrix will be denoted by $\mathbf{x}_1, \mathbf{x}_2, \ldots, \mathbf{x}_p$ for the p variables. The elements of $\mathbf{X}$ are denoted by x_{ij}, $i = 1, 2, \ldots, n; \;\; j = 1, 2, \ldots, p$.

Data Matrix

Variables

	$\mathbf{x}_1$	$\mathbf{x}_2$	$\mathbf{x}_3$	...	$\mathbf{x}_p$	Objects
	x_{11}	x_{12}	x_{13}	...	x_{1p}	1
	x_{21}	x_{22}	x_{23}	...	x_{2p}	2
$\mathbf{X} =$	x_{31}	x_{32}	x_{33}	...	x_{3p}	3
	$\vdots$					$\vdots$
	x_{n1}	x_{n2}	x_{n3}		x_{np}	n

The following four examples of data matrices are designed to show the variety of data types that can be encountered.

Example 1. The bus driver absentee records for a large city transit system were sampled in four different months of a calendar year. The purpose of the study was to determine a model to predict absenteeism. For each absentee record the variables month, day, bus garage, shift type, scheduled off days, seniority, sex and time lost were recorded. Table 1.1 below shows the obervations for 10 records.

Table 1.1. Sample From Bus Driver Absenteeism Survey

Month	Day	Garage	Shift	Days Off	Seniority	Sex	Time Lost
1	1	5	3	6	5	0	7.5
1	2	5	13	1	9	1	7.5
4	6	3	9	2	8	0	7.5
2	3	3	7	3	7	1	7.5
3	5	3	7	1	8	0	2.5
1	1	4	3	1	10	0	4.2
1	7	1	5	6	5	0	7.5
2	6	5	13	1	2	0	7.5
3	7	5	10	4	5	0	7.5
4	3	1	9	2	6	1	7.5

Example 2. The top 500 companies in Canada ranked by sales dollars in 1985 were compared using information on: percent change in sales, net income, rank on net income, percent change in net income, percent return on equity, value of total assets, rank on total assets, ratio of current assets to current liabilities (current ratio) and number of employees. Table 1.2 contains the data for the top ten companies. In this study the researcher was interested in the properties of the distributions of various quantities.

Table 1.3. Sample of Responses to R.C.M.P. Survey

Det. No.	Det. Exp.	Tot. Exp.	Age	Rank	Educ.	Stress Variables																	
						1	2	3	4	5	6	7	8	9	10	11	12	13	14	15	16	17	18
1	4	4	1	1	2	4	3	3	4	3	4	6	6	3	4	6	3	3	6	8	8	8	6
9	5	3	2	1	2	12	3	2	8	6	5	5	9	4	6	5	15	5	3	10	4	0	4
5	5	3	3	1	2	12	8	15	1	3	2	8	4	12	10	8	6	4	8	2	2	2	1
4	5	3	2	1	2	8	4	5	1	9	0	2	9	4	4	15	9	3	10	3	6	4	5
2	4	4	1	1	1	5	5	6	15	12	3	4	3	10	5	10	8	4	10	8	8	8	12
3	6	2	3	2	3	12	2	3	6	4	10	9	15	15	5	10	20	4	5	10	8	15	15
4	3	2	2	1	2	6	3	6	4	9	3	6	2	3	3	12	2	3	2	2	4	9	4
8	5	1	2	1	2	6	8	5	2	6	3	1	4	8	10	15	6	8	4	1	3	6	2
6	5	4	2	1	2	2	6	4	3	4	4	4	4	4	4	4	6	0	2	8	8	2	8
5	6	2	2	1	2	12	3	0	6	6	0	6	6	3	0	2	0	0	2	4	6	4	4
1	4	4	3	2	1	6	2	1	3	6	1	6	3	4	3	2	2	0	2	6	9	4	6
8	4	4	2	1	3	6	8	5	2	9	0	2	6	1	4	2	4	0	0	2	8	3	4
3	6	1	2	1	2	2	2	2	2	6	1	1	9	4	2	0	2	2	1	1	4	1	6
2	4	4	3	2	2	2	0	6	3	4	1	6	6	4	3	8	6	1	6	6	1	2	6
1	5	2	2	1	1	3	6	2	6	6	5	6	3	3	3	9	3	0	8	0	8	0	8
4	6	4	2	1	2	8	3	2	4	3	2	2	8	5	4	4	2	0	3	2	2	2	4
7	3	5	4	2	1	2	4	4	8	9	5	6	8	5	4	8	4	4	6	6	8	9	8
6	3	2	1	1	2	2	4	3	6	6	4	4	6	4	5	4	4	0	3	3	4	6	6

Example 3. A sample of R.C.M.P. officers were asked to respond to questions regarding the amount of stress they encounter in performing their regular duties. The officers also responded to questions seeking personal information such as age, education, rank and years of experience. The purpose of the analysis was to identify the dimensions of stress.

The data in Table 1.3 are a sample of responses obtained for eighteen stress items, and the personal variables age, education, rank and years of experience.

The 18 stress variables are measures of stress due to 1. insufficient resources, 2. unclear job responsibilities, 3. personality conflicts, 4. investigation where there is serious injury or fatality, 5. dealing with obnoxious or intoxicated people, 6. having to use firearms, 7. notifying relatives about death or serious injury, 8. tolerating verbal abuse in public, 9. unsuccessful attempts to solve a series of offences, 10. lack of availability of ambulances, doctors, etc., 11. poor presentation of a case by the prosecutor resulting in dismissal of the charge, 12. heavy workload, 13. not getting along with unit commander, 14. many frivolous complaints lodged against members of the public, 15. engaging in high speed chases, 16. becoming involved in physical violence with an offender, 17. investigating domestic quarrels, 18. having to break up fights or quarrels in bars and cocktail lounges.

Example 4. Real estate sales data pertaining to a sample of three bedroom homes sold in a calendar year in a particular area within a city were collected. The variables recorded were list price, sales price, square feet, number of rooms, number of bedrooms, garage capacity, bathroom capacity, extras, chattels, age, month sold, days to sell, listing broker, selling broker and lot type. Table 1.4 shows a sample of 12 observations, one for each month. The purpose of the study was to determine factors which influence selling price.

The four examples outlined above illustrate the variety of data matrices that may be encountered in practice. Before discussing techniques of multivariate analysis it will be useful to outline a system of classification for variables. We shall see later that the variable types influence the method of analysis that can be performed on the data. The next section outlines some terminology that is commonly applied to classify variables.

1.1.2 Measurement Scales

Variables can be classified as being *quantitative* or *qualitative*. A quantitative variable is one in which the variates differ in magnitude eg., income, age, weight, GNP, etc. A qualitative variable is one in which the variates differ in kind rather than in magnitude eg., marital status, sex, nationality, hair colour, etc.

Table 1.2. Sample of Canadian Companies 1985 Financial Data

Rank	Sales (Millions)	% Change in Sales	Net Income (Millions)	% Change in Net Income	Rank on Net Income	% Return Equity	Total Assets (Millions)	Rank on Total Assets	Current Ratio	No. of Employees (Thousands)
1	18,993	713	16.5	-19.2	2	40.5	4355	448	17	1.5
2	15,040	247	2.8	-34.5	13	5.5	21446	123	2	1.1
3	13,353	199	10.2	-33.6	17	19.2	2973	30	29	1.3
4	13,257	1051	25.4	11.8	1	16.6	20583	108	3	1.3
5	8,880	101	7.6	13.6	39	13.4	2616	57	32	1.3
6	8.667	684	1.0	28.3	3	14.9	9202	15	5	2.2
7	7,834	246	10.7	-24.9	14	6.4	9591	70	4	2.3
8	7,040	124	12.2	-29.9	30	34.2	1580	12	63	1.4
9	6,931	67	8.0	9.8	56	14.4	1530	33	68	1.2
10	6,070	146	5.9	-7.6	27	5.9	5799	7	13	2.3

Table 1.4. Sample of Real Estate Sales Data

	List Price	Selling Price	Square Feet	Rooms	Bed-Rooms	Bath	Garages	Extras	Chattels	Age	Days to Sell	Lot Type	List Broker	Selling Broker
1	79.0	75.4	1365	7	4	6	2	3	2	8	31	0	1	3
2	85.0	79.8	1170	6	3	4	2	2	1	7	48	0	5	5
3	103.8	101.5	1302	6	3	6	0	2	3	5	30	0	5	5
4	83.0	80.0	1120	5	3	6	2	1	2	5	28	0	3	1
5	109.8	107.0	1225	6	3	6	2	1	0	6	2	0	4	1
6	93.8	90.0	1160	6	3	6	2	3	2	6	44	0	2	5
7	95.8	89.8	1270	6	3	6	0	2	0	7	51	1	5	4
8	110.8	102.0	1260	6	3	6	2	3	3	7	45	0	2	2
9	97.2	94.0	1219	6	3	4	2	1	2	7	43	0	5	5
10	91.8	85.0	1170	6	3	4	0	2	3	8	25	1	5	5
11	86.0	84.0	1080	6	3	4	0	1	0	10	59	0	1	1
12	105.0	98.0	1345	6	3	6	2	1	1	7	16	0	1	1

Quantitative Scales

Obtaining values for a quantitative variable involves measurement along a scale and a unit of measure. A unit of measure may be *infinitely divisible* (eg., kilometres, metres, centimetres, millimetres, ...) or *indivisible* (eg., family size). When the units of measure are infinitely divisible the variable is said to be *continuous*. In the case of an indivisible unit of measure the variable is said to be *discrete*. A continuous variable (theoretically) can always be measured in finer units; hence, actual measures obtained for such a variable are always approximate in that they are rounded.

Analysis with discrete variables often results in summary measures or parameters taking on values which are not consistent with the scale of measurement; e.g., 1.7 children per household. Some variables which are *intrinsically continuous* are difficult to measure and hence are often measured on a discrete scale. For example the stress variable discussed in Example 3 is an intrinsically continuous variable.

Scales of measurement can also be classified on the basis of the relations among the elements composing the scale. A *ratio scale* is the most versatile scale of measurement in that it has the following properties: (a) Any two values along the scale may be expressed meaningfully as a ratio, (b) the distance between items on the scale is meaningful and (c) the elements along the scale can be ordered from low to high; eg., weight is usually measured on a ratio scale.

An *interval scale*, unlike a ratio scale, does not have a fixed origin; eg., elevation and temperature are measured relative to a fixed point (sea level or freezing point of water). The ratio between $20°C$ and $10°C$ is not preserved when these temperatures are converted to Fahrenheit. An interval scale has only properties (b) and (c) above.

An *ordinal scale* is one in which only property (c) is satisfied; eg., the grades A, B, C, D, can be ordered from highest to lowest but we cannot say that the difference between A and B is equivalent to the difference between B and C nor can we say that the ratio A/C is equivalent to the ratio B/D.

Qualitative Scales

The fourth type of scale, *nominal*, corresponds to qualitative data. An example would be the variable marital status which has the categories married, single, divorced, widowed and separated. The five categories can be assigned coded values such as 1, 2, 3, 4, or 5. Although these coded values are numerical, they must not be treated as quantitative. None of the three properties listed above can be applied to the coded data.

On occasion, variables which are quantitative are treated in an analysis as if they were nominal. In general, we shall use the term *categorical* to denote a variable which is being used as if it were nominal. The variable AGE for example can be divided into six levels and coded 1, 2, 3, 4, 5, and 6.

Measurement Scales and Analysis

We shall see throughout the remainder of this text that the scale of measurement used to measure a variable will influence the type of analysis used. The body of statistical techniques that are specially designed for ordinal data are often outlined in texts on non-parametric statistics. Variables which are measured on ordinal scales can often be handled using techniques designed for nominal data or interval data. The categories on the ordinal scale can be treated as the categories of a nominal scale by ignoring the fact that they can be ordered.

The variables in the data matrix represent the attempt by a researcher to operationalize various dimensions that are believed to be important in the research study. For dimensions such as intelligence, stress and job satisfaction, appropriate dimensions are difficult to define and measure. If there are no appropriate units of measure, dimensions are sometimes operationalized by using other variables as *surrogates* for direct measurement. The surrogate variable is usually an accessible and dependable correlate of the dimension in question; eg., a surrogate variable can be measured and is believed to be strongly correlated with the required dimension. Because surrogate variables are not in general perfectly correlated with the required dimension, a number of them are often used to measure the same dimension. The effectiveness with which a variable operationalizes a dimension is also called its validity. The measurement of validity in practice is usually complex and inadequate.

1.2 The Setting

1.2.1 Data Collection and Statistical Inference

Having decided upon the variables to be measured, an *experimental design* must be formulated which outlines how the data are to be obtained. The techniques for this are usually found under the theory and practice of *survey sampling* and the theory and practice of experimental design. In addition, texts on research methodology also discuss the issues of designs for obtaining the data. One characteristic of the quality of a research design is the reliability of the data that are obtained. The *reliability* of the design refers to the consistency of the data when the same cases are measured at some other time or by other equivalent variables, or when other samples of cases are used from the same population.

Probability Samples and Random Samples

The majority of multivariate techniques, generally employed to analyze data matrices, assume that the objects selected for the data matrix represent a *random sample* from some well defined *population* of objects. A random sample is a special case of a *probability sample.* In a probability sampling process the probability of occurrence for all possible samples is known. In some cases the sample may not be a probability sample in that the probability that any particular object will be chosen for the sample cannot be determined. *Haphazard samples* such as volunteers, *representative samples* as judged by an expert and *quota samples* where the objective is to meet certain quotas are examples of *non-probability samples* that are frequently used. On occasion the data set may represent the entire population (*a census*).

It is important to remember that without probability sampling, probability statements cannot be made about the outcomes from the multivariate analysis procedures. Since many research data sets are not obtained from probability samples, it is important to note that inference results should be stated as being conditional on the assumption of a probability sample.

In addition to the *simple random sample* there are alternative probability sampling methods that are commonly used. *Cluster sampling, stratified sampling, systematic sampling* and *multiphase sampling* are examples of more sophisticated methods which are usually used to reduce cost and improve reliability. Whenever simple random sampling is not used, adjustments have to be made to the standard inference procedures. Probability samples which are not simple random samples are called complex samples. Although modifications to some multivariate techniques have been developed for complex samples they will not be discussed here. Random sampling will be defined later in this chapter.

Exploratory and Confirmatory Analysis

The statistical techniques outlined in this text include both *exploratory analysis* and *confirmatory analysis*. In exploratory analysis, the objective is to describe the behavior of the variables in the data matrix, and to search for patterns and relationships that are not attributable to chance. In confirmatory analysis, certain hypotheses or models that have been pre-specified are to be tested to determine whether the data supports the model. The quality of the model is often measured using a *goodness of fit* criterion. In large data sets the use of goodness of fit criteria often results in the model being over fitted, i.e., a less complex model than the fitted one is sufficient to explain the variation. The use of *cross validation* techniques to further confirm the model is recommended. Cross validation involves checking the fitted model on a second data matrix that comes from the same population but was not used to estimate the model.

1.2.2 An Outline of the Techniques to be Studied

There is no widely accepted system for classifying multivariate techniques, nor is there a standard or accepted order in which the subject is presented. One useful classification is according to the number and types of variables being used, and also according to whether the focus is a comparison of means or a study of the nature of the covariance structure. Some multivariate techniques are concerned with *data analysis* and *data reduction*, while others are concerned with *models* relating various *parameters*. The presentation of topics in this text is governed by the following:

1. What topics can be assumed to be known from a typical introductory course in statistical inference?

2. How many variables in the data matrix are involved in the analysis?

3. What types of variables are involved in the analysis?

4. Is the technique a data reduction procedure?

For the most part the techniques to be studied are designed for continuous and/or categorical data. Quantitative variables, with discrete scales or ordinal scales, will sometimes be treated as if they have continuous scales, and in other cases they may be treated as categorical. For the purpose of outlining the techniques, variables are classified as either quantitative or categorical. Occasionally ordinal data techniques will be introduced to present alternative but similar procedures.

The topics in this text are split into two volumes. Volume I is primarily devoted to procedures for linear models. In addition to the linear regression model this volume also includes univariate data analysis, bivariate data analysis, analysis of variance and partial correlation. Volume II is designed to provide an overview of techniques for categorical data analysis and multivariate methods. The second volume also includes the topics of logistic regression, cluster analysis, multidimensional scaling and correspondence analysis.

Topics in Volume I

Chapter 1. The remainder of this chapter presents a review of univariate statistical inference procedures usually discussed in introductory courses in statistical inference.

Chapter 2. The second chapter provides an overview of *univariate data analysis*. This chapter is concerned with the techniques that might be applied to only one column or variable of the data matrix. The techniques of univariate analysis are designed to provide information about the distribution of the random variable being studied. The techniques include *descriptive devices* for sample distributions, *outliers, assessment of normality* and *transformations*.

Chapter 3. The third chapter is titled *quantitative bivariate analysis* and is concerned with analyses involving two quantitative variables or columns of the data matrix. The techniques employed include *regression* and *correlation.* A discussion of the theory of bivariate distributions and the *bivariate normal* is presented. An extensive survey of *simple linear regression* methodology is provided including the measurement of *outliers, residuals* and *influence diagnostics.* Chapter 3 also includes a discussion of *heteroscedasticity* and *autocorrelation.* The chapter will also discuss *partial correlation* and other procedures for determining the impact of *omitted variables* on the bivariate relationship. This discussion will assist in motivating the need for multivariate techniques.

Chapter 4. Chapter 4 provides an extensive discussion of *multiple linear regression* which is the most frequently used statistical technique. In addition to the multiple linear regression model, Chapter 4 outlines additional techniques which may be viewed as special cases or extensions of the multiple linear regression model such as the use of *dummy variables, variable selection, multicollinearity* and the measurement of *outliers* and *influence.* Chapter 4 also presents a variety of techniques designed to deal with situations which are departures from the assumptions required for the multiple linear regression model. The topics included are *curvilinear regression* models and *generalized least squares.* Curvilinear models are linear models in which the terms of the model are nonlinear in the observed variables. Generalized least squares refers to the body of techniques designed to estimate the regression parameters when the error terms in the regression model are no longer independent and identically distributed.

Chapter 5. In Chapter 5 the topics of *analysis of variance* and *experimental design* are discussed. This chapter also discusses *analysis of covariance* and shows how the analysis of variance model can be viewed as a special case of a multiple regression model. Complex design issues such as *randomization, blocking, fractional replication, random effects, repeated measures, split plots* and *nesting* are also presented.

1.3 Review of Statistical Inference for Univariate Distributions

The reader is assumed to have been exposed to the basic concepts of statistical inference commonly discussed in introductory statistics textbooks for undergraduate students. The purpose of this section is to provide an overview of these concepts and to introduce some terminology that will be required in later chapters.

1.3.1 Populations, Samples and Sampling Distributions

Statistical inference is the use of samples from a population to obtain information about properties of the population. A *population* is a set of items or individuals each of which can be assigned values of one or more characteristics. The *sample* is a subset of the population and is chosen in such a way that the degree of uncertainty in the sample values are known. The characteristic is denoted by X and the specific values of the characteristic X, assumed by the items in the population, are denoted by x. If the values of X vary over the items of the population, then X is defined to be a *random variable*.

Probability Distribution and Density

The variation in the values of the random variable X over the population is described by its *probability distribution*. The most commonly used distribution in statistics is the *normal distribution* which has the bell shape as shown in Figure 1.1 below. The probability distribution for X is usually characterized by the *distribution function* $F(x)$ where $F(x) = P[X \leq x]$ is the function which assigns a probability that X will not exceed the value x. This distribution function is related to the *density function* $f(x)$ by the equation

$$F(x) = \int_{-\infty}^{x} f(y)dy.$$

The density function $f(x)$ describes how the probability is distributed over the range of X. The density function for the *normal distribution* is given by

$$f(x; \mu, \sigma^2) = [2\pi\sigma^2]^{-1/2} \exp\{-1/2[(x - \mu)^2/\sigma^2]\}, \qquad (1.1)$$
$$-\infty < x < \infty,$$

where μ and σ^2 are known constants, $-\infty < \mu < \infty$, $\sigma^2 > 0$.

Table 1 of the Table Appendix presents the cumulative distribution function for the standard normal distribution.

Discrete Distribution

The normal density is an example of a continuous density and a normal random variable is a continuous random variable. A random variable is *discrete* if it can only assume certain specific values. These values can be itemized

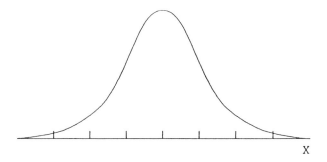

Figure 1.1. Normal Distribution

as $x_1, x_2, \ldots, x_j, \ldots$ although the list can be infinite. A *Poisson distribution* is discrete since the random variable can only assume non-negative integer values. The density is given by

$$f(x) = e^{-\lambda} \lambda^x / x! \qquad x = 0, 1, 2, 3, \ldots .\qquad (1.2)$$

In this case the distribution function has the form

$$F(x) = \sum_{y=0}^{x} f(y) \qquad x = 0, 1, 2, 3, \ldots .$$

Expectation

The *expectation* of a function of a random variable x, say $h(x)$, is the mean value of this function determined over the distribution of x and is denoted by $E[h(x)]$. For a continuous random variable the expectation is determined by

$$E[h(x)] = \int_{-\infty}^{\infty} h(x) f(x) dx.$$

For a discrete random variable the expectation is given by

$$E[h(x)] = \sum_{x} h(x) f(x)$$

where this sum is determined over all possible values of x. Special cases of the function $h(x)$ are x and $(x - E[x])^2$ which yield the mean and variance given by $E[x]$ and $E[(x - E[x])^2]$ respectively.

Parameters

The distribution of a random variable is usually characterized in terms of particular constants called *parameters*. For the normal distribution these parameters describe particular properties of the distribution; $\mu = E[x]$ is the *mean* and $\sigma^2 = E[(X-\mu)^2]$ is the *variance*. For the Poisson distribution the parameter λ has the property $E[x] = \lambda$. In general, we denote the parameters by θ and write the density function as $f(x; \theta)$. The purpose of statistical inference is to make inferences about unknown parameters, say θ, using a sample of n observations on X denoted by $(x_1, x_2, \ldots, x_n)$.

Sampling

In statistical inference the sample must be a probability sample which means that the probability of obtaining that particular sample is known. In practice the sample is usually assumed to be a *simple random sample*, which for *infinite* populations, means the items in the sample are *independent* and *identically distributed*. In other words, the selection of any item for the sample is independent of the other items selected, and also the underlying populations corresponding to the items selected are identical. For *finite* populations, containing a total of N items, simple random sampling can be described as a method of sampling which insures that all possible samples of size n have the same chance of being selected. Equivalently, when viewed as a sequence of n draws, a simple random sample from a finite population is achieved if on each draw all items remaining have the same chance of being selected. When making inferences about small finite populations the size of the population must be taken into account, but in large finite populations the population size can usually be ignored. The critical element is the *sampling fraction* n/N. For example, a random sample of 2000 voters selected from a city of one million eligible voters can be assumed to be drawn from an infinite population. If, however, a production lot of 500 bulbs is evaluated by testing a random sample of 20 bulbs (sampling fraction .04), the finite population factor may not be negligible.

1.3.2 Statistical Inference, Estimation and Hypothesis Testing

Sample observations are often used to provide information about the underlying population distribution. Any function of the sample observations $T(x_1, x_2, \ldots, x_n)$ is usually called a *statistic*. If such a function is used to provide an estimate of a parameter, θ, it is called an *estimator* and is denoted by $\hat{\theta}_n = T(x_1, \ldots, x_n)$. The choice of the function T that provides $\hat{\theta}_n$

usually depends on the variation in $\hat{\theta}_n$ over all possible samples. The distribution of values of $\hat{\theta}_n$ over all possible samples is referred to as the *sampling distribution* of $\hat{\theta}_n$ and is denoted by $g(\hat{\theta}_n; \theta)$. The sampling distribution of $\hat{\theta}_n$, $g(\hat{\theta}_n, \theta)$, usually depends on the true parameter θ and possibly other parameters. In populations with a proportionately large total number of items, N, relative to the sample size, n (e.g., $n/N < 0.01$), the properties of $\hat{\theta}_n$ that depend on the sampling fraction, n/N, can usually be ignored. We shall assume in this text that in general N is sufficiently large that such finite population corrections may be ignored.

In a random sample of n observations $(x_1, x_2, \ldots, x_n)$ from a population with mean μ and variance σ^2, the *sample mean*, $\bar{x} = \frac{1}{n} \sum_{i=1}^{n} x_i$, is a statistic which can be used to estimate μ. For infinite populations, the sampling distribution of $\bar{x}$ is known to have mean μ and variance σ^2/n. Since the expected value of $\bar{x}$ is μ, this estimator is said to be *unbiased*. The variance of $\bar{x}$, σ^2/n, measures the sample-to-sample variation in $\bar{x}$ around μ. For a finite population, the variance of $\bar{x}$ is $\left(\frac{N-n}{N-1}\right)\sigma^2/n$ and the multiplier $\left(\frac{N-n}{N-1}\right)$ is called the *finite population correction factor*.

Estimation

The "ideal" estimator $\hat{\theta}_n$ of θ is one whose sampling distribution is concentrated around θ with very little variation. This property is usually characterized by measuring the *bias*, $B(\hat{\theta}_n)$ and *mean squared error* MSE$(\hat{\theta}_n)$ defined by

$$B(\hat{\theta}_n) = E[(\hat{\theta}_n - \theta)] = E[\hat{\theta}_n] - \theta$$

and

$$\text{MSE}(\hat{\theta}_n) = E[(\hat{\theta}_n - \theta)^2],$$

where the expectations are determined over the sampling distribution of $\hat{\theta}_n$. If $B(\hat{\theta}_n) = 0$ the estimator $\hat{\theta}_n$ is unbiased for θ. An estimator $\hat{\theta}_n$ is *biased*, therefore, if its average value over all possible samples is not equal to the true parameter θ. The mean squared error measures the average value of the squared deviations from the true value over all possible samples and is often used as a measure of the *efficiency* of the estimator $\hat{\theta}_n$.

The two measures are related by the equation

$$\text{MSE}(\hat{\theta}_n) = V[\hat{\theta}_n] + [B(\hat{\theta}_n)]^2$$

where $V[\hat{\theta}_n] = E[(\hat{\theta}_n - E[\hat{\theta}_n])^2]$ is the variance of the estimator $\hat{\theta}_n$. An estimator $\hat{\theta}_n$, that has zero bias and minimizes the mean squared error, is usually referred to as a *minimum variance unbiased estimator*. This estimator is therefore simultaneously optimal with respect to both properties. Minimum variance unbiased estimators, however, do not always exist.

The *sample variance* is usually defined to be $s^2 = \dfrac{1}{(n-1)} \sum\limits_{i=1}^{n} (x_i - \bar{x})^2$ because this statistic provides an unbiased estimator of the population variance σ^2. An alternative definition for the sample variance uses the divisor n rather than $(n-1)$ and hence is biased. While s^2 is an unbiased estimator of σ^2, $s = \sqrt{s^2}$, however, is not an unbiased estimator of σ. The symbols s and σ denote the sample and population *standard deviation* respectively.

For the normal distribution the mean and median are equivalent to μ. For a sample of size n the *sample median* x_M is the central value in the sample (to be defined later) and is not necessarily equal to the mean, $\bar{x}$. The variance of the sample median for samples from a normal distribution is $\pi\sigma^2/2n$ which is larger than σ^2/n. Thus the median has greater variance than the sample mean. Though both estimators are unbiased, the sample mean has the smaller variance and is the minimum variance unbiased estimator of μ.

A weak but minimally acceptable criterion for the estimator $\hat{\theta}_n$ requires that the sampling distribution of $\hat{\theta}_n$ become more concentrated around θ as the sample size n increases without limit. This property is known as *consistency*. As a minimum $\hat{\theta}_n$ must be a *consistent* estimator of θ. The consistency property is given by

$$\underset{n \to \infty}{\text{Limit}} \, P[|\hat{\theta}_n - \theta| > \varepsilon] = 0 \quad \text{for all} \quad \varepsilon > 0.$$

The estimator $(\bar{x}+2)$ is not only a biased estimator of μ; it is also inconsistent in that the bias of 2 does not disappear as n gets large. The estimator $\left(\dfrac{n-1}{n}\right)s^2$, though biased, is still consistent since the bias disappears and the variance of s^2 goes to zero as n gets large.

The derivation of parameter estimators is most commonly approached in one of two ways; the method of moments or the method of maximum likelihood. The *method of moments* can be used if the parameter can be expressed in terms of expectations of functions of the random variable. The mean of this function determined over the sample observations then provides an estimator of the parameter. If $\theta = E[h(x)]$ then $\dfrac{1}{n}\sum\limits_{i=1}^{n} h(x_i)$ is a sample moment estimator of θ.

The *maximum likelihood estimator* begins with the definition of the likelihood function for the observation x_i as the density $f(x_i; \theta)$. For the random sample $(x_1, x_2, \ldots, x_n)$, the *likelihood function L* is given by the product of the individual likelihoods since the sample observations are independent. The product is given by

$$L = f(x_1, \theta) \, f(x_2, \theta) \ldots f(x_n, \theta) = \prod_{i=1}^{n} f(x_i, \theta),$$

where the symbol $\prod$ is used to denote the product of the elements $f(x_i, \theta)$ as i moves from 1 to n. The value of θ, say $\tilde{\theta}$, that maximizes L for a given sample is called the maximum likelihood estimator. In other words $\tilde{\theta}$ is the value of θ that maximizes the probability of the observed sample.

The likelihood function for the sample $(x_1, x_2, \ldots, x_n)$ from a *normal* population is given by

$$[2\pi\sigma^2]^{-n/2} \exp[-\frac{1}{2}\sum_{i=1}^{n}(x_i - \mu)^2/\sigma^2]$$

which is a maximum for $\mu = \bar{x}$ and $\sigma^2 = \frac{(n-1)}{n}s^2$. The estimators $\bar{x}$ and $\frac{(n-1)}{n}s^2$ are called the maximum likelihood estimators of μ and σ^2 respectively.

Confidence Intervals

Inferences for population parameters based on samples are usually made either by employing *confidence interval estimators* or by performing *tests of hypotheses*. Both procedures require that the sampling distribution of the estimator be known. The confidence interval estimator for the parameter θ, provides a range of possible values for θ say $(\hat{\theta}_L, \hat{\theta}_U)$ and also a probability that the specified range includes θ, say $(1 - \alpha)$, where α is in the range $(0, 1)$.

For the sample mean $\bar{x}$, based on a random sample of size n from a normal population with mean μ and variance σ^2, we can say that the random variable $\sqrt{n}(\bar{x} - \mu)/\sigma$ has a *standard normal* distribution ($\mu = 0$, $\sigma^2 = 1$). Denoting by $Z_{\alpha/2}$ the value of a standard normal random variable Z such that $P[Z > Z_{\alpha/2}] = \frac{\alpha}{2}$, the probability that $\sqrt{n}(\bar{x} - \mu)/\sigma$ is contained in the interval $(-Z_{\alpha/2}, Z_{\alpha/2})$ is $(1 - \alpha)$. Equivalently we can say that the probability is $(1 - \alpha)$ that the interval $(\bar{x} - Z_{\alpha/2}\sigma/\sqrt{n}, \bar{x} + Z_{\alpha/2}\sigma/\sqrt{n})$ contains μ. This latter interval is commonly referred to as a $100(1 - \alpha)\%$ confidence interval for μ based on the sample mean $\bar{x}$.

Hypothesis Testing

In hypothesis testing a specified range of values for θ is assumed to be correct and is referred to as the *null hypothesis*. The remaining possible values of θ are called the *alternative hypothesis*. Using the sample information a *test statistic* is computed, and assuming that the null hypothesis is correct, the probability that the test statistic could have that sample value or more is determined. This probability is called the *p-value*. If the *p*-value is small then the null hypothesis is rejected.

Suppose for a normal population it is believed that the mean is $\mu = \mu_0$ and that a random sample of size n from this population obtains the sample mean $\bar{x}$. Under the null hypothesis $H_0 : \mu = \mu_0$, the observed sample statistic $Z^* = \sqrt{n}(\bar{x} - \mu_0)/\sigma$ has a standard normal distribution. For the observed value of Z, Z^*, the probability $P[|Z| > |Z^*|] = \alpha$ is determined from the upper and lower tails of the standard normal distribution as shown in Figure 1.2 below. The probability α is called the p-value of the test statistic and measures the probability that the test statistic $|Z|$ could have a value at least as large as $|Z^*|$, when H_0 is true. If the value of α or the p-value is sufficiently large then the null hypothesis H_0 cannot be rejected.

If a null hypothesis is rejected when it is true a *type I error* is made. The probability of a type I error given the sample results is the p-value. If a null hypothesis is not rejected even though it is false a *type II error* occurs. The probability of rejecting the null hypothesis given that a particular alternative value of the parameter is true is called the *power* of the test at that alternative.

If the likelihood function is employed to develop an estimator it can also be used to develop a test statistic. The *likelihood ratio test* employs a ratio of likelihood function values under both the null and the alternative hypotheses to generate a test statistic.

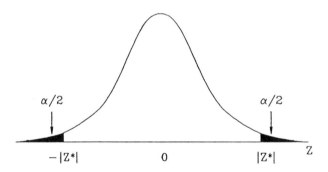

Figure 1.2. Sampling Distribution of Z

Under the null hypothesis $H_0 : \mu = \mu_0$ the likelihood function for the sample
$(x_1, x_2, \ldots, x_n)$ from the normal distribution is given by

$$L_0 = [2\pi\sigma^2]^{-n/2} \exp\left[-\frac{1}{2}\sum_{i=1}^{n}(x_i - \mu_0)^2/\sigma^2 \right]. \tag{1.3}$$

For any alternative μ the likelihood function is given by L as in (1.3) with μ_0 replaced by μ. If the likelihood L_0 is not significantly different from the likelihood L then the hypothesis $H_0 : \mu = \mu_0$ cannot be rejected. The likelihoods can be compared using the ratio L_0/L. If the ratio L_0/L is too small then H_0 must be rejected. The logarithm of this ratio can be expressed in terms of the statistic $Z^* = \sqrt{n}(\bar{x} - \mu_0)/\sigma$ discussed above and hence the likelihood ratio test is equivalent to the test for H_0 based on the distribution of $\bar{x}$. Although in this instance the test based on the distribution of $\bar{x}$ is equivalent to the likelihood ratio test, this is not always the case.

1.3.3 Three Useful Sampling Distributions

As outlined in Section 1.3.2 the sampling distribution of a sample statistic describes the distribution of the statistic over all possible samples of a given size n from some specified population. For the sample mean $\bar{x}$ based on a random sample of size n from a normal population, the sampling distribution of $\bar{x}$ is normal with mean μ and variance σ^2/n, where μ and σ^2 respectively denote the mean and variance of the population. In this section three additional useful sampling distributions are introduced. These sampling distributions will be used extensively throughout the remainder of the text.

The "χ^2" Distribution

Let $Z_1, Z_2, \ldots, Z_\nu$ be ν mutually independent standard normal random variables with common mean 0 and variance 1. The random variable $\sum_{i=1}^{\nu} Z_i^2$ has a χ^2 *distribution* with ν degrees of freedom. The distribution (see Figure 1.3) is positive valued and is skewed to the right. The mean is ν, the variance is 2ν, the index of skewness is $2\sqrt{2}/\sqrt{\nu}$ and the kurtosis coefficient is $3 + 12/\nu$. The χ^2 distribution shown in Figure 1.3 has 30 d.f.

Later in this chapter the χ^2 distribution will be used to describe the sampling distribution of the sample variance s^2 for samples from a normal population. In later chapters in volume II of this text the χ^2 distribution will be used to describe the limiting distribution of various likelihood ratio test statistics. A useful property of the χ^2 distribution is that the sum of independent χ^2 random variables is also a χ^2 random variable with degrees of freedom equal to the sum of the corresponding degrees of freedom of the components.

A table of critical values for the χ^2 distribution is given in Table 2 of the Table Appendix.

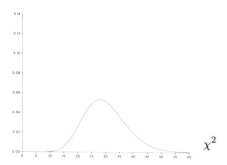

Figure 1.3. χ^2 Distribution with 30 Degrees of Freedom

The "t" Distribution

If Z is a standard normal random variable, and if χ^2_ν is a χ^2 random variable with ν degrees freedom independent of Z, then the ratio random variable $Z/(\chi^2_\nu/\nu)^{1/2}$ has a t *distribution* with ν degrees of freedom. The t distribution is symmetrical and bell shaped, and like the normal distribution, has mean 0, index of skewness 0 and kurtosis coefficient 0. The variance of the distribution is $\nu/(\nu - 2)$ which converges to 1 as $\nu \to \infty$. In large samples therefore, the t distribution is often replaced by the standard normal distribution. The t distribution for 10 degrees of freedom is compared to the standard normal distribution in Figure 1.4 below. The t distribution is the one with the wider range in Figure 1.4.

The t distribution will be used to make inferences for the population mean when sampling from a normal population, and also to make inferences about population regression coefficients. Since the t distribution is symmetrical about zero the left tail values can be obtained from the corresponding right tail values using $t_{(1-(\alpha/2));\nu} = -t_{(\alpha/2);\nu}$ where $\frac{\alpha}{2}$ and $\left(1-\frac{\alpha}{2}\right)$ denote the tail areas to the right of $t_{(\alpha/2)}$ and $t_{(1-(\alpha/2))}$ respectively.

A table of critical values of the t distribution is given in Table 3 of the Table Appendix.

The "F" Distribution

If $\chi^2_{\nu_N}$ and $\chi^2_{\nu_D}$ are two independent χ^2 random variables with degrees of freedom ν_N and ν_D respectively then the ratio random variable $(\chi^2_{\nu_N}/\nu_N)/(\chi^2_{\nu_D}/\nu_D)$ has an F distribution with ν_N and ν_D degrees of freedom. The two degrees of freedom parameters are usually referred to as the numerator and denominator degrees of freedom respectively. The F *distribution* is positive valued and is skewed to the right (see Figure 1.5). The mean is $\nu_D/(\nu_{D-2})$ and the variance is $2\nu_D^2(\nu_D + \nu_N - 2)/\nu_N(\nu_D - 2)^2(\nu_D - 4)$.

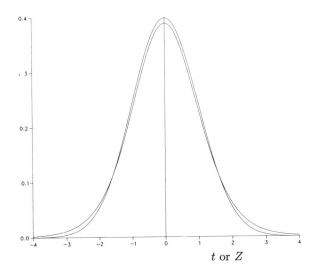

Figure 1.4.   Standard Normal Distribution (Z) and t Distribution with 10 Degrees of Freedom

The F distribution has the property that

$$F_{\alpha;\nu_N,\nu_D} = 1/F_{(1-\alpha);\nu_D,\nu_N}$$

and

$$F_{\alpha;1,\nu_D} = t^2_{(\alpha/2);\nu_D}.$$

The F distribution in Figure 1.5 has 10 and 30 degrees of freedom.

The F distribution is used extensively throughout the text. Inferences for multiple regression models, analysis of variance models and for multivariate means require the F distribution. The F distribution usually arises from the ratio of two independent sample sums of squares.

A table of critical values for the F distribution is given in Table 4 of the Table Appendix.

1.3.4 Some Inference Procedures for Normal Populations

The normal distribution is given above by (1.1) and is shown in Figure 1.1. If X is normally distributed the mean and variance are given by $\mu = E[X]$ and $\sigma^2 = E[(X - \mu)^2]$. The normal distribution is symmetric about μ and has a range of $(-\infty, \infty)$. The effective range of the distribution is 6σ which

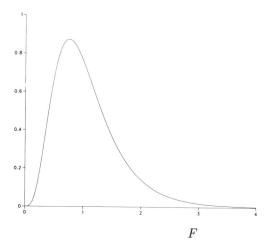

Figure 1.5. F Distribution with 10 and 30 Degrees of Freedom

encompasses 99.73% of the area. The proportion of the area within 1σ and 2σ of μ are .6826 and .9544 respectively.

A special case of a normal distribution is the *standard normal* which has parameter values $\mu = 0$ and $\sigma^2 = 1$. A normal random variable X is easily transformed to a standard normal random variable Z using the transformation $Z = (X - \mu)/\sigma$. Using this transformation, critical values and probability values for any normal distribution are easily determined from tables for the standard normal distribution.

Inference for the Mean

Given a random sample of n observations $(x_1, x_2, \ldots, x_n)$ from a normal population, the sample mean $\bar{x}$ and the sample variance s^2 provide unbiased estimators of μ and σ^2 respectively. The maximum likelihood estimators of μ and σ^2 are $\bar{x}$ and $(n - 1)s^2/n$ respectively. The sampling distribution of $\bar{x}$ is normal with mean μ and variance σ^2/n. The statistic $\sqrt{n}(\bar{x} - \mu)/\sigma$, therefore, has a standard normal distribution, and as outlined above can be used to make inferences about μ when σ is known.

If σ is unknown the statistic $\sqrt{n}(\bar{x} - \mu)/s$ has a t *distribution* with $(n - 1)$ degrees of freedom. In small samples from a normal population t is useful for making inferences about μ. In large samples the t distribution is comparable to the distribution of Z, the standard normal.

Example

Assuming the weights of boxes of breakfast cereal are normally distributed, the manufacturer weighs a random sample of 16 boxes and finds an average weight of 317.3 grams and a standard deviation of 10 grams. A 95% confidence interval for the true mean weight for boxes of cereal produced by the process is given by the interval $\bar{x} \pm t_{.025;(n-1)}s/\sqrt{n}$ which in this case is $317.3 \pm (2.131)10/\sqrt{16} = (306.0, 328.6)$. To test the null hypothesis $H_0 : \mu = 310$ grams the test statistic $t^* = (\bar{x} - \mu)/s/\sqrt{n}$ is given by $t^* = (317.3 - 310)/10/\sqrt{16} = 2.92$. The p-value for this test statistic is approximately $2(.005) = 0.010$ since $t_{.005;15} = 2.95$.

Inference for the Variance

For normal populations the statistic $(n-1)s^2/\sigma^2$ has a χ^2 *distribution* with $(n-1)$ degrees of freedom. This χ^2 statistic may be used to make inferences about the variance σ^2. A useful property of samples from a normal distribution is that $\bar{x}$ and s^2 are statistically independent.

Example

If the distribution of waiting time for customers in a bank queue is normally distributed the distribution of the sample variance s^2 is related to a χ^2 distribution. After switching to a single queue a bank manager selects a random sample of $n = 30$ customers and determines that the sample variance is 44.14 $(\text{min})^2$. A 90% confidence interval for the true variance is given by

$$\left(\frac{(n-1)s^2}{\chi^2_{.05;(n-1)}}, \frac{(n-1)s^2}{\chi^2_{.95;(n-1)}}\right) = \left(\frac{(29)(44.14)}{42.6}, \frac{(29)(44.14)}{17.7}\right) = (30.05, \ 72.32).$$

Before installing the single queue system the variance of waiting time was 80. The hypothesis $H_0 : \sigma^2 = \sigma_0^2 = 80$ that there has been no change in the variance in waiting time is tested using the test statistic $\chi^2 = \frac{(n-1)s^2}{\sigma_0^2} = \frac{(29)(44.14)}{(80)} = 16.0$ which has a p-value of 0.050 since $\chi^2_{.975;29} = 16.0$. The hypothesis therefore would be rejected at significance levels above 0.05. If this had been carried out as a one-sided test with the alternative hypothesis $H_a : \sigma^2 < 80$, the p-value would have been 0.025 since only the left tail of the χ^2 distribution is relevant to the alternative hypothesis.

1.3.5 Inference Procedures for Non-Normal Populations

If the random variable X is not normally distributed it is still possible to make inferences about the population mean μ in large samples. Given a random sample $(x_1, x_2, \ldots x_n)$ from a population with mean μ and variance σ^2, the *central limit theorem* states that as the sample size n increases the distribution of the statistic $\sqrt{n}(\bar{x} - \mu)/\sigma$ converges to a standard normal distribution. This statistic therefore can be used to make inferences about μ in large samples. If σ is unknown, inferences for μ can be made using the statistic $\sqrt{n}(\bar{x} - \mu)/s$ which also has a standard normal distribution in large samples.

To make inferences about the population variance σ^2 in non-normal populations a central limit theorem is also required. As the sample size n increases, under certain conditions the distribution of the sample variance s^2 approaches a normal distribution with mean σ^2 and variance $2\sigma^4/(n-1)$. The statistic $\sqrt{n}(s^2 - \sigma^2)/\sqrt{2s^4}$ therefore converges in distribution to a standard normal distribution as n increases. The quality of this approximation is poor in small samples and/or with highly non-normal data.

Statistical Inference for a Population Proportion

A *non-normal* population of special interest is one in which each item either has or does not have a particular *attribute*. Defining the random variable X as 1 if the item has the attribute and 0 if it does not, the parameters μ and σ^2 are given by $\mu = p$ and $\sigma^2 = p(1-p)$, where p is the *proportion* of the population that has the attribute. The sample mean $\bar{x}$ for a random sample of size n is the sample proportion and is denoted by $\hat{p}$. The variance of $\hat{p}$ is therefore $p(1-p)/n$. Inferences regarding p in large samples can be made using the statistic $(\hat{p} - p)/\sqrt{p(1-p)/n}$ or $(\hat{p} - p)/\sqrt{\hat{p}(1-\hat{p})/n}$ both of which have standard normal distributions in large samples. It is assumed that the sample size n is very small relative to the population size N. If p is close to 0 or 1, n should be at least 100 before the large sample approximation is used.

Example

A public opinion poll obtained that 60% of a random sample of 2500 voters were opposed to the death penalty for individuals convicted of first degree murder. A 95% confidence interval for the true proportion is given by

$$\hat{p} \pm Z_{.025}\sqrt{\hat{p}(1-\hat{p})/n} = 0.60 \pm 1.96\sqrt{(0.60)(0.40)/2500} = (0.581, 0.619).$$

To test the null hypothesis that the population of eligible voters is evenly divided on the issue ($H_0 : p = 0.50$) the test statistic is given by

$$Z = (\hat{p} - p)/\sqrt{p(1-p)/n} = (0.60 - 0.50)/\sqrt{(0.50)(0.50)/2500} = 10$$

which has a p-value which is much less than 0.001. The hypothesis of evenly divided opinion would therefore be rejected at conventional significance levels.

Multinomial Distribution

For *nominal* or *categorical* random variables with say g categories or cells, a random sample of size n is characterized in terms of observed frequencies n_i, $i = 1, 2, \ldots, g$, where n_i is the observed frequency in category i and $n = \sum_{i=1}^{g} n_i$. A probability distribution commonly used to describe the sample frequencies n_i for large populations is the *multinomial distribution*. Each category has an associated probability p_i where $\sum_{i=1}^{g} p_i = 1$. The probability of obtaining the observed frequencies $n_1, n_2, \ldots, n_g$ for a sample of size n is given by

$$\binom{n}{n_1 n_2 \ldots n_g} p_1^{n_1} p_2^{n_2} \ldots p_g^{n_g} = \frac{n!}{\prod_{i=1}^{g} n_i!} \prod_{i=1}^{g} p_i^{n_i}.$$

It is assumed that for a large population the sample sizes n_i are very small relative to the population sizes N_i in each category.

A test of the hypothesis that the p_i have certain values can be carried out using the *Pearson χ^2 goodness of fit test* where $\sum_{i=1}^{g} \frac{(n_i - np_i)^2}{np_i}$ has a χ^2 distribution with $(g - 1)$ degrees of freedom if the hypothesis is correct. If the p_i must be estimated from the sample, the sample proportions $\hat{p}_i = n_i/n$ provide unbiased estimators. In this case the degrees of freedom must be adjusted by subtracting from $(g - 1)$ the total number of independent parameters estimated from the sample. A likelihood ratio test of an hypothesis regarding the p_i may also be used. The likelihood ratio statistic is given by $2\sum_{i=1}^{g} n_i \ln[n_i/np_i]$ and has a χ^2 distribution with the same number of degrees of freedom as the *Pearson χ^2* above. The two χ^2 values are usually quite similar in large samples. The means, variances and covariances for the n_i are given by

$$E[n_i] = np_i, \qquad V[n_i] = np_i(1 - p_i) \qquad \text{and}$$

$$\text{Cov}(n_i, n_j) = -np_i p_j \qquad i, j = 1, 2, \ldots, g.$$

This distribution will be used more extensively in the discussion of contingency tables in Volume II.

Example

A random sample of 100 blood donors who gave blood in a particular northeastern city were distributed across the blood groups as follows

$$A: 35 \qquad B: 25 \qquad AB: 5 \qquad O: 35.$$

Previous studies have shown that for the entire nation the distribution of the population across the four categories is

$$A: 30 \qquad B: 20 \qquad AB: 10 \qquad O: 40.$$

To test the hypothesis that the northeastern city has the same blood group distribution as the entire nation, the two χ^2 statistics are given by

$$\frac{(35-30)^2}{30} + \frac{(25-20)^2}{20} + \frac{(5-10)^2}{10} + \frac{(35-40)^2}{40} = 5.208$$

and

$$2\left[35\ln\left[\frac{35}{30}\right] + 25\ln\left[\frac{25}{20}\right] + 5\ln\left[\frac{5}{10}\right] + 35\ln\left[\frac{35}{40}\right]\right] = 5.668$$

for the Pearson and likelihood ratio χ^2 statistics respectively. Comparing to the χ^2 distribution with 3 degrees of freedom the p-values for the two statistics are greater than $p = 0.10$ since $\chi^2_{.10;3} = 6.251$. The hypothesis is therefore not rejected at the 0.10 level.

Hypergeometric Distribution

For the multinomial population described above it was assumed that the population frequencies in the g categories, N_i were large relative to the sample size n_i, $i = 1, 2, \ldots, g$. If this is not the case the assumption of independence among trials required for the multinomial distribution is not acceptable. If N_i is relatively small then removal of one observation from the i-th category changes the probability p_i on the next trial. In such cases the correct expression for the probability of obtaining the sample frequencies $n_1, n_2, \ldots, n_g$ is the *hypergeometric distribution* given by

$$\frac{\binom{N_1}{n_1}\binom{N_2}{n_2}\cdots\binom{N_g}{n_g}}{\binom{N}{n}} = \frac{(N-n)!n!\prod\limits_{i=1}^{g} N_i!}{N!\prod\limits_{i=1}^{g} n_i!(N_i - n_i)!}$$

where $N = \sum_{i=1}^{g} N_i$. As the ratios n_i/N_i $i = 1, 2, \ldots, g$ go to zero this probability converges to a multinomial probability. The means, variances and covariances for the n_i are given by

$$E(n_i) = n\left(\frac{N_i}{N}\right), \qquad V(n_i) = n\left(\frac{N-n}{N-1}\right)\left(\frac{N_i}{N}\right)\left(1 - \frac{N_i}{N}\right) \qquad \text{and}$$

$$\mathrm{Cov}(n_i, n_j) = -n\left(\frac{N-n}{N-1}\right)\left(\frac{N_i}{N}\right)\left(\frac{N_j}{N}\right) \qquad i, j = 1, 2, \ldots, g.$$

This distribution will be used in the discussion of contingency tables in Volume II.

Poisson Distribution

A useful distribution for modelling discrete data for rare events is the *Poisson distribution*. For instance the number of automobile accidents experienced by an individual in a given period of time, such as a month or a year, would normally be such a rare event. The density function for X the number of event occurrences in a given period of time t has the form

$$f(x) = \lambda^x e^{-\lambda}/x!, \qquad x = 0, 1, 2, \ldots, \infty.$$

The density was also given by (1.2) in Section 1.3.1. The mean and variance are given by $E[X] = \lambda$ and $V[X] = \lambda$.

The Poisson distribution can be related to a *Poisson process* which continues over time and has certain properties. An event is a possible outcome of the process, but two events cannot occur simultaneously. In any time interval, it is assumed that the number of events that occur is independent of the number of events that have already occurred in other nonoverlapping time intervals. The Poisson process also has the property that the probability that an event occurs in a time period of length t is constant over the entire process. As a result, if λ is the expected number of event occurrences in a time period of length t, then in a period of length bt the expected number of events is $b\lambda$. The Poisson distribution is useful for describing the behavior of cell frequencies in contingency tables.

Example

In a large urban transit system the number of accidents experienced by bus drivers who had been driving for exactly ten years was found to be Poisson with $\lambda = 1$. The probabilities for $x = 1$, 2 and 3 accidents in the ten year period are therefore $e^{-1} = 0.368$, $e^{-1}/2 = 0.135$ and $e^{-1}/6 = 0.061$ respectively.

Exercises for Chapter 1

1. The distribution of taxable income for professional corporations is known to be normal with a mean of 60,000 and a standard deviation of 20,000.

   (a) What proportion of these corporations should be expected to have taxable incomes between 55,000 and 75,000?

   (b) What is the probability that in a sample of size $n = 100$ from the population the sample mean will be less than 57,000?

   (c) What sample size should be used when sampling from the population to guarantee with a probability of 0.90 the sample mean will not differ from the population mean by more than 5,000?

   (d) In a random sample of size 16 from the population a sample mean income of 55,000 and a sample standard deviation of 18,000 were obtained. Assume that the true mean μ is unknown but that σ^2 is known. Give a 95% confidence interval for μ using the sample mean obtained.

   (e) Assume both μ and σ^2 are unknown for the population and use the sample information in (d) to determine a 95% confidence interval for μ. How would you change your answer if the population distribution was not known to be normal?

   (f) Assume μ is unknown and use the sample information in (d) to test the hypothesis $H_0 : \mu \geq 58,000$, assuming σ^2 is known. How would your answer change if σ^2 is not known? What is the p-value of your test statistic in each case?

   (g) What is the power of your test in (f) against the alternative $\mu = 52,000$ if σ^2 is known to be 20,000? Assume $P[\text{Type } I \text{ error}] = 0.05$ for the test in (f).

   (h) Assume σ^2 unknown and use the sample information in (d) to give a 98% confidence interval for σ^2.

   (i) Assume σ^2 unknown and test $H_0 : \sigma \geq 20,000$ using the sample information in (d).

2. Use the central limit theorem to approximate the distribution of $\hat{p}$ the sample proportion in the following exercises.

   (a) A cable television company wishes to estimate the proportion p of cable television viewers who would be willing to rent a new 24-hour cable movie channel. If $p = 0.40$ what is the probability that in a sample of size $n = 100$ the sample proportion would differ in absolute value from $p = 0.40$ by at least 0.05?

(b) For the population in (a) if p is unknown, and no prior information is available regarding p, what sample size is necessary to insure that the sample proportion $\hat{p}$ will be within 0.05 of the true value of p with a probability of 0.99?

(c) In a random sample of 225 viewers, 125 expressed a desire to rent the new channel. Use this information to determine a 95% confidence interval for p.

(d) Use the information in (c) to test the null hypothesis $H_0 : p \leq 0.40$. What is the p-value of your computed test statistic? What is your conclusion using a type I error probability of $\alpha = 0.10$?

(e) What sample size n would be required to guarantee that the power of the test in (d) is 0.90 against the alternative $p = 0.50$? Assume in (d) that $\alpha = P[\text{Type } I \text{ error}] = 0.05$.

3. A sample of race results obtained from a local race track showed the following results relating post position to the number of winners. Use a χ^2 test to determine whether there is reason to believe that the chance of winning is related to post position.

Post Position	1	2	3	4	5	6	7	8
No. of Winners	32	28	22	19	18	13	15	13

4. An insurance company wishes to determine the premium for insurance policies for nuclear power plants. For part of the calculation it is assumed that the average number of fires per plant is one every twenty years and that the underlying process is Poisson. What is the probability that a plant could have two or more fires in a twenty-year period?

Questions for Chapter 1

1. Let $(x_1, x_2, \ldots, x_n)$ denote a random sample from a population with mean μ and variance σ^2.

   (a) Show that $E[\bar{x}] = \mu$ where $\bar{x} = \frac{1}{n}\sum_{i=1}^{n} x_i$. [NOTE: the expected value of a sum is the sum of the individual expected values.]

   (b) Show that $V(\bar{x}) = \sigma^2/n$. [NOTE: the variance of a sum of independent random variables is the sum of the variances.]

   (c) Let $s^2 = \sum_{i=1}^{n}(x_i - \bar{x})^2/(n-1)$. Show that $\sum_{i=1}^{n}(x_i - \bar{x})^2 = \sum_{i=1}^{n}(x_i - \mu)^2 - n(\bar{x} - \mu)^2$ and use this to show that $E[s^2] = \sigma^2$.

2. A hypergeometric distribution is given by

$$f(x) = \frac{\binom{N}{x}\binom{M-N}{n-x}}{\binom{M}{n}}.$$

   Show that a recursive formula for determining $f(x)$ in terms of $f(x-1)$ is given by $f(x)/f(x-1) = \dfrac{(N-x+1)(n-x+1)}{x(M-N-n+x)}$.

3. A random variable X has a mean of 0 and a variance of 1. Show that the random variable $Y = \mu + \sigma X$ has a mean of μ and a variance of σ^2. (μ and σ are constants).

4. A random variable X has a point binomial distribution with $p = P[X = 1]$ and $(1-p) = P[X = 0]$ where 0 and 1 are the only two values that X can assume. Show that $E[X] = p$ and $V(X) = p(1-p)$.

5. A random variable X has a normal density with mean μ and variance σ^2 given by

$$f(x) = \frac{1}{\sqrt{2\pi}} \frac{1}{\sigma} e^{-1/2(x-\mu)^2/\sigma^2}, \quad -\infty \le x \le \infty.$$

   (a) Show that $E[X] = \mu$ and $V[X] = \sigma^2$.

   (b) A random sample of size n, $(x_1, x_2, \ldots, x_n)$, is selected from this density. Show that the likelihood function is given by

$$L = \left(\frac{1}{\sqrt{2\pi}\,\sigma}\right)^n e^{-1/2\sum_{i=1}^{n}(x_i-\mu)^2/\sigma^2}$$

(c) Show that the logarithm of the likelihood function is given by

$$\ln L = -\frac{n}{2}\ln[2\pi\sigma^2] - \frac{1}{2}\sum_{i=1}^{n}(x_i - \mu)^2/\sigma^2.$$

(d) Show that the maximum likelihood estimators of μ and σ^2 are given by $\bar{x} = \frac{1}{n}\sum_{i=1}^{n}x_i$ and $(n-1)s^2/n = \sum_{i=1}^{n}(x_i-\bar{x})^2/n$ by maximizing $\ln L$ with respect to μ and σ^2. (HINT: Write $\sum_{i=1}^{n}(x_i - \mu)^2 = \sum_{i=1}^{n}(x_i - \bar{x})^2 + n(\bar{x} - \mu)^2$.)

6. A random sample of size n is selected from a point binomial distribution. The sample observations are denoted by $(x_1, x_2, \ldots, x_n)$. Recall that each x_i has the value 0 or 1. The density of $y = \sum_{i=1}^{n}x_i =$ (number of occurrences of $X = 1$) is given by $f(y) = \binom{n}{y}p^y(1 - p)^{n-y}$, $0 \leq y \leq n$.

(a) Show that $E[y] = np$ and $V[y] = np(1 - p)$ using the fact that $E[x] = p$ and $V[x] = p(1 - p)$. Also use the fact that since the x_i, $i = 1, 2, \ldots, n$, are mutually independent the expected value of a sum is the sum of the expected values and the variance of the sum is the sum of the variances.

(b) Show that the logarithm of the likelihood function for the sample in (a) is given by

$$\ln L = \ln\left[\binom{n}{y}\right] + y\ln p + (n - y)\ln(1 - p)$$

and that the maximum likelihood estimator of p is y/n.

(c) Use the fact that $E[y] = np$ and $V[y] = np(1 - p)$ to show that $[y^2/n^2]$ is a biased estimator of p^2. (HINT: Show that $E[y^2/n^2] = p^2 + p(1 - p)/n$ using the fact that, in general for any random variable Z, $E[Z^2] = V[Z] + (E[Z])^2$.)

7. The density for a Poisson distribution with parameter λ is given by $f(x) = e^{-\lambda}\lambda^x/x!$, $0 \leq x \leq \infty$.

(a) Show that for a random sample of size n, $(x_1, x_2, \ldots, x_n)$, the likelihood function is given by

$$L = \prod_{i=1}^{n}[e^{-\lambda}\lambda^{x_i}/x_i!].$$

(b) Show that the logarithm of the likelihood function is given by

$$\ln L = -n\lambda + (\ln \lambda)\sum_{i=1}^{n} x_i - \sum_{i=1}^{n} \ln(x_i!).$$

(c) Show that the maximum likelihood estimator of λ is given by $\bar{x} = \sum_{i=1}^{n} x_i/n$.

8. Suppose you know that a random variable X can be expressed as the sum of three *independent* random variables $X = 3V + \frac{1}{2}U + 3Z$ where V has a t distribution with 11 degrees of freedom, U has a χ^2 distribution with 16 degrees of freedom and Z has an F distribution with 2 and 8 degrees of freedom. Show that $E[X] = 12$ and $V[X] = 51$.

2
Univariate Data Analysis

As outlined in Chapter 1 the techniques of statistical inference usually require that assumptions be made regarding the sample data. Such assumptions usually include the type of sampling process that produced the data and in some cases the nature of the population distribution from which the sample was drawn. When assumptions are violated the techniques employed can lead to misleading results. Good statistical practice therefore requires that the data be studied in detail before statistical inference procedures are applied. The techniques of *exploratory data analysis* are designed to provide such a preliminary view.

Exploratory data analysis is more than a summary of data measures and various graphical displays. It also includes detecting and evaluating outliers, assessing normality, finding suitable data transformations, and finally studying residuals from fitted models. In other words, data analysis is required to ensure that the data, the assumptions and the techniques conform. Transformations of the data are usually designed to produce more normal-like distributions. The study of residuals from fitted models is done to test the adequacy of the model and its underlying assumptions. The purpose of this chapter is to provide a survey of the topics of data analysis, outliers, normality and data transformations. In later chapters the study of outliers will include the measurement of the influence such observations have on parameter estimates.

2.1 Data Analysis for Univariate Samples

Given a sample or batch of n observations on a variable X, graphical displays are useful for examining the shape of the sample distribution. Two common methods for representing sample distributions are the *stem and leaf display* and the *frequency distribution*. In both procedures the range of the variable X is first divided into non-overlapping intervals or classes. The observations are then assigned to the interval to which they belong. The result of both these procedures is a graphical display showing how the observations are distributed over the range. An alternative graphical display of a sample of observations is the *quantile plot*. The quantile plot is based on the sample order statistics and the sample p-values.

Example

The data displayed in Table 2.1 show the distribution of the number of television sets per 100 people for a selection of 30 countries. This data will be used throughout this chapter to provide examples.

Table 2.1. World Television Ownership 1975 or 1976 for Selected Countries in Northern Hemisphere

Country	No. of TV Sets per 100 people
United States	59
Canada	36
Sweden	36
Denmark	32
United Kingdom	32
East Germany	31
West Germany	30
Finland	30
Belgium	29
Switzerland	29
Netherlands	27
Norway	27
France	27
Austria	26
Czechoslovakia	25
Luxembourg	25
Hungary	23
Japan	23
Soviet Union	22
Italy	22
Spain	21
Poland	20
Israel	19
Kuwait	19
Ireland	19
Bulgaria	17
Yugoslavia	16
Lebanon	14
Rumania	14
Greece	13

Source: The World in Figures
Second Edition 1978
Facts on File Inc.
119 West 57 Street
New York, N.Y. 10019
Compiled by The Economist Newspaper Ltd., England

2.1.1 Stem and Leaf Plots

In the stem and leaf display, the range of X is divided into *intervals* of width W where $W = k10^p$, k is usually 0.2, 0.5 or 1 and p is a positive or negative integer. The observations actually appear in the display. The *boundaries* of the intervals form the stem while the observations within each interval form the leaves. The variation in the number of observations per interval over the *range* illustrates the shape of the distribution. If $k = 1$ each stem uses all ten integers for leaves, while if $k = 0.2$ each stem uses two cycles of the ten integers for leaves. If $k = 0.5$ two adjacent stems are required to cycle through all ten integers. The construction of a stem and leaf plot is best illustrated by using an example.

Example

From the data displayed in Table 2.1 we can observe that the range is [13, 59] and hence a convenient interval width is given by $5 = (0.5)(10)$. Since $k = 0.5$ two adjacent stems are required for one cycle of the ten integers. The resulting stem and leaf display is given in Figure 2.1.

Stem	Leaves
1	3 4 4
1	6 7 9 9 9
2	0 1 2 2 3 3
2	5 5 6 7 7 7 9 9
3	0 0 1 2 2
3	6 6
4	
4	
5	
5	9

Figure 2.1. Stem and Leaf Plot for Distribution of T.V. Ownership

Optimum Number of Intervals

A variety of rules have been used for determining the optimum number of intervals. For n observations Sturges (1926) suggested that the number of intervals should be determined from $L = 1 + 3.32 \log(n)$. A formula due to

Dixon and Kronmal (1965) suggests $L = 10 \log(n)$ while Velleman (1976) suggests $L = 2\sqrt{n}$. The formula due to Sturges produces the smallest value of L. For $n = 30$, the three values of L are 5.9, 14.8 and 11.0 respectively for Sturges, Dixon and Kronmal, and Velleman. The number of intervals used in the above example was 10 which is within the range suggested for L by the three criteria.

Examples

For 30 observations in the range [327, 1436] a convenient interval width is $W = (10)^2 = 100$. Using the stems 300, 400, ... , 1400 we have a total of 12 stems. The leaves provide the second digit and hence the numbers must be rounded to the nearest 10 units (i.e. 327 becomes 330). The leaves are therefore increments of 10. For 30 observations in the range [935, 3182] a convenient width is $W = (0.2)(10)^3 = 200$. Using the stems 900, 1100, ... , 3100 we have a total of 12 stems.

2.1.2 Frequency Distributions

A frequency distribution is constructed by dividing the range of X into *classes* and then assigning the observations to the appropriate class. The classes need not have equal width. The number of observations falling into each class is called the *class frequency*. The frequency distribution is commonly summarized and displayed using a *bar chart* or *histogram*.

Example

For the data displayed in Table 2.1 the classes can be chosen to conform to the intervals used in the stem and leaf display. The classes are given by 10-14, 15-19, 20-24, ... , 50-54, 55-59. The frequency distribution for this data is given in Table 2.2. The histogram for this distribution is given in Figure 2.2. The vertical scale represents the class frequency and the horizontal scale displays the range of X. The height of each bar corresponds to the class frequency.

Relative Frequency

For comparison to a probability density, the vertical scale may be transformed so that the area of each bar represents the *relative frequency* density in that class, and the total area for the 10 bars is 1. This scale is shown to

Table 2.2. Frequency Distribution for T.V. Ownership

Class Limits	10–14	15–19	20–24	25–29	30–34	35–39	40-44	45–49	50–54	55–59
Class Boundaries	9.5–14.5	14.5–19.5	19.5–24.5	24.5–29.5	29.5–34.5	34.5–39.5	39.5–44.5	44.5–49.5	49.5–54.5	54.5–59.5
Frequency/ Relative Freq.	3/ 0.10	5/ 0.17	6/ 0.20	8/ 0.27	5/ 0.17	2/ 0.07	0/ 0.00	0/ 0.00	0/ 0.00	1/ 0.03
Cumulative Frequency/ Relative Cumulative Frequency	3/ 0.10	8/ 0.27	14/ 0.47	22/ 0.73	27/ 0.90	29/ 0.97	29/ 0.97	29/ 0.97	29/ 0.97	30/ 1.00

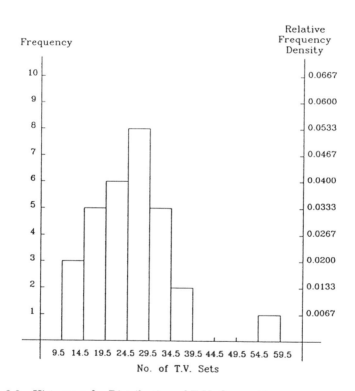

Figure 2.2. Histogram for Distribution of T.V. Ownership

the right of Figure 2.2. In the example each frequency must be divided by
30×5 or equivalently multiplied by 0.0067. Since the area of each bar is
proportional to class frequency, if the classes do not have equal width the
heights of the bars will not be proportional to class frequency.

Cumulative Frequency

The *cumulative frequency* distribution is sometimes useful for comparison to other distributions. The cumulative frequency distribution is shown in Table 2.2 above and plotted in Figure 2.3 below. By dividing the cumulative frequency by the total frequency, the vertical scale can be changed to *relative cumulative frequency*, or equivalently the *sample cumulative probability*. The relative cumulative frequency is shown on the right vertical scale of Figure 2.3.

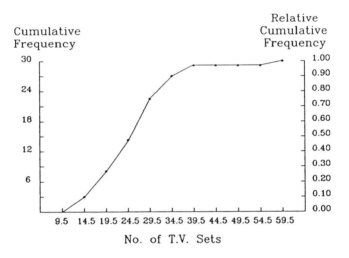

Figure 2.3. Cumulative Frequency Plot

2.1.3 Quantile Plots

Another useful graphical display of a sample distribution is provided by a *quantile plot*. Given a sample of n observations $(x_1, x_2, x_3, \ldots, x_n)$, the *quantiles* are these same observations ordered from smallest to largest and denoted by $x_{(1)}, x_{(2)}, x_{(3)}, \ldots, x_{(n)}$. These ordered observations are also called the *order statistics* for the sample. Corresponding to each *quantile value*, say $x_{(i)}$, is its *p-value* which indicates the position of the *quantile* in the data set. For the quantile value $x_{(i)}$ the p value is given by $(i - .5)/n$. A quantile plot for the data set is a plot of the quantiles, denoted by $Q(p)$ on the vertical axis, against p on the horizontal axis. A quantile of a data set is related to the *percentile* of the data set in that the p-th quantile is the $100p$-th percentile.

Motivation for the use of $(i-.5)/n$ can be obtained from Figure 2.4 below. In the figure a theoretical distribution function for the normal density is shown, along with rectangles showing the locations of the $x_{(i)}$. The heights of the rectangles corresponding to the base of $x_{(i-1)}$ to $x_{(i)}$ is shown as both $(i-1)/n$ and i/n. The use of $(i-.5)/n$ is therefore a compromise.

The quantile plot is a useful display of the data. Unlike the stem and leaf plot and the histogram, the shape of the quantile plot does not depend on the choice of class boundaries. Like the stem and leaf plot it displays all the data. Every data point on the graph corresponds to a value in the data set. Exact duplicates show as separate points. The slope of the plot in any region is an indication of the density of points in that region. If the slope is relatively flat in a region then the density is relatively high.

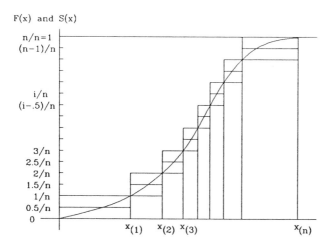

Figure 2.4.   Theoretical Distribution Function and Sample Cumulative Probability Distributions

Example

For the data set introduced in Table 2.1 the quantiles are given in Table 2.3. The quantile plot for this data is given below in Figure 2.5. In Figure 2.5 for the region between $Q(p) = 25$ and $Q(p) = 30$, p increases from .483 to .783.

The slope in this region is $5/.30 = 16.7$. For comparison the slope between $Q(p) = 13$ and $Q(p) = 17$ is $4/.133 = 30.1$. Another relatively flat region is between $Q(p) = 19$ and $Q(p) = 23$ with slope $4/.267 = 15.0$. There are two long horizontal regions in the plot corresponding to $Q(p) = 19$ and $Q(p) = 27$. In each case the change in p is .10 corresponding to a frequency of 3 out of 30. In comparison to this quantile plot, the frequency distribution hides these flat regions because it only shows frequency by classes.

Table 2.3. Quantiles for Distribution of T.V. Ownership

$Q(p)$	13	14	14	16	17	19	19	19	20	21	22	22	23	23	25
p	.017	.050	.083	.117	.150	.183	.217	.250	.283	.317	.350	.38	.417	.450	.483
$Q(p)$	25	26	27	27	27	29	29	30	30	31	32	32	36	36	59
p	.517	.550	.583	.617	.650	.683	.717	.750	.783	.817	.850	.883	.917	.950	.983

Quantile Plot for a Normal Distribution

It is useful when examining quantile plots to have in mind the shape of the quantile plot for a normal distribution as shown in Figure 2.6 below. Notice how the curve has phases of a slow build-up in p followed by a fast build up in p and finishing with a slow increase in p. These three phases correspond to the two tails and the central part of the normal distribution. The plot in Figure 2.6 provides a useful guide for studying quantile plots in general. By comparing quantile plots to this guide the shape of a distribution can be compared to the normal. This comparison will be made more formally in 2.3 where tests for normality are discussed. The quantile plot is also useful for examining the shape of a distribution which will be discussed next.

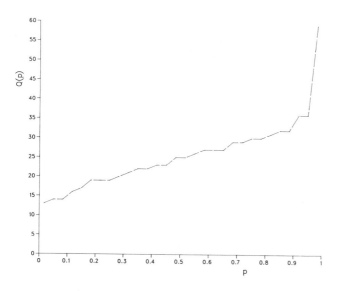

Figure 2.5.   Quantile Plot for Distribution of T.V. Ownership

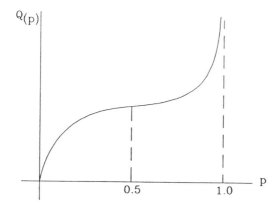

Figure 2.6.   Quantile Plot for a Normal Distribution

2.2 Characteristics of Sample Distributions

An examination of various graphical displays for sample distributions, permits one to observe various characteristics of the distribution such as location, spread and shape. The *location* or *central tendency* for the distribution indicates the middle or central value of the data, while the *spread* or *dispersion* characterizes the amount of variation in the distribution. *Shape* is usually concerned with the tails of the distribution; both tail length and the symmetry of the left and right tails are of interest. *Skewness* refers to an inequality between the two tails, and *kurtosis* refers to the length and thickness of the tails. A number of summary measures have been devised to measure these characteristics for sample distributions.

2.2.1 Measures of Location

Mean, Median and Mode

The most commonly used measure of central tendency is the (arithmetic) *mean* or *sample average*, $\bar{x} = \sum_{i=1}^{n} x_i/n$ defined in Chapter 1. As outlined in Chapter 1, this measure is commonly used in statistical inference procedures. A major disadvantage of the mean as a measure of centrality, is its sensitivity to departures from symmetry between the left and right extremes of the distribution. A preferable measure of centrality in such cases is the *median*, x_M, which is the value of the variable that divides the frequency equally into left and right halves ($Q(.50)$ or the 50th percentile). If the width of the interval containing the right half of the distribution exceeds the width of the interval containing the left half, then the distribution is skewed to the right. In such situations the mean, $\bar{x}$, usually exceeds the median, x_M. In the opposite case the distribution is skewed to the left and the mean, $\bar{x}$, is usually less than the median, x_M. The graphs in Figure 2.7 illustrate the relationship between the mean, median and the direction of the skewness.

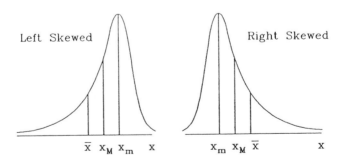

Figure 2.7. Relationship Between Mean, Median and Mode in Skewed Distributions

The two distributions shown in Figure 2.7 suggest a third measure of centrality which corresponds to the peak of the curve. The value of X corresponding to the maximum frequency is called the *mode*, x_m. Distributions may have several peaks and hence the mode is not necessarily unique. As can be seen from Figure 2.7, the mode, x_m, for a *unimodal* continuous distribution is usually on the opposite side of the median from the mean, and hence is furthest away from the direction of skewness.

Example

For the data in Table 2.1 the three measures of central tendency are given by

$$\bar{x} = 25.43;$$

$$x_M = 25 \qquad \text{since the 15th and 16th observations are both 25;}$$

and $\quad x_m = 19 \text{ or } 27 \quad$ since both 19 and 27 occur three times.

The one extreme observation to the right causes the mean to exceed the median.

Geometric Mean

A useful measure of location when the observations are ratios, such as percentages or indices is the *geometric mean*. The advantage of the geometric mean in such cases is that it preserves the inverse property. For a sample of observations $(x_1, x_2, \ldots, x_n)$ the geometric mean is given by

$$\tilde{x}_G = \sqrt[n]{\prod_{i=1}^{n} x_i},$$

which is the n-th root of the product of the n values. The logarithm of the geometric mean is the arithmetic mean of the logarithms of the observations

$$\ln(\tilde{x}_G) = \frac{1}{n} \sum_{i=1}^{n} \ln x_i,$$

and hence the geometric mean is easily computed using logarithms.

The preservation of the inverse property is useful when averaging ratios. For the inverses of the observations $1/x_1, 1/x_2, \ldots, 1/x_n$, the geometric mean is given by

$$\tilde{x}_G^* = \sqrt[n]{\prod_{i=1}^{n} 1/x_i},$$

which is equivalent to $1/\tilde{x}_G$.

Example

Given the debt/equity ratios 1.5, 0.3, 0.7 and 1.2 for four major subsidiaries of a large company, the geometric mean is given by

$$\sqrt[4]{(1.5)(0.3)(0.7)(1.2)} = 0.784.$$

The geometric mean of the equity/debt ratios for the same data is given by

$$\sqrt[4]{\left(\frac{1}{1.5}\right)\left(\frac{1}{0.3}\right)\left(\frac{1}{0.7}\right)\left(\frac{1}{1.2}\right)} = 1.2755 = 1/0.784.$$

Average Rate of Change

An alternative use of the geometric mean is the determination of the *average rate* or percentage increase. If V_n is the value of an item after n periods, and V_0 is the value at the beginning of the first period, the average rate of increase over the n periods is given by

$$\tilde{R}_G = \left(\sqrt[n]{\frac{V_n}{V_0}} - 1\right)100.$$

Example

The population of a city grew from 91,600 to 227,800 over the 10 years from Jan. 1, 1970 to Dec. 31, 1979. The average rate of growth is therefore given by

$$\tilde{R}_G = \left[\sqrt[10]{\frac{227,800}{91,600}} - 1\right] \times 100 = 9.5\%.$$

2.2.2 Measures of Spread

Variance, Standard Deviation

The amount of variation or dispersion in the distribution is most commonly measured using the *sample variance* $s^2 = \sum_{i=1}^{n}(x_i - \bar{x})^2/(n-1)$ defined in Chapter 1. The variance therefore measures dispersion relative to the mean. The square root of the variance, s, is known as the *sample standard deviation*. For the data in Table 2.1, the variance s^2 is given by 80.43 and the standard deviation s by 8.97. Like the mean, s^2 is sensitive to extreme values.

Median Absolute Deviation

An alternative measure of dispersion is the *median absolute deviation,* MAD, which is obtained by taking the median of the absolute deviations from the median. MAD = median $\{|x_i - x_M|\}$. For the example data the MAD = 5. If the extreme value 59 is removed from the example distribution, the standard deviation s becomes 6.73 which is closer to the MAD. The MAD is useful as a measure of dispersion when there are extreme values.

Range, Interquartile Range

The simplest measure of dispersion is the *range, R,* which is the difference between the largest observation U, and the smallest observation L; $R = (U - L)$. A similar measure to the range is the *interquartile range, Q,* which measures the range between the first and third quartiles. The first quartile, Q_1, is the value of X such that 25% of the observations are below Q_1, and the third quartile, Q_3, is the value of X such that 25% of the observations are above Q_3. The median x_M is also called the second quartile, Q_2. The interquartile range or *quartile deviation* is therefore $Q = (Q_3 - Q_1)$ which is the range of the central half of the data. Five useful values which show the spread of the distribution in rank order from smallest to largest are L, Q_1, x_M, Q_3 and U.

Box Plots

A useful graphical display of the summary measure information is provided by the *box plot.* A box is constructed to represent the interquartile range or quartile deviation with a vertical line somewhere in the box at the median. Horizontal lines called *"whiskers"* are extended from the edges of the box in both directions to at most a distance of 1.5 *interquartile distances,* 1.5 Q. The ends of the whisker regions are denoted by an "x". If the smallest value L and/or the largest value U are within the whisker region then the whisker stops at that value. Values which are beyond the whiskers are potential *outliers.* These extreme observations are denoted by 0 if they are within 3Q of the box and with an asterisk if they are beyond this range. The mean is plotted with a + sign along the same horizontal line as the whiskers. For most distributions the + sign appears in the box although this is not necessary.

Example

For the data in Table 2.1 $L = 13$, $Q_1 = 19$, $x_M = 25$, $\bar{x} = 25.43$, $Q_3 = 30$ and $U = 59$. The box plot for this data is shown in Figure 2.8 below. From the figure we can ascertain that the median is close to the mean, and that the median is close to the center of the interquartile range. These are indications of symmetry. The interquartile range of $Q = 11$, multipled by 1.5 yields 16.5, and hence the whiskers can extend to $19 - 16.5 = 2.5$, and to $30 + 16.5 = 46.5$. The value $L = 13$ is therefore within the lower whisker region, while the value of $U = 59$ is beyond the upper whisker region and is denoted by 0. The end of the upper whisker region is denoted by x. Since the value 59 lies beyond the upper whisker it is a potential outlier. A discussion of outliers will be provided in the next section.

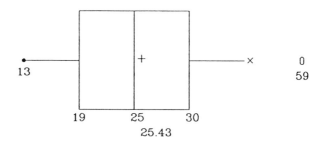

Figure 2.8. Box Plot for Distribution of T.V. Ownership (not drawn to scale)

2.2.3 Measures of Skewness

Coefficient of Skewness

The discussion of measures of central tendency and dispersion above has also introduced several measures of skewness. As shown above, the relative positions of the mean and the median can be used to measure the direction of skewness. The relationship between the median and the quartiles may also be used as a measure of skewness. In a *symmetric distribution* the distances given by $(x_M - Q_1)$ and $(Q_3 - x_M)$ should be equal. If $(x_M - Q_1)$ exceeds $(Q_3 - x_M)$ the skewness is to the left. If the opposite is true the distribution is skewed to the right. A useful *coefficient of skewness* based on these quartile distances is given by

$$Sk = [(Q_3 - x_M) - (x_M - Q_1)]/[Q_3 - Q_1].$$

In the box plot the length of the "whiskers" to the left and right of the box can be used as indicators of skewness.

A Symmetry Plot

The technique of comparing the distances $(x_M - Q_1)$ to $(Q_3 - x_M)$ to measure skewness can be extended. A *symmetric distribution* would require that the i-th lower observation from the median and the i-th higher observation from the median be the same distance from the median. Recall that the quantiles of the observed data are the ordered observations $(x_{(1)}, x_{(2)}, \ldots, x_{(n)})$. The symmetry condition requires that

$$x_M - x_{(i)} = x_{(n-i+1)} - x_M \quad \text{where} \quad i = 1, 2, \ldots, m, \qquad (2.3)$$
$$\text{and } m = (n-1)/2 \text{ if } n \text{ odd,}$$
$$\text{and } m = n/2 \text{ if } n \text{ even.}$$

A plot of the left side vs. the right of this equation for all i is called a *symmetry plot* and should yield a 45° line for a symmetrical distribution.

Example

For the data in Table 2.1, $(x_M - Q_1) = 6$ and $(Q_3 - x_M) = 5$ indicating a slight skewness to the left. The value of the coefficient of skewness Sk is $-1.0/11.0 = -0.09$. The "whiskers" of the box plot however indicate an outlier to the right which is the observation 59. With this observation included the distribution is skewed to the right. For this data the values of the left and right hand sides of the equality in (2.3) are given in Table 2.4 below. The nature of the skewness is easily recognized from a comparison of the two middle rows of the table. With the exception of the largest observation, 59, the distribution is almost perfectly symmetric about the median, $x_M = 25$.

Index of Skewness

A classical measure of skewness is based on the *third moment* about the mean, $E[(x - \mu)^3]$. To create a unit free measure this moment is divided by σ^3 where $\sigma^2 = E[(x - \mu)^2]$ is the variance. The resulting ratio $\gamma = E[(x - \mu)^3]/\sigma^3$ is commonly used as an *index of skewness*. Large negative values of γ are indications of skewness to the left while relatively large positive values suggest skewness to the right. *Left* and *right skewness* is sometimes referred to as *negative* and *positive skewness* respectively.

Table 2.4. Data for Symmetry Plot for Distribution of T.V. Ownership

i	1	2	3	4	5	6	7	8	9	10	11	12	13	14	15
$x_M - x_{(i)}$	12	11	11	9	8	6	6	6	5	4	3	3	2	2	0
$x_{(n-i+1)} - x_M$	24	11	11	7	7	6	5	5	4	4	2	2	2	1	0
$n-i+1$	30	29	28	27	26	25	24	23	22	21	20	19	18	17	16

Examples of Skewed Distributions

The four distributions in Figure 2.9 below display various degrees of skewness in comparison to a standard normal distribution. In each case the mean and variance of the skewed distribution are 0 and 1 respectively. The index of skewness, γ, is shown for each of the four examples. In panels (a) and (b), the skewness is very mild with panel (a) showing negative skewness $\gamma = -0.50$ and panel (b) showing positive skewness, $\gamma = 0.50$. In panel (c), the left tail is negligible and the right tail is lengthened to yield a skewness index of $\gamma = 1.41$. Finally in panel (d), the exponential shaped distribution has no left tail and the skewness is now $\gamma = 2.0$. The four panels also give the index of *kurtosis*, δ, for the skewed distributions. The index δ, which measures *peakedness*, will be discussed below.

Sample Form of Index of Skewness

For sample distributions the index of skewness is estimated using the sample moments $\sum_{i=1}^{n}(x_i - \bar{x})^3$ and $\sum_{i=1}^{n}(x_i - \bar{x})^2$. The index is usually estimated using the ratio of $n\sum(x_i - \bar{x})^3/(n-1)(n-2)$ to s^3 where s^2 is the sample variance. The sample index of skewness is given by

$$g = n\sum_{i=1}^{n}(x_i - \bar{x})^3/(n-1)(n-2)s^3.$$

In large samples from a normal distribution, the mean of g is 0 and the variance of g is approximately $6/n$. For the example data in Table 2.1 $g = 1.81$. The critical value for this sample index of skewness is $z = 1.81/\sqrt{6/30} = 4.05$ which is significant at conventional levels.

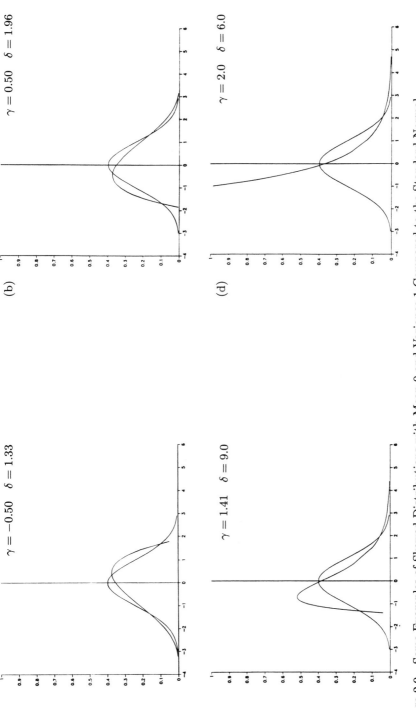

Figure 2.9.   Some Examples of Skewed Distributions with Mean 0 and Variance 1 Compared to the Standard Normal

2.2.4 Measures of Kurtosis

For symmetric distributions the proportion of the frequency that lies at the centre of the distribution is also of importance in defining shape. This characteristic is usually described by the *kurtosis* measure, which is often based on comparison to a normal distribution. If the centre of the distribution contains a larger proportion of the frequency than the normal, then the kurtosis is positive, while if the reverse is true the kurtosis is negative. A high density at the centre of the distribution is usually accompanied by long tails. The measures of kurtosis introduced here are based on a comparison of the sample distribution to a normal distribution.

Measuring Kurtosis Using the Spread

The quartile deviation, Q, measures the *spread* between the 25th and 75th percentiles. For a standard normal distribution the spread between the two quartiles is $Z_{.25} - Z_{.75} = 0.675 - (-0.675) = 1.350$, where Z_p denotes the Z value such that the area to the right is p. For any normal distribution this spread is given by 1.35σ. This value of the spread is a special case of the spread between $x_{(n-i+1)}$ and $x_{(i)}$, or equivalently the spread between the quantiles $Q(1-p)$ and $Q(p)$ where $p = (i - .5)/n$. For any normal distribution the spread between the quantiles is given by $Z_p\sigma - (-Z_{1-p})\sigma = 2Z_p\sigma$. For a standard normal distribution the standard theoretical spread is $2Z_p$. We can therefore compare the *observed spread* $Q(1-p) - Q(p)$ to the standard *theoretical spread* $2Z_p$. If the observed distribution is similar in shape to a normal density, then the ratios of $[Q(1-p) - Q(p)]$ to $2Z_p$ for various values of $p = (i - .5)/n$ should be a constant value σ.

Example

For the example data a comparison of the spreads is carried out in Table 2.5 below. The ratios of observed spread to theoretical spread fluctuate around 8, and hence there is no indication of longer tails than a normal distribution. The average ratio is 7.5 which is close to the sample standard deviation of $s = 8.97$.

An Index of Kurtosis

A classical measure of kurtosis is based on the fourth moment about the mean $E[(x-\mu)^4]$. This *index of kurtosis* is given by $\delta = E[(x-\mu)^4]/\sigma^4 - 3$. Since the value of $E[(x - \mu)^4]/\sigma^4$ is 3 for a normal distribution, this index

Table 2.5. Comparison of Spreads for Distribution of T.V. Ownership

i	1	2	3	4	5	6	7	8	9	10	11	12	12	14	15
p	.017	.050	.083	.117	.150	.183	.217	.250	.283	.317	.350	.383	.417	.450	.483
$Q(1-P)$	59	36	36	32	32	31	30	30	29	29	27	27	27	26	25
$Q(p)$	13	14	14	16	17	19	19	19	20	21	22	22	23	23	25
Observed Spread	46	22	22	16	15	12	11	11	9	8	5	5	4	3	0
Standard Theoretical/$\sigma = 2Z_p$ Spread	4.24	3.29	2.77	2.38	2.08	1.81	1.56	1.35	1.14	.96	.76	.60	.42	.26	.08
Ratio	10.8	6.7	7.9	6.7	7.2	6.6	7.1	8.1	7.9	8.3	6.6	8.3	9.5	11.5	0

measures kurtosis relative to a normal distribution. Positive values of δ indicate longer thicker tails than a normal while negative values indicate shorter thinner tails. The minimum value of δ is -2 and δ is related to the population skewness coefficient γ by the inequality $\delta \geq (\gamma^2 - 2)$. There is some dispute among statisticians regarding what properties of a distribution are measured by δ. It is generally believed that a positive δ, *positive kurtosis*, must be accompanied by a property of *peakedness* in the distribution. The peakedness property relative to a normal means that there is an excess frequency in the centre of the distribution. Peakedness is usually accompanied by relatively *long* or *fat tails*. A distribution with *negative kurtosis*, negative δ, is sometimes referred to as *platykurtic* while positive kurtosis is called *leptokurtic*. Negative kurtosis is also an indication of relatively *short* or *thin tails*.

A useful way of interpreting the kurtosis index, δ, was recently outlined by Moors (1988). For a standardized random variable $Z = (X - \mu)/\sigma$ the value of $(\delta+3) = E[(X-\mu)^4]/\sigma^4$ is equivalent to $E[(Z^2-1)^2]+1 = V[Z^2]+1$ since $E[Z^2] = 1$. Therefore $(\delta+3)$ measures the dispersion of X around the two points $\mu \pm \sigma$. This dispersion will be relatively large if the concentration of probability at the two points $\mu \pm \sigma$ is relatively low. In other words the index of kurtosis will be relatively high if

(a) There is a large concentration of probability in the centre around μ (peakedness); or

(b) There is a large concentration of probability in the tails of the distribution (fat tails).

High kurtosis is therefore a reflection of peakedness or fat tails or both.

Some Examples of Distributions with Kurtosis

Figure 2.10 below shows four non-normal symmetric distributions in comparison to the standard normal. Each of the non-normal distributions has a mean of 0 and a variance of 1. The index of kurtosis δ in each case is also indicated in the figure. In panel (a) the positive kurtosis is due to fat tails, while in panel (b) the positive kurtosis reflects the high density at the centre. In panel (c) negative kurtosis is shown. In panel (c) the density is relatively high in the neighborhood of $\mu \pm \sigma = \pm 1$. Panel (d) compares a *logistic distribution* with mean 0 and variance 1 to a normal distribution. The logistic distribution appears to display peakedness relative to the normal. The logistic distribution will be used in Volume II for logistic regression.

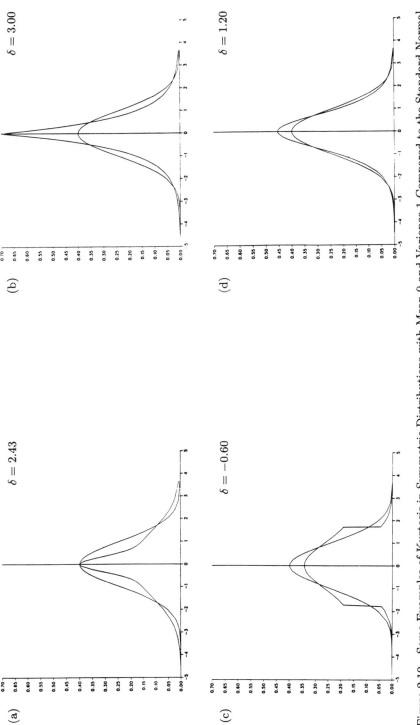

Figure 2.10.   Some Examples of Kurtosis in Symmetric Distributions with Mean 0 and Variance 1 Compared to the Standard Normal

Sample Measures of Kurtosis

For sample distributions the index of kurtosis is estimated by

$$d = \left[n(n+1) \sum_{i=1}^{n} (x_i - \bar{x})^4 / (n-1)(n-2)(n-3)s^4 \right] - 3(n-1)^2/(n-2)(n-3).$$

Under the assumption of normality the sample statistic d in large samples has a normal distribution with mean 0 variance $24/n$. For the data in Table 2.1 the kurtosis measure is $d = 2.65$. The critical value of the statistic is given by $z = 2.65/\sqrt{24/30} = 2.96$. Therefore the kurtosis value is significant at conventional levels.

A useful measure of kurtosis, due to Moors (1988), is based entirely on the quantiles of the sample distribution and is given by

$$K = [(E_7 - E_5) + (E_3 - E_1)]/(E_6 - E_2)$$

$$\text{where } E_i = Q(i/8) \qquad i = 1, 2, \ldots, 7.$$

The E measures are the quantiles which separate the distribution into eighths. The two terms in the numerator of K measure the deviation required to obtain 50 percent of the density. The first term accounts for 25 percent in the neighborhood of Q_3 and the second term accounts for 25 percent in the neighborhood of Q_1. The middle 25 percent in the neighborhood of Q_2 is omitted. This total deviation in the numerator is standardized by dividing by the quartile deviation. If the two deviations in the numerator are large relative to the quartile deviation, this is an indication of less density near Q_3 and Q_1 and greater density at the centre and/or in the two tails. For a standard normal density, K has the value of 1.233, while for a uniform density $K = 1.0$. For the *inverse triangular* distribution shown below the value of K is 0.518.

Example

For the T.V. data the values of E_1, E_3, E_5 and E_7 are 16.24, 22, 27, and 32 respectively. The quartile deviation Q determined previously was 11. The value of K in this case is therefore $\frac{10}{11} = 0.91$ which is smaller than the normal distribution measure of 1.233.

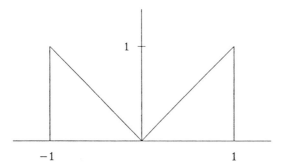

Figure 2.11. Inverse Triangular Distribution

2.2.5 Impact of Shape on Inference for μ and σ

Recall that if X is normally distributed, inferences regarding the mean μ can be made using the statistic $\sqrt{n}(\bar{x}-\mu)/s$, which has a t distribution with $(n-1)$ degrees of freedom. In large samples from non-normal distributions the central limit theorem indicates that the ratio $\sqrt{n}(\bar{x}-\mu)/s$ can be approximated by a standard normal distribution. In small samples, the impact of non-normality on the distribution of $\bar{x}$ can be partially assessed through the indices of skewness and kurtosis γ and δ. The values of the indices of skewness and kurtosis, for the distribution of $\bar{x}$, are given by $\gamma/\sqrt{n}$ and δ/n respectively. Thus as n gets large the kurtosis effect disappears more rapidly than the skewness effect. Power series expansions for the mean and variance of the t-ratio show that in large samples $E\left[\frac{\bar{x}-\mu}{s}\right] = \gamma/2\sqrt{n}$ and $\text{Var}\left[\frac{\bar{x}-\mu}{s}\right] = 1+\frac{1}{n}(2+7\gamma^2/4)$. In comparison to the skewness measure, the kurtosis measure therefore seems to have little influence on the first two moments of the t-ratio. Thus non-normality resulting in skewness seems to be more important than non-normality resulting in kurtosis.

For a normal population the numerator and denominator of $\sqrt{n}(\bar{x}-\mu)/s$ are independent. In general for any population the correlation between these two terms in large samples is $\gamma/\sqrt{\delta+2}$. Once again the magnitude of the skewness measure is more critical to the t-like behavior of the ratio. This correlation between the numerator and denominator causes the skewness in the ratio to be in the opposite direction to the skewness in the population. Monte Carlo studies have shown that in small samples if $\gamma > 1$ and $\delta > 4$, the ratio $\sqrt{n}(\bar{x}-\mu)/s$ is not t-like in its behavior. If $\gamma = 0$, however, studies have shown the distribution of the ratio to have thinner tails than the t distribution (see Miller (1986)).

Extreme observations or outliers have a major impact on the values of $\bar{x}$ and s and hence on the behavior of the t-ratio. In small samples therefore a more efficient procedure is to use a transformation of X before using a t-ratio. A study of transformations is presented in Section 2.5.

If the population is normal the statistic $(n-1)s^2/\sigma^2$ is χ^2 with $(n-1)$ d.f. As outlined in Chapter 1 this statistic is commonly used to make inferences about the population variance. If the normality assumption does not hold, the effects are described by Miller (1986) as catastrophic. The magnitude of the kurtosis index δ has a larger impact on the misbehavior of the χ^2-statistic than does the skewness index γ. For any distribution the variance of s^2 is given by $\sigma^4\left[\dfrac{2}{n-1} + \delta/n\right]$. This variance can be compared to $2\sigma^4/(n-1)$ which is the variance of s^2 under normality. Thus one impact of kurtosis, on the distribution of $(n-1)s^2/\sigma^2$, is to increase or decrease the variance depending on whether δ is positive or negative. In the case of non-normality, a correction to the distribution of s^2 can be applied using the correction factor $c = \left(1 + \tfrac{1}{2}d\right)^{-1/2}$, where d is the sample index of kurtosis. The resulting statistic $c(n-1)s^2/\sigma^2$ is approximately distributed as a χ^2 with $c(n-1)$ degrees of freedom.

An alternative approach to inference for the variance under non-normal distributions is based on the *jackknife procedure*. Denote by $s^2_{(i)}$ the sample variance determined after omitting the i-th observation. Let $\ell_{(i)} = \log s^2_{(i)}$, $i = 1, 2, \ldots, n$. The central limit theorem is then applied to obtain the distribution of $\bar{\ell} = \sum_{i=1}^{n} \ell_{(i)}/n$. In large samples the statistic $\sqrt{n}(\bar{\ell} - \log \sigma^2)/s_\ell$ has a standard normal distribution where s^2_ℓ is the sample variance for the $\ell_{(i)}$ given by $s^2_\ell = \sum_{i=1}^{n} (\ell_{(i)} - \bar{\ell})^2/(n-1)$. In order to apply the central limit theorem the observations $\ell_{(i)}$ are treated here as if they are mutually independent which they are not. Monte Carlo studies, however, have shown this procedure to be satisfactory.

2.3 Outliers

For a sample of observations, Barnett & Lewis (1978) define outliers to be observations which appear to be inconsistent with the remainder of the sample. An outlier may be the result of measurement error, execution error or inherent variability. Measurement error may be caused by inadequate measurement or incorrect recording of values, while execution error arises because of a biased sample or an improper definition of the population. Outliers can also arise naturally due to the inherent variability in the population, and may represent rare but not impossible events.

Outliers usually have a major impact on the values of estimators and test statistics which are derived from the sample. It is therefore tempting to reject such extreme observations. Because extreme observations can occur naturally, they sometimes carry information that can lead to important discoveries. Outliers may also indicate that the original assumptions need to be revised. Outliers therefore should be carefully evaluated and checked, and should only be discarded if they can be attributed to error. In the case of error they should be replaced by corrected or alternative values. Outliers which cannot be attributed to error are called *discordant* and should not be removed from the sample.

Given the presence of discordant outliers, the assumptions about the nature of the population should be re-examined. Rather than a *homogeneous* population, the sampled population may actually consist of a *mixture* of several *sub-populations*. Alternatively, the population may be composed of a major sub-population and a few minor sub-populations called *slippage* alternatives. A common model for mixtures assumes that the population is composed of k sub-populations with means and variances given by (μ_i, σ_i^2) $i = 1, 2, \ldots, k$. For the case of slippage, a common assumption is that there is one major sub-population with mean and variance (μ, σ^2) and a few very small populations with parameters $(\mu + \lambda_i, \theta_i \sigma_i^2)$ $i = 1, 2, \ldots, k$. Test statistics have been developed to judge the significance of outliers in some circumstances. The text by Barnett and Lewis (1978) contains a summary.

2.3.1 Detection of Outliers

In Section 2.2.2 a box plot was used to represent the variation in a sample of observations. The term potential outliers was attached to observations which were outside the whiskers of the box. Such observations were either less than the lower quartile Q_1 by at least 1.5 times the interquartile range, Q, or were greater than the upper quartile Q_3 by at least $1.5Q$.

For a normal distribution with mean μ and variance σ^2, the interquartile range is given by $Q = 1.35\sigma$. An additional $1.5Q$ added to both ends yields a range of $4Q = 5.40\sigma$. The outlier cutoffs are therefore $\mu \pm 2.70\sigma$ and hence 99.3% of the distribution is contained within these cutoffs.

Given a sample of size n for a normal population with order statistics $x_{(1)}, \ldots, x_{(n)}$, the sample will have no outliers if $\dfrac{Q_1 - x_{(1)}}{Q} \leq 1.5$ and if $\dfrac{x_{(n)} - Q_3}{Q} \leq 1.5$. An important question therefore is how often a sample of size n from a normal distribution will contain at least one outlier using this criterion. In a simulation study carried out by Hoaglin, Iglewicz and Tukey (1986) the proportion of samples containing at least one outlier was determined. From their study one can conclude that for a sample size n in the range [5, 20], 25% of the samples will contain at least one outlier. For $n = 50$, 100 and 200 the proportion of samples which contain at least one outlier were approximately 35%, 50% and 70% respectively.

Inference for Outliers

Given a sample of n values $(x_1, x_2, \ldots, x_n)$, with mean $\bar{x}$ and variance s^2, Shiffler (1987) has shown that the largest value of the standardized random value $|(x_i - \bar{x})/s|$ is given by $(n-1)/\sqrt{n}$. Using the critical value of 2.70 for outliers, as suggested above, implies that in samples of $n < 9$ no observations can be declared to be outliers regardless of their magnitude. If this outlier criterion is modified to $|(x_i - \bar{x})/s| > 4$, then in samples of $n < 18$ no outliers are possible. In small samples therefore it is preferable to use the quartile criterion where x is a potential outlier if $x < (Q_1 - 1.5Q)$ or $x > (Q_3 + 1.5Q)$.

Example

In Section 2.2.2 the example box plot showed one potential outlier which was the largest of the 30 observations. It would appear from the above mentioned simulation study that one such outlier might be expected to occur in at least 25% of the samples from a normal distribution. The particular outlier value, 59, corresponds to a Z value of $(59 - 25.43)/8.97 = 3.74$, which is certainly a rare occurrence.

The Z statistic value of 3.74 in the above example is a *studentized mean deviation* (SMD) statistic defined as $T = (x(n) - \bar{x})/s$. The SMD has been used to test for discordant outliers under the assumption of normality. Critical values of T for various values of n have been produced by Grubbs and Beck (1972). For the example, the T value of 3.74 may be compared to the critical values for $n = 30$ of 3.10, 2.75 and 2.56 which correspond to the p-values of 0.01, 0.05 and 0.10 respectively. The example outlier is therefore significant at the 0.01 level.

Tests for Outliers Using Order Statistics

There are a variety of other test procedures (discordancy tests) available for assessing outliers. Another useful test for normal populations is based on the sample order statistics $(x_{(1)}, x_{(2)}, \ldots, x_{(n)})$. This test was developed by Dixon (1950). For a single outlier the ratio

$$r_{10} = (x_{(n)} - x_{(n-1)})/(x_{(n)} - x_{(1)})$$

is determined and compared to tabular values based on the sample size n. For the example discussed above

$$r_{10} = (59 - 36)/(59 - 13) = 0.50.$$

Tables for r_{10} are published in Dixon (1951). From the tables for r_{10} for $n = 30$ the critical values of 0.341, 0.260 and 0.215 correspond to the p-values of 0.01, 0.05 and 0.10 respectively. The value of r_{10} is therefore significant at the 0.01 level.

The Dixon statistic r_{10} may also be used for low extremes, by replacing $x_{(n)} - x_{(n-1)}$ in the numerator of r_{10}, by $x_{(2)} - x_{(1)}$. The statistic r_{10} is only useful for situations in which there is only one extreme value. For up to two upper and two lower extreme values the statistics r_{ij} can be used where

$$r_{ij} = \left(x_{(n)} - x_{(n-i)}\right) / \left(x_{(n)} - x_{(1+j)}\right) \qquad i = 1, 2; \quad j = 0, 1, 2.$$

The subscript i denotes the number of upper extremes while the subscript j denotes the number of lower extremes. For testing the lower extremes the statistic is constructed using the corresponding lower extremes of the order statistics. If there are three potential outliers, one large and two small, the appropriate statistic for testing the large value is r_{12}, while the appropriate statistic for testing the smallest observation is the reverse of r_{21}. Thus the two statistics would be

$$r_{12} = \left(x_{(n)} - x_{(n-1)}\right) / \left(x_{(n)} - x_{(3)}\right) \qquad \text{and}$$
$$r_{21} = \left(x_{(3)} - x_{(1)}\right) / \left(x_{(n-1)} - x_{(1)}\right).$$

These statistics are designed to prevent other outliers from *masking* the differences for the outlier being tested.

Using Indices of Skewness and Kurtosis

The sample statistics g and d introduced in 2.2.3 and 2.2.4 to measure skewness and kurtosis have also been recommended for use in detecting outliers. The skewness measure is useful in detecting outliers in one extreme while the kurtosis statistic can be used to detect outliers in both extremes. If the sample consists of a small number of outliers from different normal distributions which differ from the main population in the mean but not the variance (location slippage), then g and d are ideal test statistics. Recall that our test for skewness and kurtosis for the example in the previous section showed that both g and d were significant.

2.3.2 Robust Estimation

When discordant outliers are judged to be significant it may be preferable to use alternative estimators for location (μ) and scale (σ). More *robust* estimators of the population mean μ, which are less sensitive to extreme values, are provided by the sample median and the trimmed mean. The *trimmed mean* is the mean of the observations that remain after trimming or excluding the observations in both tails of the distribution. A $p\%$ trimmed mean excludes the top $p\%$ and the bottom $p\%$ of the distribution and hence is a mean of $(100 - 2p)\%$ of the observations. The 25% trimmed mean has been called the *midmean* by Rosenberger and Gasko (1983) because it represents the mean of the middle half of the data. The midmean was shown by them to be a very efficient estimator of μ for a variety of extreme distributions.

For the example distribution, the mean without the observation 59 is 24.2 as compared to 25.43 when 59 is included. The median was 25. The 25% trimmed mean computed by omitting the lowest 7 and the highest 7 observations is 24.7. With only one extreme observation in this particular example the various measures of location are not that different.

The geometric mean, $\tilde{X}_G$, introduced in 2.2.1, is a useful measure of location when exteme observations are present because it is less sensitive to outliers in comparison to the arithmetic mean. For the observations 2, 4, 5, 8 and 29 the arithmetic mean is given by 9.6 while the geometric mean is given by 6.2. The geometric mean is most easily determined using the logarithms of the observations since

$$\log \tilde{x}_g = \frac{1}{n} \sum_{i=1}^{n} \log x_i.$$

For the T.V. data in Table 2.1, the geometric mean is given by 24.14 which is lower than the arithmetic mean of 25.43. The geometric mean is less sensitive to the extreme value of 59 for this distribution.

A commonly used robust estimator of scale is the median absolute deviation defined in the previous section. For a normal distribution the MAD is 0.6745σ, and hence the MAD divided by 0.6745 provides an estimator of σ. For the example, the MAD of 5 divided by 0.6745 yields the value of 7.4 which can be compared to the value of $s = 8.97$. Recall from the previous section that after omitting the observation 59 the value of s was reduced to 6.73. A second robust estimator of scale is provided by the quartile deviation Q divided by 1.35, which follows from the fact that for a normal distribution, $Q = 1.35\sigma$. For the sample data this estimator has the value $11/1.35 = 8.15$.

In our study of the techniques of data analysis thus far, we have seen that a few extreme observations can have a major impact on the values of various statistics that are determined from sample data. A few extreme observations may represent a rare departure from an otherwise normal distribution. It is also possible for the distribution to be non-normal-like, not because of a few extremes, but because the population distribution is truly non-normal. In the next section, procedures for the assessment of normality are outlined.

2.4 Assessing Normality

Many statistical techniques require the assumption of normality. In practical terms real data cannot be normally distributed in that all data are discrete and bounded. The normality assumption in practice simply requires that the distribution be sufficiently close to a normal density. If sample sizes are relatively large, the *normality assumption* is often a less stringent requirement. It is therefore of interest to assess the sample distribution to determine if the normality assumption is acceptable. If the normality assumption is required, but not met by the data, a transformation of the variable may be useful. Techniques for such transformations will be outlined in the next section. A discussion of procedures for testing *normal goodness of fit* is presented in this section.

   The proximity of a sample density to a normal density can be examined in a variety of ways. The procedures can be grouped into methods which compare the properties of the sample density, to the properties of a normal, and methods which compare the observed frequencies, to normal distribution expected frequencies. The latter category includes graphical as well as analytical comparisons.

2.4.1 K-S Test for Normality

In elementary textbooks the most commonly discussed procedure for *assessing normality* is the *Kolmogorov-Smirnov (K-S) test*. This test is popular partly because it can be used to compare sample distributions to any continous type theoretical distribution. It is based on a comparison of the observed cumulative probability function and a hypothesized theoretical cumulative probability function. The K-S test is an example of an *EDF (empirical distribution function)* goodness of fit procedure because it is based on a comparison of the empirical distribution function (EDF) with a theoretical distribution function. The original K-S test assumed that the parameters of the hypothesized theoretical distribution were known. More commonly in practice the required population parameters must be estimated from the sample. Lilliefors (1967) showed that in testing for normality when the population mean and variance are estimated from the sample, the tabular values of the test statistic are too conservative. A number of Monte Carlo simulation studies have shown that in assessing normality, most tests perform better if the mean and variance are estimated even when the true parameters are known. We shall assume throughout this section that the conventional unbiased estimators $\bar{x}$ and s^2 will be used to estimate μ and σ^2 for all goodness of fit tests.

To perform a K-S goodness of fit test for normality, the standardized order statistics $z_{(i)}$ are computed by transforming the sample order statistics using $z_{(i)} = (x_{(i)} - \bar{x})/s$. The theoretical cumulative probability corresponding to $z_{(i)}$ is then determined and denoted by $F_{(i)} = \Phi(z_{(i)})$, where Φ denotes the distribution function for a standard normal density. The K-S test statistic D is given by the maximum value of $|i/n - F_{(i)}|$ $i = 1, 2, \ldots, n$. Using the statistic $D(\sqrt{n} - 0.01 + 0.85/\sqrt{n})$ to denote the product of D and the factor $(\sqrt{n} - 0.01 + 0.85/\sqrt{n})$, the significance can be assessed by using the critical values 0.775 (0.15), 0.819 (0.10), 0.895 (0.05), 0.955 (0.025) and 1.035 (0.01) with the p-value shown in brackets. These critical values are available in Stephans (1974). Several other EDF tests for normality are also summarized in Stephans (1974).

Example

Using the sample distribution presented in Table 2.1 the K-S goodness of fit statistic was determined. The computed value of the test statistic was 0.59, which is not significant at the 0.15 level, and hence we cannot reject the hypothesis that the 30 observations were randomly drawn from a normal distribution. The values of i, i/n, $x_{(i)}$, $z_{(i)}$ and $F_{(i)}$ are shown in Table 2.6. Figure 2.12 shows a plot of the relative cumulative frequency i/n and the curve $F_{(i)}$, corresponding to the distribution function for the normal distribution with $\bar{x} = 25.43$, $s = 8.97$.

2.4.2 Graphical Techniques

An alternative approach to assessing normality employs graphical techniques such as a *p-p plot* or a *Q-Q plot*. The p-p plot or *probability-probability plot*, compares the sample cumulative probability distribution $S(x)$ (see Figures 2.3 and 2.4) to the theoretical cumulative probability function $F(x)$, for a set of common x values. A variation of the p-p plot is a plot of the sample cumulative probability function on specially designed normal probability graph paper. In a p-p plot on normal probability paper, a set of observations from a normal distribution will plot as a straight line with intercept 0 and slope 1. If the observations are not "normal-like" then the plot will be nonlinear. The theoretical normal density is usually estimated using the sample mean and sample variance.

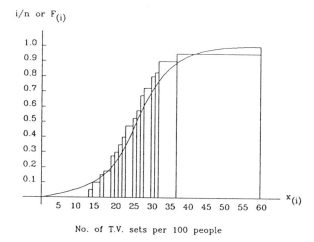

i/n or F$_{(i)}$

No. of T.V. sets per 100 people

Figure 2.12.   Goodness of Fit Test for T.V. Ownership Distribution Relative Cumulative Frequency Less Than Vs. Normal Distribution

The Q-Q Plot

A preferred graphical technique is the *quantile-quantile plot* or Q-Q plot. This plot compares the observed quantiles $Q(p)$ (X observations) defined in Section 2.1.3, to the theoretical quantiles, $Q^*(p)$, or X values corresponding to the same p-value. The sample observations determine the $x_{(i)}$ which are then used to determine $p = (i - .5)/n$. The theoretical x value is determined from the p-values. Like the p-p plot, the Q-Q plot will be a straight line through the origin with slope 1 if the data are normally distributed. If the normal density is incorrect by a scale factor and/or a location factor, the Q-Q plot will remain a straight line. This is an advantage of the Q-Q plot over the p-p plot. In the case of a location difference, the intercept in the Q-Q plot becomes non-zero, while for a scale difference, the slope will no longer be 1. As in the case of the p-p plot, the parameters for the theoretical normal density usually must be estimated using the sample mean and variance. The Q-Q plot for the example of this section is presented in Figure 2.13 below. The quantile-quantile plot may be viewed as the combination of two quantile plots. The $Q(p)$ scale from the observed data quantile plot and the $Q^*(p)$ scale from a standard normal quantile plot are plotted against each other at the values of $p = (i - .5)/n$.

If the observed sample distribution is approximately normal the Q-Q plot should be linear. The correlation coefficient between $Q(p)$ and $Q^*(p)$ can be used to test for normality (see Filliben (1975) and Johnson and Wichern (1982)). The shape of the Q-Q plot can also be used to ascertain the nature of the non-normality. A straggler at the beginning or the end of the plot, such as above in Figure 2.13, is an indication of an outlier. Curvature at

Table 2.6.   Inputs for Normality Tests for Distribution of T.V. Ownership

i/n	.033	.067	.100	.133	.167	.200	.233	.267	.300	.333	.367	.400	.433	.467	.500
$x_{(i)}$	13	14	14	16	17	19	19	19	20	21	22	22	23	23	25
$z_{(i)}$	-1.39	-1.27	-1.27	-1.05	-0.94	-0.72	-0.72	-0.72	-0.61	-0.49	-0.38	-0.38	-0.27	-0.27	-0.05
$F_{(i)}$	0.082	0.102	0.102	0.157	0.174	0.236	0.236	0.236	0.271	0.312	0.352	0.352	0.394	0.394	0.480
i/n	.533	.567	.600	.633	.667	.700	.733	.767	.800	.833	.867	.800	.933	.967	1.000
$x_{(i)}$	25	26	27	27	27	29	29	30	30	31	32	32	36	36	59
$z_{(i)}$	-0.05	0.06	0.18	0.18	0.18	0.40	0.40	0.51	0.51	0.62	0.73	0.73	1.18	1.18	3.74
$F_{(i)}$	0.480	0.524	0.571	0.571	0.571	0.655	0.655	0.695	0.695	0.732	0.767	0.767	0.881	0.881	1.000

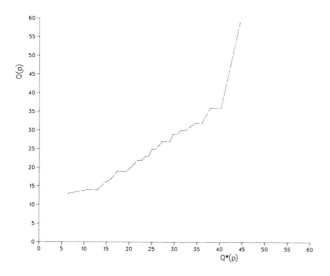

Figure 2.13.   Quantile-Quantile Plot Test of Normality for T.V. Ownership Distribution

the ends of the plot are indications of kurtosis, while convexity or concavity in the Q-Q plot are suggestive of a lack of symmetry.

Example

The data for Figure 2.13 are summarized in Table 2.7 below. The values corresponding to $p = (i - .5)/n$ for the theoretical standard normal are denoted by $Q^*(p)$. The graph in Figure 2.13 shows a straight line if the one outlier 59 is ignored.

2.4.3 Other Tests for Normality

All of the procedures discussed up to now, in this section, have involved a comparison between the EDF and the theoretical distribution function for a normal distribution. There are also a variety of procedures which are based on various properties of the normal density. Two standard test procedures, which have been used for some time, are the tests of significance for the sample measures of skewness (g) and kurtosis (d) discussed in Section 2.2. Recall that both of these measures were found to be significant at the 0.01 level for the sample data in Table 2.1.

Table 2.7.   Data for Quantile-Quantile Plot of T.V. Ownership Distribution

$p = (i - .5)/n$	.017	.050	.083	.117	.150	.183	.217	.250	.283	.317	.350	.383	.417	.450	.483
$Q(p) = x_{(i)}$	13	14	14	16	17	19	19	19	20	21	22	22	23	23	25
$Z(p)$	-2.12	-1.65	-1.39	-1.19	-1.04	-0.90	-0.78	-0.67	-0.57	-0.48	-0.39	-0.30	-0.21	-0.13	-0.04
$Q^*(p)$	6.41	10.63	12.96	14.76	16.10	17.36	18.43	19.42	20.32	21.12	21.93	22.74	23.55	24.26	25.07
$p = (i - .5)/n$	.517	.550	.583	.617	.650	.683	.717	.750	.783	.817	.850	.883	.917	.950	.983
$Q(p) = x_{(i)}$	25	26	27	27	27	29	29	30	30	31	32	32	36	36	59
$Z(p)$	0.04	0.13	0.21	0.30	0.39	0.48	0.57	0.67	0.78	0.90	1.04	1.19	1.39	1.65	2.12
$Q^*(p)$	25.79	26.60	27.31	28.12	28.93	29.74	30.54	31.44	32.43	33.50	34.76	36.10	37.90	40.23	44.45

A popular technique for testing normality is the *Wilk-Shapiro* or *W-test*. This test compares the conventional estimator of the variance, to an estimator of variance which employs the order statistics under the assumption of normality. Given the n order statistics $x_{(1)}, x_{(2)}, \ldots, x_{(n)}$ as an $n \times 1$ vector $\mathbf{x}$, the expectations of these statistics under normality are denoted by the $(n \times 1)$ vector $\mathbf{c}$. The covariance matrix relating the order statistics is given by the $(n \times n)$ matrix $\mathbf{V}$. The generalized least squares estimator of the variance under normality, is given by $\tilde{\sigma}^2 = \mathbf{c}'\mathbf{V}^{-1}\mathbf{x}/\mathbf{c}'\mathbf{V}^{-1}\mathbf{c}$. The Wilk-Shapiro test is based on the ratio of $\tilde{\sigma}^2$ and s^2. This test is much more complex than the K-S test described above. For each value of n the elements of $\mathbf{c}$ and $\mathbf{V}$ must be specified. In addition, tables for the critical value of W must also be employed. There have also been a number of extensions and approximations to the W-test.

A recent addition to the battery of normality tests is based on the property that $\bar{x}$ and s^2 are independent if the underlying population is normal. Let $\bar{x}_{(i)}$ and $s^2_{(i)}$ denote the sample mean and variance if the observation x_i is omitted. By leaving out one observation each time, a set of n pairs $(\bar{x}_{(i)}, s^2_{(i)})$, can be generated. The correlation between the two statistics over the n pairs should be zero. Since the $s^2_{(i)}$ are not normally distributed a cube root transformation is used. Defining $y_{(i)} = \left[s^2_{(i)}\right]^{1/3}$, the correlation coefficient r, between the n values of $\bar{x}_{(i)}$ and $y_{(i)}$ is determined. The conventional Fisher statistic $z = 1/2 \ln\left(\frac{1+r}{1-r}\right)$ for sample correlations is then computed (see Section 3.2). Under the normality assumption, since the true correlation is zero, in large samples from a normal population, z is approximately normal with mean 0 and variance $1/(n-3)$. Small sample versions of the statistic are outlined in Lin and Mudhulkar (1980).

The discussion presented here on goodness of fit has been limited to the assessment of normality. For the most part the techniques to be discussed in this text will require a normality assumption. Techniques are also available for judging the suitability of other distributions. The interested reader is referred to Hoaglin, Mosteller and Tukey (1985) for a survey of techniques.

The discussion presented here for the assessment of normality is but a brief overview of a vast literature that is available on this topic. The interested reader is referred to the survey article by Mardia (1980).

2.5 Transformations

In the previous section procedures for assessing normality were outlined. If normality is not a viable assumption, it may be possible to transform the scale of measurement employed to obtain a distribution shape that is more normal-like. For data that is normally distributed *linear transformations* of the data are commonly used to create a standard normal variable. The familiar equation $z = (x - \mu)/\sigma$ is a linear transformation of the variable x into the variable z. As a result of this transformation, the mean changes from μ to 0, and the variance changes from σ^2 to 1. The shape of the distribution, however, does not change as a result of this linear transformation. To change the shape of a distribution a *nonlinear transformation* is required. As discussed earlier, shape is conveniently discussed in terms of skewness and/or kurtosis. Using the normal distribution as a reference or standard, we seek to obtain "near normality" by a nonlinear transformation which yields symmetry without kurtosis.

2.5.1 Removing Skewness

It is easy to imagine examples of transformations which can be used to reduce skewness by stretching or compressing the scale. For positive values of X, the transformation $Y = X^2$ stretches the scale for $X > 1$ and compresses the scale for $X < 1$. The reverse of these properties are true for the transformation $Y = X^{1/2}$. Thus given a skewed distribution with positive X values, a transformation of the form $Y = X^k$ may be useful in eliminating skewness. Values of k exceeding 1 can be used to eliminate negative skewness, while values of k in the range $0 < k < 1$ can be used to eliminate positive skewness. This transformation is a special case of a family of transformations referred to as *power transformations*.

Box-Cox λ Method

A more general form of the power transformation is the *Box-Cox transformation* given by

$$Y = (X^\lambda - 1)/\lambda \qquad \lambda \neq 0;$$

$$= \ln X, \qquad \lambda = 0; \quad X > 0.$$

The expression $Y = \ln X$ simply reflects the limit of the transformation formula when X is positive and λ approaches zero. Given an observed data set $(x_1, x_2, \ldots, x_n)$ the transformation parameter λ must be estimated. Approaches to estimating λ are usually based on the assumption that the transformed values $y_1, y_2, \ldots, y_n$ are normally distributed.

Using a maximum likelihood approach, Box and Cox developed a technique for estimating λ. They showed that the maximum of the likelihood function (up to a constant) is

$$L_m(\lambda) = -1/2n \ln \tilde{\sigma}_z^2;$$

where

$$\tilde{\sigma}_z^2 = \sum_{i=1}^{n}(z_i - \bar{z})^2/n,$$

$z_i = x_i^\lambda/\tilde{x}_G^\lambda$ and $\tilde{x}_G$ = the geometric mean of $x_1, \ldots x_n$

$$= (x_1 \, x_2 \, \ldots \, x_n)^{1/n}.$$

Inferences for λ may be carried out using an approximation based on a χ^2 distribution with 1 degree of freedom. A $100(1 - \alpha)\%$ confidence interval for λ is given by the set of all λ satisfying $[L_m(\hat{\lambda}) - L_m(\lambda)] \leq (1/2)x_{\alpha;1}^2$, where $\hat{\lambda}$ is the maximum likelihood estimator. To test $H_0 : \lambda = \lambda_0$ the test statistic $2[L_m(\hat{\lambda}) - L_m(\lambda_0)]$ is compared to $\chi_{\alpha;1}^2$.

Example

Table 2.8 below presents the total annual crude oil production for the year 1976 for the top 50 producers in the world. A quick perusal of the table reveals a substantial positive skewness. The range of the distribution is [1.1, 520.0], the median is 10.5, and the first and third quartiles are 2.9 and 64.1 respectively. The mean of the distribution is 57.1, the standard deviation is 113.2, the skewness is 2.9 and the kurtosis is 8.3. The asymptotic standard errors for the skewness and kurtosis measures are 0.34 and 0.66 respectively, and hence both the skewness and kurtosis are highly significant. The K-S test for normality yields a p-value less than 0.00. A comparison of the quartiles to the median reveals substantial positive skewness in that $(x_M - Q_1) = 7.6$ and $(Q_3 - x_M) = 53.6$. As outlined in Section 2.2.3, the nature of the skewness can also be shown by comparing $x_M - x_{(i)}$ to $x_{(n-i+1)} - x_M$ where $i = 1, 2, \ldots, 25$. Table 2.9 below makes this comparison. The strong positive skewness in this distribution would suggest a Box-Cox transformation with a small positive or negative value of λ, or perhaps a log transformation.

Approximating λ

A common trial and error approach to the determination of the maximum likelihood estimator of λ involves determination of $L_m(\lambda)$ for a range of values of λ. An iterative process is then used to find the value of λ that maximizes $L_m(\lambda)$. A solution for λ is usually selected as an integer value such as 2, 3, 4, or if $\lambda < 1$, a common fraction such as 1/2, 1/3 or 1/4. If λ is close to zero, and X is positive $\ln X$ is often used. Once a value of λ has been determined, there is no guarantee that the solution will result in an acceptable normal-like distribution. Therefore, the distribution of the transformed data should also be carefully studied.

Table 2.8. World Crude Oil Production of Top Fifty Nations 1976
(000,000) Tonnes

Country	Oil Production	Log Oil Production	Country	Oil Production	Log Oil Production
France	1.1	0.04	Trinidad	11.0	1.04
Chile	1.1	0.04	Gabon	11.4	1.06
Italy	1.1	0.04	United Kingdom	12.0	1.08
Burma	1.2	0.08	Norway	13.7	1.14
Netherlands	1.4	0.15	Rumania	14.7	1.17
Bolivia	2.0	0.30	Egypt	16.7	1.22
Spain	2.0	0.30	Oman	18.3	1.26
Congo	2.0	0.30	Australia	20.5	1.31
Austria	2.0	0.30	Argentina	20.9	1.32
Hungary	2.1	0.32	Qatar	23.5	1.37
Albania	2.3	0.36	Mexico	40.8	1.61
Turkey	2.6	0.41	Algeria	50.0	1.70
Bahrain	2.9	0.46	Canada	64.1	1.81
Peru	3.7	0.57	Indonesia	74.0	1.87
Tunisia	3.7	0.57	China	85.0	1.93
Yugoslavia	3.9	0.59	Libya	92.8	1.97
West Germany	5.5	0.74	United Arab Emirates	93.3	1.97
Angola	6.3	0.80	Nigeria	102.7	2.01
Columbia	7.5	0.88	Kuwait	108.6	2.04
Malaysia	8.0	0.90	Iraq	112.4	2.05
Brazil	8.5	0.93	Venezuela	119.8	2.08
India	8.6	0.93	Iran	296.5	2.47
Brunei	8.6	0.93	United States	401.6	2.60
Ecuador	9.5	0.98	Saudi Arabia	424.2	2.63
Syria	10.0	1.00	Soviet Union	520.0	2.72

Source: The World in Figures
Second Edition 1978
Facts on File Inc.
119 West 57 Street
New York, N.Y. 10019
Compiled by The Economist Newspaper Ltd., England.

Example

Using an iterative procedure for the example data, the likelihood function $L_m(\lambda)$ is maximized at $\lambda = -0.13$. A 95% confidence interval for λ is given by $(-0.30, 0.04)$. Since this confidence interval contains $\lambda = 0$, the log transformation appears to be an acceptable transformation. The use of the log transformation, for distributions of oil pool sizes in various geographic basins, has been a tradition in the oil industry. The use of the log transformation in this context is therefore justified.

Table 2.8 also contains the transformed values, $Y = \log_{10} X$. The percentiles of the transformed distribution are median (1.02), first quartile

Table 2.9. Distances of Quantiles from Median for Distribution of Oil Production

i	1	2	3	4	5	6	7	8	9	10	11	12
$x_M - x_{(i)}$	9.4	9.4	9.4	9.3	9.1	8.5	8.5	8.5	8.5	8.4	8.2	7.9
$x_{(n-i+1)} - x_M$	509.5	413.7	391.1	286.0	109.3	101.9	98.1	92.2	82.8	82.3	74.5	63.5
$n - i + 1$	50	49	48	47	46	45	44	43	42	41	40	39

i	13	14	15	16	17	18	19	20	21	22	23	24	25
$x_M - x_{(i)}$	7.6	6.8	6.8	5.0	5.0	4.2	3.0	2.5	2.0	1.9	1.9	1.0	0.5
$x_{(n-i+1)} - x_M$	53.6	39.5	30.3	13.0	10.4	10.0	7.8	6.2	4.2	3.2	1.5	0.9	0.5
$n - i + 1$	38	37	36	35	34	33	32	31	30	29	28	27	26

(0.46) and third quartile (1.81). The mean is 1.13, the standard deviation is 0.75, the skewness is 0.42 and the kurtosis is -0.763. The asymptotic standard errors of the skewness and kurtosis measures are (0.34) and (0.69) respectively and hence the skewness and kurtosis measures are not significant. The K-S test for normality yields a p-value of 0.86.

The reader should note that the base used for the logarithm does not matter since changing the base involves multiplication by a constant. The base of 10 was used here for convenience and because many users are more familiar with logarithms in this form.

An Alternative Approximation

A comparison of the deviations for the transformed data, $(y_M - y_{(i)})$ to $(y_{(n-i+1)} - y_M)$, is shown in the table below. A comparison of the values of $(y_M - y_{(i)})$ to $(y_{(n-i+1)} - y_M)$ in Table 2.10 for each i reveals that the ratio is close to 1 in most instances. The determination of a suitable λ for the Box-Cox transformation, therefore, may be viewed as an attempt to make the ratios of these quantile differences equal to unity. Since the Box-Cox transformation preserves the rank order of the observations, we may write that λ must satisfy

$$(x_M^\lambda - x_{(i)}^\lambda) = (x_{(n-i+1)}^\lambda - x_M^\lambda) \qquad i = 1, 2, \ldots n;$$

for $\lambda \neq 0$. This expression can be written as

$$[x_{(i)}/x_M]^\lambda + [x_{(n-i+1)}/x_M]^\lambda = 2. \tag{2.4}$$

In the case of $Y = \ln X$, X has a lognormal distribution which has the property that

$$[x_M/x_{(i)}] = [x_{(n-i+1)}/x_M].$$

If λ is close to zero this latter equation should be used since the value $\lambda = 0$ automatically satisfies (2.4).

Hinkley (1975) used these results to suggest a "quick" estimate of λ. Hinkley's method proposes selecting a particular value of i in (2.4) and then determining the quantities $[x_{(i)}/x_M]^\lambda$ and $[x_{(n-i+1)}/x_M]^\lambda$ over a range of λ values. By trial and error, a value of λ satisfying equation (2.4) can be obtained.

Table 2.10. Distances of Quantiles from Median for Transformed Distribution of Oil Production

i	1	2	3	4	5	6	7	8	9	10	11	12
$y_M - y_{(i)}$	0.98	0.98	0.98	0.94	0.87	0.72	0.72	0.72	0.72	0.70	0.66	0.61
$y_{(n-i+1)} - y_M$	1.70	1.61	1.58	1.45	1.06	1.03	1.02	0.99	0.95	0.95	0.91	0.85
$n-i+1$	50	49	48	47	46	45	44	43	42	41	40	39

i	13	14	15	16	17	18	19	20	21	22	23	24	25
$y_M - y_{(i)}$	0.56	0.45	0.45	0.43	0.28	0.22	0.14	0.12	0.09	0.09	0.09	0.04	0.02
$y_{(n-i+1)} - y_M$	0.79	0.68	0.59	0.35	0.30	0.29	0.24	0.20	0.15	0.12	0.06	0.04	0.02
$n-i+1$	38	37	36	35	34	33	32	31	30	29	28	27	26

Table 2.11. Approximation of Transformation Parameter λ for Oil Production Data

λ	$(0.3)^\lambda$	$(6.1)^\lambda$	$(0.3)^\lambda + (6.10)^\lambda$
1	0.3	6.1	6.4
0.5	0.5	2.5	3.0
0.1	0.9	1.2	2.1
-0.1	1.1	0.8	1.9
-0.5	2.0	0.4	2.4
-1	3.3	0.2	3.5

Example

For the oil production data the first and third quartiles were used along with the x_M to estimate λ. The values of x_M, Q_1 and Q_3 were used in Table 2.11 to approximate λ. The ratios $Q_1/x_M = 0.3$ and $Q_3/x_M = 6.1$ were the starting points for the iteration.

From the table it would appear that λ should lie between 0.1 and −0.1, which would suggest that the log transformation should prove useful. Notice that this solution is comparable to the solution obtained using the likelihood method. The advantage of the Hinkley approximation is that it can be done quickly and easily without computing software. For the log transformation we should have that $x_M/Q_1 = Q_3/x_M$, which does not hold in this case since 3.6 ≠ 6.1. It would appear that this is close enough however for practical purposes since λ is close to zero.

Negative Observations

We have assumed that the observations in the original data set are positive. When this is the case the impact of the Box-Cox transformation is either a stretching or shrinking of the entire scale in one direction. If, however, the data set contains negative observations, the impact of this transformation on the negative values will be different from the positive values. In addition the transformation would be undefined for some values of λ. For data sets with negative values, a preliminary transformation of the form $Y = (X+L)$, L a positive constant, is recommended to ensure that all the values are positive.

2.5.2 Removing Kurtosis

While the power transformation is useful for removing skewness, it does not necessarily yield a distribution without kurtosis. The distribution may require an additional transformation to remove kurtosis. In some cases the original distribution may be symmetrical but contains positive or negative kurtosis. Removing kurtosis from a symmetric distribution involves shrinking or stretching both tails of the distribution simultaneously, and hence for symmetric distributions with positive values the simple power transformation cannot be used to remove kurtosis. A *modified power transformation* is outlined next.

A Modified Power Transformation

To eliminate kurtosis in a symmetric distribution, the same power type transformation must be carried out on both sides of the centre of the distribution. This could be achieved by first subtracting the median x_M from the observations, to obtain a new set of translated observations with median at zero. To ensure that a power transformation can be used on both sides, absolute values of the translated observations can be used. The transformation is given by

$$Y = \text{SIGN} \frac{(|X - x_M| + 1)^\lambda - 1}{\lambda} \qquad \lambda \neq 0;$$

$$Y = \text{SIGN} \ln(|X - x_M| + 1) \qquad \lambda = 0;$$

where SIGN is the sign of the original value of $(X - x_M)$ before taking absolute values. A form of this transformation was introduced by John and Draper (1980).

The value of λ required, can be approximated by comparing various deviations between pairs of quantiles $y_{(p)}$ and $y_{(1-p)}$, to the corresponding quantiles for the normal distribution. The distance between $y_{(p)}$ and $y_{(1-p)}$ can be expressed as a function of σ, say $k(p)\sigma$, where $k(p)$ is obtained from standard normal tables. The ratio of any pair of distances can be written as

$$[y_{(p)} - y_{(1-p)}]/[y_{(q)} - y_{(1-q)}] = k(p)/k(q).$$

Replacing $y_{(p)}$ by $(x_{(p)} - x_M)^\lambda$ the equation becomes,

$$[(x_{(p)} - x_M)^\lambda - (x_{(1-p)} - x_M)^\lambda]/[(x_{(q)} - x_M)^\lambda - (x_{(1-q)} - x_M)^\lambda] = k(p)/k(q).$$

By computing the left hand side of the equality for a variety of values of λ, a value of λ may be determined that yields the predetermined ratio on the right hand side.

Example

Table 2.13 displays the trade balances in 1976 for the 140 largest importing countries. A perusal of the table reveals a distribution which is symmetric but with substantial kurtosis. The median is given by $x_M = 148.5$. The observed quantiles corresponding to $p = 0.05$, $p = 0.95$, $p = 0.15$ and $p = 0.85$, yield values of $(x_{(p)} - x_M)$ given by -3157, 4939, -1291 and 547 respectively. For a normal distribution $X(.95) - X(.05) \simeq 2\sigma$ and $X(.85) - X(.15) \simeq \sigma$. The ratio of these two deviations therefore should be approximately 2. Table 2.12 below shows the values of the ratio of these two deviations for several values of λ. From the table, it appears that a square root transformation applied to the two tails of the distribution may be adequate to remove kurtosis.

The transformation corresponding to $\lambda = 0.5$ was carried out, and the moments were determined. The magnitudes for the coefficients of skewness and kurtosis for the transformed data were $g = 0.31$ and $d = 1.24$. For comparison, the value of these statistics for the untransformed data were $g = 0.56$ and $d = 3.31$ respectively. For the untransformed data the Z value corresponding to d is $3.31/\sqrt{24/140} = 7.99$ indicating highly significant kurtosis relative to a normal distribution. The Z value for d for the transformed data is $1.24/\sqrt{24/140} = 2.99$. The p-values for the K-S test of normality were 0.230 and 0.000 for the transformed and untransformed data respectively.

Table 2.12. Approximation of Transformation Parameter λ for World Trade Data

λ	$[X(.05) - x_M]^\lambda$	$[X(.95) - x_M]^\lambda$	Deviation
1	-3157	4939	8096
0.5	-56.2	70.3	126.5
0.4	-25.1	30.0	55.1

λ	$[X(.15) - x_M]^\lambda$	$[X(.85) - x_M]^\lambda$	Deviation	Ratio of Deviations
1	-1291	695	1986	4.1
0.5	-35.9	26.4	62.3	2.0
0.4	-17.6	13.7	31.3	1.8

Table 2.13. World Balance of Trade

Rank	Country	$mn	Rank	Country	$mn
1	Saudi Arabia	27431	36	Rumania	43
2	Germany, West	14354	37	Macao	28
3	Iran	10626	38	New Caledonia	24
4	Kuwait	6509	39	El Salvador	16
5	Iraq	5787	40	Ghana	16
6	UtdArabEmirates	5224	41	Surinam	15
7	Libya	4791	42	Nicaragua	10
8	Indonesia	2874	43	Liechtenstein	0
9	Nigeria	2366	44	Mauritania	-2
10	Japan	2365	45	Swaziland	-3
11	Venezuela	2317	46	Burma	-8
12	Qatar	1379	47	Sri Lanka	-13
13	Malaysia	1367	48	Guinea	-20
14	China	1100	49	Madagascar	-37
15	Brunei	1040	50	India	-38
16	Argentina	882	51	Paraguay	-41
17	Oman	850	52	Bolivia	-42
18	Australia	684	53	Malawi	-46
19	South Africa	654	54	Albania	-50
20	Angola	599	55	Honduras	-61
21	Taiwan	547	56	Uruguay	-63
22	Chile	399	57	Nepal	-65
23	Gabon	357	58	Haiti	-66
24	Ivory Coast	338	59	Iceland	-67
25	Zaire	249	60	Togo	-69
26	Trinidad & Tobago	231	61	Somalia	-71
27	South-West Africa	230	62	Ethiopia	-72
28	Colombia	172	63	Botswana	-74
29	Papua New Guinea	158	64	Guyana	-83
30	Ecuador	117	65	Mauritius	-91
31	Rhodesia	111	66	Cameroon	-98
32	Liberia	61	67	Senegal	-119
33	Switzerland	56	68	Fiji	-124
34	Korea, North	50	69	Jersey	-134
35	Lebanon	50	70	Tanzania	-148

Table 2.13. World Balance of Trade (continued)

Rank	Country	$mn	Rank	Country	$mn
71	Algeria	-149	106	Bahamas	-681
72	Barbados	-150	107	Tunisia	-742
73	Dominican Rep	-162	108	Peru	-772
74	Afghanistan	-163	109	Jordan	-815
75	Yemen, South	-165	110	Ireland	-858
76	Kenya	-183	111	Sweden	-869
77	Malta	-194	112	Soviet Union	-941
78	Guatemala	-200	113	Pakistan	-967
79	Mozambique	-200	114	Finland	-1049
80	Cuba	-203	115	Korea, South	-1056
81	Costa Rica	-217	116	Neth Antilles	-1142
82	Guadeloupe	-227	117	Syria	-1310
83	Cyprus	-230	118	Morocco	-1352
84	Guam	-239	119	Philippines	-1439
85	Bulgaria	-243	120	Germany, East	-1835
86	Martinique	-258	121	Puerto Rico	-2086
87	French Polynesia	-274	122	Egypt	-2287
88	Jamaica	-280	123	Portugal	-2421
89	Bahrain	-324	124	Mexico	-2455
90	Zambia	-328	125	Singapore	-2484
91	Mongolia	-330	126	Jugoslavia	-2489
92	Netherlands	-332	127	Belgium-Lux	-2681
93	Hong Kong	-356	128	Poland	-2820
94	Reunion	-356	129	Austria	-3002
95	Yemen, North	-404	130	Norway	-3153
96	Canada	-405	131	Turkey	-3251
97	Sudan	-426	132	Israel	-3300
98	New Zealand	-454	133	Denmark	-3305
99	Bangladesh	-458	134	Brazil	-3313
100	Vietnam	-600	135	Greece	-3499
101	Thailand	-604	136	Italy	-6496
102	Hungary	-608	137	France	-7297
103	Panama	-612	138	Spain	-8773
104	Czechoslovakia	-647	139	Utd Kingdom	-9910
105	Virgin Islands, US	-669	140	United States	-14568

Source: The World in Figures
Second Edition 1978
Facts on File Inc.
119 West 57 St.
New York, N.Y. 10019
Compiled by The Economist Newspaper Ltd., England.

2.5.3 Transformations in the Presence of Other Variables

This section on transformations has been concerned with procedures to achieve more normal-like sample distributions. These transformations have been performed in the absence of any other prescribed purpose. Although many statistical inference procedures require the assumption of normality, transformations to achieve normality can affect the outcome of other procedures that are carried out following the transformation. In analyses involving several variables such as analysis of variance, transformations to achieve normality can have an important impact on the equality of group means across the groups being compared. Transformations must therefore be used with caution, and must take into account the end use of the data analysis. A related data analysis technique which is used in the assessment of the performance of fitted models is residual analysis. Suitable transformations on sample distributions are sometimes determined on the basis of residual analysis. The transformation is selected only after the relationship among the variables is taken into consideration. These issues will be explored in later chapters.

Cited Literature for Chapter 2

1. Barnett, V. and Lewis, T. (1978). *Outliers in Statistical Data*, New York: John Wiley and Sons Inc.
2. Dixon, W.J. (1950). "Analysis of Extreme Values," *Annals of Mathematical Statistics* 21, 488–506.
3. Dixon, W.J. (1951). "Ratios Involving Extreme Values," *Annals of Mathematical Statistics* 22, 68–78.
4. Dixon, W.J. and Kronmal (1965). "The Choice of Origin and Scale for Graphs," *Journal of the Association for Computing Machinery* 12, 259–261.
5. Filliben, J.J. (1975). "The Probability Plot Correlation Coefficient Test for Normality," *Technometrics* 17, 111–117.
6. Grubbs, F.E. and Beck, G. (1972). "Extension of Sample Sizes and Percentage Points for Significance Tests of Outlying Observations," *Technometrics* 14, 847–854.
7. Hinkley, D.V. (1975). "On Power Transformations to Symmetry," *Biometrika* 62, 101–111.
8. Hoaglin, C., Mosteller, F. and Tukey, J.W. (1985). *Exploring Data, Tables, Trends and Shapes*. New York: John Wiley and Sons Inc.
9. Hoaglin, David C., Iglewicz, Boris and Tukey, John W. (1986). "Performance of Some Resistant Rules for Outlier Labelling," *Journal of the American Statistical Association* 81, 991–999.
10. Hogg, Robert V. (1974). "Adaptive Robust Procedures: A Partial Review and Some Suggestions for Future Applications and Theory," *Journal of the American Statistical Association* 69, 909–923.
11. John, J.A. and Draper, N.R. (1980). "An Alternative Family of Transformations," *Applied Statistics* 29, 190–197.
12. Johnson, R.A. and Wichern, D.W. (1982). *Applied Multivariate Statistical Analysis*. Englewood Cliffs, N.J.: Prentice–Hall.
13. Lilliefors, H.W. (1967). "On the Kolmogorov–Smirnov Test for Normality with Mean and Variance Unknown," *Journal of the American Statistical Association* 62, 399–402.
14. Lin, C. and Mudhulkar, G.S. (1980). "A Simple Test for Normality Against Asymmetric Alternatives," *Biometrika* 67, 455–461.
15. Mardia, K.V. (1980). "Tests of Univariate and Multivariate Normality" in P.R. Krishnaiah ed., *Handbook of Statistics, Vol. 1*. New York: North Holland.
16. Miller, Rupert G. (1986). *Beyond Anova, Basics of Applied Statistics*. New York: John Wiley and Sons Inc.
17. Moors, J.J.A. (1988). "A Quantile Alternative for Kurtosis," *Statistician* 37, 25–32.
18. Rosenberger, James L. and Gasko, Miriam (1983). "Comparing Location Estimators: Trimmed Means, Medians and Trimean" in Hoaglin,

Mosteller and Tukey ed., *Understanding Robust and Exploratory Data Analysis*. New York: John Wiley.

19. Shiffler, Ron E. (1987). "Bounds for the Maximum Z-Score," *Teaching Statistics* 9, 80–81.

20. Stephans, M.A. (1974). "EDF Statistics for Goodness of Fit and Some Comparisons," *Journal of the American Statistical Association* 69, 730–737.

21. Sturges, H.A. (1926). "The Choice of a Class Interval," *Journal of the American Statistical Association* 21, 65–66.

22. Velleman, P.F. (1976). "Interactive Computing for Exploratory Data Analysis I: Display Algorithms," 1975 Proceedings of the Statistical Computing Section. Washington, D.C.: American Statistical Association.

Exercises for Chapter 2

1. This exercise is based on Table D.1 in the Data Appendix.

   (a) Use a computer software package to study the distributions for the variables SMAX, PMAX and PM2. Obtain various measures of location, dispersion, skewness and kurtosis. Examine a box plot, a stem and leaf diagram and a normal distribution quantile–quantile plot. Also carry out a normal goodness of fit test. Identify any outliers and discuss the impact that these outliers have on the various measures obtained above.

   (b) Examine the sample distribution for the variable

   $$LPOP = LOG_{10}(POP)$$

   and discuss the normal goodness of fit. Carry out the exponential transformation to get the variable POP, examine the sample distribution and discuss the normal goodness of fit. Compare the distributions for POP and LPOP and discuss the impact of the log transformation on the distribution.

2. This exercise is based on Table D.2 in the Data Appendix.

   (a) Use a computer software package to study the distributions for the variables RETCAP, PAYOUT, GEARRAT and CURRAT. Obtain various measures of location, dispersion, skewness and kurtosis. Examine a box plot, stem and leaf diagram, and a normal distribution quantile–quantile plot. Also carry out a normal goodness of fit test. Identify any outliers and discuss the impact that these outliers have on the various measures obtained above.

   (b) Examine the sample distributions for the variables LOGSALE and LOGASST and discuss the normal goodness of fit in each case. Carry out the exponential transformation to get the variables SALE and ASST and once again examine the two sample distributions. In each case discuss the normal goodness of fit. Compare the untransformed and transformed distributions and discuss the impact of the transformation. Repeat the above comparisons using a square root transformation on SALE and ASST. Which transformation (log or square root seems more appropriate).

Questions for Chapter 2

1. Using only the values 0.2, 0.5 or 1.0 for k select a suitable value for k for the following sample distributions. Recall that the interval width for the stem and leaf display is $W = k10^p$. In each case indicate whether the leaves for the stem and leaf plot will contain half cycles, one cycle or two cycles of the digits 0, 1, 2, 3, 4, 5, 6, 7, 8, and 9.

   (a) range of distribution $(392, 764)$, $n = 49$;

   (b) range of distribution $(1041, 9368)$, $n = 36$;

   (c) range of distribution $(13, 879)$, $n = 100$.

2. Assuming a normal distribution with mean $\mu = 10$ and variance $\sigma^2 = 100$ determine the quantiles $Q(p)$ corresponding to $n = 20$ using $p = (i - 0.5)/n$. Use this information to construct a quantile plot for the distribution.

3. Given a sample of n observations $x_1, x_2, \ldots, x_n$ denote the mean for all n observations by $\bar{x}$ and let the variance be denoted by s^2. Let $\bar{x}(i)$ and $s^2(i)$ denote the mean and variance for the $(n - 1)$ observations excluding x_i. Answer the following:

   (a) Show that $\bar{x}(i) = \bar{x} + \dfrac{1}{(n-1)}(\bar{x} - x_i)$ and $\bar{x} = \bar{x}(i) + \frac{1}{n}(x_i - \bar{x}(i))$.

   (b) Show that $s^2(i) = \dfrac{(n-1)}{(n-2)}s^2 - \dfrac{n}{(n-1)(n-2)}(\bar{x} - x_i)^2$ and $s^2 = \dfrac{(n-2)}{(n-1)}s^2(i) + \frac{1}{n}(\bar{x}(i) - x_i)^2$.

   (c) Use the relationships in (a) and (b) to discuss the impact on $\bar{x}$ and s^2 of removing or adding an outlier.

   (d) Define the Z score $z_i = (x_i - \bar{x})/s$ and show using (b) that in large samples
   $$s^2(i)/s^2 = 1 - \frac{1}{n}z_i^2$$
   and hence if $|z_i| > 3$, $s^2 > s^2(i)\left[\frac{n}{n-9}\right]$.

   (e) Compute the inequality in (d) for $n = 10$, 30 and 50.

4. For a sample distribution a useful method for determining the quantile $Q(p)$ corresponding to a specified value of p, say p^*, is given by the interpolation formula
   $$Q(p^*) = (1 - f)Q(p_i) + fQ(p_{i+1})$$

where p_i and p_{i+1} are the values of p immediately below and above p^* for which observed quantiles exist and $f = (p^* - p_i)/(p_{i+1} - p_i)$. Draw a graph illustrating this interpolation. Use the equation to determine $Q(.85)$ if $Q(.865) = 41$ and $Q(.844) = 40$. Also determine n using $p = (i - 0.5)/n$.

5. Given a standard normal distribution determine Q the quartile deviation. What interval of values lie in the range $0.5Q$ to $2.0Q$? What interval of values lie in the range $2.0Q$ to $3.5Q$? What is the probability of a random selection from this distribution lying outside a range of $7Q$ ($3.5Q$ either side of 0)?

6. Show that for a normal distribution with variance σ^2 the quartile deviation Q is approximately 1.35σ.

7. Chebyshev's Inequality states that for any random variable X
$$P[|X - \mu| \geq k] \leq \sigma^2/k^2$$
where $E[X] = \mu$ and $V[X] = \sigma^2$.

(i) Determine this probability for $k = 1$, 2 and 3 and compare this probability to the normal distribution case.

(ii) Suppose that the distribution of X has the quartile deviation $Q = c\sigma$. Use Chebyshev's Inequality to determine expressions for $P[|X - \mu| \geq 3.5Q]$ and $P[|X - \mu| \geq 2.0Q]$ at $c = 1$ and $c = 2$. How do these probabilities compare to the normal distribution case? Comment on the application of the Chebyshev Inequality to the detection of outliers.

8. Explain why the coefficients of skewness and kurtosis given by
$$s_k = [(Q_3 - x_M) - (x_M - Q_1)]/[Q_3 - Q_1]$$
and $\quad k = [(E_7 - E_5) + (E_3 - E_1)]/[E_6 - E_2]$
are relatively insensitive to outliers.

9. What impact would the removal of a single outlier have on a symmetry plot?

10. The uniform distribution has the density function
$$f(x) = \frac{1}{(b - a)} \qquad -\infty < a < x < b < \infty.$$

(a) Use calculus to show that
$$E[X] = \int_a^b \frac{x}{(b-a)} dx = (b + a)/2$$

$$E[X^2] = \int_a^b \frac{x^2}{(b-a)} dx = (b^2 + ab + a^2)/3$$

$$V[X] = \frac{(b-a)^2}{12}.$$

(b) Show that the median $x_M = (b+a)/2$ is obtained from x_M in the equation

$$\int_a^{x_M} \frac{1}{(b-a)} dx = 0.5.$$

(c) Use the fact that

$$\int_a^{Q(p)} \frac{1}{(b-a)} dx = p$$

to show that $Q(p) = a + p(b-a)$.

(d) Show graphically why the formula for $Q(p)$ in (c) makes sense.

(e) Use the result in (d) to determine the quartile deviation and the coefficient of skewness based on the quartiles.

(f) Use the result in (c) to determine a coefficient of kurtosis based on the quantiles.

11. Let $x_1, x_2, \ldots x_n$ denote a sample of observations and let $\bar{x}$ and s^2 denote the sample mean and variance. Let $\bar{x}(i)$ and $s^2(i)$ denote the sample mean and variance if observation x_i is omitted. Assume that x_i is an outlier and that $\bar{x}(i) = 0$.

(a) Show that $\bar{x} = x_i/n$ and $s^2 = [(n-2)/(n-1)]s^2(i) + x_i^2/n$.

(b) Show that the Z score corresponding to x_i given by $z_i = (x_i - \bar{x})/s$ is equal to $x_i(n-1)/ns$.

(c) Show that s^2 is minimized if $s^2(i) = 0$ and hence that $(n-1)/\sqrt{n}$ is an upper bound for z_i.

(d) Compute the maximum value of z_i for samples of size $n = 5, 10, 15, 20$ and 25 and comment on the use of z scores to detect outliers in small samples.

12. An alternative measure of kurtosis which is designed to measure tail weight can be based on averages of various parts of the distribution. Denote by $\bar{U}(\alpha)$ the average of the largest $100\alpha\%$ of the observations and by $\bar{L}(\alpha)$ the average of the smallest $100\alpha\%$ of the observations. A measure of kurtosis proposed by Hogg (1974) is given by

$$K = [\bar{U}(0.20) - \bar{L}(0.20)]/[\bar{U}(0.50) - \bar{L}(0.50)].$$

For a normal distribution this measure is approximately 1.75. Compute this measure for the distribution of T.V. ownership in Section 2.2 and for the uniform distribution in question 10.

13. Use Hinkley's approximation method to determine a suitable value of λ for the power transformation for the data given below. Use x_M, $Q(0.20)$ and $Q(0.80)$.

25·	5·	2·	22·	67·
7·	3·	12·	11·	5·
143·	4·	6·	8·	2·
9·	32·	3·	13·	16·
12·	3·	11·	27·	6·
23·	3·	7·	9·	10·

14. For the distribution given in Question 13 determine the following:

(a) The mean, variance, standard deviation, range, median, first quartile, third quartile and quartile deviation.

(b) A quantile-quantile plot to determine goodness of fit to normality.

(c) A box-plot for the distribution.

(d) A symmetry plot.

(e) Compare the spreads $Q(.90)-Q(.10)$ and $Q(.80)-Q(.20)$ to a normal distribution spread assuming $\sigma = s$ as determined in (a). Comment on the results.

3
Bivariate Analysis for Quantitative Random Variables

In this chapter we study relationships between two columns of the data matrix which contain observations on quantitative random variables. The first section of the chapter will be devoted to a review of theory for bivariate distributions for both discrete and continuous random variables. This section will also include an introduction to the theory of correlation and regression and an introduction to the bivariate normal distribution. The second and third sections of the chapter will be devoted to the techniques of inference for correlation and regression respectively. The last section of the chapter will discuss bivariate inference in the presence of other variables including the topics of partial correlation and lurking variables. In each of the four sections a review of techniques normally presented in an introductory course will be combined with additional material.

3.1 Joint Distributions

In *bivariate analysis* we are concerned with the relationship between two random variables X and Y. The relationship between X and Y can be described by the *joint probability distribution* which assigns probabilities to all possible outcomes (x, y). The joint distribution is usually characterized by the *joint density function* $f(x, y)$. The joint densities for both discrete and continuous random variables will be discussed in this Chapter. A discussion of *bivariate distributions* when one of the variables is qualitative will be discussed in Chapter 5. The discussion of discrete bivariate distributions begins in 3.1.1 below. Bivariate distributions with both variables qualitative will appear in the contingency table chapter of Volume II.

3.1.1 Discrete Bivariate Distributions

For discrete random variables X and Y, each of which can take on only a finite number of values, it is possible to construct a *joint probability table* which contains a cell for each possible combination of X and Y. The *joint density* $f(x, y)$ in this case is the probability of an observation occurring in the cell corresponding to $X = x$ and $Y = y$. Since X and Y are discrete and finite we may denote the population values by $(x_1, x_2, \ldots, x_r)$ and $(y_1, y_2, \ldots, y_c)$ respectively. The density $f(x_i, y_j)$ denotes the probability for cell (i, j) and $\sum_{i=1}^{r} \sum_{j=1}^{c} f(x_i, y_j) = 1$. Table 3.1 shows a *bivariate density* for a *discrete bivariate distribution*.

Table 3.1. Bivariate Density $f(x, y)$

				Y		
$X \backslash Y$	1	2	3		c	Totals
1	$f(x_1, y_1)$	$f(x_1, y_2)$	$f(x_1, y_3)$	$\ldots$	$f(x_1, y_c)$	$f_x(x_1)$
2	$f(x_2, y_1)$	$f(x_2, y_2)$	$f(x_2, y_3)$	$\ldots$	$f(x_2, y_c)$	$f_x(x_2)$
X 3	$f(x_3, y_1)$	$f(x_3, y_2)$	$f(x_3, y_3)$	$\ldots$	$f(x_3, y_c)$	$f_x(x_3)$
	$\vdots$	$\vdots$	$\vdots$		$\vdots$	$\vdots$
r	$f(x_r, y_1)$	$f(x_r, y_2)$	$f(x_r, y_3)$	$\ldots$	$f(x_r, y_c)$	$f_x(x_r)$
Totals	$f_y(y_1)$	$f_y(y_2)$	$f_y(y_3)$	$\ldots$	$f_y(y_c)$	1

The row totals and column totals in Table 3.1 provide the marginal densities of X and Y respectively. The *marginal density* of X describes the probability density for X ignoring Y. This density is given by $f_x(x_i) = \sum_{j=1}^{c} f(x_i, y_j)$ and is shown in the row totals. Similarly the marginal density for Y is given by $f_y(y_j) = \sum_{i=1}^{r} f(x_i, y_j)$ and is shown in the column totals.

Example

The example presented in Table 3.2 illustrates a joint density function $f(x, y)$ for the returns on two financial portfolios. The random variable X represents returns on a portfolio of domestic stocks while the random variable Y represents returns on a portfolio of foreign stocks. Both portfolios can be purchased from a large mutual fund company. The levels of return have been coded into six categories for each of X and Y. The row and column totals provide the marginal densities $f_x(x)$ and $f_y(y)$ for the two random variables. From the joint density table we can see that the probability that both X and Y are 0 is 0.10. The probability that X is less than or equal to 1 is 0.79 and for Y this probability is 0.80.

Table 3.2. Joint Distribution of Returns From a Domestic and a Foreign Portfolio

| | | Y | | | | | | |
| | | Returns on the Foreign Portfolio | | | | | | |
		-2	-1	0	1	2	3	Total
	-2	.00	.01	.02	.02	.01	.00	.06
Returns	-1	.02	.03	.05	.04	.03	.01	.18
X on the	0	.03	.06	.10	.06	.05	.02	.32
Domestic	1	.03	.04	.07	.05	.03	.01	.23
Portfolio	2	.01	.03	.05	.03	.02	.01	.15
	3	.00	.01	.03	.01	.01	.00	.06
	Total	.09	.18	.32	.21	.15	.05	1.00

Moments for Joint Distributions

The first two moments of the random variables X and Y can be calculated as follows:

$$\mu_x = E[X] = \sum_{i=1}^{r} x_i f_x(x_i); \qquad \mu_y = E[Y] = \sum_{j=1}^{c} y_j f_y(y_j);$$

$$E[X^2] = \sum_{i=1}^{r} x_i^2 f_x(x_i); \qquad E[Y^2] = \sum_{j=1}^{c} y_j^2 f_y(y_j).$$

The variances are given by

$$\sigma_x^2 = V[X] = E[(X - \mu_x)^2] = E[X^2] - \mu_x^2$$

and

$$\sigma_y^2 = V[Y] = E[(Y - \mu_y)^2] = E[Y^2] - \mu_y^2.$$

A measure of the strength and direction of association between the variables X and Y is provided by the *covariance* which is defined by

$$\text{Cov}(X, Y) = \sigma_{xy} = E[(X - \mu_x)(Y - \mu_y)] = E[XY] - \mu_x \mu_y$$

where

$$E[XY] = \sum_{i=1}^{r} \sum_{j=1}^{c} x_i y_j f(x_i, y_j).$$

The *covariance matrix* for the joint distribution is given by

$$\Sigma = \begin{bmatrix} \sigma_x^2 & \sigma_{xy} \\ \sigma_{xy} & \sigma_y^2 \end{bmatrix}.$$

A relatively large value of σ_{xy} indicates a strong relationship between X and Y. If the covariance is positive, X and Y tend to move in the same direction (the covariation is direct) while a negative covariance indicates that X and Y tend to move in opposite directions (an inverse relationship between X and Y). Like the variances the covariance depends on the overall magnitude of X and Y.

An index of covariation between X and Y is provided by the *correlation coefficient* $\rho = \sigma_{xy}/\sigma_x \sigma_y$. This index has the range -1 to $+1$. When compared to 1, the absolute value of ρ indicates the strength of linear association between X and Y. For quantitative variables $|\rho| = 1$ is equivalent to a perfect linear relationship. The slope of the linear relationship is indicated by the sign of ρ. If $\rho = 0$ no linear relationship exists. The correlation matrix for the joint distribution is given by $\boldsymbol{\rho} = \begin{bmatrix} 1 & \rho \\ \rho & 1 \end{bmatrix}$.

Example

The means, variances and covariances for the joint distribution of portfolio returns in Table 3.2 are given by

$$\mu_x = (.06)(-2) + (.18)(-1) + (.32)(0) + (.23)(1)$$
$$+ (.15)(2) + (.06)(3) = 0.41$$

$$\mu_y = (.09)(-2) + (.18)(-1) + (.32)(0) + (.21)(1)$$
$$+ (.15)(2) + (.05)(3) = 0.30$$

$$\sigma_x^2 = (.06)(-2)^2 + (.18)(-1)^2 + (.32)(0)^2 + (.23)(1)^2$$
$$+ (.15)(2)^2 + (.06)(3)^2 - (0.41)^2 = 1.6219$$

$$\sigma_y^2 = (.09)(-2)^2 + (.18)(-1)^2 + (.32)(0)^2 + (.21)(1)^2 + (.15)(2)^2$$
$$+ (.05)(3)^2 - (0.30)^2 = 1.7100$$

and

$$\sigma_{xy} = (.00)(-2)(-2) + (.01)(-2)(-1) + (.02)(-2)(0) + (.02)(-2)(1)$$
$$+ (.01)(-2)(2) + (.00)(-2)(3) + (.02)(-1)(-2) + (.03)(-1)(-1)$$
$$+ (.05)(-1)(0) + (.04)(-1)(1) + (.03)(-1)(2) + (.01)(-1)(3)$$
$$+ (.03)(0)(-2) + (.06)(0)(-1) + (.10)(0)(0) + (.06)(0)(1)$$
$$+ (.05)(0)(2) + (.02)(0)(3) + (.03)(1)(-2) + (.04)(1)(-1)$$
$$+ (.07)(1)(0) + (.05)(1)(1) + (.03)(1)(2) + (.01)(1)(3)$$
$$+ (.01)(2)(-2) + (.03)(2)(-1) + (.05)(2)(0) + (.03)(2)(1)$$
$$+ (.02)(2)(2) + (.01)(2)(3) + (.00)(3)(-2) + (.01)(3)(-1)$$
$$+ (.03)(3)(0) + (.01)(3)(1) + (.01)(3)(2) + (.00)(3)(3)$$
$$- (.41)(.30) = -0.043.$$

The correlation coefficient is given by

$$\rho = (-0.043)/(1.6219)^{1/2}(1.7100)^{1/2} = -0.0258$$

which suggests a very weak correlation between the returns on the two portfolios.

Linear Combinations

The expectation and variance for a linear combination of the two random variables is given by

$$E[aX + bY] = aE[X] + bE[Y] \quad \text{and}$$

$$V[aX + bY] = a^2 V[X] + b^2 V[Y] + 2ab \, \text{Cov}[X, Y]$$

where a and b are two constants. These expressions may be used to obtain the mean and variance for a portfolio consisting of a combination of the two portfolios discussed above.

Example

An investment consisting of $1/3$ of the foreign portfolio and $2/3$ of the domestic portfolio would have expected return and return variance given by

$$E\left[\frac{2}{3}X + \frac{1}{3}Y\right] = \frac{2}{3}(0.41) + \frac{1}{3}(0.30) = 0.3733 \quad \text{and}$$

$$V\left[\frac{2}{3}X + \frac{1}{3}Y\right] = \frac{4}{9}(1.6219) + \frac{1}{9}(1.7100) + \frac{4}{9}(-0.043) = 0.8917.$$

Thus an investor can reduce the return variance by almost 50% with this investment as compared to an investment entirely in X. The mean return for the combined portfolio is only marginally less than for X alone. This reflects the potential benefit of diversification.

Conditional Distributions and Independence

If the variables X and Y are *statistically independent*, the joint density is given by

$$f(x_i, y_j) = f_x(x_i)f_y(y_j) \quad \text{for all} \quad (x_i, y_j), \ i = 1, 2, \dots, r; \quad j = 1, 2, \dots, c.$$

In general, if X and Y are not independent the probability density for X at $Y = y_j$ differs from the density for X at $Y = y_k$.

Independence implies that the covariance σ_{xy} and the correlation coefficient, ρ, are necessarily zero. It is possible for σ_{xy} and ρ to be zero however, even if X and Y are not independent. Independence and zero correlation are therefore not equivalent. We shall see later in this chapter that for the bivariate normal density, independence and zero correlation are equivalent.

Conditional Densities

In the absence of independence the probability for X is conditional on the value of Y. Given that $Y = y_j$, the probability that $X = x_i$ is given by $f(x_i, y_j)/f_y(y_j)$ and is called the *conditional probability* of x_i given $Y = y_j$. In general the *conditional density* for $X = x_i$ given $Y = y_j$ is given by

$$f_x(x_i \mid Y = y_j) = f(x_i, y_j)/f_y(y_j).$$

Similarily the conditional density for $Y = y_j$ given $X = x_i$ is given by

$$f_y(y_j \mid X = x_i) = f(x_i, y_j)/f_x(x_i).$$

Example

From the joint distribution for the two portfolios in Table 3.2, the conditional density for Y given $X = 0$ can be determined as

Y	-2	-1	0	1	2	3
$f_y(y \mid X = 0)$	.03/.32	.06/.32	.10/.32	.06/.32	.05/.32	.02/.32 .
	.094	.188	.312	.188	.156	.062

In comparison to the marginal density for Y we can conclude that when $X = 0$ the conditional density for Y is almost identical to the marginal density. The largest difference occurs for the value $Y = 1$ which is less likely at 0.188 for $X = 0$ than the joint probability which is 0.21.

Under independence the joint density is given by the product of the marginals and would have the values given in Table 3.3. A comparison of Tables 3.2 and 3.3 shows how close to independence the two portfolios are.

Table 3.3. Joint Distribution of the Two Portfolios Under Independence

			Y					
			Returns on the Foreign Portfolio					
		-2	-1	0	1	2	3	Total
	-2	.0054	.0108	.0192	.0126	.0090	.0030	.0600
Returns	-1	.0162	.0324	.0576	.0378	.0270	.0090	.1800
on	0	.0288	.0576	.1024	.0672	.0480	.0160	.3200
X the	1	.0207	.0414	.0736	.0483	.0345	.0115	.2300
Domestic	2	.0135	.0270	.0480	.0315	.0225	.0075	.1500
Portfolio	3	.0054	.0108	.0192	.0126	.0090	.0030	.0600
	Total	.0900	.1800	.3200	.2100	.1500	.0500	1.0000

Moments for Conditional Distributions

The moments for the *conditional distributions* can be obtained by employing the conditional densities. The *conditional means* are given by

$$\mu_{x \cdot y_j} = E[X \mid Y = y_j] = \sum_{i=1}^{r} x_i f_x(x_i \mid Y = y_j)$$

$$\mu_{y \cdot x_i} = E[Y \mid X = x_i] = \sum_{j=1}^{c} y_j f_y(y_j \mid X = x_i).$$

The *conditional variances* can also be obtained from the conditional second moments

$$E[X^2 \mid Y = y_j] = \sum_{i=1}^{r} x_i^2 f_x(x_i \mid Y = y_j);$$

$$E[Y^2 \mid X = x_i] = \sum_{j=1}^{c} y_j^2 f_y(y_j \mid X = x_i);$$

yielding the conditional variances

$$\sigma_{x \cdot y_j}^2 = E[X^2 \mid Y = y_j] - \mu_{x \cdot y_j}^2;$$

$$\sigma_{y \cdot x_i}^2 = E[Y^2 \mid X = x_i] - \mu_{y \cdot x_i}^2.$$

Example

The mean and variance for the distribution of Y given $X = 0$ are given by

$$E[Y \mid X = 0] = (.094)(-2) + (.188)(-1) + (.312)(0) + (.188)(1)$$
$$+ (.156)(2) + 3(.062) = 0.3074 \quad \text{and}$$

$$V[Y \mid X = 0] = (.094)(-2)^2 + (.188)(-1)^2 + (.312)(0)^2 + (.188)(1)^2$$
$$+ (.156)(2)^2 + (.062)(3)^2 - (.3074)^2 = 1.839.$$

As expected because of near independence the conditional mean and variance are very similar to the marginal distribution mean (0.30) and variance (1.70) for Y.

3.1.2 Continuous Bivariate Distributions and the Bivariate Normal

For the bivariate distribution discussed above in 3.1.1 the random variables X and Y could each assume only a finite number of values. For continuous random variables there are infinitely many possible values and hence the summation notation used in the definitions above must be replaced by integral notation. These definitions will be introduced using the *bivariate normal distribution* as an example.

The Bivariate Normal

The *bivariate normal density* is given by

$$f(x, y) = c_1^{-1} \exp\{c_2[(x-\mu_x)^2/\sigma_x^2 + (y-\mu_y)^2/\sigma_y^2 - 2\rho(x-\mu_x)(y-\mu_y)/\sigma_x\sigma_y]\}$$
(3.1)

where $c_1 = 2\pi\sigma_x\sigma_y(1 - \rho^2)^{1/2}$ and $c_2 = -1/[2(1 - \rho^2)]$. The bivariate normal density contains five parameters μ_x, μ_y, σ_x, σ_y and ρ or equivalently the parameters are given by the mean vector $\boldsymbol{\mu} = \begin{bmatrix} \mu_x \\ \mu_y \end{bmatrix}$ and the covariance matrix $\boldsymbol{\Sigma} = \begin{bmatrix} \sigma_x^2 & \sigma_{xy} \\ \sigma_{xy} & \sigma_y^2 \end{bmatrix}$ where $\rho = \sigma_{xy}/\sigma_x\sigma_y$. The bivariate normal density can also be written in matrix notation as

$$f(x, y) = c_1^{-1} \exp\left\{ -\frac{1}{2} \begin{bmatrix} x - \mu_x \\ y - \mu_y \end{bmatrix}' \boldsymbol{\Sigma}^{-1} \begin{bmatrix} x - \mu_x \\ y - \mu_y \end{bmatrix} \right\}.$$

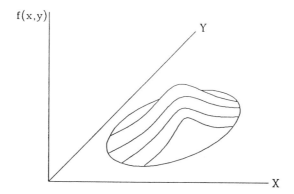

Figure 3.1. Bivariate Normal Density

Figure 3.1 shows a bivariate normal surface. The three dimensional picture shows that the density at (x, y) is given by the height $f(x, y)$. The probability for any region of values $(x_0 \leq X \leq x_1, \ y_0 \leq Y \leq y_1)$ is given by the volume under the normal surface.

The marginal density for X is obtained from the integral expression

$$f_x(x) = \int_{-\infty}^{\infty} f(x, y)dy = \frac{1}{\sqrt{2\pi\sigma_x^2}} \ \exp\left[(-1/2)\left(\frac{x - \mu_x}{\sigma_x}\right)^2\right]$$

and similarly for Y

$$f_y(y) = \int_{-\infty}^{\infty} f(x, y)dx = \frac{1}{\sqrt{2\pi\sigma_y^2}} \ \exp\left[(-1/2)\left(\frac{y - \mu_y}{\sigma_y}\right)^2\right].$$

These densities are the familiar univariate normal densities. The marginal densities for the bivariate normal therefore are also normal. It is not necessarily true however that X and Y will be bivariate normal if X and Y are both univariate normal.

If $\rho = 0$ the joint density for the bivariate normal can be written as the product of the marginal densities and hence X and Y are independent. Thus for the bivariate normal, independence and zero correlation are equivalent.

Example

As an example, suppose that for a large population of fathers and oldest sons (all over 21) the distributions of height (in inches), X for fathers, Y for sons are normal with identical means $\mu_x = \mu_y = 72$ and variances $\sigma_x^2 =$

$\sigma_y^2 = 6$. A useful way of graphically displaying a sample of observations from a bivariate distribution is the *scatterplot*. The four scatterplots in Figure 3.3 show random samples of 200 points for correlation coefficients $\rho = 0.0$, 0.8, 0.4 and -0.5 for panels (a), (b), (c) and (d) respectively.

In panel (a) we would conclude that there is no linear relationship between father's height and son's height. In panels (b) and (c) there is positive linear association with panel (b) illustrating a stronger association than (c). In panel (d) the linear association is negative indicating that the heights of fathers and sons tend to move in opposite directions.

From the four scatterplots we can conclude that the correlation coefficient has an important impact on the bivariate scatter. For $\rho = 0$, the points are scattered in a circle with centre (72, 72), while for $\rho \neq 0$, the scatterplots have elliptical shapes with centre (72, 72). The larger the magnitude of ρ the greater the length of the major axis relative to the minor axis. Also, the sign of ρ determines the sign of the slope of the major axis.

Elliptical Contours For the Bivariate Normal Distribution

Recall that for a univariate distribution, a plot of the density of the standardized random variable $\left(\frac{X - \mu_x}{\sigma_x}\right)$ is useful for making comparisons with other densities such as the normal distribution. For a normal random variable X we know that $100(1 - \alpha)\%$ of the distribution is contained within the range $[-Z_{\alpha/2}, Z_{\alpha/2}]$. This region could be described by the equation $\left(\frac{X - \mu_x}{\sigma_x}\right)^2 \leq Z_{\alpha/2}^2$. The region $\left|\left(\frac{X - \mu_x}{\sigma_x}\right)\right| > c$ is shown by the shaded area in the tails in Figure 3.2. Recall that for a normal density the effective range for c is the interval (0,3).

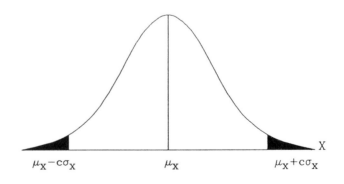

Figure 3.2. Univariate Normal Density

(a) $\rho = 0.0$

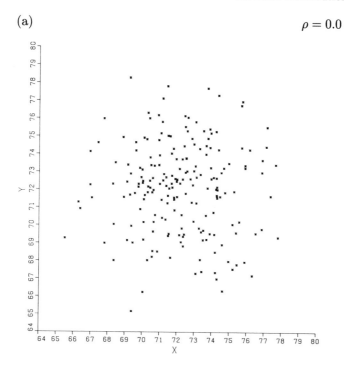

(b) $\rho = 0.80$

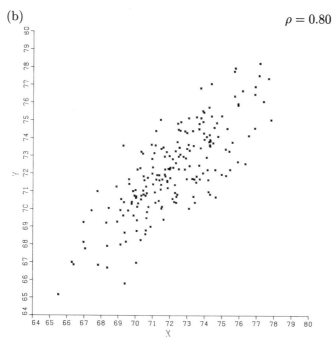

Figure 3.3. Scatterplots for Various Bivariate Normal Samples

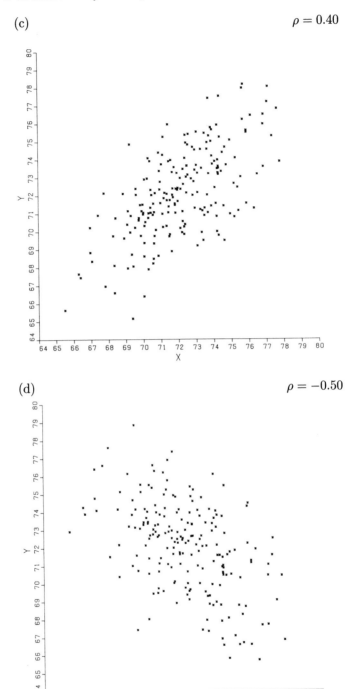

Figure 3.3. Scatterplots for Various Bivariate Normal Samples (continued)

For bivariate distributions *elliptical contours* can be used to study the distribution of the density over the $X - Y$ plane. The equation

$$\left(\frac{1}{1-\rho^2}\right)\left[\left(\frac{X-\mu_x}{\sigma_x}\right)^2 + \left(\frac{Y-\mu_y}{\sigma_y}\right)^2 - 2\rho\left(\frac{X-\mu_x}{\sigma_x}\right)\left(\frac{Y-\mu_y}{\sigma_y}\right)\right] = k \quad (3.2)$$

describes an ellipse in the $X - Y$ plane with centre at (μ_x, μ_y), as shown in Figure 3.4. As k increases the area of the ellipse increases. The equation for the elliptical contour can also be written in matrix notation as

$$\begin{bmatrix} X-\mu_x \\ Y-\mu_y \end{bmatrix}' \Sigma^{-1} \begin{bmatrix} X-\mu_x \\ Y-\mu_y \end{bmatrix} = k.$$

For the bivariate normal density, the constant k on the right-hand side of (3.2) is equal to $\chi^2_{\alpha;2}$, where $\chi^2_{\alpha;2}$ denotes the value of a χ^2 random variable with 2 degrees of freedom and a p-value of α in the upper tail. The elliptical contour will on average contain $100(1-\alpha)\%$ of the sample points.

Example

The ellipse shown in Figure 3.5 corresponds to $k = \chi^2_{.05;2}$ for the joint distribution discussed in the above example with $\rho = 0.67$. The plotted points represent a random sample of 400 points from this bivariate normal population. Approximately 20 points should therefore be outside the elliptical boundary. Any sample bivariate scatterplot can be compared to elliptical contours of the bivariate normal, in order to judge the acceptability of a bivariate normal assumption and also to check for outliers. This concept will be employed in a discussion of inference in Section 3.2.

Some Geometry for Elliptical Contours

A transformation of axes, so that the origin is at (μ_x, μ_y), is given by $Z = (X - \mu_x)$ and $W = (Y - \mu_y)$ as in Figure 3.4. A rotation of the axes $Z - W$ through an angle θ yields new axes $V - U$ corresponding to the major and minor axes of the ellipse shown in Figure 3.4.

The major axis of the ellipse makes an angle θ with the X axis where

$$\tan\theta = \left[\frac{\sigma_y}{2\rho\sigma_x} - \frac{\sigma_x}{2\rho\sigma_y}\right] \pm \sqrt{\left[\frac{\sigma_y}{2\rho\sigma_x} - \frac{\sigma_x}{2\rho\sigma_y}\right]^2 + 1}, \qquad \rho \neq 0.$$

The slope, $\alpha = \tan\theta$, of the major axis is a solution to the quadratic equation $\alpha^2\sigma_{xy} + \alpha[\sigma_x^2 - \sigma_y^2] - \sigma_{xy} = 0$.

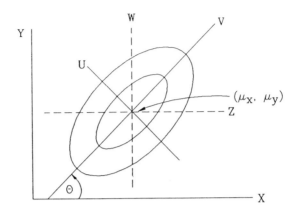

Figure 3.4. Elliptical Contours for Bivariate Normal Density

If $\sigma_x = \sigma_y$ and $\rho > 0$ the angle $\theta = 45°$, and if $\rho < 0$, $\theta = 135°$. If $\rho = 0$, the angle $\theta = 0$ or $180°$, and the contour is an ellipse centered at (μ_x, μ_y) with the equation given by

$$\left(\frac{X - \mu_x}{\sigma_x}\right)^2 + \left(\frac{Y - \mu_y}{\sigma_y}\right)^2 = k.$$

If $\sigma_x = \sigma_y$ and $\rho = 0$ the contour is a circle centered at (μ_x, μ_y).

For a given σ_x and σ_y, the magnitude of ρ affects the relative lengths of the major and minor axes. The squares of the lengths of the major and minor axes are given by

$$\frac{(1 - \rho^2)\sigma_x^2\sigma_y^2 k}{4(\sigma_x^2 + \sigma_y^2) \pm 4[(\sigma_x^2 - \sigma_y^2)^2 + 4\rho^2\sigma_x^2\sigma_y^2]^{1/2}}.$$

The plus sign in the denominator is used for the minor axis and the minus sign for the major axis. If $\sigma_x = \sigma_y$, the correlation coefficient ρ can be expressed as $\rho = (A^2 - B^2)/(A^2 + B^2)$ where A and B denote the lengths of the major and minor axes respectively.

Mahalanobis Distance

For every point (X, Y) on an elliptical contour, the equation

$$\left(\frac{1}{1 - \rho^2}\right)\left[\left(\frac{X - \mu_x}{\sigma_x}\right)^2 + \left(\frac{Y - \mu_y}{\sigma_y}\right)^2 - 2\rho\left(\frac{X - \mu_x}{\sigma_x}\right)\left(\frac{Y - \mu_y}{\sigma_y}\right)\right] = k$$

measures the square of the *Mahalanobis distance* between (X, Y) and the centre (μ_x, μ_y). Therefore the elliptical contour traces the path of points with equal Mahalanobis distance from the centre. This distance is the radius of a circle, which could be obtained by standardizing the two variables and rotating the axes through the angle θ (hence transforming X and Y to X^* and Y^* so that $\rho = 0$, $\sigma_{x^*} = \sigma_{y^*} = 1$). The Mahalanobis distance is therefore equivalent to the *Euclidean distance* for a standardized bivariate normal under independence. The squared Mahalanobis distance can also be written in matrix notation as

$$k = \begin{bmatrix} X - \mu_x \\ Y - \mu_y \end{bmatrix}' \begin{bmatrix} \sigma_x^2 & \sigma_{xy} \\ \sigma_{xy} & \sigma_y^2 \end{bmatrix}^{-1} \begin{bmatrix} X - \mu_x \\ Y - \mu_y \end{bmatrix}.$$

The square of the Mahalanobis distance between any pair of points (x_1, y_1), (x_2, y_2) is given by

$$\left(\frac{1}{1 - \rho^2} \right) \left[\frac{(x_1 - x_2)^2}{\sigma_x} + \frac{(y_1 - y_2)^2}{\sigma_y} - 2\rho \frac{(x_1 - x_2)(y_1 - y_2)}{\sigma_x \sigma_y} \right],$$

where (x_1, y_1) and (x_2, y_2) are two points in the bivariate normal distribution with density $f(x, y)$ given by (3.1).

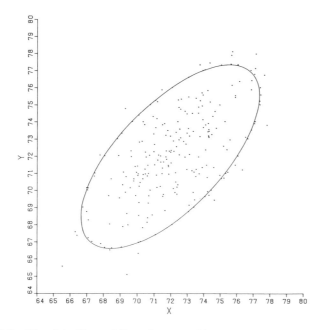

Figure 3.5. Bivariate Normal Sample and 95% Elliptical Contour

Conditional Density and Regression Functions

The conditional density of Y given $X = x$, for the bivariate density $f(x, y)$, is obtained from the expression

$$f_y(y \mid X = x) = f(x, y)/f_x(x).$$

The conditional expectation of Y given $X = x$, is given by

$$\mu_{y \cdot x} = E[Y \mid X = x] = \int_{-\infty}^{\infty} y f_y(y \mid X = x) dy,$$

and is called the *regression function of Y on X*. Note that the regression function is a function of x. The variance of the conditional distribution of Y given $X = x$, is given by

$$\sigma_{y \cdot x}^2 = V[Y \mid X = x] = E[Y^2 \mid X = x] - \mu_{y \cdot x}^2,$$

where

$$E[Y^2 \mid X = x] = \int_{-\infty}^{\infty} y^2 f_y(y \mid X = x) dy.$$

For the bivariate normal distribution, the conditional density of Y given X is normal, and has the form

$$f_y(y \mid X = x) = \frac{1}{\sqrt{2\pi\sigma_{y \cdot x}^2}} \exp\left[-\frac{1}{2}\left(\frac{y - \beta_0 - \beta_1 x}{\sigma_{Y \cdot x}} \right)^2 \right],$$

where

$$\beta_0 = \mu_y - \beta_1 \mu_x$$

$$\beta_1 = \sigma_{xy}/\sigma_x^2$$

$$\sigma_{y \cdot x}^2 = \sigma_y^2(1 - \rho^2).$$

The mean and variance of this conditional distribution are given by $\mu_{y \cdot x} = \beta_0 + \beta_1 x$ and $\sigma_{y \cdot x}^2$ respectively. The regression function of Y on X is therefore given by

$$\mu_{y \cdot x} = \beta_0 + \beta_1 x.$$

For the bivariate normal distribution the regression function derived above is a <u>linear</u> function of x. Figure 3.6 shows a plot of this regression function. At each value of X, say x_i, the conditional density of Y is normal with mean lying on the regression line. The variances of the conditional distributions, $\sigma_{y \cdot x_i}^2$ are equal for all x_i and will be denoted by $\sigma_{y \cdot x}^2$. The conditional density of Y given $X = x$ is therefore a normal density with mean $\mu_{y \cdot x}$ and variance $\sigma_{y \cdot x}^2$.

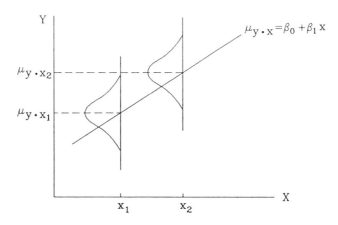

Figure 3.6. Regression Function Y on X

Partitioning the Variance σ_y^2

The equation for $\sigma_{y \cdot x}^2$ given above can be written as $\sigma_{y \cdot x}^2 = \sigma_y^2 - \sigma_y^2 \rho^2 = \sigma_y^2 - \beta_1^2 \sigma_x^2$, and hence $\sigma_y^2 = \beta_1^2 \sigma_x^2 + \sigma_{y \cdot x}^2$. The variance of the random variable Y therefore, has been partitioned into the *variance around the regression line* given by $\sigma_{y \cdot x}^2$ and the *variance explained* by the points along the regression line, $\beta_1^2 \sigma_x^2$. The ratio of $\beta_1^2 \sigma_x^2$ to σ_y^2 is called the *coefficient of determination* and is equivalent to ρ^2. The square of the correlation coefficient, ρ^2, is equivalent therefore to the coefficient of determination.

Regression Function for X on Y

The conditional density of X given $Y = y$ can be studied in a manner similar to the above. The *regression function for X given $Y = y$* is given by

$$\mu_{x \cdot y} = \gamma_0 + \gamma_1 y \qquad \text{and} \qquad \sigma_{x \cdot y}^2 = \sigma_x^2 (1 - \rho^2),$$

$$\text{where} \qquad \gamma_1 = \sigma_{xy} / \sigma_y^2 \qquad \text{and} \qquad \gamma_0 = \mu_x - \gamma_1 \mu_y.$$

The two regression lines are therefore necessarily not the same, although they both pass through the point (μ_x, μ_y) and both regression lines have the same coefficient of determination ρ^2. The correlation coefficient ρ can also be expressed as $\rho = \sqrt{\beta_1 \gamma_1}$ where β_1 and γ_1 are the slopes in the regressions of Y on X and X on Y respectively. The correlation coefficient is therefore the geometric mean of the two slopes. If the point (μ_x, μ_y) is taken as the origin, a line of slope 1 through this origin bisects the angle between these two regression lines.

Regression Functions

The term "regression" in the context of a linear relationship between two random variables X and Y, was first suggested by the English scientist Sir Francis Galton in the late nineteenth century. Galton was concerned with the study of hereditary traits, and in particular, the relationship between the heights of parents and the heights of their adult offspring. In fitting a linear least squares relationship between height of offspring, Y, and height of parent, X, Galton discovered that the slope was much less than 1. As a result of the analysis he concluded that sons of tall fathers are on average shorter than their fathers and that sons of short fathers tend to be taller than their fathers. Galton labelled this phenomenon as "regression to mediocrity". It is from this label that the technique of least squares became known as regression.

Example

For the joint distribution of fathers and sons discussed above the parameters were $\mu_x = \mu_y = 72$ and $\sigma_x^2 = \sigma_y^2 = 6$. We assume that $\sigma_{xy} = 4$ and hence $\rho = 0.67$.

The regression function for sons on fathers is given by

$$\mu_{y \cdot x} = \beta_0 + \beta_1 X = 24 + 0.67X$$

and $\mu_y = 24 + 0.67\mu_x$. The variance in Y explained by the regression is $\beta_1^2 \sigma_x^2 = 2.67$ and the variance around the regression line is $\sigma_{y \cdot x}^2 = \sigma_y^2 - \beta_1^2 \sigma_x^2 = 3.33$. The proportion of variation in Y explained by the regression line is 0.44.

Similarily the regression function for fathers on sons is given by

$$\mu_{x \cdot y} = \gamma_0 + \gamma_1 Y = 24 + 0.67X$$

and $\mu_x = 24 + 0.67\mu_y$. The variance in X explained by the regression is $\gamma_1^2 \sigma_y^2 = 2.67$ and the variance around the regression line is $\sigma_{x \cdot y}^2 = \sigma_x^2 - \gamma_1^2 \sigma_y^2 = 3.33$. The proportion of variation in X explained by the regression line is 0.44. Solving the second regression equation for Y in terms of X we obtain

$$\mu_y = -36 + 1.5\mu_x.$$

Thus the two regression relationships are quite different although both lines pass through the mean $(\mu_x, \mu_y) = (72, 72)$. Figure 3.7 shows a graph which includes the two regression lines along with the line $\mu_y = \mu_x$ and a random sample of 200 observations from this bivariate normal population. The regression line for Y on X has slope less than 1, while the regression line

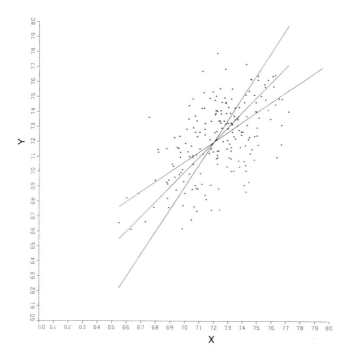

Figure 3.7.   Bivariate Normal Sample and True Regression Lines For Y on X and X on Y

for X on Y has slope greater than 1. The line with slope 1 is the line $\mu_y = \mu_x$ which lies in between the two regression lines.

For a relatively tall father say $X = 78$, we have $\mu_{Y \cdot X} = 24 + 0.67(78) = 76$, and hence the sons of very tall fathers are expected to be shorter than their fathers but still taller than average. Among this group of tall fathers in comparison to their fathers (son's grandfathers) more of them can be expected to be unusually tall than can be expected to be unusually short given their genetic make-ups. These unusual traits are not likely to be passed on to the sons and hence many of the sons revert back to more normal heights. A similar argument can be made for a group of short fathers and their sons. Because by assumption the two marginal distributions are equal there must be a regression toward the mean from both extremes in order to preserve the equality of the two marginal distributions. The means for the conditional distributions for Y given $X = 78$; Y given $X = 66$; X given $Y = 78$; and X given $Y = 66$ are respectively $\mu_y = 76.26$ (less than 78), $\mu_y = 68.22$ (greater than 66), $\mu_x = 76.26$ (less than 78) and $\mu_x = 68.22$ (greater than 66). Thus if this bivariate distribution with the assumed properties is to remain stable from generation to generation, then the regression toward the mean must continue.

A second example which helps to clarify the regression phenomenon is to assume that X represents the batting average of a major league ball player at the all-star break, and that Y represents the batting average at the end of the season. Assume once again that the joint distribution is bivariate normal and that the two marginal distributions are identical. Assuming a positive correlation between X and Y the regression function would imply that the year-end batting average for players with high first half averages will tend to be lower while the reverse is true for players with low first half averages. This phenomenon can be explained by assuming that at the all-star break some ball players are batting above their potential and some below their potential. Assuming that batting averages are normally distributed there are fewer players in the tails of the distribution. Among the players who attained a relatively high first-half average will be some who are below their potential and some who are above their potential. At the relatively high first-half average there will be more who attained an average greater than their potential than the number who attained an average below their potential. We must have regression to the mean if the two marginal distributions are to be the same. For example, if those with high averages tended to get better and those with lower averages tended to get worse the distribution of second half averages would have a larger variance than the distribution of first half averages. In other words the marginal distributions would not be the same.

Control Variables

Another way of viewing the regression phenomenon is to imagine that we are trying to predict the outcome Y for a variety of processes in which a *control variable* X is being manipulated. X is assumed to be error free whereas Y is assumed to be a random variable at each X. The regression function is an expression for the conditional mean of Y at each X. In regression analysis we are concerned with determining the variation in Y that can be explained by the control variable X. It is this explained variation that is used to predict Y. The prediction for Y at X is the mean of the conditional distribution of Y at X.

Alternative Linear Relationships

For the bivariate normal distribution given by (3.1) two different linear relationships relating X and Y have been derived using the regression functions of Y on X and X on Y. Both of the regression relationships pass through the point (μ_x, μ_y) but have different slopes $\beta = \sigma_{xy}/\sigma_x^2$ and $\gamma = \sigma_{xy}/\sigma_y^2$ for Y on X and X on Y respectively. Other linear relationships between X and Y which also pass through (μ_x, μ_y) can be derived under an alternative model.

Assume that the bivariate distribution is generated as follows. An underlying relationship between X and Y is the *exact functional relation*

$Y^* = \beta_0 + \beta_1 X^*$. For each observation in a random sample of size n, (x_i, y_i), $i = 1, 2, \ldots, n$ the values of x_i and y_i are generated by

$$x_i = x_i^* + v_i \qquad \text{and} \qquad y_i = y_i^* + u_i$$

where

$$\begin{bmatrix} V \\ U \end{bmatrix} \sim N\left(\begin{bmatrix} 0 \\ 0 \end{bmatrix}, \begin{bmatrix} \sigma_v^2 & 0 \\ 0 & \sigma_u^2 \end{bmatrix} \right) \tag{3.3}$$

independently of $\begin{bmatrix} X^* \\ Y^* \end{bmatrix}$. The variable X^* is assumed to be normally distributed with mean μ_x and variance $\sigma_{x^*}^2$. The joint distribution of $\begin{bmatrix} X^* \\ Y^* \end{bmatrix}$ is therefore bivariate normal with mean vector $\begin{bmatrix} \mu_x \\ \mu_y \end{bmatrix}$ and covariance matrix

$\begin{bmatrix} \sigma_{x^*}^2 & \sigma_{x^*y^*} \\ \sigma_{x^*y^*} & \sigma_{y^*}^2 \end{bmatrix}$ where $\mu_y = \beta_0 + \beta_1\mu_x$, $\sigma_{y^*}^2 = \beta_1^2\sigma_{x^*}^2$ and $\sigma_{x^*y^*} = \beta_1\sigma_{x^*}^2$.

The joint distribution of $\begin{bmatrix} X \\ Y \end{bmatrix}$ is given by

$$\begin{bmatrix} X \\ Y \end{bmatrix} \sim N\left(\begin{bmatrix} \mu_x \\ \mu_y \end{bmatrix}, \begin{bmatrix} \sigma_x^2 & \sigma_{xy} \\ \sigma_{xy} & \sigma_y^2 \end{bmatrix} \right)$$

where

$$\mu_x = E[X^*], \qquad \mu_y = E[Y^*],$$

$$\sigma_x^2 = V(X^*) + \sigma_v^2, \qquad \sigma_y^2 = V(Y^*) + \sigma_u^2, \qquad \sigma_{xy} = \text{Cov}(X^*, Y^*).$$

Figure 3.8 shows the location of sample points relative to the true line and underlying true point. The observed points are denoted by $\odot$ and the underlying generating points on the line are denoted by $\otimes$.

Example

An example application of this model is provided by an assumed linear relationship between work related stress, Y, experienced by police officers and X, the level of crime in the community. For each community the average level of stress for the police officers is denoted by Y^* and the average level of crime as perceived by the citizens is denoted by X^*. To estimate the linear relationship over a population of n communities, samples of citizens are surveyed to determine average crime level and samples of officers are used to determine average stress level. For the i-th community the sample average levels of stress and crime are related to the true mean values by the equations

$$x_i = x_i^* + v_i \quad \text{and} \quad y_i = y_i^* + u_i \qquad i = 1, 2, \ldots, n.$$

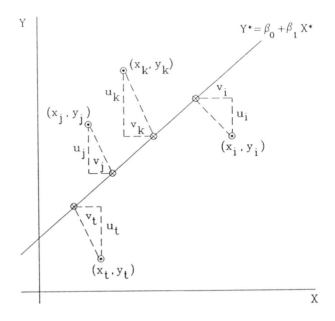

Figure 3.8. Scatterplot of Sample Points and True Line

Assuming that the bivariate distribution of $\begin{bmatrix} U \\ V \end{bmatrix}$ is given by (3.3), and that the distribution of X^* is normal with mean μ_{x^*} and variance $\sigma_{x^*}^2$, the model outlined above can be used to describe the observations (x_i, y_i) on average crime level and average stress level for the n communities. In this case the true relationship is between the true means, and the sample observations are sample means which are subject to sampling error.

Determining the True Parameters

In practice since we do not observe (X^*, Y^*) we must determine the relationship between the parameters β_0, β_1, and the parameters of the distribution of $\begin{bmatrix} X \\ Y \end{bmatrix}$. From the equation $Y^* = \beta_0 + \beta_1 X^*$ we obtain

$$\mu_y = \beta_0 + \beta_1 \mu_x,$$

$$(\sigma_y^2 - \sigma_u^2) = \beta_1 \sigma_{xy},$$

$$\sigma_{xy} = \beta_1 (\sigma_x^2 - \sigma_v^2),$$

therefore the slope β_1 is given by the quadratic equation

$$\beta_1^2 \sigma_{xy} \sigma_v^2 + \beta_1 [\sigma_x^2 \sigma_u^2 - \sigma_y^2 \sigma_v^2] - \sigma_{xy} \sigma_u^2 = 0.$$

If $\sigma_v^2 = 0$, X is observed precisely, and the solution for β_1 provides the regression function of Y on X. If $\sigma_u^2 = 0$, Y is observed precisely, and the solution for β_1 provides the regression function of X on Y. In general if both σ_v^2 and σ_u^2 are non-zero and have the ratio $b = \sigma_v^2 / \sigma_u^2$ then the slope β_1 is a solution to the quadratic

$$\beta_1^2 \sigma_{xy} b + \beta_1 [\sigma_x^2 - b\sigma_y^2] - \sigma_{xy} = 0. \tag{3.4}$$

If $b = 1$ then the quadratic becomes

$$\beta_1^2 \sigma_{xy} + \beta_1 [\sigma_x^2 - \sigma_y^2] - \sigma_{xy} = 0$$

and hence the linear relationship between Y and X is the major axis of the ellipsoid given by (3.2). In this case β_1 is the slope of the major axis of the ellipsoid. The quadratic equation (3.4) will be discussed in Section 3.3.9 in connection with measurement error.

3.2. Statistical Inference for Bivariate Random Variables

In this section techniques useful for the study of bivariate distributions are introduced. The most commonly used technique, simple linear regression, will be discussed separately in the next section. The study of the bivariate distribution should begin with the examination of the two marginal distributions using the techniques of Chapter 2. In some applications a comparison of the two marginal distributions may also be of interest if the two random variables are measuring similar quantities. Also of interest is the detection of outliers, which are over and above the outliers that would be determined in a univariate analysis of each variable separately. Bivariate analysis almost always includes the study of the correlation between the two random variables. This topic is discussed first.

3.2.1 The Correlation Coefficient

The relationship between two quantitative random variables X and Y is usually characterized by the covariance σ_{xy} and/or the correlation coefficient ρ. These parameters can be used to indicate both a lack of independence and a strength of linear association between X and Y. Given a random sample of n observations (x_i, y_i), $i = 1, 2, \ldots, n$, for the random variables X and Y, it is often of interest to make inferences about ρ. The sample observations can be used to determine unbiased estimators of the parameters μ_x, μ_y, σ_x^2 and σ_y^2 using the estimators $\bar{x}$, $\bar{y}$, s_x^2 and s_y^2 respectively. The covariance parameter σ_{xy} has the unbiased estimator s_{xy} called the *sample covariance* where

$$s_{xy} = \sum_{i=1}^{n} (x_i - \bar{x})(y_i - \bar{y})/(n-1).$$

Pearson Correlation Coefficient

The *Pearson correlation coefficient* estimator r is obtained from $\rho = \sigma_{xy}/\sigma_x\sigma_y$ by replacing the parameters σ_{xy}, σ_x and σ_y by s_{xy}, s_x and s_y respectively; $r = s_{xy}/s_x s_y$. This estimator r of ρ is biased.

Example

The following example of a bivariate sample comes from a study of garbage disposal costs in a large municipality. A sample of 70 neighborhoods was observed; the variables of interest are cost per tonne (CPT) and number of loads (LOADS). The 70 observations are displayed in Table 3.4. Figure 3.9 shows a scatterplot of the data. The sample statistics are given by

$$\overline{\text{LOADS}} = 1352.07 \qquad \overline{\text{CPT}} = 24.69$$

$$s_{\text{LOADS}}^2 = 76448.80 \qquad s_{\text{CPT}}^2 = 13.90$$

$$s_{\text{CPT.LOADS}} = -572.40 \qquad r_{\text{CPT.LOADS}} = -0.555.$$

The correlation between CPT and LOADS is therefore of moderate size and negative. We can see from Figure 3.9 that there is an inverse relationship between the two variables and that the relationship is somewhat linear.

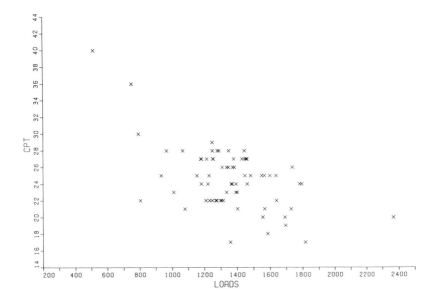

Figure 3.9. Scatterplot of CPT vs. LOADS

Inference when $\rho = 0$

To make inferences about ρ using r we require the *sampling distribution of r*
which is complex. When $\rho = 0$ and (X, Y) is bivariate normal, the statistic
$r\sqrt{n-2}/\sqrt{1-r^2}$ has a t distribution with $(n-2)$ degrees of freedom and
can be used to test the hypothesis that $\rho = 0$. The distribution of this
statistic is also the same if one of the variables say Y is normal and the
other variable say X is independent of Y. In the case of a bivariate normal
a test of $\rho = 0$ would also be a test for independence. If X and Y are not
bivariate normal and $\rho = 0$, the statistic $r\sqrt{n-2}/\sqrt{1-r^2}$ has a standard
normal distribution in large samples. Thus in large samples this statistic
can be used to test for zero correlation even if the joint distribution is not
bivariate normal.

If X and Y are not bivariate normal but are still independent the t-
statistic performs very well. If however, with $\rho = 0$, X and Y are not
bivariate normal and are also not independent then the t-statistic is not
well behaved and should be used primarily as a test for independence rather
than for zero correlation. Thus a rejection of the hypothesis $H_0 : \rho = 0$
should be viewed as a rejection of a hypothesis of independence.

Table 3.4. Garbage Disposal Cost Data Cost per Tonne and Number of Loads

OBSERV.NO	1	2	3	4	5	6	7	8	9	10	11	12	13	14
CPT	23	25	21	17	22	20	24	25	22	24	22	27	22	21
LOADS	1336	933	1729	1359	1305	1691	1370	1482	1275	1221	1248	1213	1270	1080

OBSERV.NO	15	16	17	18	19	20	21	22	23	24	25	26	27	28
CPT	26	25	25	24	22	22	24	22	19	25	27	27	25	26
LOADS	1308	1548	1598	1181	1298	804	1365	1315	1695	1636	1252	1460	1564	1736

OBSERV.NO	29	30	31	32	33	34	35	36	37	38	39	40	41	42
CPT	24	27	28	27	21	23	25	28	27	28	29	27	18	26
LOADS	1793	1177	1066	1250	1403	1393	1228	965	1445	1248	1247	1253	1585	1383

OBSERV.NO	43	44	45	46	47	48	49	50	51	52	53	54	55	56
CPT	21	24	28	27	36	24	23	24	40	17	30	28	22	27
LOADS	1567	1780	1288	1454	748	1393	1401	1461	513	1815	792	1278	1639	1181

OBSERV.NO	57	58	59	60	61	62	63	64	65	66	67	68	69	70
CPT	25	26	26	26	25	22	23	28	20	20	22	27	28	27
LOADS	1447	1347	1336	1372	1154	1231	1010	1348	2366	1555	1209	1429	1444	1379

Example

For the sample data in Table 3.4, the t-statistic for testing $H_0 : \rho = 0$ is given by $t = -5.50$ which has lower p-value of less than 0.000. The null hypothesis of zero correlation between CPT and LOADS is therefore rejected at conventional levels.

Fisher Transformation for $\rho \neq 0$

For $\rho \neq 0$ the *Fisher transformation* $\frac{1}{2} \ln \left(\frac{1+r}{1-r} \right)$ can be used to characterize the distribution of r. In large samples from a bivariate normal distribution the statistic $\frac{1}{2} \ln \left(\frac{1+r}{1-r} \right)$ is approximately normally distributed with mean $\frac{1}{2} \ln \left(\frac{1+\rho}{1-\rho} \right)$ and variance $\left(\frac{1}{n-3} \right)$. This statistic therefore can be used to make inferences about ρ in large samples from a bivariate normal. If (X, Y) are not bivariate normal the large sample distribution of r is unknown.

Example

An approximate 95% confidence interval for the true correlation coefficient ρ between CPT and LOADS can be obtained under the assumption of bivariate normality. Using the expression

$$\frac{1}{2} \ln \left(\frac{1+r}{1-r} \right) \pm 1.96 \sqrt{\frac{1}{n-3}}$$

the data in Table 3.4 yields the 95% confidence interval $(-.6256 \pm 0.24)$ for $\frac{1}{2} \ln \left(\frac{1+\rho}{1-\rho} \right)$ and hence the 95% confidence interval $(-.70, -.37)$ for ρ.

Other Measures of Correlation

The Pearson product moment correlation coefficient r measures the degree of linear association between X and Y. In the event of nonlinear but monotonic relationships, a useful correlation coefficient is provided by the *Spearman correlation coefficient*, r_s, which is a Pearson type correlation computed on the ranks of X and Y, ranked separately. In large samples the statistic $r_s \sqrt{n-1}$ has a standard normal distribution if there is no correlation between X and Y. This statistic can be used to test the null hypothesis of zero correlation.

Unlike the Pearson correlation coefficient, Spearman's rank correlation, r_s, does not measure the degree of linear association but instead measures *monotonicity* either increasing or decreasing. If r_s is close to plus 1, we know that for any two points (x_i, y_i) and (x_j, y_j) if $x_i > x_j$ then in general $y_i > y_j$ and the two points are said to be *concordant*. Similarily if r_s is close to minus 1 we would expect that if $x_i > x_j$ then $y_i < y_j$ and the two points (x_i, y_i), (x_j, y_j) are called *discordant*. The strength of the monotonicity is indicated by how close $|r_s|$ is to 1 while the sign of r_s indicates whether the relationship is increasing or decreasing.

An alternative measure of association, *Kendall's tau*, is based on this property of monotonicity. For any sample of n observations there are $\binom{n}{2}$ possible paired comparisons of points (x_i, y_i) and (x_j, y_j). Let C be the number of pairs that are concordant and let D be the number of discordant pairs. Kendall's tau is given by

$$\tau = (C - D)/\binom{n}{2}.$$

Clearly τ has the range $-1 \leq \tau \leq +1$. Kendall's tau may be viewed as a sample estimator of the difference between the probability of concordance and the probability of discordance.

If in the comparison of (x_i, y_i) and (x_j, y_j); $x_i = x_j$, or $y_i = y_j$ or both, the comparison is called a 'tie' and is not counted as concordant or discordant. If there are a large number of ties, the denominator of τ should be corrected by replacing $\binom{n}{2}$ by the quantity

$$\left(\left[\binom{n}{2} - n_x \right] \left[\binom{n}{2} - n_y \right] \right)^{1/2},$$

where n_x and n_y are the number of ties involving X and Y observations respectively.

In large samples the statistic $3\tau\sqrt{n(n-1)}/\sqrt{2(2n+5)}$ has a standard normal distribution if X and Y are independent. The distribution of this statistic approaches normality more rapidly than the distribution of $r_s\sqrt{n-1}$ for Spearman's correlation r_s.

Example

From the scatterplot of the data in Figure 3.9 there appears to be some nonlinearity in the relationship between CPT and LOADS. The Spearman correlation coefficient is -0.383 and the Kendall correlation coefficient is -0.282. Both of these correlation coefficients indicate a much weaker association between CPT and LOADS than the Pearson correlation of -0.555 obtained above.

The scatterplot in Figure 3.9 also suggests that observation numbers 47, 51 and 65 with data points (748, 36), (513, 48) and (2366, 20) may be outliers. These observations would also appear to have some influence on the strength of the correlation coefficient. The table below summarizes the three correlation coefficients when one or more of the extreme points is eliminated from the calculation. The correlations in Table 3.5 show that both the Spearman and Kendall coefficients are less sensitive to the extreme observations. After all three extreme observations have been deleted the Pearson correlation is much closer to the other two correlations. The difference between the two rank type correlations appears to be constant over the entire table.

These calculations were performed using the SAS procedure PROC CORR.

Table 3.5. Comparison of Correlation Coefficients with Extreme Observations Deleted

Observations Deleted	Pearson	Spearman	Kendall
None	-0.555	-0.383	-0.282
No. 47	-0.510	-0.355	-0.260
No. 51	-0.461	-0.355	-0.260
No. 65	-0.550	-0.357	-0.264
Nos. 47, 51	-0.388	-0.326	-0.236
Nos. 47, 65	-0.498	-0.328	-0.241
Nos. 51, 65	-0.439	-0.328	-0.241
Nos. 47, 51, 65	-0.351	-0.298	-0.217

3.2.2 Goodness of Fit and Outlier Detection For a Bivariate Normal

Elliptical Contours

Given n observations (x_i, y_i), $i = 1, 2, \ldots, n$ from a bivariate population we compute the sample quantities $\bar{x}$, $\bar{y}$, s_x^2, s_y^2, s_{xy} and r. For several choices of α and using the sample form of (3.2) we can determine $\chi^2_{\alpha;2}$ and determine elliptical contours

$$\left(\frac{1}{1-r^2}\right)\left[\left(\frac{X-\bar{x}}{s_x}\right)^2 + \left(\frac{Y-\bar{y}}{s_y}\right)^2 - 2r\left(\frac{X-\bar{x}}{s_x}\right)\left(\frac{Y-\bar{y}}{s_y}\right)\right] \leq \chi^2_{\alpha;2}.$$

The proportion of sample observations lying inside the contour can then be compared to $(1 - \alpha)$. The left hand side of this expression can be written in matrix notation as

$$\begin{bmatrix} X-\bar{x} \\ Y-\bar{y} \end{bmatrix}' \begin{bmatrix} s_x^2 & s_{xy} \\ s_{xy} & s_y^2 \end{bmatrix}^{-1} \begin{bmatrix} X-\bar{x} \\ Y-\bar{y} \end{bmatrix}$$

which is the general form for multivariate elliptical contours.

A useful approach would be to compute

$$c_i = \left(\frac{(x_i - \bar{x})^2}{s_x^2} + \frac{(y_i - \bar{y})^2}{s_y^2} - 2r \frac{(x_i - \bar{x})}{s_x} \frac{(y_i - \bar{y})}{s_y} \right) \left(\frac{1}{1 - r^2} \right)$$

for all observations $i = 1, 2, \ldots, n$. In large samples the values of c_i should be distributed as a χ^2 with 2 degrees of freedom. A quantile plot can be used to compare the quantiles for the observed c_i values to the quantiles for the χ^2 distribution. Observations which appear to be too far outside the elliptical contour may indicate outliers.

Example

Using the sample data for CPT and LOADS shown in Table 3.4 the value of c_i was computed using the sample statistics for this data presented at the beginning of Section 3.2. Comparison of the c_i values to $\chi^2_{.05;2} = 5.99$ shows that the c_i values which exceed 5.99 correspond to observation numbers 4, 20, 47, 51, 52, 53 and 65, which have c_i values of 6.186, 6.458, 20.194, 37.665, 10.235, 8.871, 21.749 respectively. Once again the three observations 47, 51 and 65 appear to be outliers. Observations 4, 20, 52 and 53 are closer to the bound of 5.99 and hence will assumed to be larger due to chance. The critical χ^2 of 5.99 should only include 95% of the 70 observations.

An Alternative Method

Another useful way to detect outliers is to use the Pearson correlation coefficient which is very sensitive to outliers. Let $r_{(i)}$ denote the correlation coefficient between X and Y determined after omitting the observation (x_i, y_i). Clearly for most observations if the sample is large the values of $r_{(i)}$ should not differ very much. The quantities

$$Z_{(i)} = \left(\frac{1}{2} \right) \ln \left(\frac{1 + r_{(i)}}{1 - r_{(i)}} \right) \qquad i = 1, 2, \ldots, n$$

obtained using the Fisher transformation should be approximtely normal and hence should be plotted using a quantile plot. Values of $Z_{(i)}$ which are too large or too small should be evaluated as outliers.

Example

For the data in Table 3.4 the correlation coefficient $r_{(i)}$ between CPT and LOADS after deleting observation i was determined for all observations $i = 1, 2, \ldots, 70$. The values of $r_{(i)}$ which differed most from $r = -0.555$ were observations 47 and 51. As shown in Table 3.5 deleting each of these observations separately yields correlations of -0.510 and -0.461 respectively. As can be seen from Table 3.5 deleting both of these observations reduces the correlation further to -0.388.

Robust Estimation of Covariance and Correlation

The sample covariance s_{xy} provides an unbiased estimator of σ_{xy}. As in the case of the variance however there are situations where a small number of outliers can yield a poor estimator of the covariance. In Chapter 2 robust estimators for the mean and variance were introduced. The use of trimming to obtain robust estimators was exemplified. For a bivariate distribution the observations trimmed on X and Y to determine robust estimators, $\tilde{s}_x^2$ and $\tilde{s}_y^2$, may not be the same. In such situations, a robust estimator of the covariance is given by

$$\tilde{s}_{xy} = [\tilde{s}_{(x+y)}^2 - \tilde{s}_{(x-y)}^2]/4,$$

where $\tilde{s}_{(x+y)}$ and $\tilde{s}_{(x-y)}$ are robust estimators of the variances of $(X + Y)$ and $(X - Y)$ respectively. This estimator uses the property that

$$\text{Cov}(X, Y) = [V(X + Y) - V(X - Y)]/4.$$

This method avoids the difficulty of determining bivariate outliers before determining an estimator of the covariance.

A robust estimator of the correlation coefficient can also be obtained from robust estimators $\tilde{s}_x^2$, $\tilde{s}_y^2$ and $\tilde{s}_{xy}$ of the variances and covariance respectively by computing

$$\tilde{r} = \tilde{s}_{xy}/\tilde{s}_x \tilde{s}_y.$$

If $\tilde{s}_{xy}$ is used as an estimator of σ_{xy} there is no guarantee that $|\tilde{r}|$ will lie in the range $[0, 1]$. To ensure that the correlation coefficient lies in the proper range an alternative estimator is given by

$$\tilde{r}^* = [\tilde{s}_{(z+w)}^2 - \tilde{s}_{(z-w)}^2]/[\tilde{s}_{(z+w)}^2 + \tilde{s}_{(z-w)}^2]$$

where $Z = X/\tilde{s}_x$ and $W = Y/\tilde{s}_y$ are standardized values of X and Y respectively. This estimator uses the property that

$$\frac{V(X/\sigma_x + Y/\sigma_y) - V(X/\sigma_x - Y/\sigma_y)}{V(X/\sigma_x + Y/\sigma_y) + V(X/\sigma_x - Y/\sigma_y)} = \rho.$$

A robust estimator of the covariance can then be obtained from $\tilde{r}^*$ using $\tilde{r}^* \tilde{s}_x \tilde{s}_y$.

3.2.3 Comparison of the Marginal Distributions

Up to this point our study of bivariate analysis has been concerned with the relationships between two quantitative variables X and Y. Correlation analysis for example is concerned with the tendency for X and Y to move together or in opposite directions as we compare observations on objects or individuals. On occasion we may be interested in comparing the probability distributions for X and Y. Are the means equal? Do the distributions have the same shape? The observations on the two random variables may represent observations made on n individuals under two different conditions and hence it might be of interest to compare the two distributions.

Graphical Comparisons

A useful way to compare two distributions is to place two stem and leaf diagrams back to back using the same stem. The leaves for one distribution would be to the right of the stem and for the other distribution, to the left of the stem.

The two empirical distributions can also be compared using the quantile-quantile plots discussed in Chapter 2. The quantile-quantile plots were used previously to compare observed data to a theoretical distribution. In a similar fashion a quantile-quantile plot can be used to compare two empirical distributions. If the two distributions are similar we would expect to see scatter about a 45° line through the origin. Other patterns might suggest a transformation of X and/or Y that would result in the transformed distributions being similar. If the median of X exceeds the median of Y by an amount c the scatter would appear as a 45° line but displaced c units from the origin on the X axis. If $Y = cX$ then $\log Y = \log c + \log X$ and hence a quantile-quantile plot of $\log Y$ vs $\log X$ should appear as a 45° line displaced $\log c$ units from the origin on the $\log Y$ axis.

Example

For a random sample of 50 employees in a large organization the distribution of current salary was compared to the distribution of starting salary. The salary data were first transformed using a natural log transformation to obtain more normal shaped distributions. A comparison of the two distributions using a stem and leaf plot is shown in Figure 3.10. A comparison using a quantile-quantile plot is provided in Figure 3.11. From both figures we can see that the means of the two distributions are different. The stem and leaf plots also indicate that the distribution of current salary is more spread out and less skewed than the distribution of starting salary. A comparison of the various descriptive measures for the two distributions is given in Table 3.6.

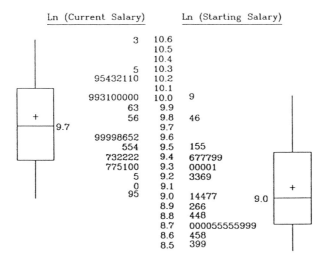

Figure 3.10. Comparison of Salary Distributions Using Stem and Leaf Plot

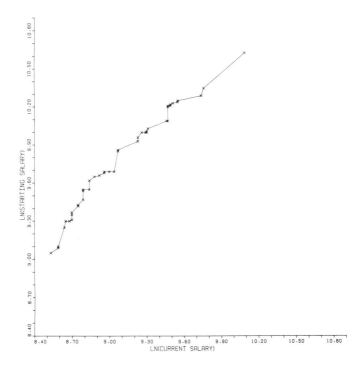

Figure 3.11. Quantile-Quantile Plot Comparison of Two Salary Distributions

Table 3.6. Comparison of Descriptive Measures for Two Salary Distributions

	Ln (Current Salary)	Ln (Starting Salary)
Mean $\bar{x}$	9.755	9.076
Median x_M	9.685	9.021
$Q1$ (first quarter)	9.417	8.748
$Q3$ (third quarter)	10.047	9.344
Range R	1.581	1.561
Quartile Deviation Q_3-Q_1	0.630	0.596
Standard Deviation s	0.384	0.371
Skewness Coefficient g	0.085	0.535
Kurtosis Coefficient d	-0.913	-0.498

Paired Comparisons

A common approach to the comparison of the mean of the two distributions is the *paired comparison test*. A paired comparison test which compares the weights of individuals before and after a particular diet would be an example. A second example would be the comparison of the distributions of scores obtained by individuals who wrote two different mathematics aptitude tests.

To compare the means μ_x and μ_y a new variable $d_i = X_i - Y_i$, $i = 1, 2, \ldots, n$ is defined. Under the null hypothesis $H_0 : \mu_x = \mu_y$ the variable d has the mean $\mu_d = 0$. A test of H_0 can be carried out by testing $H_0 : \mu_d = 0$ using standard univariate techniques. In large samples the statistic $(\sqrt{n})\bar{d}/s_d$ has a standard normal distribution if H_0 is true where s_d^2 is the sample variance of the d_i, $i = 1, 2, \ldots, n$. If X and Y are normal then d is normal and the statistic $(\sqrt{n})\bar{d}/s_d$ has a t distribution with $(n-1)$ degrees of freedom if H_0 is true.

When the normality assumption is troublesome the *Wilcoxon signed-rank test* can be used if the distribution of d is symmetric. The absolute values of the observations $d_i = X_i - Y_i$ are ranked, and then the sign of d_i is applied to the rank and the result is denoted by R_i, [hence $R_i = \text{SIGN}(d_i)\text{RANK}(|d_i|)$.] The null hypothesis H_0 : Median of d is zero, may be tested using

$$T = \sum_{i=1}^{n} R_i \bigg/ \sqrt{\sum_{i=1}^{n} R_i^2}$$

which has a standard normal distribution in large samples if H_0 is true. Exact tables for T are also available for small samples in textbooks on nonparametric statistics such as Conover (1980).

Example

For the example comparing salary distributions, the mean value of $d =$ ln(current salary) $-$ ln(starting salary) is 0.678 with a standard deviation of 0.193. The t-statistic for testing a zero mean hypothesis is 24.867 which is highly significant indicating that the means for the two distributions are different. The Wilcoxon signed-rank-statistic in this case has a value of 637.5 which is also highly significant. The observed distribution of the differences d_i was quite comparable to a normal distribution (K-S p-value 0.661).

3.3 The Simple Linear Regression Model

The most common type of bivariate analysis for quantitative random variables is *simple linear regression*. In the simple linear regression model one variable, say Y the *dependent variable*, is assumed to be dependent on the other variable, X, which is called the *explanatory variable*. The purpose is usually to explain the variation of Y in terms of the variation in the variable X, which is assumed to be linearly related to Y. The data set in Table 3.7 gives the net operating income (NOI) and the number of suites (SUITES) for a sample of 47 apartment buildings in a major city. A bivariate scatterplot of the 47 observations is shown in Figure 3.12. The scatterplot suggests a linear relationship between NOI and SUITES. The data will be used as an example throughout this section. A review of the theory for the simple linear regression model is presented next.

3.3.1 The Theoretical Model and Estimation

Theory

The regression function of Y on X is defined to be the conditional expection $E[Y \mid X] = f(X)$, which is a function of X. For the simple linear regression model, $f(X)$ is assumed to be linear (recall that for the bivariate normal this function is linear). For the simple linear regression model we write $f(X) = \beta_0 + \beta_1 X$. The function is a line, and each point on the line is the mean of a conditional density function $f(Y \mid X)$. See Figure 3.6 in the previous section.

For a given value of X say $X = x$, we have the conditional mean of Y given by $E[Y \mid X = x] = \beta_0 + \beta_1 x$ is a point on the regression line. At $X = x$, Y is distributed around the regression line. For any particular Y, say y, at $X = x$, we write $y - \beta_0 - \beta_1 x = u$. The distribution of u at $X = x$ has a mean of 0, and the variance of u given $X = x$ is given by

Table 3.7.   Net Operating Income and Number of Suites for a Sample of Apartment Buildings

Apt. Bldg.	NOI - Net Operating Income	SUITES No. of Suites	Apt. Bldg.	NOI - Net Operating Income	SUITES No. of Suites
1	119202	58	25	226375	44
2	50092	30	26	247203	78
3	33263	22	27	28519	69
4	18413	21	28	154278	150
5	26641	12	29	157332	62
6	32628	20	30	171305	86
7	19877	15	31	109461	44
8	196500	29	32	159245	104
9	63200	28	33	34057	21
10	43484	23	34	15392	18
11	26424	14	35	60791	24
12	81413	27	36	48008	15
13	153284	52	37	42299	21
14	187993	48	38	145998	65
15	33869	20	39	54357	24
16	562942	205	40	17288	12
17	10217	17	41	24058	12
18	26712	26	42	12397	12
19	48721	22	43	9882	12
20	51282	24	44	13713	12
21	31572	20	45	12782	15
22	107169	33	46	24020	12
23	345608	104	47	36187	20
24	350633	140			

$V[y \mid X = x]$. We assume that this variance is constant in that it does not depend on the choice of $X = x$. Therefore we write

$$V[y \mid X = x] = V[u \mid X = x] = \sigma_u^2,$$

and hence the variance of u at each point on the regression line is constant. (Recall that this is true for the bivariate normal.)

To estimate the unknown parameters β_0, β_1 and σ_u^2 a random sample of observations (y_i, x_i), $i = 1, 2, \ldots, n$ is selected from the bivariate population. The values of X, x_i, $i = 1, 2, \ldots, n$ are assumed to be fixed, or if not, the estimation results are conditional on the particular X values observed. The assumptions for the simple linear regression model are usually written in the format summarized below.

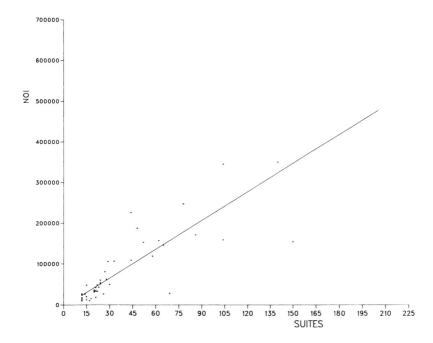

Figure 3.12. Scatterplot of NOI vs. SUITES

Assumptions

Given a random sample of n observations (x_i, y_i) $i = 1, 2, \ldots, n$ from the joint distribution for (X, Y) the observations are assumed to satisfy the linear model

$$y_i = \beta_0 + \beta_1 x_i + u_i$$

with the following properties

1. The distribution of the *error terms* $(u_1, u_2, \ldots, u_n)$ is independent of $(x_1, \ldots, x_n)$ OR the X observations $(x_1, \ldots, x_n)$ are fixed; hence the regression function is linear and is given by $E[Y \mid X = x] = \beta_0 + \beta_1 x$;

2. $E[u_i] = 0$ and $V[u_i] = \sigma_u^2$, $i = 1, 2, \ldots, n$; hence $V[Y \mid X = x] = $ a constant, $\sigma_{y \cdot x}^2 = \sigma_u^2$ for all $(x_1, \ldots, x_n)$;

3. The unknown parameters β_0 and β_1 are constants.

Since the observations $(y_1, \ldots, y_n)$ are a random sample they are mutually independent and hence the error terms are necessarily mutually independent. The regression function for a sample from the bivariate normal discussed in 3.1.2 satisfies these linear model assumptions.

Least Squares Estimation

The random sample of observations (x_i, y_i) $i = 1, 2, \ldots, n$ can be used to estimate the unknown parameters β_0, β_1 and σ_u^2. The *least squares estimators* of β_0 and β_1 are obtained by minimizing the expression $\sum_{i=1}^{n} (y_i - \beta_0 - \beta_1 x_i)^2$ with respect to β_0 and β_1. Differentiation of this expression with respect to β_0 and β_1 respectively yields two expressions which are set equal to zero to obtain a minimum. The resulting equations are called the *normal equations* and are given by

$$\sum_{i=1}^{n} y_i = n\beta_0 + \beta_1 \sum_{i=1}^{n} x_i$$

$$\sum_{i=1}^{n} x_i y_i = \beta_0 \sum_{i=1}^{n} x_i + \beta_1 \sum_{i=1}^{n} x_i^2.$$

The solutions to these equations are given by

$$b_1 = \left[n \sum_{i=1}^{n} x_i y_i - \sum_{i=1}^{n} x_i \sum_{i=1}^{n} y_i \right] \bigg/ \left[n \sum_{i=1}^{n} x_i^2 - \left(\sum_{i=1}^{n} x_i \right)^2 \right]$$

and

$$b_0 = \left[\sum_{i=1}^{n} y_i - b_1 \sum_{i=1}^{n} x_i \right] \bigg/ n,$$

which can also be written as

$$b_1 = s_{xy}/s_x^2$$

and

$$b_0 = \bar{y} - b_1 \bar{x},$$

where $\bar{x}$, $\bar{y}$, s_{xy} and s_x^2 are the usual unbiased estimators of μ_x, μ_y, σ_{xy} and σ_x^2 respectively. The estimators b_0 and b_1 are *unbiased* and are *minimum variance* among all estimators that are linear functions of the Y observations. Under the assumption of normality for the u_i they are also the maximum likelihood estimators. An unbiased estimator of $\sigma_{y \cdot x}^2 = \sigma_u^2$ is provided by

$$s_{y \cdot x}^2 = \left(\frac{n-1}{n-2} \right) \left[s_y^2 - s_{xy}^2/s_x^2 \right] = s_y^2 (1 - r^2) \left(\frac{n-1}{n-2} \right) = s^2 \text{ say,}$$

where s_y^2 is an unbiased estimator of σ_y^2 and r is the Pearson correlation coefficient.

Also, under the normality assumption the estimator s^2 of σ_u^2 can be used to make inferences about σ_u^2. The statistic $\dfrac{(n-2)s^2}{\sigma_u^2}$ has a χ^2 distribution with $(n-2)$ degrees of freedom.

Residuals

For each of the X values $(x_1, \ldots, x_n)$ the *predicted, fitted* or estimated Y values are the points on the regression line

$$\hat{y}_i = b_0 + b_1 x_i.$$

The differences between the observed values and the fitted values defined by $e_i = (y_i - \hat{y}_i)$ are called the *residuals*. The *residual variation* $\sum_{i=1}^{n} e_i^2/(n-2)$ is equivalent to s^2.

Example

Using the data in Table 3.7 the means and covariance matrix parameters are given by $\bar{y} = 9.223 \times 10^4$, $\bar{x} = 4.132 \times 10$, $s_y^2 = 1.191 \times 10^{10}$, $s_x^2 = 1.690 \times 10^3$ and $s_{xy} = 3.972 \times 10^6$. The least squares estimators for the parameters β_0 and β_1 in the model relating net operating income (NOI) to number of suites (SUITES) are $b_0 = -4872.0$ and $b_1 = 2350.7$. The estimated variance of the error term is given by $s^2 = 2.626 \times 10^9$. A plot of the fitted least squares line

$$\text{NOI} = -4,872.0 + 2,350.7 \text{ SUITES}$$

is shown with the scatterplot in Figure 3.12.

Throughout this section all of the regression estimation calculations for this data set have been carried out using SAS PROC REG.

3.3.2 Inference For The Regression Coefficients

Confidence Intervals

The covariance matrix for the estimators b_0 and b_1 is given by

$$\begin{bmatrix} V(b_0) & \text{Cov}(b_0 b_1) \\ \text{Cov}(b_0 b_1) & V(b_1) \end{bmatrix} = \sigma_u^2 \begin{bmatrix} \dfrac{1}{n} + \dfrac{\bar{x}^2}{(n-1)s_x^2} & \dfrac{-\bar{x}}{(n-1)s_x^2} \\ \dfrac{-\bar{x}}{(n-1)s_x^2} & \dfrac{1}{(n-1)s_x^2} \end{bmatrix}$$

which is estimated by replacing σ_u^2 by the unbiased estimator s^2.

Under the assumption of normality the statistics

$$(b_0 - \beta_0)/\left(s^2 \left[\frac{1}{n} + \frac{\bar{x}^2}{(n-1)s_x^2} \right] \right)^{1/2}$$

and

$$(b_1 - \beta_1) / \left(s^2 \left[\frac{1}{(n-1)s_x^2} \right] \right)^{1/2}$$

are both distributed as t random variables with $(n-2)$ degrees of freedom. In large samples without the normality assumption these two statistics have a standard normal distribution. These statistics may be used to make inferences about β_0 and β_1. Confidence interval estimators for β_0 and β_1 are therefore given by

$$b_0 \pm t_{\alpha/2;(n-2)} s \left[1/n + \frac{\bar{x}^2}{(n-1)s_x^2} \right]^{1/2}$$

$$b_1 \pm t_{\alpha/2;(n-2)} s \left[\frac{1}{(n-1)s_x^2} \right]^{1/2}$$

(3.5)

Example

The standard errors of the regression coefficients b_0 and b_1, for the estimated relationship between NOI and SUITES, are given by 10655.3 and 183.8 respectively. Using these standard errors, the t-statistics for testing $H_0 : \beta_0 = 0$ and $H_1 : \beta_1 = 0$ are given by $t = -0.5$ and $t = 12.8$ respectively. The p-values for these t-statistics which have 45 degrees of freedom are 0.65 and 0.0001 respectively. We can conclude therefore that the slope is significantly greater than zero but that the intercept is not. For the range of the data studied the regression line can be assumed to pass through the origin. (The reader may wish to ponder the practical implications of a zero intercept in this example.) Confidence interval estimates (95%) for the true parameters β_1 and β_0 are given by $2350.7 \pm (2.016)(183.8)$ and $-4872.0 \pm (2.016)(10655.3)$, respectively. The resulting confidence intervals for β_0 and β_1 are given by $(-26{,}353,\ 16{,}609)$ and $(1980.2,\ 2721.2)$ respectively.

Joint Inferences

From the expression for the covariance matrix for b_0 and b_1 we can see that b_0 and b_1 are negatively correlated if $\bar{x}$ is positive. Joint inferences about β_0 and β_1 therefore cannot be obtained from the separate inferences for β_0 and β_1. A $100(1-\alpha)\%$ *confidence region* for (β_0, β_1) is given by

$$n(b_0 - \beta_0)^2 + 2(b_0 - \beta_0)(b_1 - \beta_1) \sum_{i=1}^{n} x_i + (b_1 - \beta_1)^2 \sum_{i=1}^{n} x_i^2 = 2s^2 F_{\alpha;2,(n-2)}$$

(3.6)

which describes an ellipse in (β_0, β_1) space with centre at (b_0, b_1). The quantity $F_{\alpha;2,(n-2)}$ denotes the upper $(100\alpha)\%$ point in an F distribution with 2 and $(n-2)$ degrees of freedom.

Example

For the example the 95% confidence region is given by the ellipse

$$47(-4872 - \beta_0)^2 + 2(1942)(-4872 - \beta_0)(2350.7 - \beta_1)$$

$$+ 157,985(2350.7 - \beta_1)^2 = 2(2.626)(2.43) \times 10^9.$$

A plot of this ellipse is shown in Figure 3.13. The negative slope of the ellipse reflects the negative correlation between b_0 and b_1. Figure 3.13 also illustrates the individual 95% confidence interval regions for β_0 and β_1 obtained above (see the horizontal and vertical lines defined by H_L, H_U, and V_L, V_U). The joint region defined by these horizontal and vertical lines tends to be too large in most of the (β_0, β_1) plane and too small at the upper left and lower right extremes of the ellipse.

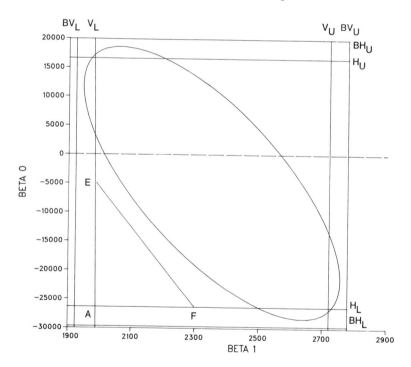

Figure 3.13. Joint Confidence Region for β_0 and β_1

It is important to recognize that the joint confidence region for (β_0, β_1) excludes a large number of values of (β_0, β_1) that would be included by the rectangular region obtained from the individual confidence intervals. For example in a study of profitability it may be that the values of (β_0, β_1) in the region EAF in Figure 3.13 represent non-profitable situations. Use of the rectangular region could therefore lead to incorrect decisions.

Bonferroni Approximation

A conservative approach is to choose a rectangular region which contains the elliptical region. This is the approach used in the *Bonferroni approximation* described next.

To obtain an approximate joint confidence region the addition theorem of probability can be used. The *Venn diagram* in Figure 3.14 can be used to represent the various probabilities associated with the covering of the point (β_0, β_1) by the two intervals (3.5). The area (1) outside both circles corresponds to intervals that simultaneously cover (β_0, β_1). The area (2) represented by the intersection of the two circles represents the probability that both intervals simultaneously do not cover (β_0, β_1). The areas of the two circles (3) and (4) respectively represent the probabilities that β_1 is not covered and the probability that β_0 is not covered. The familiar addition theorem of probability is given by

$$(1) = 1 - [(3) + (4) - (2)] = 1 - (3) - (4) + (2).$$

We can conclude therefore that

$$(1) \geq 1 - [(3) + (4)].$$

The probability $1 - [(3) + (4)]$ is a lower bound for the probability (1). If (3) and (4) have the conventional values of α, then we can conclude that the joint region formed by the two intervals covers (β_0, β_1) with a probability of at least $(1 - 2\alpha)$. Therefore an approximate $100(1 - \alpha)\%$ interval is obtained by using $100(1 - \alpha/2)\%$ intervals for β_0 and β_1. This approximation is commonly referred to as the Bonferroni aproximation. Note that in the case of independence between b_0 and b_1 (i.e. $\bar{x} = 0$) the joint probability of coverage is simply the product of the two marginal probabilities

$$(1 - \alpha)^2 = 1 - 2\alpha + \alpha^2.$$

In the case of independence therefore the error in the Bonferroni approximation is $\alpha^2/4$ since

$$(1 - \alpha/2)^2 = 1 - \alpha + \alpha^2/4.$$

Example

Using the Bonferroni approximation a joint 95% confidence region for (β_0, β_1) for the model relating NOI and SUITES is given by

$$-4872.0 \pm (t_{.0250/2;45})(10655.3) \quad \text{for } \beta_0$$

and

$$2350.7 \pm (t_{.0250/2;45})(183.8) \quad \text{for } \beta_1.$$

The resulting joint region is given by the rectangle $(-29,677 \leq \beta_0 \leq 19,933)$ and $(1922.8 \leq \beta_1 \leq 2778.6)$. This region is shown in Figure 3.13 and is defined by the rectangle described by the horizontal and vertical lines (BH_L, BH_U, BV_L, BV_U). As can be seen from the Figure this rectangle contains the entire elliptical region.

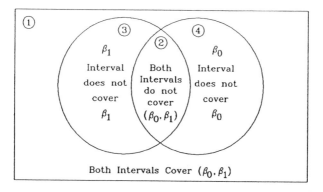

Figure 3.14. Venn Diagram Illustrating Possible Values of Intervals for (β_0, β_1)

3.3.3 Goodness of Fit and Prediction

The goodness of fit for the simple linear regression model is usually expressed in terms of the sample coefficient of determination

$$R^2 = r^2 = \frac{s_{xy}^2}{s_x^2 s_y^2} = 1 - \left(\frac{n-2}{n-1}\right)s^2/s_y^2$$

which measures the proportion of the variation in Y explained by the fitted linear regression relationship. The regression relationship has divided the *total variance* in Y, s_y^2, into the *explained variance* s_{xy}^2/s_x^2 and the unexplained variance $s^2\left(\frac{n-2}{n-1}\right)$. The coefficient of determination is the ratio of the explained variance to the total variance.

Analysis of Variance

The *total sum of squares* for Y, SST, the *regression sum of squares*, SSR, and the *error sum of squares*, SSE, are given by $(n-1)s_y^2$, $(n-1)s_{xy}^2/s_x^2$ and $(n-2)s^2$ respectively. The *analysis of variance* relationship between these sums of squares is given by SST = SSR + SSE. The coefficient of determination can also be written as $R^2 = $ SSR/SST.

Under the normality assumption if $\beta_1 = 0$, the statistic $(n-2)R^2/(1 - R^2) = (n-2)$ SSR/SSE has an F *distribution* with 1 and $(n-2)$ degrees of freedom. The F-statistic here is the square of the t-statistic $b_1/(s^2[1/(n-1)s_x^2])^{1/2}$ and may also be used to test $H_0 : \beta_1 = 0$. The t-statistic for testing $H_0 : \rho = 0$, $r\sqrt{n-2}/\sqrt{1-r^2}$, given in 3.2.1 is also equivalent to the t-statistic for testing $H_0 : \beta_1 = 0$.

The Pearson correlation coefficient between the observation values y and the $\hat{y}$ values is given by $r = \sqrt{R^2}$. Thus the correlation coefficient also measures goodness of fit.

Example

For the regression model fitted to the data in Table 3.7, SSR $= 4.2951 \times 10^{11}$, SSE $= 1.1185 \times 10^{11}$, SST $= 5.4766 \times 10^{11}$ and $R^2 = 0.7843$. The F-statistic for testing $H_0 : \beta_1 = 0$ is given by $F = 163.59$. Under H_0, the F-statistic has an F distribution with 1 and 45 degrees of freedom. The p-value of F is 0.0001. The Pearson correlation coefficient between NOI and SUITES is $r = 0.89$.

Prediction Intervals

Confidence intervals for Y or $E[Y]$, at a specific value of X, $X = x_0$, are called *prediction intervals*.

Confidence Interval for $E[Y]$

A $100(1 - \alpha)\%$ confidence interval for the true value of $E[Y]$ at $X = x_0$ for the sampled population is provided by

$$\hat{y}_{x_0} \pm t_{\alpha/2;(n-2)} s \left(\frac{1}{n} + \frac{(x_0 - \bar{x})^2}{(n-1)s_x^2} \right)^{1/2},$$

where $\hat{y}_{x_0} = \hat{\beta}_0 + \hat{\beta}_1 x_0$ is a point on the least squares regression line at $X = x_0$.

Confidence Interval for a Particular Y

If the true regression line was known, a $100(1 - \alpha)\%$ confidence interval for the true Y value for a given $X = x_0$ would be

$$y_{x_0} \pm Z_{\alpha/2} \sigma_u$$

where $y_{x_0} = \beta_0 + \beta_1 x_0$. If σ_u is unknown the interval becomes

$$y_{x_0} \pm t_{\alpha/2;n-2} s.$$

This interval uses the fact that the variance around the regression line is σ_u^2. Since y_{x_0} is unknown the total variance is $\sigma_u^2 + V(\hat{y}_{x_0})$ which is estimated by $s^2 + s^2 \left[\frac{1}{n} + \frac{(x_0 - \bar{x})^2}{(n-1)s_x^2} \right]$. Thus for an individual selected at random from this population with X value at $X = x_0$, a $100(1-\alpha)\%$ confidence interval for Y for this individual is given by

$$\hat{y}_{x_0} \pm t_{\alpha/2;(n-2)} s \left(1 + \frac{1}{n} + \frac{(x_0 - \bar{x})^2}{(n-1)s_x^2} \right)^{1/2}$$

This interval is wider than the interval for $E[Y]$, reflecting the fact that individuals at $X = x_0$ are distributed around the regression line with variance σ_u^2.

Example

For the apartment data a 95% confidence interval for the $E[\text{NOI}]$ is given by

$(-4872 + 2350.7 \text{ SUITES})$

$$\pm (2.016)(2.626 \times 10^9)^{1/2} \left[\frac{1}{47} + \frac{(\text{SUITES} - 41.32)^2}{(46)(1690)} \right]^{1/2}$$

A plot showing this confidence region is shown in panel (a) of Figure 3.15. At the value SUITES= 100, the 95% confidence interval is given by $(231,998 \pm 6,776.5)$.

A 95% confidence interval for the value of NOI for any particular apartment is given by

$(-4872 + 2350.7 \text{ SUITES})$

$$\pm (2.016)(2.626 \times 10^9)^{1/2} \left[1 + \frac{1}{47} + \frac{(\text{SUITES} - 41.32)^2}{(46)(1690)} \right]^{1/2}$$

A plot showing this confidence region is shown in panel (b) of Figure 3.15. This confidence region is much wider than the confidence region for the expected value of NOI because this region must allow for variation in NOI around its expectation. At the value SUITES $= 100$, the 95% confidence interval is given by $(231,998 \pm 110,076.5)$.

The predicted values of NOI at each observed value of SUITES are given in Table 3.8. The table also contains the 95% confidence interval values, for the expected value of NOI, and for a particular value of NOI, evaluated at the 47 values of SUITES. The residuals (observed NOI — predicted NOI) are also shown in Table 3.8. The residuals corresponding to NOI values which are outside the 95% confidence interval for NOI have been marked with an asterisk in the table. The three observations falling outside the 95% confidence interval for NOI have been circled in panel (b) of Figure 3.15. The influence that these extreme values have on the linear fit will be studied later in this section.

(a)

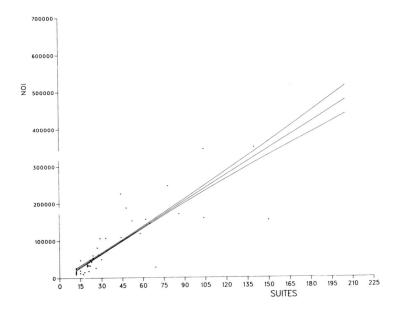

(b)

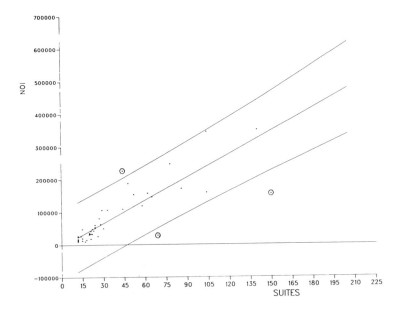

Figure 3.15. Confidence Regions for $E[\text{NOI}]$ and NOI

Table 3.8. Predicted Values and Residuals for Least Squares Fit of NOI vs. SUITES

Obs. No.	SUITES	NOI	Predicted NOI	95% Interval For	Mean	95% Interval For	NOI	Residuals
1	58	119202	131469	115198	147740	26990	235948	-12267
2	30	50092	67649	50023	81275	-38731	170029	-15557
3	22	33263	46844	30177	63510	-57697	151384	-13581
4	21	18413	44493	27664	61321	-60074	149060	-26080
5	12	26641	23336	4778	41895	-81523	128196	3305
6	20	32628	42142	25145	59139	-62452	146736	-9514
7	15	19877	30389	12457	48320	-74362	135139	-10512
8	29	106500	63298	47569	79028	-41097	167694	43202
9	28	63200	60948	45107	76788	-43465	165360	2252
10	23	43484	49194	32684	65705	-55322	153710	-5710
11	14	26424	28038	9903	46173	-76747	132823	-1614
12	27	81413	58597	42637	74557	-45834	163028	22816
13	52	153284	117365	101800	132929	12994	221736	35919
14	48	187993	107962	92706	123217	3637	212287	80031
15	20	33869	42142	25145	59139	-62452	146736	-8273
16	205	562942	477023	414590	539455	356404	597641	85919
17	17	10217	35090	17550	52630	-69594	139774	-24873
18	26	26712	56246	40160	72333	-48204	160696	-29534
19	22	48721	46844	30177	63510	-57697	151384	1877
20	24	51282	51545	35183	67907	-52948	156038	-263
21	20	31572	42142	25145	59139	-62452	146736	-10570
22	33	107169	72701	57336	88067	-31640	177043	34468
23	104	345608	239601	211943	267260	132756	346447	106007
24	140	350633	324227	284717	363736	213719	434735	26406
25	44	226375	98559	83473	113646	-5742	202860	127816*
26	78	247203	178483	158210	198756	73307	283659	68720
27	69	28519	157327	139116	175537	52529	262125	-128808*
28	150	154278	347734	304779	390689	235948	459520	-193456*
29	62	157332	140872	123983	157760	36295	245448	16460
30	86	171305	197289	174924	219653	91689	302888	-25984
31	44	109461	98559	83473	113646	-5742	202860	10902
32	104	159245	239601	211943	267260	132756	346447	-80356
33	21	34057	44493	27664	61321	-60074	149060	-10436
34	18	15392	37441	20088	54794	-67212	142093	-22049
35	24	60791	51545	35183	67907	-52948	156038	9246
36	15	48008	30389	12457	48320	-74362	135139	17619
37	21	42299	44493	27664	61321	-60074	149060	-2194
38	65	145998	147924	130504	165344	43260	252588	-1926
39	24	54357	51545	35183	67907	-52948	156038	2812
40	12	17288	23336	4778	41895	-81523	128196	-6048
41	12	24058	23336	4778	41895	-81523	128196	722
42	12	12397	23336	4778	41895	-81523	128196	-10939
43	12	9882	23336	4778	41895	-81523	128196	-13454
44	12	13713	23336	4778	41895	-81523	128196	-9623
45	15	12782	30389	12457	48320	-74362	135139	-17607
46	12	24020	23336	4778	41895	-81523	128196	684
47	20	36187	42142	25145	59139	-62452	146736	-5955

Confidence Band for the Entire Regression Line

For each (b_0, b_1) point in the joint $100(1 - \alpha)\%$ region for (β_0, β_1) there is a corresponding regression line

$$\hat{y} = b_0 + b_1 x.$$

The region generated by all such regressions provides a *confidence band* for the regression line. The regression band is given by

$$\hat{y} \pm s \sqrt{2 F_{\alpha;2,(n-2)}} \sqrt{\frac{1}{n} + \frac{n(x - \bar{x})^2}{n \sum_{i=1}^{n} x_i^2 - \left(\sum_{i=1}^{n} x_i \right)^2}}.$$

The probability is $(1 - \alpha)$ that this confidence region covers the entire true regression given the fixed X values.

In comparison to the equation for the confidence interval for $E[Y]$ at $X = x$, the statistic $t_{\alpha/2;(n-2)}$ has been replaced by the statistic $\sqrt{2 F_{\alpha;2,(n-2)}}$. The confidence band for the entire regression line is therefore wider by a factor of

$$\sqrt{2 F_{\alpha;2,(n-2)}} / t_{\alpha/2;(n-2)} = \sqrt{2 F_{\alpha;2,(n-2)} / F_{\alpha;1,(n-2)}}.$$

At 10 d.f. and $\alpha = 0.05$ this factor is $\sqrt{2(4.10)/(4.96)} = 1.286$ and at 30 d.f. the factor is $\sqrt{2(3.32)/(4.17)} = 1.592$. For the NOI vs SUITES example, the expression for the confidence band for the entire regression line at SUITES $= 100$ is given by $231,198 \pm 8,503.6$.

Using Prediction Intervals

Use of these intervals for predictive purposes assumes that the predictions are being made about the sampled population that was used to estimate the linear relationship. This assumption is often violated in practice. Other assumptions such as linearity and independent, identically distributed error terms should also be valid.

An example of incorrect use of an estimated regression to make predictions is when the sample is not random but has been restricted to only a subset of the population. The results should not be presumed to apply to the whole population. A study of a sample of volunteers is an excellent example of hidden restrictions being placed on the sample. Another example would be the firm which uses its own employees to study a relationship and then applies the fitted model to individuals who have not been hired. Suppose a large sales organization uses sales aptitude tests to identify potential sales personnel. To study the relationship between aptitude and performance a sales aptitude test is given to all sales personnel and the aptitude scores are then related to sales performance using regression. These results should not be used to predict sales performance for potential employees using the aptitude scores of the potential employees. Presumably the firm is placing more than just a sales aptitude restriction on the sample of its own employees. If there are other factors that influence sales performance the sample will be restricted with respect to these other factors. If the firm is a middle of the road firm with respect to salary, presumably it does not have very many super salespeople nor would it be expected to have many poor salespeople either. A similar example could be constructed for the study of the relationship between college admission test scores and college performance.

Errors due to extrapolation beyond the range of the sample data can also result when nonlinear relationships exist over larger ranges. In Figure 3.16 the relationship between Y and X is a curve but between A and B a straight line is adequate.

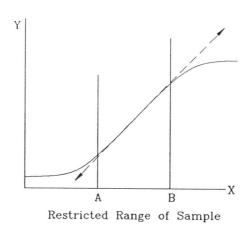

Restricted Range of Sample

Figure 3.16. Extrapolation Beyond the Range of Estimation

3.3.4 Repeated Observations and a Test for Lack of Fit

In experimental situations where the observations on the X variable are fixed, it is often possible to repeat the experiment so that for each of the X values several observations on Y have been obtained. In this case of *repeated observations* it is possible to measure variation in Y that is due to a *lack of fit* of the linear model. The data is assumed to satisfy the linear model

$$y_{ij} = \beta_0 + \beta_1 x_j + u_{ij}$$

where $j = 1, 2, \ldots, g$; $i = 1, 2, \ldots, n_j$, and $n = \sum_{j=1}^{g} n_j$ is the total number of observations. Thus for each of the pre-selected *design values* of X; $(x_1, x_2 \ldots, x_g)$, the experiment is repeated to produce n_j observations on Y.

The total variation in Y is given by SST $= \sum_{j=1}^{g} \sum_{i=1}^{n_j} (y_{ij} - \bar{y}_{...})^2$ where $\bar{y}$ denotes the mean of all the Y observations $\left(\bar{y} = \frac{1}{n} \sum_{j=1}^{g} \sum_{i=1}^{n_j} y_{ij} \right)$. This variation can be separated into two parts. The variation around $\bar{y}_j$, the mean of Y for the design value $X = x_j$, is given by $\sum_{i=1}^{n_j} (y_{ij} - \bar{y}_j)^2$. For all g design values, the total variation around the design means is given by SSPE $= \sum_{j=1}^{g} \sum_{i=1}^{n_j} (y_{ij} - \bar{y}_j)^2$ which is a measure of *pure error*. This variation is due only to variation in Y since the X values are held fixed. The usual error sum of squares for the regression is given by SSE $= \sum_{j=1}^{g} \sum_{i=1}^{n_i} (y_{ij} - \hat{y}_{ij})^2$, where $\hat{y}_{ij}$ denotes the predicted values of Y obtained from the least squares fit. The difference between SSE and SSPE, SSE $-$ SSPE is called the *variation due to lack of fit* and is denoted by SSLF. It can be shown that

$$\text{SSLF} = \sum_{j=1}^{g} \sum_{i=1}^{n_j} (\hat{y}_{ij} - \bar{y}_j)^2 = \sum_{j=1}^{g} n_j (\hat{y}_j - \bar{y}_j)^2 \text{ since } \hat{y}_{ij} = \hat{y}_j, \ i = 1, 2, \ldots, n_j$$

are equal value predictions at x_j. As usual the regression sum of squares is given by SSR $= \sum_{j=1}^{g} \sum_{i=1}^{n} (\hat{y}_{ij} - \bar{y})^2 = \sum_{j=1}^{g} n_j (\hat{y}_j - \bar{y})^2$.

For this repeated observations model, the total variation is divided into the three parts

$$\text{SST} = \text{SSR} + \text{SSPE} + \text{SSLF}$$

where SSE $=$ SSPE $+$ SSLF is the former error sum of squares. The advantage of replication in the design is therefore a further breakdown of error variation into two parts. An *F-test for lack of fit* can be carried out using the statistic $\dfrac{\text{SSLF}/(g-2)}{\text{SSPE}/(n-g)}$ which has an F distribution with $(g-2)$ and $(n-g)$ degrees of freedom if the true model is linear. A significant F-statistic is an indication that a nonlinear model would be more appropriate.

Where possible, the F-test for lack of fit is a useful preliminary test before the conventional *F-test for linearity*, $F = \dfrac{\text{SSR}/1}{\text{SSE}/(n-2)}$. If the F-test for lack of fit indicates nonlinearity then the F-test for linearity should not be carried out. It is important to keep in mind that the conventional F-test for zero slope coefficient is a test of linear association rather than a test of association. In the case of severe nonlinearity this F-test would indicate no linear association (nonsignificant F) while the F-test for lack of fit would yield a significant F.

Example

In a study of consumer expenditures on milk products the relationship between family expenditures on milk products (EXPEND) and family size (SIZE) was determined using a simple linear regression model. For each family size, SIZE = 1, 2, 3, 4, 5, and 6, twenty families were observed. Table 3.9 summarizes this data and the scatterplot appears in Figure 3.17. The simple linear regression model relating EXPEND to SIZE is given by

$$\text{EXPEND} = \underset{(0.290)}{2.869} + \underset{(0.000)}{10.813} \text{ SIZE}$$

with $R^2 = 0.673$. The regression sum of squares is SSR $= 40926.016$ and the error sum of squares is SSE $= 19847.109$. The error sum of squares can be split into the *lack of fit sum of squares* SSLF $= 39.744$ and the *sum of squares for pure error* SSPE $= 19807.365$. The F ratio for the lack of fit test is given by $F = (39.744/4)/(19807.365/114) = 0.06$. Therefore we cannot reject the hypothesis of linearity. We shall see in the study of analysis of variance in Chapter 5 that other factors can be used to account for the pure error in this example. The scatterplot in Figure 3.17 shows the variation in milk product expenditures by family size. At each level of family size there seems to be about the same amount of variation in expenditure.

3.3.5 An Alternative Geometric Interpretation

The most common geometric representation of a bivariate sample of n observations (x_i, y_i), $i = 1, 2, \ldots, n$ is a scatterplot of the n points in a *two-dimensional space* with coordinate axes X and Y. An alternative representation of this sample can be obtained using two points in an n-*dimensional space*. The n observations on each of X and Y can be viewed as n-dimensional vectors

$$\mathbf{x} = \begin{bmatrix} x_1 \\ x_2 \\ \vdots \\ x_n \end{bmatrix} \quad \text{and} \quad \mathbf{y} = \begin{bmatrix} y_1 \\ y_2 \\ \vdots \\ y_n \end{bmatrix}.$$

The two vectors locate two points in an n-dimensional space and are represented by rays drawn from the origin to the respective points. See Figure 3.18 panel (a).

Table 3.9. Expenditures on Milk Products by Test Families in Second Month of Advertising

Family Expenditures on Milk Products				Family Size
12.35	21.86	14.43	21.44	1
28.26	13.76	14.44	30.78	1
10.97	0.00	2.90	6.46	1
0.00	11.90	4.48	27.62	1
13.11	8.00	10.90	14.36	1
20.52	42.17	22.26	31.21	2
37.67	24.59	29.63	45.75	2
26.70	2.41	17.28	18.61	2
4.52	27.75	18.01	42.63	2
16.89	18.27	28.22	26.37	2
30.85	49.61	23.99	40.09	3
44.70	37.30	38.27	56.37	3
36.81	16.10	19.62	30.14	3
13.71	42.22	21.96	59.20	3
27.99	27.72	38.62	34.15	3
39.35	63.65	36.98	55.68	4
57.54	49.53	51.59	70.19	4
51.34	22.71	29.53	39.12	4
27.91	56.06	34.42	74.92	4
36.35	42.04	48.31	54.02	4
48.87	73.75	42.13	65.81	5
67.57	59.25	59.09	79.81	5
62.69	30.19	38.57	51.15	5
38.57	66.16	40.14	92.37	5
48.85	48.50	60.23	59.90	5
58.01	85.95	54.19	76.61	6
77.70	67.68	71.69	94.23	6
42.71	78.71	57.06	98.02	6
72.68	41.64	48.20	59.11	6
61.97	59.92	71.39	74.79	6

The vectors **y** and **x** can also be re-expressed in deviation form by subtracting the means $\bar{y}$ and $\bar{x}$ from the y values and x values respectively. The *mean corrected vectors* are denoted by

$$\mathbf{y}^* = \begin{bmatrix} y_1 - \bar{y} \\ y_2 - \bar{y} \\ \vdots \\ y_n - \bar{y} \end{bmatrix} \quad \text{and} \quad \mathbf{x}^* = \begin{bmatrix} x_1 - \bar{x} \\ x_2 - \bar{x} \\ \vdots \\ x_n - \bar{x} \end{bmatrix}.$$

The *predicted values* of y_i, $\hat{y}_i$, $i = 1, 2, \ldots, n$ are summarized by the prediction vector

$$\widehat{\mathbf{y}} = \begin{bmatrix} \hat{y}_1 \\ \hat{y}_2 \\ \vdots \\ \hat{y}_n \end{bmatrix}$$

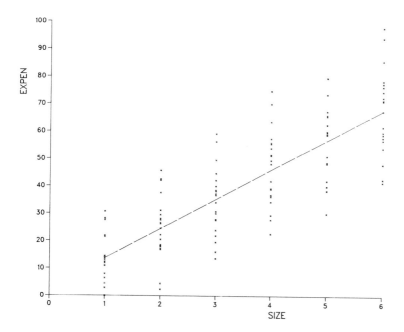

Figure 3.17. Fitted Relationship Between EXPEND and SIZE

and the *mean corrected prediction vector*

$$\widehat{\mathbf{y}}^* = \begin{bmatrix} \hat{y}_1 - \bar{y} \\ \hat{y}_2 - \bar{y} \\ \vdots \\ \hat{y}_n - \bar{y} \end{bmatrix}.$$

The mean corrected prediction vector $\widehat{\mathbf{y}}^*$ can be viewed as the *projection* of the mean corrected sample vector $\mathbf{y}^*$ on the mean corrected sample vector $\mathbf{x}^*$. Figure 3.18 panel (b) shows the relationship between $\mathbf{y}^*$, $\widehat{\mathbf{y}}^*$ and $\mathbf{x}^*$. The residuals $e_i = y_i - \hat{y}_i$ can also be summarized in vector form by

$$\mathbf{e} = \begin{bmatrix} e_1 \\ e_2 \\ \vdots \\ e_n \end{bmatrix} = \begin{bmatrix} y_1 - \hat{y}_1 \\ y_2 - \hat{y}_2 \\ \vdots \\ y_n - \hat{y}_n \end{bmatrix} = (\mathbf{y} - \hat{\mathbf{y}}).$$

The vector $\mathbf{e}$ is also shown in Figure 3.18 panel (b).

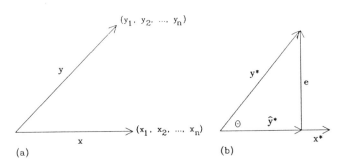

Figure 3.18.   Geometric Representation of a Bivariate Sample

The lengths of the vectors $\mathbf{y}^*$, $\mathbf{x}^*$, $\widehat{\mathbf{y}}^*$ and $\mathbf{e}$ are given by

$$\|\mathbf{y}^*\| = \left[\sum_{i=1}^{n}(y_i - \bar{y})^2 \right]^{1/2},$$

$$\|\mathbf{x}^*\| = \left[\sum_{i=1}^{n}(x_i - \bar{x})^2 \right]^{1/2},$$

$$\|\widehat{\mathbf{y}}^*\| = \left[\sum_{i=1}^{n}(\hat{y}_i - \bar{y})^2 \right]^{1/2}$$

and

$$\|\mathbf{e}\| = \left[\sum_{i=1}^{n}e_i^2 \right]^{1/2} = \left[\sum_{i=1}^{n}(y_i - \hat{y}_i)^2 \right]^{1/2}.$$

The right-angled triangle relationship is given by

$$\|\mathbf{y}^*\|^2 = \|\widehat{\mathbf{y}}^*\|^2 + \|\mathbf{e}\|^2$$

or

$$\sum_{i=1}^{n}(y_i - \bar{y})^2 = \sum_{i=1}^{n}(\hat{y}_i - \bar{y})^2 + \sum_{i=1}^{n}(y_i - \hat{y}_i)^2$$

or

$$\text{SST} = \text{SSR} + \text{SSE}.$$

This triangle shows the relationship between SST, SSR and SSE discussed earlier.

The angle θ between $\mathbf{y}^*$ and $\mathbf{x}^*$ can be obtained from the relationship

$$\text{Cos } \theta = \sum_{i=1}^{n}(y_i - \bar{y})(x_i - \bar{x}) \Big/ \left\{ \left[\sum_{i=1}^{n}(y_i - \bar{y})^2 \right]\left[\sum_{i=1}^{n}(x_i - \bar{x})^2 \right] \right\}^{1/2}.$$

Thus the Pearson correlation coefficient r is equivalent to Cos θ. If the angle between the vectors $\mathbf{y}^*$ and $\mathbf{x}^*$ is 0 we say that the two vectors are collinear and $r = \text{Cos } 0 = 1$ (Y can be written precisely as a linear function of X). If the angle between $\mathbf{y}^*$ and $\mathbf{x}^*$ is 90 we say that the vectors are orthogonal and $r = \text{Cos } 90 = 0$ (Y and X are uncorrelated). Thus the magnitude of the Pearson correlation coefficient indicates the degree of closeness of the vectors $\mathbf{y}^*$ and $\mathbf{x}^*$ as measured by the cosine of the angle between them.

An extensive summary of the simple linear regression model is available in Neter, Wasserman and Kutner (1983).

3.3.6 Scatterplots, Transformations and Smoothing

Scatterplots

As outlined above the simple linear regression model requires several assumptions regarding the joint distribution of X and Y. The most critical assumption is that the regression function $E[Y \mid X]$ is linear. Before carrying out an ordinary least squares fit the data should be plotted and the resulting scatterplot should be studied. The scatterplot of Y (vertical) versus X (horizontal) should appear as a random scatter of points about a straight line.

Figure 3.19 illustrates a variety of plots. Although for the first four plots the correlation coefficients have similar magnitudes these scatterplots reveal a variety of situations. If the scatter appears to be a single nonlinear shaped cluster, a transformation of one or both of the variables X and Y may be of value. In addition to the linearity property, the scatterplot is also useful for observing whether there are outliers in the data, and also whether there are two or more separate and distinct clusters of points.

Before a bivariate analysis is performed on a sample of observations (x_i, y_i) $i = 1, 2, \ldots, n$, the individual sample distributions for X and Y should be studied separately following the procedures suggested in Chapter 2. The analyst should be aware of distribution outliers, lack of symmetry and kurtosis before studying the relationship between X and Y.

Smoothing

If there are a large number of data points in the scatterplot it is sometimes difficult to determine the shape simply by eye. In such cases a *smoothing function* is useful for showing the shape. In a smoothed plot, each data point (x_i, y_i) is replaced by the value $(x_i, \tilde{y}_i)$, where $\tilde{y}_i$ is a *smoothed value* of y_i determined from all the points in the neighborhood of y. An example of a smoothing technique commonly used with time series data is the simple moving average. More generally the observations are ordered in terms of the X values to obtain $x_{(1)}, x_{(2)}, \ldots, x_{(n)}$. For notational purposes we denote the corresponding Y values as $y_{(1)}, \ldots, y_{(n)}$ although these values are not necessarily ordered.

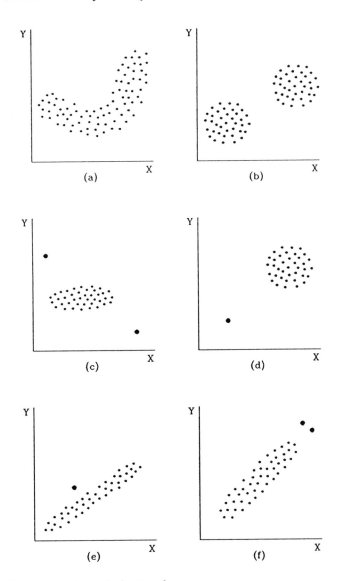

Figure 3.19. Some Example Scatterplots

Simple Moving Average

A *simple moving average* of length r (r is odd) replaces $y_{(k)}$ by a smoothed value obtained from the average of the r values, $y_{(k)}$, the $\left(\frac{r-1}{2}\right)$ values preceeding $y_{(k)}$, and the $\left(\frac{r-1}{2}\right)$ values following $y_{(k)}$. A simple moving average of length 5 would replace $y_{(7)}$ by the *smoothed value* $\tilde{y}_{(7)} = \left(y_{(5)} + y_{(6)} + y_{(7)} + y_{(8)} + y_{(9)}\right)/5$. The length of the moving average, r, governs the degree of smoothness. The longer the average the smoother the curve. The disadvantage of a long moving average is the loss of smoothed values at the beginning and end of the scatterplot. In a 5-point moving average there are no smoothed values for the first two and the last two points.

Example

The sample bivariate distribution of CPT vs LOADS was introduced in the study of correlation in the previous section. A study of the scatterplot in Figure 3.9 suggests that there may be some nonlinearity in the relationship between CPT and LOADS. To obtain an indication of the nature of the nonlinearity a five point moving average of the CPT values was determined and plotted in Figure 3.20. The smoothed scatterplot does suggest that there is some curvature in the shape of the function (particularly at the left). Below this data will be fitted using a transformation.

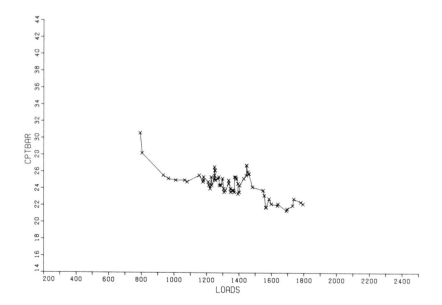

Figure 3.20. Five-Point Moving Average of CPT (CPTBAR) vs. LOADS

Weighted Moving Average

The simple moving average is sometimes replaced by a *weighted moving average* which gives greater weight to the values closest to the centre. This type of a moving average is more sensitive to curvature than the simple moving average. A common set of weights are the binomial weights based on the binomial theorem. For a 5-point *binomially weighted moving average* the binomial coefficients are [1, 4, 6, 4, 1] which total 16 and hence the weights are (1/16, 4/16, 6/16, 4/16, 1/16). The binomially weighted moving average will tend to have more curvature than the simple moving average.

Using Medians

Like the simple univariate mean discussed in Chapter 2, moving averages are sensitive to extreme values and hence to outliers. To avoid this effect medians can be used in place of means. A 5-point running median therefore replaces the value of $y_{(k)}$ by the median of the five points $y_{(k-2)}, y_{(k-1)}, y_{(k)}, y_{(k+1)}, y_{(k+2)}$.

Other Plots

In addition to the scatterplot it is useful to recognize that there are two marginal distributions (one for each of X and Y). Box plots can be plotted adjacent to the two axes to provide information about the two marginal distributions.

   In the regression of Y on X we are interested in the distribution of Y at each X. It is sometimes of value to divide the range of X into equal intervals and to study the variation of Y in each interval. At the center of each X interval a vertical box plot can be used to show the distribution of Y. It is easy then for the eye to follow the pattern of successive quantiles and the range for the Y distributions.

Transformations

From the scatterplot and/or smoothed scatterplot the relationship between Y and X may appear to be nonlinear. The technique of linear regression can still be applied if the model can be written as a linear function of variables which themselves are nonlinear. Simple linear models involving variables such as $\log X$, $\log Y$, $1/Y$, $1/X$, X^2 and Y^2 are examples. There are a number of nonlinear functions relating Y to X which may be converted to a linear function by suitable transformations on Y and/or X. This class of models is called *intrinsically linear*. Table 3.10 and Figure 3.21 summarize five classes of nonlinear functions which are intrinsically linear. From the graphs shown in Figure 3.19 the general shape of the various nonlinear functions can be observed. It may be that one of this set of nonlinear models is consistent with the theory underlying the data.

Table 3.10. Some Intrinsically Linear Functions

Linear Form	Function in Terms of Y	Graph
$Y = \beta_0 + \beta_1 \log X$	$Y = \beta_0 + \beta_1 \log X$	(a)
$\log Y = \beta_0 + \beta_1 X$	$Y = AB^X \quad \beta_0 = \log A, \ \beta_1 = \log B$	(b)
$\log Y = \beta_0 + \beta_1 \log X$	$Y = AX^{\beta_1} \quad \beta_0 = \log A$	(c)
$Y = \beta_0 + \beta_1(1/X)$	$Y = \beta_0 + \beta_1(1/X)$	(d)
$Y = \beta_0 + \beta_1 X^2$	$Y = \beta_0 + \beta_1 X^2$	(e)

For the intrinsically linear model the ordinary least squares procedure is applied to the transformed data to obtain a fit for the nonlinear model. It is important to recognize that this nonlinear fit is not an ordinary least squares fit for the original data. The determination of the goodness of fit should be carried out by determining the correlation coefficient between the original y values and the fitted $\hat{y}$ values after transforming back from the fitted linear model. The square of this correlation gives the proportion of variation in y explained by $\hat{y}$. We shall use a intrinsically linear function to fit the relation between CPT and LOADS discussed above.

Example

From a comparison of the plot in Figure 3.20 to the graphs of the intrinsically linear functions shown in Figure 3.21 it would appear that a function of the form $CPT = \beta_0 + \beta_1/LOADS$ may provide a superior fit to the data. A simple linear regression of CPT on the variable 1/LOADS yielded the regression relationship

$$CPT = 16.06 + 11091.85/LOADS.$$

A plot of this fitted curve is shown in Figure 3.22. For comparison the equation obtained from an ordinary least squares fit yielded the equation $CPT = 34.80 - 0.0075\ LOADS$. This equation is plotted in Figure 3.23. The correlation coefficient between the observed values of CPT and the predicted values of CPT from the nonlinear model was 0.632. This correlation is larger than the correlation of 0.555 between CPT and the predicted

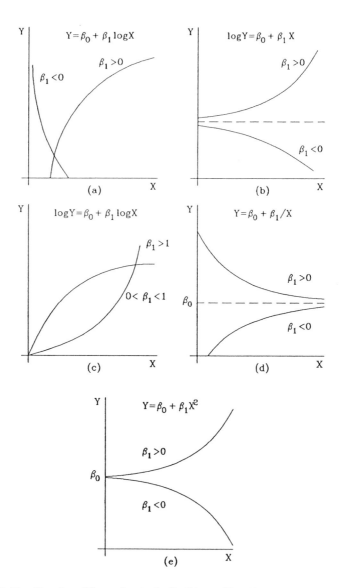

Figure 3.21. Graphs of Some Intrinsically Linear Functions

values from the linear fit. A useful measure of goodness of fit for nonlinear relationships is the correlation coefficient between y and $\hat{y}$. It is important to note however that $\bar{y}$ and $\bar{\hat{y}}$ are not necessarily equal as in the case of ordinary least squares.

An examination of the fitted nonlinear relationship suggests that little improvement could be obtained from any other nonlinear function. It would appear that there are other factors in addition to LOADS which are contributing to the variation in CPT. A discussion of the possible impact on the regression results of other explanatory variables will be discussed later in this chapter.

Estimating a Power Transformation Using the Box-Cox Family

In Chapter 2 the Box-Cox family of transformations was introduced for transforming a skewed distribution to normality. This family of power transformations may also be used to transform Y in a linear regression model. Given X and Y the assumption is that the transformed value of Y, say Z, is given by

$$Z(\lambda) = \frac{Y^\lambda - 1}{\lambda} \qquad \lambda \neq 0$$

$$= \log Y \qquad \lambda = 0,$$

and that $Z = \beta_0 + \beta_1 X + U$ where the error term U is assumed to be well-behaved. For a given λ, the likelihood function is maximized with respect to β_0, β_1 and $\sigma^2 = V(U)$, by minimizing the residual sum of squares $\sum_{i=1}^{n} \left(z_i^*(\lambda) - \hat{z}_i^*(\lambda)\right)^2 = \mathrm{SSE}_Z^*(\lambda)$, where $z^*(\lambda) = z(\lambda)/\tilde{y}_G$ and $\tilde{y}_G$ is the geometric mean of the Y observations. The minimum residual sum of squares $\mathrm{SSE}_Z^*(\lambda)$ can then be compared for various λ values. The value of λ which minimizes the value of $\mathrm{SSE}_Z^*(\lambda)$ yields the optimum power transformation.

For variables that are constrained to the interval $[0, b]$ the family of folded power transformations can be used. The transformed variable Z is given by

$$Z(\lambda) = \frac{Y^\lambda - (b - Y)^\lambda}{\lambda} \qquad \lambda \neq 0$$

$$= \log(Y/(b - Y)) \qquad \lambda = 0.$$

An example is a sample proportion p which is constrained to the interval $[0, 1]$. For $\lambda = 0$ this case is called a logit transformation and will be useful for logistic regression to be discussed later in this text.

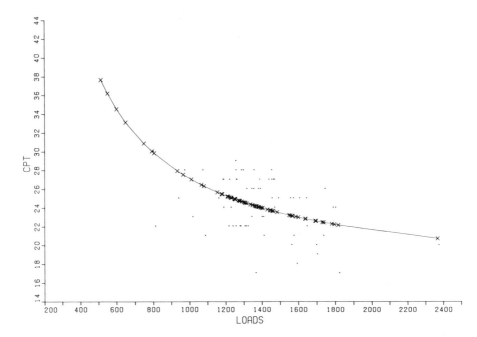

Figure 3.22. Scatterplot of CPT vs. LOADS Showing Curvilinear Fit

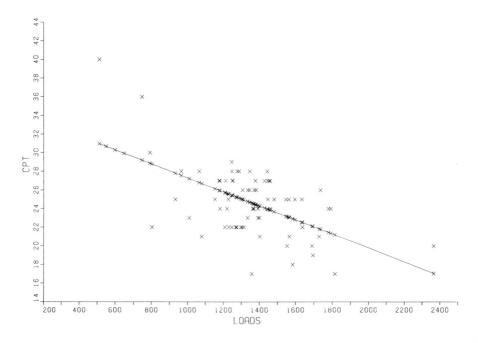

Figure 3.23. Scatterplot of CPT vs. LOADS Showing Linear Fit

The equation for the power transformation assumes that the Y values are positive. If this is not the case the formula can be modified by adding a constant θ to Y so that $(Y + \theta)$ is positive. The formula becomes

$$Z(\lambda) = \frac{(Y + \theta)^\lambda - 1}{\lambda} \qquad \lambda \neq 0$$

$$= \log(Y^\lambda + \theta) \qquad \lambda = 0.$$

3.3.7 Influence, Outliers and Leverage

Leverage

A major difficulty with the least squares fit of a straight line is its sensitivity to outliers or extreme values. The graphs in Figure 3.19 illustrate that outliers can have a strong influence on the estimates of the fitted parameters. For graphs (c) and (d) one can see that removal of the outlier points would markedly change the estimates of the various parameters. A useful approach to the assessment of outliers therefore is to determine the impact on the least squares fit for each of the data points, and also to determine how well the least squares relationship would fit a given data point if that point was not used in determining the estimates of the parameters.

For a particular data point (x_i, y_i), the impact of this point on the least squares fit can be determined from the expression for the predicted value of Y at x_i given by

$$\hat{y}_i = b_0 + b_1 x_i = \sum_{j=1}^{n} h_{ij} y_j,$$

where

$$h_{ij} = \frac{1}{n} + \frac{(x_i - \bar{x})(x_j - \bar{x})}{\sum\limits_{j=1}^{n}(x_j - \bar{x})^2},$$

and b_0 and b_1 are the least squares estimators defined in 3.3.1. Each $\hat{y}_i$ value is related to all y_j values through the h_{ij} values.

The $n \times n$ matrix of values h_{ij}, $i, j = 1, 2, \ldots, n$ is usually called the *'hat'* or *"H"* *matrix*. From this expression we can see that each fitted value $\hat{y}_i$ is a weighted linear combination of the Y observations $(y_1, y_2, \ldots, y_n)$ where h_{ij} is the weighting value for each y_i. From this expression it is clear that the further the observation x_j is from the center of the data, $\bar{x}$, the greater the weight placed on the corresponding y_j observation in the determination of $\hat{y}_i$. For points whose X observations are close to the mean, $\bar{x}$, the corresponding Y observations have much less impact on the determination of the $\hat{y}$ values. It is also clear that each y_j observation has an impact on each value of $\hat{y}_i$ and thus an extreme value of y_j influences all $\hat{y}_i$. The least squares fit is thus very sensitive to extreme points.

The most useful elements of the $\boldsymbol{H}$ matrix in the above expression are the diagonal elements h_{ii}, $i = 1, 2, \ldots, n$, which measure the impact that y_i has on $\hat{y}_i$. The quantity h_{ii} given by

$$h_{ii} = \frac{1}{n} + \frac{(x_i - \bar{x})^2}{\sum\limits_{j=1}^{n} (x_j - \bar{x})^2}$$

is referred to as the *leverage* or *h hat value* of the data point (x_i, y_i), and is a measure of how far the observation x_i is from the centre of the X values. When the leverage h_{ii} is large, $\hat{y}_i$ is more sensitive to changes in y_i than when h_{ii} relatively small. A calculation of the values of h_{ii}, $i = 1, 2, \ldots, n$ can show which Y observations will have the most leverage. High leverage Y observations correspond to points where $(x_i - \bar{x})$ is relatively large. Outliers therefore tend to have a large leverage value and hence a major impact on the predicted values $\hat{y}_i$.

The scatterplot in Figure 3.24 shows three points N, O and D which are somewhat distant from the main cluster of points which form an ellipse with the center at M. The points O and D are distant from the center of the data in both the X and Y directions. The point N is at the mean of the X values but is distant from the Y mean. The points O and D will have large values of h-hat while the point N will yield an h-hat value of zero. Fitting a least squares line to the main cluster in Figure 3.24 and then comparing the fit to the least squares fit including only one of the points O, D and N would yield differences in the slope and intercept. From the figure we can observe that N and O will have a much larger influence on the estimated regression parameters than the point D. The leverage value h_{ii} is therefore only a partial measure of the impact of observation (x_i, y_i) on the slope and intercept parameters. Better measures of the influence of (x_i, y_i) on the fitted regression relationship will be introduced later in this section.

To effectively use the h_{ii} value as a measure of leverage we need to know something about its range. The elements h_{ii} are contained in the interval $[0, 1]$ and for a simple linear regression, $\sum\limits_{i=1}^{n} h_{ii} = 2$. The average size of the leverage measure is therefore $2/n$. In practice h_{ii} is considered to be relatively large if it exceeds $4/n$. If $x_i = \bar{x}$ then $h_{ii} = 1/n$. In the extreme case that $x_j = a^*$ a constant, for all $j \neq i$ then all observed data points are located at a^* except x_i and hence $h_{ii} = 1$. In this case $\hat{y}_i$ only depends on y_i, the other y_j do not affect $\hat{y}_i$; that is $h_{ij} = 0$ $j \neq i$. Also in this case $h_{jj} = \left(\frac{1}{n-1}\right)$ for all $j \neq i$. The straight line passes through the points (x_i, y_i) and $(a^*, \bar{y}_{(i)})$ where $\bar{y}_{(i)}$ is the average of the y values excluding y_i.

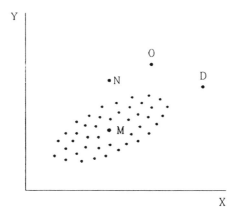

Figure 3.24. Scatterplot With Outliers

Example

For the NOI data presented earlier in Table 3.7 the values of h_{ii} are presented in Table 3.11 along with the values of the variable SUITES. The relatively large values of h_{ii} are 0.366, 0.072, 0.147, 0.173, 0.072 corresponding to the five largest values of the variable SUITES, given by 205, 104, 140, 150 and 104 respectively. The h-hat values thus indicate the distance of the observations on the variable SUITES from the centre of the data. Comparing these observations to the observations with large residuals in Table 3.8 we can conclude that only one of the large h-hat values also corresponds to a value of NOI outside the 95% confidence interval, namely the NOI value of 154,278 for the 150 suite building.

Table 3.11. Leverage Values for Suites Variable

SUITES	58	30	22	21	12	20	15	29	28	23	14	27
h_{ii}	0.025	0.023	0.026	0.027	0.032	0.027	0.030	0.023	0.024	0.026	0.031	0.024
SUITES	52	48	20	205	17	26	22	24	20	33	104	140
h_{ii}	0.023	0.022	0.027	0.366	0.029	0.024	0.026	0.025	0.027	0.022	0.072	0.147
SUITES	44	78	69	150	62	86	44	104	21	18	24	15
h_{ii}	0.021	0.039	0.031	0.173	0.027	0.047	0.021	0.072	0.027	0.028	0.025	0.030
SUITES	21	65	24	12	12	12	12	12	15	12	20	
h_{ii}	0.027	0.029	0.025	0.032	0.032	0.032	0.032	0.032	0.030	0.032	0.027	

It is worth noting here that points with high leverage values can have a large impact on R^2, the coefficient of determination, if they are close to the fitted line as in the case of point D in Figure 3.24. Omitting the point D from a linear fit will reduce R^2 by a large amount. In the example the point corresponding to SUITES $= 205$ will have a large influence on R^2. Omitting this observation from the fit reduces R^2 from 0.784 to 0.669. Therefore observations with large leverage values that are close to the fitted line can bring about large increases in R^2.

Residuals and Deleted Residuals

A useful way of judging outliers is to determine whether the extreme point (x_i, y_i) is within a predictable range, given the linear relationship that would be fitted without this extreme point. Denote by e_i the residual obtained by taking the difference between the observed y_i and the predicted $\hat{y}_i$ using all the data points. Denote by $e_{(i)i}$ the residual obtained by taking the difference between the observed y_i and the least squares prediction $\hat{y}_{(i)i}$ of y_i obtained after omitting the point (x_i, y_i) from the least squares fit. The residual $e_{(i)i}$ will be referred to as a *deleted residual* since it represents the residual at y_i when (x_i, y_i) was omitted from the fit. The two residuals are given by

$$e_i = y_i - \hat{y}_i \qquad \text{and} \qquad e_{(i)i} = y_i - \hat{y}_{(i)i}.$$

In Figure 3.25 a scatterplot of values (x_j, y_j) is shown with an ordinary least squares line $\hat{y}$. A second least squares line $\hat{y}_{(i)}$ denotes the fitted line obtained after omitting observation (x_i, y_i) at point A. The corresponding values of $\hat{y}_i$, $\hat{y}_{(i)i}$, e_i and $e_{(i)i}$ are also shown in the figure. In this case the point A appears to have an impact on the magnitude of the residual.

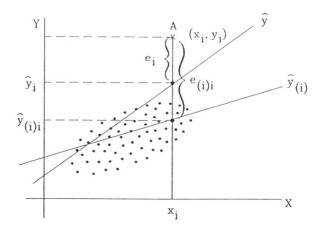

Figure 3.25. Residuals and Deleted Residuals

These two residuals can be related by the equation

$$e_{(i)i} = e_i/(1 - h_{ii})$$

where

$$h_{ii} = \frac{1}{n} + \frac{(x_i - \bar{x})^2}{\displaystyle\sum_{j=1}^{n}(x_j - \bar{x})^2}$$

is the leverage defined above. When the leverage is relatively large (closer to 1) the two residuals will tend to be quite different. When the leverage is relatively large the deleted residual $e_{(i)i}$ is large relative to the ordinary residual e_i.

PRESS Statistic

In large data sets the omission of a single data point from a least squares fit should have little impact on the residual for the omitted data point. The two residuals $e_{(i)i}$ and e_i should therefore be similar. By the same argument the sum of squared residuals $\sum_{i=1}^{n} e_i^2$ and $\sum_{i=1}^{n} e_{(i)i}^2$ in each case should also be of similar magnitude. The statistic $\sum_{i=1}^{n} e_{(i)i}^2$ which is the sum of squares of the deleted residuals is called the *PRESS* (predicted residual sum of squares) statistic. The ratio of PRESS to the ordinary sum of squared residuals $\sum_{i=1}^{n} e_i^2$, gives an indication of the sensitivity of the fit to omitted observations. Several outliers can often cause this ratio to be much larger than unity. The contribution of each point to the PRESS statistic $e_{(i)i}^2 = e_i^2/(1 - h_{ii})^2$ can be determined, and hence dominant values of $e_{(i)i}^2$ can be identified. Such dominant values can therefore be related to the leverage values.

Example

For the fitted relationship between NOI and SUITES the sum of squared residuals was 118.15×10^9 while the PRESS statistic was 152.65×10^9. The ratio of these is therefore 1.29 which indicates a small number of outliers might be present.

Standardized Residuals

Under the usual assumptions for the simple linear regression model the residuals $e_i = (y_i - \hat{y}_i)$ have the properties

$$E[e_i] = 0 \quad \text{and} \quad V[e_i] = \sigma_u^2(1 - h_{ii}) \quad i = 1, 2, \ldots, n.$$

Thus even though the variances of the true errors u_i are assumed to be homogeneous $V[u_i] = \sigma_u^2$, $i = 1, 2, \ldots, n$ this is not true for the e_i. The transformed statistics $e_i/\sqrt{1 - h_{ii}}$, $i = 1, 2, \ldots, n$ therefore have the common variance σ_u^2. The residuals can be standardized therefore by dividing by $s\sqrt{1 - h_{ii}}$ where $E[s^2] = \sigma_u^2$. The *standardized residuals* $r_i = e_i/s\sqrt{1 - h_{ii}}$, $i = 1, 2, \ldots, n$ have mean 0 and common variance near 1, and are useful for studying the behavior of the true error terms.

The standardized residuals can be used to test for a single outlier using the distribution of the maximum absolute value of the standardized residuals

$$R_n = \max_i |r_i|.$$

Upper bounds for the α critical values of R_n are given by

$$(n - 2)F_{\alpha;1,(n-2)}/(n - 3 + F_{\alpha;1,(n-2)}).$$

Critical values of R_n have been approximated by Tietjen, Moore and Beckman (1973) using Monte Carlo simulation. Prescott (1975) has shown these values to be very similar to the upper bound values determined from the above expression. Table 3.12 summarizes the upper bound values available in Lund (1975).

Example

The standardized residuals for the linear fit between NOI and SUITES are presented in Table 3.13. Using the critical value of 2.95 for $N = 45$ obtained from Table 3.12 we can conclude that the largest standardized residual -4.1522 corresponding to observation no. 28 represents an outlier. There are four additional observations in which the standardized residual exceeds $|2|$. Observations 16, 23, 25 and 27 have the standardized residuals 2.1058, 2.1473, 2.5215 and -2.5538 respectively. These standardized residuals should not be compared to the critical value of R_n since this statistic is designed for the maximum. In the evaluation of the h-hat statistics in the previous section observations 16, 23 and 28 also exhibited relatively large values of h-hat.

The quantities displayed in Table 3.13 were generated using SAS PROC REG.

Table 3.12. Upper Bound for Critical Values for R_n

n	$\alpha = 0.10$	$\alpha = 0.05$	$\alpha = 0.01$
5	1.87	1.92	1.98
6	2.00	2.07	2.17
7	2.10	2.19	2.32
8	2.18	2.28	2.44
9	2.24	2.35	2.54
10	2.30	2.42	2.62
12	2.39	2.52	2.76
14	2.47	2.61	2.86
16	2.53	2.68	2.95
18	2.58	2.73	3.02
20	2.63	2.78	3.08
25	2.72	2.89	3.21
30	2.80	2.96	3.30
35	2.86	3.03	3.37
40	2.91	3.08	3.43
45	2.95	3.13	3.48
50	2.99	3.17	3.52
60	3.06	3.23	3.60
70	3.11	3.29	3.65
80	3.16	3.33	3.70
90	3.20	3.37	3.74
100	3.23	3.41	3.78

Studentized Residuals

The standardized residual, r_i, given above is sometimes referred to as an *internally studentized residual* because s^2 is not independent of e_i. An alternative residual is the *externally studentized residual* which employs an alternative estimate of $\sigma_u^2, s_{(i)}^2$, which is determined from the estimated regression of all points excluding (x_i, y_i). The estimator $s_{(i)}^2$ is therefore independent of e_i. The externally studentized residual is often referred to more generally as the studentized residual and is given by

$$t_i = \frac{e_i}{s_{(i)}\sqrt{1 - h_{ii}}} \qquad i = 1, 2, \ldots, n.$$

This studentized residual is distributed as a t distribution with $(n - 3)$ d.f. if the error terms are normal, independent, mean zero and with constant variance σ_u^2. The estimators $s_{(i)}^2$ $i = 1, 2, \ldots, n$ can be determined without performing additional regressions by using the relationship

$$s_{(i)}^2 = [(n - 2)s^2 - e_i^2/(1 - h_{ii})]/(n - 3).$$

Table 3.13. Influence Diagnostic Statistics for NOI vs. SUITES Relationship

Obs. No.	Standard Residual	Student Residual	H Hat	Cook's D	COVRATIO	DFFITS	DFBETA Intercept	DFBETA Suites
1	-0.2424	-0.2399	0.0249	0.001	1.0698	-0.0383	-0.0145	-0.0145
2	-0.3072	-0.3040	0.0229	0.001	1.0660	-0.0466	-0.0404	0.0125
3	-0.2686	-0.2658	0.0261	0.001	1.0705	-0.0435	-0.0409	0.0187
4	-0.5159	-0.5116	0.0266	0.004	1.0619	-0.0846	-0.0800	0.0378
5	0.0656	0.0648	0.0323	0.000	1.0807	0.0119	0.0117	-0.0069
6	-0.1882	-0.1862	0.0271	0.000	1.0734	-0.0311	-0.0296	0.0144
7	-0.2083	-0.2061	0.0302	0.001	1.0765	-0.0364	-0.0355	0.0198
8	0.8531	0.8504	0.0232	0.009	1.0365	0.1311	0.1151	-0.0380
9	0.0445	0.0440	0.0236	0.000	1.0711	0.0068	0.0061	-0.0021
10	-0.1129	-0.1116	0.0256	0.000	1.0728	-0.0181	-0.0169	0.0074
11	-0.0320	-0.0316	0.0309	0.000	1.0792	-0.0056	-0.0055	0.0031
12	0.4507	0.4467	0.0239	0.002	1.0619	0.0699	0.0628	-0.0232
13	0.7091	0.7051	0.0227	0.006	1.0465	0.1076	0.0535	0.0273
14	1.5792	1.6067	0.0219	0.028	0.9541	0.2401	0.1385	0.0389
15	-0.1637	-0.1619	0.0271	0.000	1.0739	-0.0270	-0.0257	0.0126
16	2.1058	2.1931	0.3660	1.280	1.3406	1.6662	-0.8706	1.6170
17	-0.4926	-0.4884	0.0289	0.004	1.0655	-0.0842	-0.0815	0.0432
18	-0.5835	-0.5792	0.0243	0.004	1.0559	-0.0914	-0.0830	0.0322
19	0.0371	0.0367	0.0261	0.000	1.0730	0.0060	0.0056	-0.0026
20	-.005197	-0.0051	0.0251	0.000	1.0729	-0.0008	-0.0008	0.0003
21	-0.2091	-0.2069	0.0271	0.001	1.0730	-0.0345	-0.0329	0.0160
22	0.6802	0.6761	0.0222	0.005	1.0478	0.1018	0.0845	-0.0204
23	2.1473	2.2413	0.0718	0.178	0.9078	0.6235	-0.1347	0.5230
24	0.5578	0.5535	0.1466	0.027	1.2087	0.2294	-0.0898	0.2121
25	2.5215	2.6906	0.0214	0.069	0.7881	0.3976	0.2597	0.0262
26	1.3678	1.3815	0.0386	0.038	0.9994	0.2768	0.0120	0.1854
27	-2.5538	-2.7310	0.0311	0.105	0.7893	-0.4896	-0.0876	-0.2755
28	-4.1522	-5.2275	0.1732	1.806	0.4814	-2.3929	1.0090	-2.2411
29	0.3256	0.3224	0.0268	0.001	1.0697	0.0535	0.0162	0.0242
30	-0.5194	-0.5152	0.0470	0.007	1.0844	-0.1144	0.0063	-0.0846
31	0.2151	0.2128	0.0214	0.001	1.0666	0.0314	0.0205	0.0021
32	-1.6278	-1.6592	0.0718	0.103	0.9981	-0.4615	0.0997	-0.3872
33	-0.2064	-0.2042	0.0266	0.001	1.0725	-0.0338	-0.0319	0.0151
34	-0.4365	-0.4326	0.0283	0.003	1.0673	-0.0738	-0.0711	0.0367
35	0.1828	0.1808	0.0251	0.000	1.0713	0.0290	0.0168	-0.0114
36	0.3492	0.3457	0.0302	0.002	1.0727	0.0610	0.0595	-0.0331
37	-0.0434	-0.0429	0.0266	0.000	1.0745	-0.0071	-0.0067	0.0032
38	-0.0381	-0.0377	0.0285	0.000	1.0766	-0.0065	-0.0016	-0.0032
39	0.0556	0.0550	0.0251	0.000	1.0728	0.0088	0.0082	-0.0035
40	-0.1200	-0.1187	0.0323	0.000	1.0802	-0.0217	-0.0214	0.0127
41	0.0143	0.0142	0.0323	0.000	1.0809	0.0026	0.0026	-0.0015
42	-0.2170	-0.2147	0.0323	0.001	1.0787	-0.0393	-0.0387	0.0230
43	-0.2669	-0.2642	0.0323	0.001	1.0075	-0.0483	-0.0476	0.0282
44	-0.1909	-0.1889	0.0323	0.001	1.0792	-0.0345	-0.0340	0.0202
45	-0.3489	-0.3455	0.0302	0.002	1.0727	-0.0610	-0.0595	0.0331
46	0.0136	0.0134	0.0323	0.000	1.0809	0.0025	0.0024	-0.0014
47	-0.1178	-0.1165	0.0271	0.000	1.0745	-0.0195	-0.0185	0.0090

The t-statistic for evaluating the point (x_i, y_i) can therefore be expressed as

$$t_i = \frac{e_i\sqrt{n-3}}{[(n-2)s^2(1-h_{ii}) - e_i^2]^{1/2}}.$$

This t-statistic is a measure of the *influence* of (x_i, y_i) on the least squares fit. From this expression for the t-statistic we can see that for data points with relatively large values of e_i and/or with relatively large values of h_{ii} the t-value will tend to be large. The leverage, h_{ii}, of an observation x_i measures the distance from x_i to the centre of the data while the residual e_i measures the difference $(y_i - \hat{y}_i)$ which is the distance between y_i and $\hat{y}_i$.

If the analyst believes a priori that the i-th observation is an outlier then the t_i-statistic can be compared to a critical t with $(n-3)$ d.f. More commonly however the studentized residuals are used for outlier detection. In this case the largest value of t_i, $i = 1, 2, \ldots, n$ is used to test the null hypothesis that there are no outliers. Since there are n values of t_i we are testing the hypothesis that all n observations are not outliers. Using a Bonferroni inequality, the type I error is less than or equal to $n\alpha$ if t_α is used as the critical value of t_i.

Example

The studentized residuals for the NOI vs SUITES linear fit are presented in Table 3.13 above. The studentized residuals that exceed $|2|$ are given by 2.1931, 2.2413, 2.6906, -2.7310 and -5.2275 corresponding to observations 16, 23, 25, 27 and 28 respectively. If a t-statistic corresponding to $47 - 3 = 44$ d.f. is used with $\alpha = 0.05$ the critical value is ± 2.015. If a Bonferroni approximation is used to allow for multiple tests, the adjusted p-value is $0.05/47 = 0.001$ which is an upper bound for the correct p-value, hence the critical value of t to use is ± 3.52. Using the more conservative Bonferroni approach there is one significant outlier corresponding to observation 28. This conclusion is consistent with the results obtained from the standardized residuals.

The tests for outliers discussed in this chapter have been tests for a single outlier. These tests are also used in practice to identify additional outliers although the type I error probability is no longer valid. Techniques for identifying more than one outlier will be discussed in Chapter 4.

Influence Diagnostics

In our earlier discussion of the h-hat statistic it was shown that h-hat measures the distance of an observation from the centre of the data along the X axis. In Figure 3.24 it was shown that an observation with a relatively large value of h-hat may or may not have much influence on the fitted parameters. Figure 3.24 also displayed an observation, point N, that had a relatively low h-hat value but had a large influence on the fitted parameters. This influential observation was not far from the centre of the data along the X axis but was influential because its Y value was distant relative to the line that would fit the remaining points. The use of residuals from a fitted linear relationship to identify outliers has the disadvantage that an outlier can mask itself by drawing the line toward itself, as a result the residual is considerably smaller than it would have been had this point been omitted from the data set. In this section we approach the measure of *influence* more directly by comparing estimates of parameters determined with and without the observation of interest being included in the data. If the estimates do not change very much as a result of the deletion than the observation is said to be non-influential.

Cook's D

A variety of diagnostic statistics have been developed to measure the influence of a particular observation on a linear regression fit. *Cook's D statistic* measures the influence of (x_i, y_i) by computing

$$D_i = \sum_{j=1}^{n} (\hat{y}_{(i)j} - \hat{y}_j)^2 / 2s^2 \quad i = 1, 2, \ldots, n,$$

where $\hat{y}_{(i)j}$ and $\hat{y}_j$ are the predictions for y at $x = x_j$ without and with the point (x_i, y_i), respectively. The statistic can also be written in the more convenient computational form

$$D_i = \left(\frac{h_{ii}}{1 - h_{ii}} \right) \frac{e_i^2}{2s^2(1 - h_{ii})} = \frac{h_{ii}}{2(1 - h_{ii})} r_i^2 \quad i = 1, 2, \ldots, n.$$

Thus D_i will be large if the standardized residual is large and/or if the leverage is large. There can be trade offs between the two parts in that a large leverage effect can be dampened by a small residual. Another way of writing the D_i statistic can be obtained by substituting for $(\hat{y}_{(i)j} - \hat{y}_j)$ in D_i. We may write

$$(\hat{y}_{(i)j} - \hat{y}_j) = (b_{(i)0} - b_0) + (b_{(i)1} - b_1)x_j,$$

where $b_{(i)0}$, $b_{(i)1}$, b_0 and b_1 are the least squares estimators of β_0 and β_1 without and with observation i. Therefore D_i is given by

$$\left[n(b_{(i)0} - b_0)^2 + (b_{(i)1} - b_1)^2 \sum_{j=1}^{n} x_j^2 + s \sum_{j=1}^{n} x_j(b_{(i)0} - b_0)(b_{(i)1} - b_1)\right]\Big/2s^2.$$

In this form D_i can be recognized as the expression for a confidence ellipsoid for (β_0, β_1) given by (3.6) with $(b_{(i)0}, b_{(i)1})$ replacing (β_0, β_1). Thus if $D_i \leq F_{\alpha;1,(n-2)}$ then $(b_{(i)0}, b_{(i)1})$ is contained in the $100(1-\alpha)\%$ confidence ellipse for (β_0, β_1). One common rule of thumb is to investigate observations with D_i values which exceed one. This corresponds to an approximate 50% confidence ellipse for the parameters β_0 and β_1. See Cook and Weisburg (1982) for further discussion.

The DF Family

Other measures related to D_i developed by Belsey, Kuh and Welsch (1980) are also used. The measure

$$\text{DFFITS}_i = (\hat{y}_i - \hat{y}_{(i)i})/s_{(i)}\sqrt{h_{ii}}$$

also measures the difference between the two predicted values. This measure can also be written as

$$\text{DFFITS}_i = \left[\frac{h_{ii}}{1 - h_{ii}}\right]^{1/2} \frac{e_i}{s_{(i)}\sqrt{1 - h_{ii}}} \qquad i = 1, 2, \ldots, n.$$

Thus the measure $(\text{DFFITS}_i)^2$ is comparable to D_i except that $s_{(i)}^2$ is used in place of $2s^2$.

Similarly, the differences in any particular regression coefficient can be measured by

$$\text{DFBETA}_{1i} = (b_1 - b_{(i)1})/c_1 s_{(i)} \qquad i = 1, 2, \ldots, n,$$

where

$$c_1^2 = \left(\frac{n}{n \sum_{j=1}^{n} x_j^2 - \left(\sum_{j=1}^{n} x_j\right)^2}\right)$$

is the variance of b_1 divided by σ_u^2 and

$$\text{DFBETA}_{0i} = (b_0 - b_{(i)0})/c_0 s_{(i)} \qquad i = 1, 2, \ldots, n,$$

where

$$c_0^2 = \left(\frac{\sum\limits_{j=1}^{n} x_j^2}{n \sum\limits_{j=1}^{n} x_j^2 - \left(\sum\limits_{j=1}^{n} x_j \right)^2} \right)$$

is the variance of b_0 divided by σ_u^2.

Thus $DFFITS_i$ measures the overall influence of observation i on the fit while $DFBETA_{0i}$ and $DFBETA_{1i}$ show the separate influence of observation i on the intercept and slope of the fitted regression line respectively. Figure 3.26 shows the values of $DFBETA_{0i}$ and $DFBETA_{1i}$ for various situations regarding the influence of observation i.

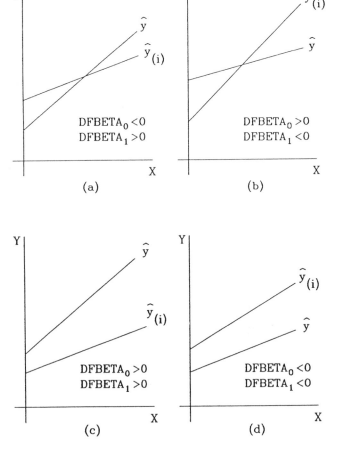

Figure 3.26. Illustration of Various Combinations of $DFBETA_0$ and $DFBETA_1$

The impact of observation i on the variances and covariances among the coefficients b_0 and b_1 is also of interest. The elements of the estimated covariance matrix can be written as

$$\frac{s^2}{n\sum_{j=1}^{n} x_j^2 - \left(\sum_{j=1}^{n} x_j\right)^2} \begin{bmatrix} \sum_{j=1}^{n} x_j^2 & -\sum_{j=1}^{n} x_j \\ -\sum_{j=1}^{n} x_j & n \end{bmatrix}.$$

If x_i is omitted from the regression it must be omitted from the summations in the covariance matrix and hence $(n-1)$, $s_{(i)}^2$, $\sum_{\substack{j=1 \\ j\neq i}}^{n} x_j$ and $\sum_{\substack{j=1 \\ j\neq i}}^{n} x_j^2$ must replace n, s^2, $\sum_{j=1}^{n} x_j$ and $\sum_{j=1}^{n} x_j^2$, respectively. The ratio of the determinants of this new covariance matrix to the original covariance matrix was labelled COVRATIO by Belsey, Kuh and Welsch (1980) and can be written as

$$\frac{(n-1)}{[(n-2)+t_i](1-h_{ii})}.$$

For h_{ii} values in the range $\frac{1}{n} \leq h_{ii} \leq \frac{2}{n}$ and studentized residuals $|t_i| \leq 2$ the COVRATIO lies approximately in the interval $\left(1-\frac{3}{n}\right) \leq \text{COVRATIO} \leq \left(1+\frac{3}{n}\right)$. If $|\text{COVRATIO}-1| > \frac{3}{n}$ the point (x_i, y_i) should be investigated. The COVRATIO statistic can be written as

$$\left[\frac{1}{\dfrac{(n-2)}{(n-1)} + \dfrac{t_i}{(n-1)}}\right] \frac{1}{(1-h_{ii})}$$

which shows that if h_{ii} is close to zero the ratio tends to be close to 1. The term involving t_i can be written approximately as

$$\frac{1}{1+t_i/(n-1)},$$

which suggests that if t_i is large negatively this term will tend to increase the ratio, while if t_i is large positively the ratio will tend to decrease. Thus large negative studentized residuals which are distant from the X mean will yield large values of COVRATIO while large positive studentized residuals which are close to the X mean will yield the small values of COVRATIO. For the remaining two cases there is a trade off between the impact of the two measures on COVRATIO.

Example

The various *influence diagnostic statistics* outlined above have been determined for the NOI vs SUITES data and are presented in Table 3.13 above. An examination of the Cook's D statistics reveals that observations 16 and 28 have Cook's D values of 1.280 and 1.806 respectively. The remaining values of Cook's D are much less than 1.00.

The DFFITS statistic also identifies the same influential observations as Cook's D. Observations 16 and 28 have the largest values of DFFITS 1.6662 and -2.3929 respectively. The DFFITS statistic also contains a sign which indicates the direction of the influence. A positive sign indicates that the observation increases the predicted value of Y. Figure 3.27 identifies the influential observations on a scatterplot. The two most influential points are shown with the symbol $\bigcirc$.

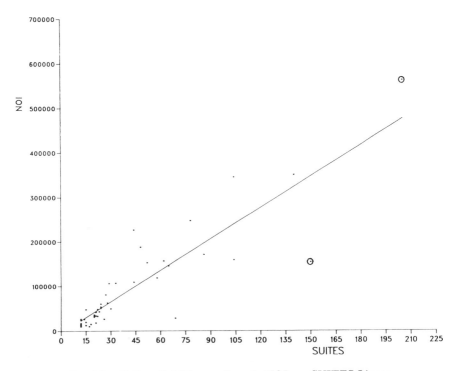

Figure 3.27.   Most Influential Observations in NOI vs. SUITES Linear Relationship

The DFBETA values for the intercept and slope for observation 16 indicate that adding observation 16 to the data set results in an increase in the slope and a decrease in the intercept. This can also be seen by looking at Figure 3.27 and imagining how least squares would adjust if observation 16 were omitted from the data. The reverse is true of observation 28.

The COVRATIO statistic indicates that observations 16 and 28 have a large influence on the covariance matrix for the estimated parameters. The values of COVRATIO are 1.3406 and 0.4814 respectively. For observation 16 the generalized variance is decreased when this observation is added while for observation 28 the generalized variance is increased when observation 28 is added.

A least squares fit was carried out after omitting both observations 16 and 28. The resulting equation was

$$\text{NOI} = -9645.8 + 2553.8 \text{ SUITES} \quad \text{with } R^2 = 0.77.$$

This equation can be compared to the least squares fit obtained in Section 3.3.1 for all 47 points which was

$$\text{NOI} = -4872.0 + 2350.7 \text{ SUITES} \quad \text{with } R^2 = 0.78.$$

The standard errors of the intercept and slope for the complete data set were 10655.3 and 183.8 respectively. For the data set omitting observations 16 and 28 these standard errors were 9601.5 and 210.7.

A Second Example

A second example which illustrates the measurement of outliers and influence can be provided using the scatterplot of CPT vs LOADS discussed earlier in Section 3.2. Table 3.14 presents the values of the various statistics for the simple linear regression relationships. Both the standardized and studentized residuals show four observations (4, 20, 47, 51) with a value larger than $|2|$. The h-hat values indicate that the two most extreme observations are 51 and 65 which have h-hat values of 0.1478 and 0.2092 respectively. The Cook's D statistic indicates that the most influential observation is 51 with a Cook's D value of 0.851. The DFFITS values corresponding to observations 20, 47, 51 and 65 are -0.6440, 0.7078, 1.399 and 0.5371 respectively. The DFFITS value for observation 51 indicates that this observation tends to increase the predicted value. The values of DFBETA for this outlier indicate that for this observation, exclusion results in a smaller value of the intercept and a larger value for the slope. Observation 51 is marked $\bigcirc$ in Figure 3.28.

The calculations for this example were performed using SAS PROC REG.

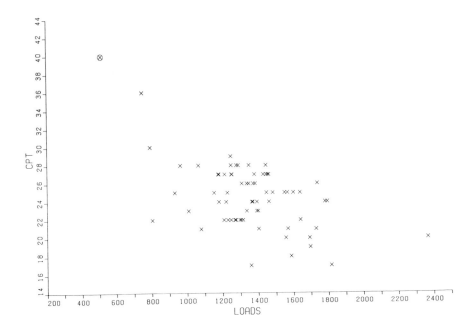

Figure 3.28. Influential Observations in CPT vs. LOADS Scatterplot

Robust Linear Fit

In the presence of outliers, a more *robust linear fit* that is less sensitive
to extremes can be obtained by *grouping* observations. This method is
outlined in Emerson and Hoaglin (1983). Given n pairs of observations
(x_i, y_i), $i = 1, 2, \ldots, n$ the X's are ordered to obtain the order statistics
$(x_{(1)}, x_{(2)}, \ldots, x_{(n)})$. The n points are divided into thirds, a lower third,
a middle third and an upper third. For each of the thirds the median X
value and median Y value are determined and denoted by x_L, x_M, x_U and
y_L, y_M, y_U. Note that the points (x_L, y_L), (x_M, y_M) and (x_U, y_U) are not
necessarily data points. These three points are used to estimate the slope
and the intercept.

The slope is obtained from the expression $b_1^* = (y_U - y_L)/(x_U - x_L)$
while the intercept is estimated using

$$b_0^* = \frac{1}{3}[(y_L - b_1^* x_L) + (y_M - b_1^* x_M) + (y_U - b_1^* x_U)].$$

The estimated equation is given by

$$\hat{y}^* = b_0^* + b_1^* x.$$

Table 3.14. Influence Diagnostic Statistics for CPT vs. LOADS Relationship

Obs. No.	Standard Residual	Studentized Residual	H Hat	Cook's D	DFFITS	DFBETA Intercept	DFBETA LOADS
1	-0.5822	-0.5793	0.0143	0.002	-0.0699	-0.0179	0.0041
2	-0.9251	-0.9241	0.0476	0.021	-0.2065	-0.1918	0.1728
3	-0.2830	-0.2811	0.0412	0.002	-0.0583	0.0393	-0.0471
4	-2.4609	-2.5593	0.0143	0.044	-0.3082	-0.0537	-0.0078
5	-0.9795	-0.9792	0.0147	0.007	-0.1196	-0.0433	0.0202
6	-0.7009	-0.6983	0.0361	0.009	-0.1351	0.0859	-0.1050
7	-0.1778	-0.1765	0.0143	0.000	-0.0213	-0.0029	-0.0014
8	0.4153	0.4128	0.0175	0.002	0.0551	-0.0132	0.0236
9	-1.0522	-1.0531	0.0154	0.009	-0.1318	-0.0601	0.0356
10	-0.5380	-0.5352	0.0175	0.003	-0.0715	-0.0430	0.0308
11	-1.1179	-1.1200	0.0163	0.010	-0.1443	-0.0770	0.0512
12	0.4114	0.4089	0.0180	0.002	0.0553	0.0343	-0.0250
13	-1.0644	-1.0655	0.0156	0.009	-0.1340	-0.0631	0.0384
14	-1.8756	-1.8925	0.0283	0.050	-0.3231	-0.2685	0.2274
15	0.3174	0.3153	0.0147	0.001	0.0385	0.0135	-0.0061
16	0.5760	0.5731	0.0216	0.004	0.0851	-0.0347	0.0494
17	0.6985	0.6958	0.0258	0.006	0.1131	-0.0572	0.0755
18	-0.6354	-0.6326	0.0198	0.004	-0.0900	-0.0618	0.0476
19	-0.9965	-0.9964	0.0148	0.007	-0.1223	-0.0470	0.0236
20	-2.2536	-2.3255	0.0712	0.195	-0.6440	-0.6217	0.5758
21	-0.1899	-0.1885	0.0143	0.000	-0.0227	-0.0035	-0.0011
22	-0.9553	-0.9547	0.0145	0.007	-0.1160	-0.0381	0.0155
23	-1.0174	-1.0177	0.0366	0.020	-0.1983	0.1271	-0.1548
24	0.7956	0.7900	0.0296	0.010	0.1379	-0.0781	0.0991
25	0.5052	0.5024	0.0162	0.002	0.0644	0.0337	-0.0221
26	1.0075	1.0076	0.0165	0.009	0.1305	-0.0226	0.0477
27	0.6151	0.6122	0.0228	0.004	0.0935	-0.0413	0.0571
28	1.3691	1.3782	0.0422	0.041	0.2894	-0.1972	0.2354
29	0.8585	0.8569	0.0511	0.020	0.1989	-0.1446	0.1689
30	0.3248	0.3226	0.0201	0.001	0.0462	0.0321	-0.0248
31	0.3815	0.3791	0.0298	0.002	0.0664	0.0561	-0.0479
32	0.5004	0.4976	0.0163	0.002	0.0640	0.0338	-0.0223
33	-1.0656	-1.0667	0.0148	0.009	-0.1306	-0.0022	-0.0238
34	-0.4448	-0.4421	0.0146	0.001	-0.0538	-0.0028	-0.0079
35	-0.1982	-0.1968	0.0172	0.000	-0.0260	-0.0152	0.0107

Table 3.14. Influence Diagnostic Statistics for CPT vs. LOADS Relationship (continued)

Obs. No.	Standard Residual	Studentized Residual	H Hat	Cook's D	DFFITS	DFBETA Intercept	DFBETA LOADS
36	0.1369	0.1359	0.0427	0.000	0.0287	0.0262	-0.0234
37	0.9710	0.9705	0.0159	0.008	0.1235	-0.0155	0.0396
38	0.8183	0.8163	0.0163	0.006	0.1052	0.0561	-0.0373
39	1.1386	1.1411	0.0164	0.011	0.1472	0.0789	-0.0526
40	0.5076	0.5048	0.0161	0.002	0.0647	0.0336	-0.0220
41	-1.6019	-1.6209	0.0246	0.032	-0.2573	0.1241	-0.1664
42	0.4983	0.4955	0.0145	0.002	0.0600	0.0053	0.0067
43	-0.6728	-0.6701	0.0230	0.005	-0.1029	0.0460	-0.0634
44	0.8256	0.8237	0.0490	0.018	0.1870	-0.1341	0.1574
45	0.9142	0.9131	0.0151	0.006	0.1129	0.0470	-0.0257
46	0.9928	0.9927	0.0163	0.008	0.1276	-0.0197	0.0444
47	2.2716	2.3456	0.0835	0.235	0.7078	0.6898	-0.6444
48	-0.1224	-0.1215	0.0146	0.000	-0.0148	-0.0008	-0.0022
49	-0.4255	-0.4229	0.0147	0.001	-0.0517	-0.0012	-0.0091
50	0.0417	0.0414	0.0165	0.000	0.0054	-0.0009	0.0020
51	3.1330	3.3620	0.1478	0.851	1.3999	1.3905	-1.3305
52	-1.3901	-1.3999	0.0549	0.056	-0.3374	0.2502	-0.2902
53	0.3739	0.3715	0.0738	0.006	0.1048	0.1014	-0.0941
54	0.8902	0.8889	0.0153	0.006	0.1109	0.0496	-0.0289
55	-0.1752	-0.1739	0.0299	0.000	-0.0305	0.0174	-0.0221
56	0.3344	0.3322	0.0198	0.001	0.0473	0.0325	-0.0250
57	0.3305	0.3284	0.0160	0.001	0.0419	-0.0055	0.0137
58	0.4114	0.4089	0.0143	0.001	0.0491	0.0107	-0.0009
59	0.3849	0.3825	0.0143	0.001	0.0461	0.0118	-0.0027
60	0.4718	0.4690	0.0144	0.002	0.0566	0.0072	0.0041
61	-0.3778	-0.3754	0.0217	0.002	-0.0559	-0.0411	0.0327
62	-1.1594	-1.1624	0.0171	0.012	-0.1532	-0.0885	0.0618
63	-1.3840	-1.3936	0.0365	0.036	-0.2711	-0.2410	0.2115
64	1.0586	1.0595	0.0143	0.008	0.1276	0.0272	-0.0019
65	1.0436	1.0443	0.2092	0.144	0.5371	-0.4801	0.5184
66	-1.0252	-1.0256	0.0221	0.012	-0.1541	0.0651	-0.0916
67	-1.2132	-1.2175	0.0182	0.014	-0.1656	-0.1042	0.0765
68	0.9321	0.9312	0.0154	0.007	0.1165	-0.0085	0.0314
69	1.2912	1.2976	0.0159	0.013	0.1649	-0.0202	0.0524
70	0.8110	0.8090	0.0144	0.005	0.0979	0.0100	0.0096

It is sometimes useful to write the equation in terms of deviations from x_M

$$\hat{y}^* = (b_0^* + b_1^* x_M) + b_1^* (x - x_M)$$
$$= b^* + b_1^* (x - x_M) \text{ where } b^* = b_0^* + b_1^* x_M.$$

The residuals from this fitted line are given by $e_i^* = (y_i - \hat{y}_i^*)$.

Example

Application of this robust fitting procedure to the NOI vs SUITES scatterplot yields the equation

$$\text{NOI} = -17,173.4 + 2650.4 \text{ SUITES}.$$

The robust fit therefore yields a higher slope and lower intercept than the least squares fit

$$\text{NOI} = -4872.0 + 2350.7 \text{ SUITES}.$$

The linear fit which was obtained after omitting the two major outliers (observations 16 and 28) was

$$\text{NOI} = -9645.8 + 2553.8 \text{ SUITES}.$$

Omitting the two major outliers from a least squares fit therefore results in intercept and slope estimates within the range of the estimates obtained from the robust linear fit and the least squares fit to the entire data set.

3.3.8 Residual Plots, Assumption Violations and Some Remedies

Residual Plots

In previous sections of this chapter scatterplots were employed to assist in determining the suitability of a linear fit to an observed data set. While scatterplots are useful for assessing linearity and the presence of outliers, they are not useful for assessing the validity of the assumptions required for the error term. To study the behavior of the error term the residuals determined in the previous section will be plotted. In particular, assumptions such as normality, homogeneity of variance and independence of the error terms need to be evaluated. *Residual plots* may also be used for checking for mis-specification of the functional form and for outliers, both of which have already been discussed in the previous section. Since the techniques already discussed are more suited to the study of mis-specification or outliers they will not be discussed here.

The distribution of the residuals can be studied using the techniques discussed in Chapter 2. Histograms and stem and leaf plots are useful for representing the distribution of residuals. The quantile-quantile plot can be used to assess normality. [The tests for normality outlined in Chapter 2 can be applied to the residuals from the linear fit. In large samples these tests tend to perform as if the residuals were the true error terms.] Departures from normality are not usually serious as long as the distribution of residuals is bell-shaped without outliers. If the Y observations are normally distributed the residuals should also be, unless there are other problems such as outliers, *autocorrelation, heteroscedasticity*, etc. Transformations to normality for the Y variables have already been discussed earlier in this chapter.

A variety of residual plots can be used to study the properties of the residuals. Usually the studentized residuals on the vertical axis are plotted against either the predicted values $\hat{y}$ or the explanatory variable X on the horizontal axis. Figure 3.29 illustrates several patterns of residual plots that might be obtained using the predicted values $\hat{y}$. Panels (b), (c) and (e) show heteroscedasticity. In (b) the variance is increasing with $\hat{y}$ or X and in (e) it is decreasing as X increases. In (c) the variance increases and then decreases. In Panels (a) and (d) a nonlinear relationship between Y and X is evident. In (d) there is also heteroscedasticity. In Panel (f) the residuals are plotted against time and the residual plot shows a cyclical pattern.

Heteroscedasticity

If heteroscedasticity exists among the error terms, the ordinary least squares estimators will still be unbiased but will no longer be minimum variance. More importantly, the estimates of the standard errors for the coefficients will be understated, and hence test statistics will tend to be too large. In some situations heteroscedasticity can be eliminated using weighted least squares, which is equivalent to a special transformation applied to both Y and X before determining the ordinary least squares fit. If for example the heteroscedasticity can be described by $v(u_i) = \sigma_u^2 = \sigma_v^2 X$ (similar to (b) in Figure 3.29), then by multiplying through the simple linear model by $1/\sqrt{X}$, the new error term will have variance σ_v^2. The linear model becomes

$$y_i/\sqrt{x_i} = \beta_0/\sqrt{x_i} + \beta_1 x_i/\sqrt{x_i} + u_i/\sqrt{x_i}.$$

Applying ordinary least squares to the transformed data is equivalent to weighting the squared deviations $(\hat{y}_i - y_i)^2$ by $1/x_i$. If the heteroscedasticity is of the form $v(u_i) = \sigma^2/x^2$ (similar to (e) in Figure 3.29) then we should multiply through by x.

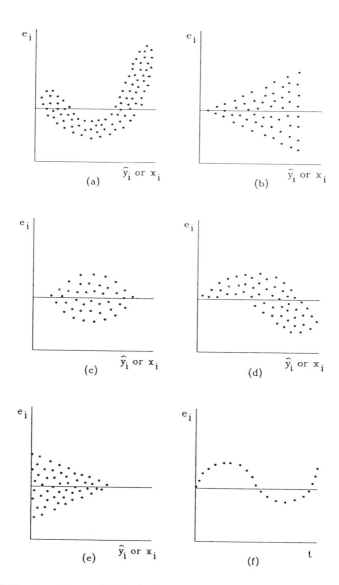

Figure 3.29. Some Example Residual Plots

Example

Data based on a sample of 116 real estate sales transactions was collected over a one year period in a particular region of a large city. Table 3.15 and Figure 3.30 show the relationship between observations on selling price (SELLP) and number of square feet (SQF) for 116 bungalows. The ob-

servations on RMS will be used in Section 3.4. From Figure 3.30 we can
see that the scatter of points is much larger at larger values of SQF. The
scatterplot suggests that the variance of the error term is related to X.
The least squares estimate of the regression line has the equation

$$SELLP = 2306.54 + 75.90 \, SQF,$$

with $R^2 = 0.50$, $s = 8972.56$ and coefficient standard errors of 8689.94 for
the intercept and 7.11 for the slope.

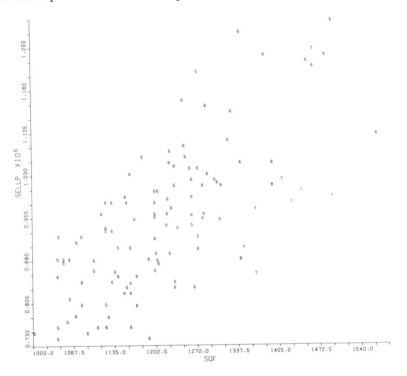

Figure 3.30. Scatterplot of SELLP vs. SQF

The residuals e_i, $i = 1, 2, \ldots, 116$ from the least squares fit were computed and the correlation coefficients between e_i^2 and each of SQF and
SQF^2 were obtained. These correlations were 0.33 in both cases. A correction for heteroscedasticity was carried out by using the two weighting schemes $1/\sqrt{SQF}$ and $1/SQF$ corresponding to the model assumptions
$v(u_i) = \sigma_v^2 SQF$ and $v(u_i) = \sigma_v^2 SQF^2$ respectively.

The regression with the weights $1/\sqrt{SQF}$ yielded the equation

$$SELLP = 1903.72 + 76.23 \, SQF,$$

Table 3.15. Real Estate Sales Data

OBS	SELLP	SQF	RMS	OBS	SELLP	SQF	RMS	OBS	SELLP	SQF	RMS	OBS	SELLP	SQF	RMS
1	85400	1365	7	30	98600	1221	6	59	90000	1140	5	88	109000	1320	6
2	79800	1170	6	31	100000	1205	6	60	103000	1160	6	89	99000	1490	7
3	82000	1160	6	32	85000	1170	6	61	75000	1089	6	90	130000	1491	6
4	95000	1306	6	33	98000	1200	6	62	76000	1040	5	91	83000	1154	6
5	80000	1120	5	34	93500	1120	6	63	96000	1200	6	92	89000	1224	6
6	85000	1040	6	35	93000	1130	6	64	104000	1257	6	93	96000	1220	6
7	84000	1130	6	36	92000	1270	5	65	94000	1260	5	94	99000	1152	6
8	84000	1232	6	37	98000	1120	6	66	76000	1106	6	95	101000	1392	8
9	96800	1364	7	38	91000	1072	6	67	76000	1120	6	96	124000	1480	6
10	95800	1260	6	39	98000	1130	6	68	74000	1190	6	97	92000	1042	5
11	101500	1302	6	40	84000	1080	5	69	83000	1264	6	98	95000	1167	6
12	74000	1040	5	41	88000	1060	6	70	83000	1232	6	99	116000	1246	6
13	101000	1278	6	42	101000	1232	6	71	83800	1160	6	100	100000	1440	7
14	102000	1408	7	43	100000	1200	6	72	88000	1190	6	101	110000	1564	8
15	107000	1225	6	44	92000	1080	6	73	105000	1392	6	102	105000	1224	6
16	90000	1160	6	45	122000	1460	6	74	96000	1220	6	103	114000	1325	6
17	88000	1040	5	46	87800	1100	6	75	123000	1450	6	104	121000	1270	5
18	98000	1424	7	47	85000	1140	6	76	87200	1206	6	105	88000	1340	8
19	104400	1232	6	48	108000	1248	6	77	82000	1150	6	106	104000	1270	6
20	89800	1270	6	49	77400	1050	5	78	87800	1204	6	107	125000	1460	7
21	102000	1298	6	50	103000	1286	6	79	105000	1340	6	108	75000	1001	5
22	128000	1340	6	51	106000	1250	6	80	76000	1160	6	109	97000	1227	6
23	124000	1380	6	52	77000	1056	5	81	86000	1200	5	110	77800	1124	6
24	91600	1200	6	53	85800	1135	6	82	89000	1202	6	111	78000	1070	6
25	106000	1180	6	54	96000	1280	6	83	95200	1278	6	112	93000	1120	5
26	91600	1200	6	55	86000	1100	5	84	99000	1260	6	113	95400	1200	6
27	102000	1260	6	56	81000	1060	5	85	115000	1284	6	114	101000	1308	6
28	94000	1219	6	57	84000	1080	6	86	93500	1237	6	115	96000	1113	6
29	98000	1154	6	58	80000	1080	6	87	88000	1050	5	116	90000	1345	7

with intercept and slope standard errors given by 8654.34 and 7.14 respectively. For the regression with the weights 1/SQF the equation obtained was

$$\text{SELLP} = 1565.84 + 76.51 \text{ SQF},$$

with standard errors 8637.83 and 7.19 for the intercept and slope respectively. For all three regressions the coefficient estimates are necessarily unbiased. In the presence of heteroscedasticity however the weighted least squares estimates of the standard error should also be unbiased estimates of the true standard error. In this case the estimates of the standard error of the coefficients did not vary greatly over the three different regressions. The use of the two weighted least squares procedures has simply placed different weights on the sample observations. The SAS procedure PROC REG was used to determine the three estimated lines.

The particular examples of heteroscedasticity discussed above involve simple functions of the explanatory variable. A more general approach to heteroscedasticity will be outlined in Chapter 4.

Autocorrelation

In time series data autocorrelation can exist, reflecting correlation among the error terms. Autocorrelation is a common occurrence with economic time series data. When autocorrelation exists, the ordinary least squares estimators are unbiased, but inefficient in that the variances of the estimators are no longer minimum. In addition, the standard errors for the ordinary least squares regression coefficients are incorrectly stated if the usual formulae are used. Alternative methods of estimation are therefore preferred when autocorrelation is present.

First Order Autocorrelation

The simplest form of autocorrelation is the first order model, where for the linear relationship

$$y_t = \beta_0 + \beta_1 x_t + u_t \qquad t = 1, 2, \ldots, n,$$

the error term u_t satisfies

$$u_t = \rho u_{t-1} + \varepsilon_t, \quad |\varepsilon_t| < 1, \quad t = 2, 3, \ldots, n,$$

where the ε_t are mutually independent and identically distributed with mean 0 and constant variance σ^2.

A statistical test for first order autocorrelation is provided by the *Durbin-Watson test* which is based on the ordinary least squares residuals e_t, $t = 1, 2, \ldots, n$, where $e_t = y_t - \hat{\beta}_0 - \hat{\beta}_1 x_t$. The test statistic is given by

$$d = \sum_{t=2}^{n} (e_t - e_{t-1})^2 \Big/ \sum_{t=1}^{n} e_t^2.$$

If there is no first order autocorrelation then d should in general be within a specific range. For a given α critical values of d can be obtained from tables. Tables for the Durbin-Watson statistic provide two critical points d_L and d_U. For positive alternative values of ρ if $d < d_L$ the hypothesis $H_0 : \rho = 0$ is rejected while if $d > d_U$ the hypothesis is accepted. If $d_L \leq d \leq d_U$ the test is inconclusive. For negative ρ, H_0 is rejected if $d > 4 - d_L$ and is accepted if $d < 4 - d_U$. If $4 - d_U \leq d \leq 4 - d_L$ the test is inconclusive.

If first order autocorrelation is present, the coefficient ρ can be estimated and a transformation can be used to remove the autocorrelation. The estimator of ρ is given by

$$r = \sum_{t=2}^{n} e_t e_{t-1} \Big/ \sum_{t=1}^{n} e_t^2.$$

This estimate of ρ is used to determine the transformations in the model (3.7) below.

$$\sqrt{1 - \rho^2}\, y_1 = \beta_0 \sqrt{1 - \rho^2} + \beta_1 \sqrt{1 - \rho^2}\, x_1 + \varepsilon_1 \tag{3.7}$$

$$(y_t - \rho y_{t-1}) = \beta_0 (1 - \rho) + \beta_1 (x_t - \rho x_{t-1}) + \varepsilon_t \qquad t = 2, \ldots, n.$$

This is easily carried out by defining new variables

$$z_1 = \sqrt{1 - r^2}\, y_1; \qquad z_t = y_t - r y_{t-1} \qquad t = 2, \ldots, n$$

$$w_1 = \sqrt{1 - r^2}\, x_1; \qquad w_t = x_t - r x_{t-1} \qquad t = 2, \ldots, n$$

$$v_1 = \sqrt{1 - r^2}; \qquad v_t = (1 - r) \qquad t = 2, \ldots, n.$$

New estimators $\tilde{b}_0$ and $\tilde{b}_1$ are obtained by fitting the multiple regression model (two explanatory variables and no intercept)

$$z_t = \beta_0 v_t + \beta_1 w_t + \varepsilon_t.$$

This method is often called the *Prais-Winsten* method. A better estimator can often be obtained by repeating the process. The estimators $\tilde{b}_0$ and $\tilde{b}_1$ are used to determine a new estimator of $r_{(1)}$, which is then employed to determine new estimators $\tilde{b}_{0(1)}$ and $\tilde{b}_{1(1)}$ using the new transformed equation with $r_{(1)}$ replacing r. This Prais-Winsten iterative process usually converges very quickly.

Example

An example of a simple linear regression model with autocorrelation is provided by the data given in Table 3.16. The dependent variable is the Canadian monthly treasury bill interest rate (TRSBILL) and the explanatory variable is the Canadian monthly consumer price index (CPI) for the period January 1971 to December 1973. The estimated simple linear regression model for this data is

$$\text{TRSBILL} = -12.176 + 0.366 \text{ CPI},$$

with $R^2 = 0.684$, $s = 0.615$ and estimated standard errors for intercept and slope 1.909 and 0.043 respectively. The Durbin-Watson statistic is 0.217 indicating significant positive autocorrelation. The residuals from this linear fit are also shown in Table 3.16. From the residuals shown in Table 3.16 it is clear that there is a strong tendency for adjacent residuals to have the same sign. The estimator of ρ based on these residuals is $r = 0.775$.

Using the Prais-Winsten procedure the estimated equation is given by

$$\text{TRSBILL} = -10.224 + 0.326 \text{ CPI},$$

with $s = 0.313$ and estimated standard errors for intercept and slope given by 3.236 and 0.072 respectively. The standard errors of the regression coefficients have almost doubled in size.

Cochrane-Orcutt Procedure

In data sets with a large number of observations the first equation of the transformation (3.7) can often be ignored. The model can then be estimated using the simple linear regression model. Ordinary least squares yields estimators of the intercept $\beta_0(1 - \rho)$ and the slope β_1. This method is usually referred to as the *Cochrane-Orcutt procedure*.

Higher Order Autoregressive Models

The autocorrelation present in the error term may be more complex than the first order model assumed above. A more general model assumes the error terms u_t satisfy

$$u_t = \rho_1 u_{t-1} + \rho_2 u_{t-2} + \cdots + \rho_k u_{t-k} + \varepsilon_t$$

which is called *k-th order autocorrelation*. For a given k the transformation used in the Prais-Winsten procedure can be generalized. Estimators of

Table 3.16.   Monthly Returns on Canadian Treasury Bills (TRSBILL) and Canadian Consumer Price Index (CPI) – January 1971 - December 1973

Year	Month	TRSBILL	CPI	Residuals
1971	Jan	4.59	41.2	1.43
	Feb	4.51	41.4	1.45
	Mar	3.30	41.5	0.42
	Apr	3.05	41.8	0.05
	May	3.06	42.0	0.10
	June	3.15	42.1	-0.03
	July	3.58	42.4	0.05
	Aug	3.88	42.7	0.48
	Sept	3.93	42.6	0.55
	Oct	3.79	42.7	0.30
	Nov	3.31	42.8	-0.15
	Dec	3.25	43.1	-0.50
1972	Jan	3.29	43.3	-0.54
	Feb	3.48	43.4	-0.24
	Mar	3.51	43.5	-0.51
	Apr	3.65	43.7	-0.42
	May	3.68	43.8	-0.19
	June	3.58	43.8	-0.62
	July	3.48	44.4	-0.76
	Aug	3.47	44.7	-0.58
	Sept	3.57	44.9	-0.77
	Oct	3.57	44.9	-0.69
	Nov	3.61	45.0	-0.92
	Dec	3.66	45.3	-1.20
1973	Jan	3.79	45.7	-0.76
	Feb	3.92	46.0	-0.76
	Mar	4.29	46.1	-0.51
	Apr	4.73	46.6	-0.04
	May	5.08	47.4	0.23
	June	5.40	47.4	0.18
	July	5.65	47.8	0.43
	Aug	6.03	48.4	0.74
	Sept	6.41	48.7	1.01
	Oct	6.51	48.8	1.31
	Nov	6.46	49.2	0.96
	Dec	6.38	49.5	0.52

the k autoregressive parameters $\rho_1, \rho_2, \ldots, \rho_k$ can be determined from the residuals from a simple linear regression fit. This procedure is employed by the SAS procedure AUTOREG. A more detailed discussion of this topic will be presented in the section on generalized least squares in Chapter 4. The topics of heteroscedasticity and autocorrelation are usually discussed more extensively in texts devoted to Econometrics. The reader may wish to consult Fomby, Hill and Johnson (1984) or Judge, Griffiths, Hill, Lütkepohl and Lee (1985).

Example

In the above example relating TRSBILL to CPI the AUTOREG procedure was fit assuming autocorrelation models with parameters varying from $k = 1, 2, \ldots, 8$. For $k = 1$ the results are equivalent to the Prais-Winsten procedure outlined above. For $k = 2, \ldots, 8$ the results are summarized in Table 3.17. There appears to be very little change in the model parameter estimates as a result of the change in the order of the autoregressive model.

Table 3.17. Estimated Parameters for Linear Relationship Between TRSBILL and CPI for Various Autoregressive Models for the Error Term

Order of Autocorrelation	Intercept		Slope	
k	Estimate	Standard Error	Estimate	Standard Error
1	-10.224	3.236	0.326	0.072
2	-10.956	2.616	0.340	0.058
3	-11.309	2.960	0.349	0.066
4	-11.103	2.932	0.344	0.065
5	-11.509	3.068	0.354	0.068
6	-11.417	3.112	0.352	0.069
7	-11.257	3.153	0.348	0.070
8	-10.872	3.125	0.339	0.070

3.3.9 Measurement Error in the Regression Model

Measurement Error

In some situations it is unrealistic to assume that a random variable X is measured without error. If the *measurement error* in X is a random variable V with mean zero, and is independent of the true value X^*, then the sample mean remains an unbiased estimator of the population mean. Denoting the observed value by x and the true value by x^* we have the relationship

$$x_i = x_i^* + v_i.$$

The reader should note here that v_i and x_i are correlated even though x_i^* and v_i are independent. The observation x_i will tend to be larger (smaller) when v_i is positive (negative). Denoting the means and variances of X^*

and V by $(\mu_{x^*}, \sigma_{x^*}^2)$ and $(0, \sigma_v^2)$ respectively, the mean and variance of X are therefore given by (μ_x, σ_x^2) where $\mu_x = \mu_{x^*}$ and $\sigma_x^2 = (\sigma_{x^*}^2 + \sigma_v^2)$. The measurement error in X therefore causes the variance of X to be larger than the variance of X^*.

The *reliability ratio* ϕ is the proportion of the total variance not due to measurement error and is given by $\sigma_{x^*}^2/(\sigma_{x^*}^2 + \sigma_v^2)$. If the ratio, ϕ, is known or if σ_v^2 is known then an unbiased estimator of $\sigma_{x^*}^2$ can be obtained from the sample variance s_x^2. The estimator is given by $\hat{\sigma}_{x^*}^2 = \phi s_x^2$ or $\hat{\sigma}_{x^*}^2 = (s_x^2 - \sigma_v^2)$.

In the bivariate case a second random variable Y may also contain measurement error U. If so, we assume the relationship is given by

$$y_i = y_i^* + u_i,$$

where y_i^* is an observed value of Y^* the true value, with mean μ_{y^*} and variance $\sigma_{y^*}^2$, and u_i is an observed value of U which is independent of Y^* with mean 0 and variance σ_u^2. The mean and variance of Y are given by $\mu_y = \mu_{y^*}$ and $\sigma_y^2 = (\sigma_{y^*}^2 + \sigma_u^2)$. The reliability ratio for Y is denoted by

$$\delta = \sigma_{y^*}^2/(\sigma_{y^*}^2 + \sigma_u^2).$$

Measurement Error in Bivariate Analysis

In the bivariate case it is usually assumed that the measurement errors V and U are mutually independent of each other and as well are independent of both X^* and Y^*. The joint distribution of $\begin{bmatrix} X \\ Y \end{bmatrix}$ therefore has mean vector $\begin{bmatrix} \mu_x \\ \mu_y \end{bmatrix}$ and covariance matrix

$$\begin{bmatrix} \sigma_{x^*}^2 + \sigma_v^2 & \sigma_{x^*y^*} \\ \sigma_{x^*y^*} & \sigma_{y^*}^2 + \sigma_u^2 \end{bmatrix},$$

since by the independence of U and V, $\sigma_{xy} = \sigma_{x^*y^*}$. The sample covariance s_{xy} therefore provides an unbiased estimator of σ_{xy}.

The correlation coefficient between X and Y is given by $\rho_{xy} = \sigma_{xy}/\sigma_x\sigma_y = \sigma_{x^*y^*}/\sqrt{(\sigma_{x^*}^2 + \sigma_v^2)(\sigma_{y^*}^2 + \sigma_u^2)}$ and hence $\rho_{xy} < \rho_{x^*y^*} = \sigma_{x^*y^*}/\sigma_{x^*}\sigma_{y^*}$ where $\rho_{x^*y^*}$ is the correlation between X^* and Y^*. The estimator $s_{xy}/s_x s_y$ is therefore a consistent estimator of ρ_{xy} but is not a consistent estimator of $\rho_{x^*y^*}$. Using the sample correlation coefficient of the measured variables therefore leads to an underestimate of the true correlation coefficient. If the reliability ratios ϕ and δ are known a consistent estimator of $\rho_{x^*y^*}$ is given by $s_{xy}/s_x s_y \sqrt{\phi\delta}$. Thus $\rho_{x^*y^*}$ tends to be understated by a factor $\sqrt{\phi\delta}$ which is the geometric mean of the two reliability ratios.

Estimator for the Simple Linear Regression Model With Measurement Error

For the simple linear regression model in 3.3.1 the relationship between Y and X was defined by

$$Y = \beta_0 + \beta_1 X + U$$

where the error term U is assumed to have mean 0 variance σ_u^2 and to be independent of X. If the random variable Y contains measurement error as outlined above a second error term independent of X is added on to the right-hand side of the model. These two error terms can without loss of generality be conveniently represented by a single error term. The presence of measurement error in Y therefore increases the value of σ_u^2, but does not have an impact on the consistency or bias of the conventional estimators of β_0 and β_1.

If there is measurement error in X as defined above then the model becomes

$$Y = \beta_0 + \beta_1 X + (U - \beta_1 V)$$

and since V is correlated with X the error term in the model is no longer independent of X. The conventional ordinary least squares estimator $b_1 = s_{xy}/s_x^2$, is therefore an inconsistent downward biased estimator of β_1, since $E[s_{xy}] = \sigma_{x^*y^*}$, $E[s_x^2] = (\sigma_{x^*}^2 + \sigma_v^2)$ and $\beta_1 = \sigma_{x^*y^*}/\sigma_{x^*}^2$. Similarly, $b_0 = \bar{y} - b_1\bar{x}$, will be inconsistent and upward biased for β_0. If σ_v^2 is known, however, the estimators $\tilde{b}_1 = s_{xy}/(s_x^2 - \sigma_v^2)$ and $\tilde{b}_0 = \bar{y} - \tilde{b}_1\bar{x}$ are consistent estimators of β_1 and β_0 respectively.

Although the estimator $b_1 = s_{xy}/s_x^2$ is a biased estimator of β_1, its bias is equal to a multiplicative constant times β_1. It is therefore possible to test the hypothesis $H_0 : \beta_1 = 0$ using a t distribution based on the observed data. Other inferences regarding β_1 or β_0 cannot be made using the conventional methods.

If the reliability ratio $\phi = \sigma_{x^*}^2/(\sigma_{x^*}^2 + \sigma_v^2)$ is known the estimators $\tilde{b}_1 = s_{xy}/\phi s_x^2 = \hat{\beta}_1/\phi$ and $\tilde{b}_0 = (\bar{y} - \tilde{b}_1\bar{x}) = (\bar{y} - \hat{\beta}_1\bar{x}/\phi)$ are unbiased for β_1 and β_0 respectively. Inferences for β_1 can be made using the statistic $(\tilde{b}_1 - \beta_1)/[s_e^2/ns_x^2]^{1/2}$ which has a t distribution with $(n-2)$ d.f. Inferences for β_0 can be made using the statistic

$$(\tilde{b}_0 - \beta_0)/\left[\frac{1}{n} s_{e^*}^2 + \bar{x}^2 s_e^2/ns_x^2\right]^{1/2}$$

which has a standard normal distribution in large samples. The mean square error variances are given by

$$s_e^2 = \sum_{i=1}^{n}(y_i - b_0 - b_1 x_i)^2/(n-2) \quad \text{and}$$

$$s_{e^*}^2 = \sum_{i=1}^{n}(y_i - \tilde{b}_0 - \tilde{b}_1 x_i)^2/(n-2).$$

If the ratio $\lambda = \sigma_v^2/\sigma_u^2$ between the two error variances is known, a consistent estimator, b_1, of β_1 is obtainable by solving the quadratic equation

$$\beta_1^2 \lambda s_{xy} + \beta_1(s_x^2 - \lambda s_y^2) - s_{xy} = 0$$

where the sign of b_1 must be the sign of s_{xy}. The intercept β_0 is then estimated using $b_0 = \bar{y} - \hat{\beta}_1 \bar{x}$. Expressions for the standard errors of b_0 and b_1 are quite complex and are given in Chapter 1 of Fuller (1987).

The alternative linear model introduced in 3.1.2 can be viewed as a special case of the simple linear regression model with measurement error. The true relation $Y^* = \beta_0 + \beta_1 X^*$ is called a functional relation and is assumed to generate the model $Y = \beta_0 + \beta_1 X^* + U$ and $X = X^* + V$. A more detailed discussion of the bivariate measurement error model is provided in Chapter 1 of Fuller (1987) and also in Kmenta (1986).

Example

In a study of the performance of the R.C.M.P. force a population of 18 detachments was surveyed. The police officers were asked a number of questions regarding the STRESS they felt in carrying out their duties. The scores on eighteen stress items were added together to determine a STRESS index. Samples of residents of each community were asked to respond to questions regarding the nature of the crime situation in their community. The 9 items were added together to form a CRIME index. Table 3.18 reports the sample size, means and variances on the two indices for the eighteen detachments.

To determine the relationship between STRESS experienced by officers and the CRIME situation as perceived by the residents the sample mean vector and covariance marix were determined to be

Covariance Matrix

	Means			STRESS	CRIME		
STRESS	$\begin{bmatrix} 5.30 \\ 2.01 \end{bmatrix}$	and	$\begin{bmatrix}$	0.715	0.096	$\end{bmatrix}$	STRESS
CRIME				0.096	0.032		CRIME

respectively.

Using this data the estimated linear relationship between average stress, $\overline{\text{STRESS}}$, and average crime, $\overline{\text{CRIME}}$, was determined to be $\overline{\text{STRESS}} = \underset{(1.84)}{0.672} + \underset{(0.91)}{2.967}\ \overline{\text{CRIME}}$ with $R^2 = 0.40$ (the numbers in brackets are the coefficient standard errors).

Since the linear relationship being estimated is between true detachment means the use of sample means introduces a problem of measurement error. To estimate the detachment variances for the means, the detachment variances were divided by the sample size. The estimate of the measurement error variance was then obtained by averaging the eighteen variances

Table 3.18. Police Detachment Data

Det. No.	Officer Data			Community Data		
	Sample Size	Stress Mean	Stress Variance	Sample Size	Crime Mean	Crime Variance
1	8	6.28	2.45	64	2.19	0.39
2	12	4.05	1.16	67	1.94	0.23
3	8	5.30	3.38	80	2.00	0.27
4	16	6.86	7.33	57	2.37	0.52
5	8	5.97	2.18	70	1.87	0.20
6	13	4.62	0.93	70	1.82	0.19
7	11	6.79	3.42	76	2.02	0.36
8	4	5.21	1.61	70	2.00	0.35
9	5	4.59	1.89	75	1.72	0.16
10	19	5.69	3.93	73	2.16	0.27
11	6	5.43	1.62	71	2.40	0.43
12	6	4.50	0.77	80	1.85	0.22
13	5	5.94	3.47	64	2.07	0.33
14	6	4.81	1.08	79	1.96	0.34
15	5	4.28	0.54	72	1.84	0.30
16	5	5.01	3.45	67	1.98	0.27
17	10	5.73	3.70	75	2.09	0.36
18	15	4.41	3.96	67	1.97	0.27

for the means. For the STRESS mean the estimate of measurement error variance was 0.32 and for CRIME the estimated measurement error variance was 0.004. Using these estimates of measurement error variance the estimated linear relationship was revised by using

$$b_1 = 0.096/(0.032 - 0.004) = 3.429$$

$$b_0 = 5.30 - 3.429(2.01) = -0.41.$$

The estimate of ρ^2 was revised to $(0.096)^2/(0.032 - 0.004)(0.715 - 0.32) = 0.84$. The estimate of the coefficient of determination has more than doubled. The effect of the measurement error is a downward bias on the estimate of the strength of the linear association. The true linear association between average STRESS and average CRIME has a larger slope than the expected value of the slope obtained from the measurement data using ordinary least squares.

Grouping Method

An alternative approach to the least squares procedure for estimating a linear relationship in the presence of measurement error is to use a *method of grouping* similar to the method introduced earlier for robust regression. The values of X are ordered to obtain $(x_{(1)}, x_{(2)}, \ldots, x_{(n)})$ and the X's divided into three groups. For the points in the lowest group, the means $(\bar{x}_L, \bar{y}_L)$ are determined and similarily, for the highest group, $(\bar{x}_U, \bar{y}_U)$ are the means. The slope is estimated by

$$b_1^+ = (\bar{y}_U - \bar{y}_L)/(\bar{x}_U - \bar{x}_L)$$

and the intercept by

$$b_0^+ = \bar{y} - b_1^+ \bar{x}.$$

These estimates should eliminate bias due to measurement error in X.

If the x values are approximately bell-shaped the lower and upper groups should be allocated approximately 25% of the data each, leaving 50% of the data for the left out middle group. In general, if the distribution of X has more observations in the tail relative to a normal, then a larger percentage of the data should be included in the group corresponding to that tail. For a rectangular distribution an equal three-way split is recommended, while for a U-shaped X distribution the two extreme groups should contain 80% of the data.

3.4 Regression and Correlation in a Multivariate Setting

In most applications of simple linear regression and correlation the variables X and Y being studied represent only a subset of the total number of variables in the data matrix. In most settings there are a number of other variables that are believed to have an influence on both X and Y and hence may have an impact on the relationship between X and Y. A variable that has an important effect and yet is not included in the analysis is called a *lurking variable*. Some lurking variables may be in the data matrix and some may not. Some lurking variables may have been controlled by the researcher either intentionally or unintentionally when collecting the sample. In some cases it may be possible to argue from some underlying theory that the relationship between X and Y is due only to the relationship that both X and Y have with some other variable Z. In other words the *partial correlation* between X and Y is negligible after controlling for Z. In the case of the ratio variables $W = X/Z$ and $V = Y/Z$ the correlation between W and V is necessarily positive when the variables X, Y and Z are mutually uncorrelated. In other cases it is possible for the interaction between X and the missing Z to significantly change the relationship between Y and X. The degree of linear association between the variable Y and the two variables X and Z as measured by the *multiple correlation coefficient* also reflects on the relationship between Y and X. It is possible for the joint relationship described by the multiple correlation to be stronger than the sum of the individual correlations. In this section we shall outline procedures for measuring the impact of a third variable Z on both regression and correlation relationships between X and Y.

3.4.1 Partial Correlation Coefficient

A useful approach to the study of the relationship between two variables X and Y, in the presence of a third variable Z, is to determine the partial correlation between X and Y after controlling for the impact of Z. Before introducing this concept some notation from simple linear regression must be introduced.

Joint Distribution of Three Variables and Partial Correlation

Assume that the three variables X, Y and Z are jointly normally distributed with mean vector, covariance matrix and correlation matrix given by

$$
\begin{bmatrix} \mu_x \\ \mu_y \\ \mu_z \end{bmatrix}, \quad
\begin{bmatrix} \sigma_x^2 & \sigma_{xy} & \sigma_{xz} \\ \sigma_{xy} & \sigma_y^2 & \sigma_{yz} \\ \sigma_{xz} & \sigma_{yx} & \sigma_z^2 \end{bmatrix} \quad \text{and} \quad
\begin{bmatrix} 1 & \rho_{xy} & \rho_{xz} \\ \rho_{xy} & 1 & \rho_{yz} \\ \rho_{xz} & \rho_{yz} & 1 \end{bmatrix} \quad \text{respectively.}
$$

To study the impact of Z on the linear relationship between the variables X and Y we first examine the simple linear relationship between Y and Z and X and Z separately. From the theory presented earlier in this chapter, the simple linear regression of Y on Z yields a regression function $\beta_0 + \beta_1 Z$, where β_0 and β_1 are given by

$$\beta_0 = \mu_y - \beta_1 \mu_z$$

$$\beta_1 = \sigma_{zy}/\sigma_z^2.$$

The regression model relating Y to Z is given by

$$Y = \beta_0 + \beta_1 Z + U_{y \cdot z} \tag{3.8}$$

where $U_{y \cdot z}$ is assumed to be uncorrelated with Z and represents the part of Y that is not linearly related to Z. By changing Z by 1 unit the expected change in Y is β_1 units. The variance of Y, σ_y^2, has been divided into two components: $\beta_1^2 \sigma_z^2$, the variation explained by Z, and $\sigma_{y \cdot z}^2 = V[U_{y \cdot z}] = \sigma_y^2(1 - \rho_{yz}^2)$, the variation in Y not explained by Z. Recall that $\rho_{yz} = \beta_1 \sigma_z/\sigma_y$ is the correlation coefficient relating Y and Z.

In a similar fashion the simple linear regression of X on Z yields the regression function $\gamma_0 + \gamma_1 Z$, where γ_0 and γ_1 are given by

$$\gamma_0 = \mu_x - \gamma_1 \mu_z$$

$$\text{and} \quad \gamma_1 = \sigma_{xz}/\sigma_z^2$$

and the regression model for X on Z by

$$X = \gamma_0 + \gamma_1 Z + U_{x \cdot z}. \tag{3.9}$$

The variance of $U_{x \cdot z} = \sigma^2_{z \cdot x} = \sigma^2_z (1 - \rho^2_{xz})$ where $\rho_{xz} = \gamma_1 \sigma_x / \sigma_z$ is the correlation between X and Z. The variance in Z can be partitioned as $\sigma^2_z = \gamma^2_1 \sigma^2_x + \sigma^2_{z \cdot x}$.

The error terms $U_{x \cdot z}$ and $U_{y \cdot z}$ from the above two regressions represent the parts of X and Y respectively that are uncorrelated with Z. The correlation coefficient between $U_{x \cdot z}$ and $U_{y \cdot z}$ is called the partial correlation between Y and X controlling for Z and is denoted by $\rho_{yx \cdot z}$. This correlation represents the degree of linear association between Y and X after removing the effect of Z. This *partial correlation coefficient* can be written explicitly in the form

$$\rho_{yx \cdot z} = \frac{\mathrm{Cov}(U_{x \cdot z} U_{y \cdot z})}{\sigma_{y \cdot z} \sigma_{x \cdot z}} = \frac{\rho_{yx} - \rho_{yz} \rho_{xz}}{\sqrt{1 - \rho^2_{yz}} \sqrt{1 - \rho^2_{xz}}}. \tag{3.10}$$

Similarly the partial correlation between Y and Z controlling for X is given by

$$\rho_{yz \cdot x} = \frac{\rho_{yz} - \rho_{yx} \rho_{zx}}{\sqrt{1 - \rho^2_{yx}} \sqrt{1 - \rho^2_{zx}}}. \tag{3.11}$$

Sample Partial Correlation

Given a sample of n observations on the variables X, Y and Z the regression relationships can be estimated using least squares as outlined in Section 3.3.1. The sample mean vector, covariance matrix and correlation matrix are denoted by

$$\begin{bmatrix} \bar{x} \\ \bar{y} \\ \bar{z} \end{bmatrix}, \quad \begin{bmatrix} s^2_x & s_{yx} & s_{xz} \\ s_{yx} & s^2_y & s_{yz} \\ s_{xz} & s_{yz} & s^2_z \end{bmatrix} \quad \text{and} \quad \begin{bmatrix} 1 & r_{yx} & r_{xz} \\ r_{yx} & 1 & r_{yz} \\ r_{xz} & r_{yz} & 1 \end{bmatrix},$$

where the elements of the mean vector are the usual sample means and the elements of the covariance matrix are the sums of squares and sums of cross products divided by $(n-1)$.

The estimates for the regression coefficients are given by

$$b_0 = \bar{y} - b_1 \bar{z},$$
$$b_1 = s_{yz} / s^2_z$$
$$\text{and} \quad g_0 = \bar{x} - g_1 \bar{z},$$
$$g_1 = s_{xz} / s^2_z.$$

The residuals for the two regressions are

$$e_{y \cdot z_i} = y_i - b_0 - b_1 z_i, \tag{3.12}$$

$$e_{x \cdot z_i} = x_i - g_0 - g_1 z_i, \tag{3.13}$$

and both sets of residuals are by construction uncorrelated with the z_i. The error sums of squares in each of the regressions

$$s_{y \cdot z}^2 = \frac{(n-1)}{(n-2)} s_y^2 (1 - r_{yz}^2)$$

and $$s_{x \cdot z}^2 = \frac{(n-1)}{(n-2)} s_x^2 (1 - r_{xz}^2)$$

provide unbiased estimators of $\sigma_{y \cdot z}^2$ and $\sigma_{x \cdot z}^2$ respectively.

The *sample partial correlation* between Y and X after controlling for Z is determined from the Pearson correlation between $e_{y \cdot z}$ and $e_{x \cdot z}$. This correlation coefficient can be determined from the sample correlation coefficients r_{yx}, r_{yz} and r_{xz} using

$$r_{yx \cdot z} = \frac{r_{yx} - r_{yz} r_{xz}}{\sqrt{(1 - r_{yz}^2)(1 - r_{xz}^2)}}. \tag{3.14}$$

This expression for $r_{yx \cdot z}$ can be obtained from (3.10) by replacing each population correlation coefficient with the corresponding sample correlation coefficient.

Similarly the sample partial correlation coefficient between Z and Y controlling for X is given by

$$r_{zy \cdot x} = \frac{r_{zy} - r_{yx} r_{zx}}{\sqrt{(1 - r_{yx}^2)(1 - r_{zx}^2)}}.$$

Inference for Partial Correlation

The hypothesis $H_0 : \rho_{yx \cdot z} = 0$ can be tested using the statistic

$$t = \frac{r_{yx \cdot z} \sqrt{n - 3}}{\sqrt{1 - r_{yx \cdot z}^2}}$$

which under joint normality has a t distribution with $(n - 3)$ d.f. if H_0 is true. The similarity between this statistic and the t-statistic used for the simple correlation coefficient earlier in this chapter should be noted.

Example

For the sample of 116 real estate sales introduced in Section 3.3.8, Table 3.15, a sample correlation matrix was obtained relating the variables selling price, SELLP; square feet, SQF; and number of rooms, RMS. The sample correlation matrix is given by

$$\begin{array}{c} \\ \text{SELLP} \\ \text{SQF} \\ \text{RMS} \end{array} \begin{array}{ccc} \text{SELLP} & \text{SQF} & \text{RMS} \\ \begin{bmatrix} 1.000 & 0.707 & 0.262 \\ 0.707 & 1.000 & 0.620 \\ 0.262 & 0.620 & 1.000 \end{bmatrix} \end{array}. \tag{3.15}$$

Applying the expression (3.14) the partial correlation coefficient relating SELLP and SQF after controlling for RMS is given by 0.719 while the partial correlation coefficient between SELLP and RMS after controlling for SQF is -0.318. While the correlation between SELLP and SQF was not affected by the variable RMS the correlation between SELLP and RMS was changed substantially after controlling for SQF. The positive correlation of 0.262 between SELLP and RMS if SQF is not held constant indicates a weak positive relationship. If however SQF is held fixed the correlation between SELLP and RMS becomes weak and negative. It would seem that a house with more rooms does not necesarily mean a higher price unless the overall size of the house is also larger. Small rooms appear to detract from the value of the house.

Some Relationships Among ρ_{xz}, ρ_{yz} and ρ_{xy}

It is of interest to examine the expression (3.10) for $\rho_{yx \cdot z}$ to determine how the three correlations ρ_{xz}, ρ_{yz} and ρ_{yx} interact to produce $\rho_{yx \cdot z}$. The numerator of the expression determines the sign of $\rho_{yx \cdot z}$ while the denominator has a large impact on the magnitude of $\rho_{yx \cdot z}$. The figure below compares ρ_{yx} to the product $\rho_{yz} \rho_{xz}$ and shows the impact on the sign of $\rho_{yx \cdot z}$.

If ρ_{yx} is positive and ρ_{xz} and ρ_{yz} are opposite in sign then $\rho_{yx \cdot z}$ must be positive. If ρ_{yx} is negative and ρ_{xz} and ρ_{yz} have the same signs then $\rho_{yx \cdot z}$ is negative. If ρ_{yx} is positive and ρ_{xz} and ρ_{yz} have the same sign then the sign of $\rho_{yx \cdot z}$ depends on the relative magnitudes of ρ_{yx} and the product $\rho_{xz} \rho_{yz}$. Similarly for the case when $\rho_{yx} < 0$ and ρ_{xz} and ρ_{yz} are opposite in sign.

The denominator of the expression for $\rho_{yx \cdot z}$ decreases as ρ_{xz}^2 and ρ_{yz}^2 increase. For a given ρ_{yx} the magnitude of $\rho_{yx \cdot z}$ increases with the magnitude of ρ_{xz} and ρ_{yz}. Thus if the correlations between Z and X and Y and Z are relatively large then the partial correlation between Y and X after controlling for Z will also tend to be relatively large.

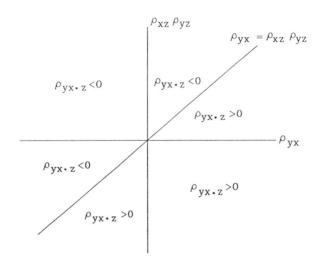

Figure 3.31. Relationship Between $\rho_{yx\cdot z}$ and ρ_{yx} and $\rho_{xz}\rho_{yz}$

Example

In the real estate example discussed above let $Y = \text{SELLP}$, $X = \text{RMS}$ and $Z = \text{SQF}$. Replacing the population correlations by their sample counterparts the values of r_{yx}, r_{xz} and r_{yz} are given by 0.262, 0.620 and 0.707 respectively. Since $0.262 = r_{yx} < r_{xz}r_{yz} = 0.438$ then $r_{yx\cdot z}$ is negative.

In the garbage disposal cost study discussed in Section 3.2.1 the sample correlations relating total cost, COST, number of apartments, APTS, and number of homes, HOMES, yielded the following correlation matrix

	COST	HOMES	APTS
COST	1.000	0.582	-0.216
HOMES	0.582	1.000	0.254
APTS	-0.216	0.254	1.000 .

The partial correlation between COST and HOMES controlling for APTS is 0.675 while the partial correlation between COST and APT controlling for HOMES is -0.463. In this case the result of controlling for the other variable results in an increase in the magnitude of the correlation.

In studies of adults it is not uncommon to find a positive correlation between income and both education and age but a negative correlation between age and education. As a result the partial correlations between income and age and income and education will be larger than the simple correlations between these variables. In this case the negative sign on the correlation between age and education causes the partial correlations between income and education controlling for age and between income and age controlling for education to increase.

A Graphical Illustration

In Figure 3.32 the two panels (a) and (b) show the overall linear relationship between Y and X, and also show the linear relationship between Y and X when Z is held fixed at $Z = z_1$, $Z = z_2$, and $Z = z_3$. In panel (a) the partial correlation between Y and X has the same sign as the simple correlation. In panel (b) the simple correlation between Y and X is negative however the partial correlation is positive when Z is fixed.

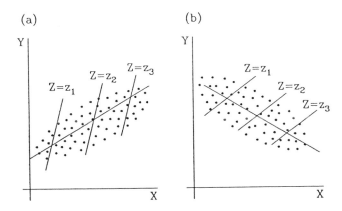

Figure 3.32. Impact of Variable Z on Linear Relationship Between Y and X

Suppressor Variables

The expression (3.10) for $\rho_{yx \cdot z}$ indicates that if ρ_{xz} is zero the expression for $\rho_{yx \cdot z}$ is given by

$$\rho_{yx \cdot z} = \rho_{yx} / \sqrt{1 - \rho_{yz}^2}.$$

Similarly if $\rho_{yz} = 0$, $\rho_{yx \cdot z} = \rho_{yx} / \sqrt{1 - \rho_{xz}^2}$. In this case $\rho_{yx \cdot z}$ is necessarily larger in magnitude than ρ_{yx}.

   In general if $\rho_{yx \cdot z}$ is larger in magnitude than ρ_{yx} we say that the variable Z is a *suppressor variable* because its relationship with Y and X affects the correlation between Y and X. The variable Z suppresses the variation in Y related to Z. This concept will be discussed further in connection with regression in the next section.

Possible Range for ρ_{xy} Given ρ_{xz} and ρ_{yz}

An interesting question is the potential range of values of ρ_{yx} given ρ_{xz} and ρ_{yz}. The expression for $\rho_{yx \cdot z}$ permits us to examine this question. Since the range of $\rho_{yx \cdot z}$ is $[-1, 1]$ we have that

$$-1 \leq \frac{\rho_{yx} - \rho_{yz}\rho_{xz}}{\sqrt{1 - \rho_{yz}^2}\sqrt{1 - \rho_{xz}^2}} \leq 1.$$

For simplicity assume $\rho_{yz} = \rho_{xz} = \rho$. From this expression we have the range of values for ρ_{yx} given by

$$2\rho^2 - 1 \leq \rho_{yx} \leq 1.$$

For instance if $\rho_{yx} = 0$, $\rho = \sqrt{0.5} = 0.71$ and hence it is possible for $\rho_{yz} = \rho_{xz} = 0.71$ and still have $\rho_{yx} = 0$. Therefore a relatively large positive correlation between both Z and Y and X and Z does not guarantee that there will be any positive correlation between Y and X. This is often a suprising result for users of correlation coefficients.

As an example of this surprising phenomenon suppose you are told that the correlation between income and education level is 0.50 and that the correlation between income and age is 0.50. What does this imply about the relationship between age and education? Nothing! From the above we can see that the correlation between age and education could be zero. In fact, as suggested above, in some populations this correlation is negative.

This relationship can be seen geometrically using the n-dimensional space representation introduced in 3.3.5. The three variables can be shown in mean corrected form as three vectors in n-dimensional space. The angles between $\mathbf{x}^*$ and $\mathbf{z}^*$ and $\mathbf{x}^*$ and $\mathbf{y}^*$ must be $45°$ and the angle between $\mathbf{y}^*$ and $\mathbf{z}^*$ must be $90°$.

From Figure 3.33, given that the correlation between X and Z is $\rho_{xz} = \text{Cos } \theta$, and between Y and Z is $\rho_{yz} = \text{Cos } \theta$, the angle between $\mathbf{y}^*$ and $\mathbf{x}^*$ can range between $0°$ and $2\theta°$. The correlation between Y and X can therefore range between $\rho_{yx} = \text{Cos } 0° = 1$ and $\rho_{yx} = \text{Cos } 2\theta$. If $\theta = 45°$ the correlation between Y and X can be 0. A similar diagram is given in Mulaik (1972).

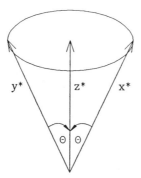

Figure 3.33. Correlation Relationships Among X, Y and Z

3.4.2 Partial Correlation and Regression

Partial Regression Coefficients

In this section we are interested in determining the impact of adding a second explanatory variable Z to a regression of Y on X. Consider the regression model

$$Y = \beta_0^* + \beta_1^* X + \beta_2^* Z + U_{y \cdot xz}.$$

The regression coefficient for Z, β_2^*, in the regression of Y on X and Z, is related to the partial correlation coefficient $\rho_{yz \cdot x}$ and hence to the regression of $U_{y \cdot x}$ on $U_{z \cdot x}$. The regression coefficient β_2^* is given by

$$\beta_2^* = \rho_{yz \cdot x} \frac{\sigma_{y \cdot x}}{\sigma_{z \cdot x}} = \frac{\rho_{yz} - \rho_{yx}\rho_{zx}}{\sqrt{1 - \rho_{yx}^2}\sqrt{1 - \rho_{zx}^2}} \cdot \frac{\sigma_y \sqrt{1 - \rho_{yx}^2}}{\sigma_z \sqrt{1 - \rho_{zx}^2}}$$

$$= \frac{\rho_{yz} - \rho_{yx}\rho_{zx}}{(1 - \rho_{zx}^2)} \frac{\sigma_y}{\sigma_z}. \tag{3.16}$$

If $\rho_{yz \cdot x}$ is relatively large it would be of interest to know the contribution that could be made if the variable Z was added to the model relating Y to X. If X, Y and Z are jointly normal it can be shown that the coefficients β_0^* and β_1^* are given by

$$\begin{bmatrix} \beta_0^* \\ \beta_1^* \end{bmatrix} = \begin{bmatrix} \beta_0 \\ \beta_1 \end{bmatrix} - \begin{bmatrix} \gamma_0 \\ \gamma_1 \end{bmatrix} \beta_2^*, \tag{3.17}$$

where β_2^* is defined above by (3.16) and β_0, β_1, γ_0 and γ_1 are obtained from (3.8) and (3.9) by interchanging the roles of X and Z. Therefore the marginal impact on Y of a one unit change in Z is given by β_2^* when X is assumed to be fixed. Similarly the marginal impact on Y of a one unit change in X is β_1^* when Z is fixed. The regression coefficients β_1^* and β_2^* are therefore *partial regression coefficients*. The marginal impact of X on Y in this model can be compared to β_1 which is the impact of X on Y in the simple linear regression model. The difference between the two coefficients β_1^* and β_1 represents the effect on the coefficient of X of holding Z fixed. Similar comments can be made regarding the difference between β_0^* and β_0.

As in the case of β_2^* the regression coefficient β_1^* can also be expressed in terms of a partial correlation coefficient

$$\beta_1^* = \frac{\rho_{xy} - \rho_{xz}\rho_{yz}}{\sqrt{1 - \rho_{xz}^2}\sqrt{1 - \rho_{yz}^2}} \cdot \frac{\sigma_{y \cdot z}}{\sigma_{x \cdot z}}$$

$$= \frac{\rho_{xy} - \rho_{xz}\rho_{yz}}{(1 - \rho_{zx}^2)} \cdot \frac{\sigma_y}{\sigma_x}.$$

Sample Relationships

For the sample of n observations on X, Y and Z the partial regression coefficient β_2^* defined in (3.16) is estimated by regressing the residuals $e_{y \cdot x}$ on the residuals $e_{z \cdot x}$ where $e_{y \cdot x}$ and $e_{z \cdot x}$ are determined from (3.12) and (3.13) by interchanging the roles of X and Z. The formula for the resulting estimated regression coefficient b_2 is obtained by replacing population parameters by their sample counterparts in (3.16) and is given by

$$b_2 = \frac{r_{yz} - r_{yx}r_{zx}}{(1 - r_{zx}^2)} \frac{s_y}{s_x} = r_{yz \cdot x} \frac{\sqrt{1 - r_{yx}^2}}{\sqrt{1 - r_{zx}^2}} \frac{s_y}{s_x}$$

$$= r_{yz \cdot x} \frac{s_{y \cdot x}}{s_{z \cdot x}}.$$

Testing the hypothesis $H_0 : \beta_2^* = 0$ in the two variable model is equivalent to testing the hypothesis $H_0 : \rho_{yz \cdot x} = 0$ already discussed above. A more detailed discussion of inference precedures for regression models containing two or more explanatory variables will be discussed in Chapter 4.

Example

For the previous example relating SELLP to SQF and RMS the simple linear regression results are

$$\text{SELLP} = \underset{(11928.18)}{60255.33} + \underset{(1990.42)}{5772.95} \text{ RMS} \quad R^2 = 0.07 \qquad (3.18)$$

$$\text{SQF} = \underset{(90.35)}{458.10} + \underset{(15.08)}{127.24} \text{ RMS} \qquad R^2 = 0.38 \qquad (3.19)$$

$$\text{SELLP} = \underset{(9211.95)}{16743.87} + \underset{(8.63)}{94.98} \text{ SQF}$$
$$- \underset{(1769.98)}{6312.89} \text{ RMS} \qquad\qquad R^2 = 0.55 \qquad (3.20)$$

The revised values of the intercept and coefficient of RMS after adding SQF are related by the equation

$$\begin{bmatrix} 16,743.87 \\ -6,312.89 \end{bmatrix} = \begin{bmatrix} 60,255.33 \\ 5,772.95 \end{bmatrix} - \begin{bmatrix} 458.10 \\ 127.24 \end{bmatrix} [94.98]$$

as shown in equation (3.17). The intercept decreased substantially and the coefficient of RMS changed sign. Thus adding RMS while holding SQF fixed results in a decrease in the value of the house.

In a similar fashion the impact of adding RMS to the regression model relating SELLP to SQF can also be studied. The regressions are given by

$$\text{SELLP} = \underset{(8689.94)}{2306.54} + \underset{(7.11)}{75.90} \text{ SQF} \qquad R^2 = 0.50$$

$$\text{RMS} = \underset{(0.44)}{2.287} + \underset{(0.00036)}{0.003} \text{ SQF} \qquad R^2 = 0.38.$$

The revised intercept and coefficient of SQF in (3.20) are given by substituting into equation (3.17).

$$\begin{bmatrix} 16,743.87 \\ 94.98 \end{bmatrix} = \begin{bmatrix} 2,306.54 \\ 75.90 \end{bmatrix} - \begin{bmatrix} 2.287 \\ 0.003 \end{bmatrix} [-6312.89]$$

Thus after adding the variable RMS to the model the intercept increases and the coefficient of SQF also increases. Adding more SQF to the house while holding RMS fixed adds more to the value of the house than if RMS is not held fixed.

Multiple Correlation

When there are two explanatory variables it is possible to obtain a measure of linear association between Y and the two variables. In the simple linear regression of Y on X the variation in Y explained by X is $\beta_1^2 \sigma_x^2$. For the residual variation in Y, $U_{y \cdot x}$, the amount $\beta_2^2 \sigma_{z \cdot x}^2 = \beta_2^2 \sigma_z^2 (1 - \rho_{zx}^2)$ is the variation explained by Z. The total variation in Y explained by the model with both X and Z is equivalent to the sum of these two parts

$$\beta_1^2 \sigma_x^2 + \beta_2^2 \sigma_z^2 (1 - \rho_{zx}^2).$$

The proportion of variation in Y explained by both X and Z is denoted by $\rho_{y \cdot xz}^2$ and is given by

$$\rho_{y \cdot xz}^2 = \frac{\beta_1^2 \sigma_x^2}{\sigma_y^2} + \frac{\beta_2^2 \sigma_z^2 (1 - \rho_{zx}^2)}{\sigma_y^2} = \rho_{xy}^2 + \rho_{yz \cdot x}^2 (1 - \rho_{zx}^2)$$

$$= \rho_{xy}^2 + \frac{[\rho_{yz} - \rho_{yx} \rho_{zx}]^2}{[1 - \rho_{zx}^2]}. \tag{3.21}$$

The quantity $\rho_{y \cdot xz}$ is called the *coefficient of multiple correlation* between the variable Y and the variables X and Z.

If $\rho_{zx} = 0$ we can see from (3.21) that

$$\rho_{y \cdot xz}^2 = \rho_{xy}^2 + \rho_{yz}^2$$

and hence that the squared multiple correlation between Y and both X and Z is the sum of the squared correlations between Y and X and between Y and Z. If $\rho_{zx} \neq 0$ then $\rho_{y \cdot xz}^2$ can be larger <u>or</u> smaller than $(\rho_{yx}^2 + \rho_{yz}^2)$ depending on ρ_{xz}. From (3.21) $\rho_{y \cdot xz}^2 > (\rho_{xy}^2 + \rho_{zy}^2)$ requires that

$$\frac{[\rho_{yz} - \rho_{yx} \rho_{zx}]^2}{(1 - \rho_{zx}^2)} > \rho_{yz}^2.$$

From this expression we can conclude that if ρ_{zx} is close to 1 the condition should be met particularly if ρ_{yz} is small. This is the suppressor variable effect discussed above.

Example

For the population correlation matrix

$$
\begin{array}{c|ccc}
 & X & Y & Z \\
\hline
X & 1.00 & 0.60 & -0.50 \\
Y & 0.60 & 1.00 & 0.20 \\
Z & -0.50 & 0.20 & 1.00
\end{array}
$$

$$\frac{[\rho_{yz} - \rho_{yx}\rho_{zx}]^2}{(1 - \rho_{zx}^2)} = \frac{[0.20 - (0.60)(-0.50)]^2}{(1 - (-.50)^2)} = 0.33$$

which is greater than $\rho_{yz}^2 = 0.04$. In this case the partial correlation between Y and Z after controlling for X is

$$(\rho_{yz} - \rho_{yx}\rho_{zx}) \big/ \sqrt{1 - \rho_{zx}^2}\sqrt{1 - \rho_{yx}^2} = \frac{0.20 - (0.60)(-0.50)}{\sqrt{1 - (.20)^2}\sqrt{1 - (-.50)^2}} = 0.72$$

which is substantially larger than $\rho_{yz} = 0.20$.

Suppressor Variables and Multiple Correlation

An examination of the above relationship between $\rho_{y\cdot xz}^2$ and ρ_{yx}^2 and ρ_{yz}^2 reveals that it is possible for $\rho_{y\cdot xz}^2$ to be close to 1 even though ρ_{yx}^2 and ρ_{yz}^2 are close to 0. This can happen if ρ_{zx}^2 is close to one. This means that the multiple correlation between Y and the variables X and Z can be relatively large even though the simple correlation between Y and each of the variables X and Z is relatively small. In this case the simple linear regression of Y on either X or Z yields a weak relationship but the regression of Y on both X and Z yields a strong relationship. The variable Z in this case is called a suppressor variable since adding it to the regression after X increases the importance of X. The variable Z is said to be masking the true relationship between Y and X. Similarly the variable X is a suppressor variable in the relation between Y and Z.

Example

The population correlation matrix given by

$$
\begin{array}{c|ccc}
 & X & Y & Z \\
\hline
X & 1.00 & 0.30 & 0.95 \\
Y & 0.30 & 1.00 & 0.05 \\
Z & 0.95 & 0.05 & 1.00
\end{array}
$$

provides an interesting example of the suppressor variable phenomenon. The correlation coefficient $\rho_{zy} = 0.05$ is very weak indicating that the proportion $(0.05)^2 = 0.0025$ of variation in Y is explained by Z. The simple correlation between Y and X, $\rho_{yx} = 0.30$ also indicates a relatively weak relationship between that pair. The partial correlation between Y and X after controlling for Z is given by

$$\rho_{yx \cdot z} = \frac{0.30 - (0.05)(0.95)}{\sqrt{1 - (0.05)^2}\sqrt{1 - (0.95)^2}} = 0.8098,$$

suggesting that the variation in Y not explained by Z is strongly correlated with the part of X that is not related to Z. The additional variation in Y explained by X (after Z) is given by

$$\rho_{yx \cdot z}^2(1 - \rho_{yz}^2) = 0.6541$$

The total variation in Y explained by both X and Z is therefore

$$\rho_{y \cdot xz}^2 = \rho_{yz}^2 + \rho_{yx \cdot z}^2(1 - \rho_{yz}^2) = 0.6566.$$

Suppose that in the above correlation matrix, Z = rate of change in consumer price index (CPI), Y = rate of return on a stock market index and X = interest rate on treasury bills. The correlations between the market return and each of the CPI ($\rho_{yz} = 0.05$) and the treasury bill rate ($\rho_{yx} = 0.30$) are assumed to be relatively weak, but a substantial part of the variation in treasury bill return is related to the change in the CPI ($\rho_{xz} = 0.95$). When CPI is held fixed, however, the partial correlation between market return and treasury bills ($\rho_{yx \cdot z} = 0.8098$) is quite strong. The small part of market return variation that is unrelated to CPI is strongly related to the part of the treasury bill return that is not related to CPI.

Geometric Explanation of Suppressor Variable Effect

Geometrically the multiple correlation is equal to Cos θ where θ is the angle between $\mathbf{y}^*$ (the mean corrected $\mathbf{y}$) and the plane formed by the vectors $\mathbf{x}^*$ and $\mathbf{z}^*$. Hence θ is close to $0°$ if $\rho_{y \cdot xz}$ is close to 1. If, however, ρ_{yx} and ρ_{yz} are close to zero then the angles between $\mathbf{y}^*$ and the individual vectors $\mathbf{x}^*$ and $\mathbf{z}^*$ are both close to $90°$. It must be true therefore that the angle between $\mathbf{x}^*$ and $\mathbf{z}^*$ is close to $0°$. Thus the vector $\mathbf{y}^*$ almost lies in the plane of $\mathbf{x}^*$ and $\mathbf{z}^*$ but $\mathbf{y}^*$ is almost orthogonal to both $\mathbf{x}^*$ and $\mathbf{z}^*$ which are almost co-linear.

Partial Coefficient of Determination

The variation in Y explained by X and Z is given by $\sigma^2_{y \cdot xz}$ and the residual or unexplained variation in Y is given by $(\sigma^2_y - \sigma^2_{y \cdot xz}) = \sigma^2_y(1 - \rho^2_{y \cdot xz})$. In the simple linear regression model of Y on X the variation in Y left unexplained is $\sigma^2_y(1 - \rho^2_{xy})$. The proportion of this residual variation now explained by adding Z is given by

$$\frac{\beta^2_2 \sigma^2_z(1 - \rho^2_{zx})}{\sigma^2_y(1 - \rho^2_{xy})} = \frac{[\rho_{yz} - \rho_{yx}\rho_{zx}]^2}{(1 - \rho^2_{xy})(1 - \rho^2_{zx})} \tag{3.22}$$

which is the square of the partial correlation coefficient $\rho_{yz \cdot x}$. Thus $\rho^2_{yz \cdot x}$ represents the proportion of the variation in Y explained by Z after removing the effect of X. The quantity $\rho^2_{yz \cdot x}$ is sometimes called the *partial coefficient of determination*.

Examples

In the discussion of suppressor variables and multiple correlation an example relating stock market returns Y to CPI X, and treasury bill rate Z was discussed. The partial correlation between X and Y controlling for Z was given by 0.8098. The partial coefficient of determination for X after Z is therefore $\rho^2_{yx \cdot z} = (0.8098)^2 = 0.66$. Therefore by adding the variable X to the regression after Z, 66% of the residual variation in Y is explained by X.

For the addition of the variable Z after X the partial coefficient of determination using (3.22) is given by

$$\rho^2_{yz \cdot x} = \frac{[(0.05) - (0.30)(0.95)]^2}{[1 - (0.05)^2][1 - (0.95)^2]} = 0.82.$$

Therefore by adding the variable Z after X, 82% of the residual variation in Y is explained by Z.

The total variation in Y explained by both X and Z is given by

$$\rho^2_{y \cdot xz} = (0.30)^2 + (0.66)[1 - (0.30)^2] = (0.05)^2 + (0.82)[1 - (0.05)^2] = 0.8103.$$

In the relationship between SELLP and the explanatory variables SQF and RMS given by (3.15) the partial correlations squared given by $(-0.318)^2$ = 0.10 and $(0.719)^2 = 0.52$ yield the partial coefficients of determination for RMS after SQF and for SQF after RMS respectively.

Sample Inference and Variance Inflation

For a sample regression it is important to note that the addition of the variable Z to the model relating Y to X can also have an important impact on the statistical significance of the estimators b_0 and b_1. By replacing population parameters by sample estimators in (3.17) we can see that b_0 and b_1 can change through the magnitudes of g_0, g_1 and b_2. In addition the standard errors of the coefficients b_0 and b_1 can also change as a result of the addition of Z.

Before entering the variable Z the standard error of b_1 is given by

$$\sqrt{\frac{s_{y\cdot x}^2}{\Sigma x^2 - (\Sigma x)^2/n}} = \sqrt{\frac{s_y^2(1 - r_{yx}^2)}{(n-2)s_x^2}}.$$

After Z has been included the standard error becomes

$$\sqrt{\frac{s_y^2(1 - r_{y\cdot xz}^2)}{(n-3)s_x^2(1 - r_{xz}^2)}}$$

where $r_{y\cdot xz}^2$ is the sample multiple correlation between Y and both X and Z. This sample multiple correlation can be obtained from (3.21) by replacing population parameters by their sample counterparts. The revised standard error now includes the term $\dfrac{1}{(1 - r_{xz}^2)}$ which is called the *variance inflation factor* since it measures the impact on the standard error of the correlation between X and Z. Note that if r_{xz} is close to 1 the standard error of b_1 can increase without limit. Similarily the standard error of b_2 is given by

$$\sqrt{\frac{s_y^2(1 - r_{y\cdot xz}^2)}{(n-3)s_z^2(1 - r_{xz}^2)}}$$

and can also increase without limit under the same circumstances.

The sample expression for the proportion of variation in Y explained by both X and Z is given by the sample form of (3.21) which is given by

$$r_{y\cdot xz}^2 = r_{xy}^2 + r_{yz\cdot x}^2\,(1 - r_{zx}^2).$$

Therefore the proportion of variation in Y explained by both X and Z can approach 1 even though the regression coefficients individually may not be significant. This topic will be discussed under the heading of multicollinearity in Chapter 4.

Examples

In the example above relating the stock market index return, Y, to CPI change, X, and treasury bill return, Z, the treasury bill return acted as a suppressor variable for the relationship between market return and CPI. Using the population correlation as a sample correlation recall that ρ_{xz} was 0.95. The variance inflation factor because of this high correlation is given by $1/[1 - (0.95)^2] = 10.26$. As a result of regressing Y on both X and Z the standard errors for the two regression coefficients would be very large. The individual regression coefficients therefore will tend to be insignificant even though the overall regression may be quite significant.

In the example relating SELLP to SQF and RMS adding the second variable to the model does not have as large an impact on the standard errors of the coefficients. The sample correlation coefficient between SQF and RMS is 0.62 and hence the variance inflation factor is $1/[1 - (0.62)^2] = 1.62$. An examination of the standard errors of the coefficients in equations (3.18), (3.19), (3.20) illustrates the changes in the standard errors. Notice that when SQF is added after the variable RMS, the standard errors decrease. This results from the increase in overall goodness of fit from $R^2 = 0.07$ to $R^2 = 0.55$. The change in R^2 reflects the difference between $r_{y\cdot x}^2$ and $r_{y\cdot xz}^2$. Large differences between $(1 - r_{y\cdot x}^2)$ and $(1 - r_{y\cdot xz}^2)$ therefore can also have an impact on the change in standard errors.

3.4.3 Missing Variable Plots

The impact of another variable Z on the relationship between X and Y can also be studied using the techniques of residual analysis discussed in 3.3.8. We have already seen that the residuals $e_{y\cdot x}$ and $e_{z\cdot x}$ from the two regressions Y on X and Z on X can be used to determine the potential benefit of including Z as an additional explanatory variable. It should be expected therefore that a plot of $e_{y\cdot x}$ vs $e_{z\cdot x}$ should yield a relationship. This is the principle behind the *added variable plot* or *partial plot*. The slope of the linear relationship between these two residuals is equivalent to the regression coefficient that would result if Z was added to the model. The residuals around the fitted line would be equivalent to the residuals from the regression that includes both X and Z. In addition to these properties, the partial plot also permits one to detect violations such as potential nonlinear relationships and heteroscedasticity, as well as potential outliers and points with significant influence and/or leverage. Sometimes a useful Z variable to study would be the order in which the observations were taken. In other cases an unknown Z variable may be discovered by examining clusters of similar residual values to determine if the clustering is due to some other variables.

An alternative type of plot that has been suggested is the *partial residual plot* or *parres plot* which is based on the residual from the larger model $e_{y \cdot xz} = \hat{y} - b_0 - b_1 X - b_2 Z$. In the partial residual plot $W = e_{y \cdot xz} + b_2 Z$ is plotted against Z rather than $e_{z \cdot x}$. The residual $e_{y \cdot xz}$ is uncorrelated with Z and X and hence the correlation between W and Z is due to the relation between $b_2 Z$ and Z. The partial residual plot is sometimes more easily obtained from statistical software.

Example

The first panel, (a), of Figure 3.34 shows the partial plot of the residuals from the regression of SELLP on RMS against the residuals from the regression of SQF on RMS. From the plot we can see that there is a strong positive linear relationship. Adding the variable SQF to the model should therefore provide a substantial increase in R^2. The partial residual plot of the SELLP on RMS residuals against SQF is shown in the second panel, (b), of Figure 3.34. This plot also indicates a strong linear relationship. Notice how in the second plot the points are more scattered because the variable RMS has not been removed from SQF.

The partial plot and partial residual plot in the lower two panels (c) and (d) of Figure 3.34 show the importance of the variable RMS after SQF has been included first. The linear relationships indicated are quite weak compared to the previous plots for SQF after RMS.

The residuals for these plots were obtained from the SAS procedure PROC REG.

3.4.4 Empirical vs. Structural Relationships and Omitted Variable Bias

Section 3.4 has been concerned with the impact of a third variable Z on the relationship between two variables X and Y. In a linear regression model relating Y to X, if a variable Z is added to the model the regression coefficients will change if Z is correlated with X. The question therefore arises as to what is the meaning of a regression coefficient in a simple linear regression relationship if there are other variables that are correlated with both Y and X.

For discussion purposes, we assume that the random variable Y is related to the random variables X and Z by the linear model

$$Y = \beta_0 + \beta_1 X + \beta_2 Z + U, \qquad (3.23)$$

where the error term U has $E[U] = 0$, $V[U] = \sigma_U^2$, and U is independent of X and Z over some population. This model describes the marginal

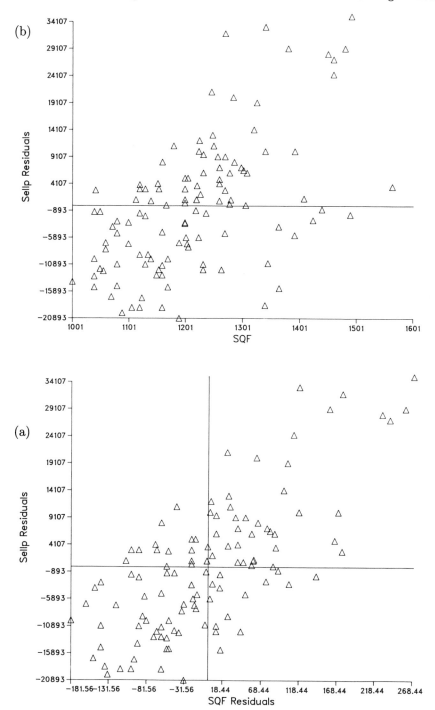

Figure 3.34. Partial Plots and Partial Residual Plots for Regression SELLP on SQF and RMS

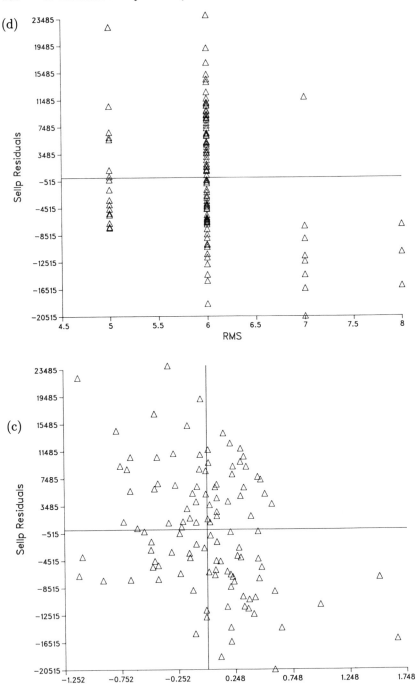

Figure 3.34. Partial Plots and Partial Residual Plots for Regression SELLP on SQF and RMS (continued)

impact of X on Y when Z is fixed and of Z on Y when X is fixed. All other variables which have an influence on Y are included in U and are by assumption independent of X and Z. Equivalently, the conditional expectation of Y holding X and Z fixed, and the conditional expectation of Y holding X, Z and any other variables fixed, yields the same regression coefficients β_1 and β_2 without regard for the other variables. If some other variable W, which is correlated with Y, is held fixed or limited in some way, the regression coefficients β_1 and β_2 will remain the same.

While the above model describes a *structural relationship* between Y and the explanatory variables X and Z, we can also discuss an *empirical relationship* between Y and X. Assuming that the model (3.23) still holds, the regression function for Y on X is given by

$$E[Y \mid X] = \beta_0 + \beta_1 X + \beta_2 E[Z \mid X]$$

since U is independent of X and $E[U] = 0$. If Z and X are correlated we may write the regression function of Z on X as

$$E[Z \mid X] = \gamma_0 + \gamma_1 X.$$

The regression function of Y on X is therefore given by

$$E[Y \mid X] = (\beta_0 + \beta_2 \gamma_0) + (\beta_1 + \beta_2 \gamma_1) X.$$

The regression function of Y on X therefore provides an alternative value $(\beta_1 + \beta_2 \gamma_1)$ for the impact of X on Y.

The linear model relating Y and X can be written

$$Y = \alpha_0 + \alpha_1 X + V$$

where $\alpha_0 = (\beta_0 + \beta_2 \gamma_0)$ and $\alpha_1 = (\beta_1 + \beta_2 \gamma_1)$. The error term $V = (\beta_2 Z + U - \beta_2 \gamma_0 - \beta_2 \gamma_1 X)$ is uncorrelated with X but is not uncorrelated with Z. If $V[Z \mid X] = \sigma^2$ then $V[V \mid X] = \sigma^2 + \beta_2 \sigma_U^2$. Since the error term V is correlated with Z the regression function relating Y and X depends on whether Z is controlled. If Z is not controlled, then α_1 is the impact of X on Y, while if Z is held fixed, β_1 is the impact of X on Y.

Given a random sample of n observations $(x_i, y_i \mid i = 1, 2, \ldots, n)$ where no restrictions were placed on z_i in the sampling process, the least squares estimators provide unbiased estimators of the empirical relationship between Y and X. These estimators however are not unbiased estimators of β_0 and β_1 which are the true regression coefficients if Z is also included in the model.

As an example, a builder of single family dwellings is considering the addition of more overall floor space to his three-bedroom designs. In order to determine the market value of such increases he collects a sample of sales data for the past year. The data were used to estimate the regression function relating selling price to size. The relationship being estimated is an empirical one. Many other variables influencing selling price could be included as explanatory variables. How is the coefficient of size to be interpreted? We must ask what restrictions if any were placed on the sample. Was the sample restricted to three-bedroom bungalows? Clearly the regression coefficient for size will depend on such restrictions. If the data represents many different designs such as two storey, and single storey with more than three bedrooms, then the coefficient of size will reflect that some of the houses were bigger and hence the coefficient is an average over the various types. If, however, the regression model includes number of bedrooms and number of storeys as additional explanatory variables then these variables will be controlled and the coefficient of size will be more realistic.

It is important when employing regression models that the distinction between these two types of models be considered. In Chapter 4 the issue of omitted variable bias will be discussed in connection with variable selection techniques.

3.4.5 Qualitative Lurking Variables

In some applications a third variable Z which has an impact on the relation between X and Y can be a classification variable. For simplicity we shall restrict our discussion here to the case where Z has only two possible values which we shall code as $Z = 0$ or $Z = 1$. Figure 3.35 suggests several possible relationships among the three variables. In panel (a) of the figure the two lines are parallel, while in panel (b) the two lines have a common intercept. In panel (c) the two lines have different slopes and different intercepts. In panel (d) the two lines are located in distinct ranges of X with a known joint point at $X = X_0$. In all cases fitting a single line to describe the relationship between Y and X represents a compromise which could be somewhat misleading.

An example of a lurking variable Z might be the variable sex in the study of the relationship between income (Y) and years of work experience (X) for a particular professional group. Classifying sex using the variable $Z = 0$ or 1 for males and females respectively, we might expect to see one of the relationships in panels (a), (b) or (c). A second example might be provided by the relationship between annual observations on aggregate consumption and aggregate income. This relationship has been found to be different in the United States during the Second World War than it was before and after the war. The relationship in panel (a) was found to be useful with the variable Z having the value 0 for peacetime years and 1 for the years 1941 to 1945. The reason for the lower intercept during the war was claimed to be the shortage of consumer goods which resulted in forced saving.

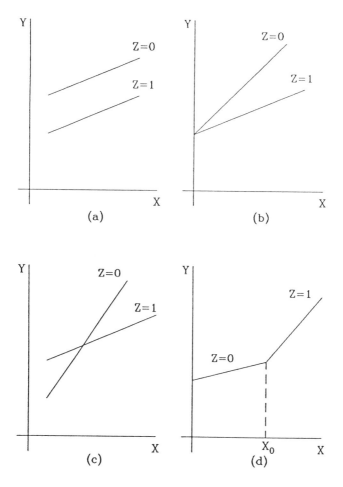

Figure 3.35. Impact of a Qualitative Lurking Variable Z on Linear Relationship Between Y and X

The relationship shown in panel (d) could represent the relationship between two variables where there is a structural change in the system at the point $X = X_0$. A linear cost function can have this shape if at output levels beyond $X = X_0$ there must be a change in the system in order to increase output, for instance, overtime pay for workers or use of a less efficient backup process.

A linear regression model can be used to describe the relationships illustrated in Figure 3.35. If the Z variable is added as a second explanatory variable we have the model

$$Y = \beta_0 + \beta_1 X + \beta_2 Z + U.$$

Since Z only takes on the values 0 or 1 we may write the model as

$$Y = \beta_0 + \beta_1 X + U \qquad \text{when } Z = 0$$

$$\text{and} \quad Y = (\beta_0 + \beta_2) + \beta_1 X + U \quad \text{when } Z = 1.$$

Thus we obtain a pair of parallel lines as shown in panel (a) of Figure 3.35.

We can also add a third variable to the model which will permit the two lines to have different slopes and different intercepts as in panel (c). Define the new variable $XZ = (X)(Z)$ which is the product of the two variables. A model which contains all three explanatory variables is given by

$$Y = \beta_0 + \beta_1 X + \beta_2 Z + \beta_3 XZ + U.$$

Writing separate models for the two values of Z we obtain

$$Y = \beta_0 + \beta_1 X + U$$

$$\text{and} \quad Y = (\beta_0 + \beta_2) + (\beta_1 + \beta_3)X + U.$$

In this case we obtain two distinct lines with different intercepts and different slopes.

If the variable Z is omitted from the previous model we obtain

$$Y = \beta_0 + \beta_1 X + \beta_3 XZ + U$$

which results in two lines with common intercept β_0 as in panel (b).

To describe the relationship in panel (d) a new variable Z^* is defined as follows

$$Z^* = 0 \qquad\qquad X \leq X_0$$

$$Z^* = (X - X_0) \qquad X > X_0.$$

The model
$$Y = \beta_0 + \beta_1 X + \beta_2 Z^* + U$$

results in the two lines

$$Y = \beta_0 + \beta_1 X + U \qquad\qquad\qquad X \leq X_0$$

$$\text{and} \quad Y = (\beta_0 - \beta_2 X_0) + (\beta_1 + \beta_2)X + U \qquad X > X_0.$$

In this case the intercepts and slopes for the two lines are different subject to the condition that the two lines intersect at $X = X_0$.

Techniques for dealing with these models and more complex models involving additional categories and variables will be outlined in Chapter 4 in the discussion of dummy variable techniques.

Cited Literature for Chapter 3

1. Belsey, D.A., Kuh, E., and Welsch, R.E. (1980). *Regression Diagnostics: Identifying Influential Data and Sources of Collinearity.* New York: John Wiley and Sons, Inc.

2. Conover, W.J. (1980). *Practical Nonparametric Statistics* (Second Edition). New York: John Wiley and Sons Inc.

3. Cook, R. Dennis and Weisberg, Sanford (1982). *Residuals and Influence in Regression.* New York: Chapman and Hall.

4. Emerson, John D. and Hoaglin, David C. (1983). "Resistant Lines for *y* versus *x*," in Hoaglin, David C., Mosteller, Frederick and Tucker, John W. eds., *Understanding Robust and Exploratory Data Analysis.* New York: John Wiley and Sons Inc.

5. Fomby, T.B., Hill, R.C. and Johnson, S.R. (1984). *Advanced Econometric Methods.* New York: Springer–Verlag.

6. Fuller, Wayne A. (1987). *Measurement Error Models.* New York: John Wiley and Sons Inc.

7. Judge, G.G., Griffiths, W.E., Hill, R.C., Lütkepohl, H. and Lee, T. (1985). *The Theory and Practice of Econometrics* (Second Edition). New York: John Wiley and Sons Inc.

8. Kmenta, Jan (1986). *Elements of Econometrics* (Second Edition). New York: Macmillan Publishing Co.

9. Lund, Richard E. (1975). "Tables for an Approximate Test for Outliers in Linear Models." *Technometrics* 17, 473–476.

10. Mulaik, Stanley A. (1972). *The Foundations of Factor Analysis.* New York: McGraw–Hill Book Co.

11. Neter, J., Wasserman, W. and Kutner, M.H. (1983). *Applied Linear Regression Models.* Homewood, Illinois: Richard D. Irwin Inc.

12. Prescott, P. (1975). "An Approximate Test for Outliers in Linear Models." *Technometrics* 17, 129–132.

13. Tietjen, G.L., Moore, R.H. and Beckman, R.J. (1973). "Testing for a Single Outlier in Simple Linear Regression." *Technometrics* 15, 717–721.

Exercises for Chapter 3

1. This exercise is based on Table D1 in the Data Appendix.

   (a) Obtain three scatterplots relating the variables PMAX, SMAX and TMR in pairs. Discuss the relationships and comment on the presence of outliers. Obtain the Pearson, Kendall and Spearman correlation coefficients and compare them. Omit any outliers and repeat the three correlation coefficients. Discuss the impact of the outliers on the correlations.

   (b) Compute the two ratio variables SMAX/SMIN and PMAX/PMIN. Compare the distributions of these two variables both graphically and using various analytical measures. Also examine the scatterplot relating the two ratios and discuss the results. Are outliers present?

   (c) Fit simple linear regression models relating TMR to each of PM2 and SMAX. Discuss the fitted relationship and the goodness of fit. Examine the residuals and influence diagnostics and comment. Plot the two fitted relationships on separate graphs. Also examine the residual plots and comment on the residual behavior. Remove any influential observations and fit the simple linear regression models. Discuss the fitted relationship, the residuals and influence diagnostics and compare to the previous results. Plot the fitted relationships on the same two graphs as the previous fitted relationships and compare the results.

   (d) Determine the correlation matrix relating the variables GE65, TMR and PERWH and comment on the correlations. Obtain the partial correlations between TMR and each of GE65 and PERWH controlling for the other variable. Compare these first order partial correlations to the zero order correlations. Discuss the relationships. Regress TMR on each of the variables separately and together. Compare the regression coefficients and R^2 values and comment on the impact of adding the second variable to the regression.

2. This exercise is based on Table D2 in the Data Appendix.

   (a) Obtain three scatterplots relating the variables RETCAP, QUIKRAT and CURRAT in pairs. Discuss the relationships and comment on the presence of outliers. Obtain the Pearson, Kendall and Spearman correlation coefficients and compare them. Omit any outliers and repeat the three correlation coefficients. Discuss the impact of the outliers on the correlations.

   (b) Repeat the steps in exercise (a) for the variables RETCAP, GEAR-RAT and WCFTCL.

(c) Fit simple linear regression models relating RETCAP to each of QUIKRAT and CURRAT. Discuss the fitted relationships and the goodness of fit. Examine the residuals and influence diagnostics and comment. Plot the two fitted relationships on separate graphs. Also examine the residual plots and comment on the residual behavior. Remove any influential observations and fit the simple linear regression models. Discuss the fitted relationships, the residuals and influence diagnostics and compare to previous results. Plot the fitted relationships on the same two graphs as the previous fitted relationship and compare the results.

(d) Repeat exercise (c) for the regressions of RETCAP on GEARRAT and WCFTCL.

(e) Determine the correlation matrix relating the variables RETCAP, WCFTCL and CURRAT and comment on the correlations. Obtain the partial correlations between RETCAP and each of WCFTCL and CURRAT controlling for the other variable. Compare the first order partial correlations to the zero order correlations. Discuss the relationships. Regress RETCAP on each of the variables separately and together. Compare the regression coefficients and R^2 values, and comment on the impact of adding the second variable to the regression.

3. The exercises in this part are based on the data in Table D5 in the Data Appendix.

(a) Regress TRSBILL on CPI and carry out the Durbin–Watson test for first order autocorrelation. Discuss the results.

(b) Assuming there is first order autocorrelation use a procedure which takes this into account and refit the model. Compare the results to the ordinary least squares results in (a).

4. This exercise is based on the data in Table D6 in the Data Appendix. Carry out a regression of the variable EXPEN on the variable SIZE using a procedure which permits calculation of the component of SSE that is pure error (SSPE) and perform a test for lack of fit. If a regression program is not available to obtain SSPE, a one-way ANOVA program comparing EXPEN across the six size groups can be used. The sum of squares within group SSW can be used for SSPE required above.

5. This exercise is based on the data in Table D3 in the Data Appendix.

(a) Examine a scatterplot relating the variable LCURRENT to the variable EDUC. Comment on the relationship between the two variables. Comment on the scatter in LCURRENT over the range of EDUC. Divide the range of EDUC into groups and determine the variance in LCURRENT for each of your EDUC groups and comment on the existence of heteroscedasticity.

(b) Fit a regression model relating LCURRENT to EDUC using ordinary least squares. Develop a weighting scheme which takes into account the heteroscedasticity determined in (a). Fit a weighted least squares model relating LCURRENT to EDUC using the weighting scheme. Compare the two estimated models.

(c) Repeat the analyses carried out in (a) and (b) for the explanatory variables AGE, EXPER and SENIOR.

Questions for Chapter 3

1. (a) Given that X and Y are random variables and a and b are constants, use the fact that $E[a + X] = a + E[X]$, $E[b + Y] = b + E[Y]$, $V[a + X] = V[X]$ and $V[b + Y] = V[Y]$ to show that the correlation between $(a + X)$ and $(b + Y)$ is equivalent to the correlation between X and Y.

   (b) Use the fact that $E[aX] = aE[X]$, $E[bY] = bE[Y]$, $V[aX] = a^2 V[X]$ and $V[bY] = b^2 V[Y]$ to show that the correlation between aX and bY is equivalent to the correlation between X and Y.

   (c) Use the definitions of expectation, variance, conditional expectation and conditional variance to show that $V[Y] = V[E[Y \mid X]] + E[V[Y \mid X]]$.

2. Let $R = aX + bY$ denote the return on a portfolio of two stocks with returns X and Y where $(a + b) = 1$. Denote the mean return vector and covariance matrix by

$$\boldsymbol{\mu} = \begin{bmatrix} \mu_x \\ \mu_y \end{bmatrix}, \qquad \boldsymbol{\Sigma} = \begin{bmatrix} \sigma_x^2 & \sigma_{xy} \\ \sigma_{xy} & \sigma_y^2 \end{bmatrix}$$

and denote the vector

$$\begin{bmatrix} a \\ b \end{bmatrix}$$

by $\mathbf{w}$.

Show that $E[R] = \mathbf{w}'\boldsymbol{\mu}$ and $V[R] = \mathbf{w}'\boldsymbol{\Sigma}\mathbf{w}$.

3. Assume X and Y are independent discrete random variables and assume that each assume only a finite number of values. Use the definition of independence to show that $\text{Cov}(X, Y) = 0$.

4. Let $\mathbf{z} = \begin{bmatrix} X \\ Y \end{bmatrix}$, $\boldsymbol{\mu} = \begin{bmatrix} \mu_x \\ \mu_y \end{bmatrix}$, $\boldsymbol{\Sigma} = \begin{bmatrix} \sigma_x^2 & \sigma_{xy} \\ \sigma_{xy} & \sigma_y^2 \end{bmatrix}$, $\rho = \dfrac{\sigma_{xy}}{\sigma_x \sigma_y}$.

   (a) Show that the Mahalanobis distance between $\mathbf{z}$ and $\boldsymbol{\mu}$ given by

$$(\mathbf{z} - \boldsymbol{\mu})'\boldsymbol{\Sigma}^{-1}(\mathbf{z} - \boldsymbol{\mu}),$$

   can be written as

$$\left(\frac{1}{1 - \rho^2}\right)\left[\left(\frac{X - \mu_x}{\sigma_x}\right)^2 + \left(\frac{Y - \mu_y}{\sigma_y}\right)^2 - 2\rho\left(\frac{X - \mu_x}{\sigma_x}\right)\left(\frac{Y - \mu_y}{\sigma_y}\right)\right].$$

(b) Show that the expression for the density of the bivariate normal density given in 3.1 can be written as

$$\frac{1}{2\pi|\Sigma|^{1/2}} \exp\left[-\frac{1}{2}(\mathbf{z} - \boldsymbol{\mu})'\Sigma^{-1}(\mathbf{z} - \boldsymbol{\mu}) \right]$$

where

$$\mathbf{z} = \begin{bmatrix} X \\ Y \end{bmatrix}, \quad \boldsymbol{\mu} = \begin{bmatrix} \mu_x \\ \mu_y \end{bmatrix} \quad \text{and} \quad \Sigma = \begin{bmatrix} \sigma_x^2 & \sigma_{xy} \\ \sigma_{xy} & \sigma_y^2 \end{bmatrix}.$$

5. Given that the slope of the regression of Y on X is 0.9 and the correlation between X and Y is 0.6, determine the slope of the regression of X on Y. Assume X and Y are bivariate normal.

6. (a) Show that σ_{xy} can be written as

$$\sigma_{xy} = [V(X + Y) - V(X - Y)]/4.$$

(b) Show that ρ_{xy} can be written as

$$\rho_{xy} = \frac{V\left(\frac{X}{\sigma_x} + \frac{Y}{\sigma_y}\right) - V\left(\frac{X}{\sigma_x} - \frac{Y}{\sigma_y}\right)}{V\left(\frac{X}{\sigma_x} + \frac{Y}{\sigma_y}\right) + V\left(\frac{X}{\sigma_x} - \frac{Y}{\sigma_y}\right)}.$$

7. (a) Let (x_i, y_i) denote a random sample of observations on the random variables (X, Y) and define $d_i = (x_i - y_i)$. Let $\bar{d}$ and s_d^2 be the sample mean and sample variance for d_i based on n observations. Show that $\bar{d} = \bar{x} - \bar{y}$ and $s_d^2 = s_x^2 + s_y^2 - 2s_{xy}$ where $\bar{x} = \frac{1}{n}\sum_{i=1}^{n} x_i$,

$$\bar{y} = \frac{1}{n}\sum_{i=1}^{n} y_i, \quad s_x^2 = \frac{1}{(n-1)}\sum_{i=1}^{n}(x_i-\bar{x})^2, \quad s_y^2 = \sum_{i=1}^{n}(y_i-\bar{y})^2/(n-1),$$

$$s_{xy} = \frac{1}{(n-1)}\sum_{i=1}^{n}(x_i - \bar{x})(y_i - \bar{y}).$$

(b) The paired comparison test employs the test statistic $\bar{d}/\sqrt{s_d^2/n}$ to test $H_0 : \mu_x = \mu_y$. Show that for a given $\bar{x}$, $\bar{y}$, s_x^2, and s_y^2 the statistic is maximized when $r = +1$ where r is the Pearson correlation between x and y given by $r = s_{xy}/\sqrt{s_x^2 s_y^2}$. Use this result to suggest why identical twins make ideal pairs in such a test.

8. Assume that X and Y have common mean zero and common variance 1 and correlation coefficient ρ. Let the regression function for Y on X be given by $E[Y \mid X = x] = f(x)$. Under the assumption of bivariate normality $f(x) = \rho x$. Using Chebyshev's Inequality it can be shown that for any bivariate distribution $P[|f(x) - \rho x| < \varepsilon] \geq 1 - \frac{(1 - \rho^2)}{\varepsilon^2}$.

Use this result to justify the notion that for any bivariate distribution as $\rho^2 \to 1$ the regression function approaches linearity.

9. Assume that X and Y are bivariate normal with zero means, unit variances and positive correlation coefficient ρ. Denote by μ_{x_0} the expected value of X **given** that $X > X_0$ where $X_0 > 0$. From the regression of Y on X the expected value of Y given $X > X_0$ is given by $\rho\mu_{x_0}$. The difference between these two expectations is $\mu_{x_0}(1 - \rho)$. Explain how this result can be used to explain the regression to the mean phenomenon.

10. The ordinary least squares estimators for β_0 and β_1 are given by

$$b_0 = \bar{y} - b_1\bar{x}, \qquad b_1 = \frac{s_{xy}}{s_x^2} = \frac{rs_y}{s_x} = \frac{\sum\limits_{i=1}^{n}(x_i - \bar{x})(y_i - \bar{y})}{\sum\limits_{i=1}^{n}(x_i - \bar{x})^2}$$

$$= \frac{n\sum\limits_{i=1}^{n} x_i y_i - \sum\limits_{i=1}^{n} x_i \sum\limits_{i=1}^{n} y_i}{n\sum\limits_{i=1}^{n} x_i^2 - (\sum\limits_{i=1}^{n} x_i)^2}$$

and are obtained by minimizing $S = \sum\limits_{i=1}^{n}(y_i - \beta_0 - \beta_1 x_i)^2$ with respect to β_0 and β_1.

(a) Show that S can be written as

$$S = \sum_{i=1}^{n}[(y_i - \bar{y}) - \beta_1(x_i - \bar{x}) + (\bar{y} - \beta_0 - \beta_1\bar{x})]^2$$

$$= \left[s_y^2(1 - r^2) + (\beta_1 s_x - rs_y)^2 + \frac{n}{(n-1)}(\bar{y} - \beta_0 - \beta_1\bar{x})^2\right](n - 1)$$

and hence that S is minimized by b_0 and b_1.

(b) Use calculus to derive the normal equations for b_0 and b_1.

(c) Assume $\beta_0 = 0$ and show the least squares estimator of β_1 is given by $b_1 = \sum\limits_{i=1}^{n} x_i y_i / \sum\limits_{i=1}^{n} x_i^2$.

11. Show that $\sum\limits_{i=1}^{n}(x_i - \bar{x})(y_i - \bar{y})$ is equivalent to $\sum\limits_{i=1}^{n} y_i(x_i - \bar{x})$ and hence show that b_1 can be written as a linear function of the y_i.

12. (a) Use the expression for the $\text{Cov}(b_0\ b_1)$ given in 3.3.2 to show that this covariance is zero if $\bar{x} = 0$.

(b) Use the expression for the $100(1-\alpha)\%$ confidence ellipsoid given in 3.3.2 to show that if $\bar{x} = 0$ then the ellipse is given by

$$n(b_0 - \beta_0)^2 + \sum_{i=1}^{n} x_i^2 (b_1 - \beta_1)^2 = 2s^2 F_{\alpha;2,(n-2)}.$$

(c) A Bonferroni approximation to this ellipse is provided by the rectangle

$$b_0 \pm t_{\frac{\alpha}{4};(n-2)} \frac{s}{\sqrt{n}}$$

$$b_1 \pm t_{\frac{\alpha}{4};(n-2)} \frac{s}{\sqrt{\sum x_i^2}}.$$

Show this using the formulae given in Chapter 3.

(d) Assume that $n = 17$, $\sum_{i=1}^{n} x_i^2 = 68$, $s^2 = 12.59$. Show that the 90% confidence ellipse is given by $\dfrac{(b_0 - \beta_0)^2}{4} + (b_1 - \beta_1)^2 = 1$ and plot the ellipse. Use (b_0, b_1) as the origin.

Hint: $\dfrac{(X - \alpha)^2}{a^2} + \dfrac{(Y - \beta)^2}{b^2} = 1$ is an ellipse with centre at (α, β) and axes of length $2a$ and $2b$.

(e) Show that the Bonferroni rectangle is given by

$$b_0 \pm 1.84; \qquad b_1 \pm 0.92.$$

Plot the rectangle on the same graph as the ellipse and hence show how the Bonferroni approximation compares to the ellipse.

13. In the case of repeated observations y_{ij}, $i = 1, 2, \ldots, n$, for each x_j, $j = 1, 2, \ldots, g$, the total error sum of squares $\text{SSE} = \sum\limits_{j=1}^{g} \sum\limits_{i=1}^{n_j} (y_{ij} - \hat{y}_{ij})^2$ can be written as $\text{SSE} = \text{SSPE} + \text{SSLF}$ where SSPE is the sum of squares for pure error

$$\text{SSPE} = \sum_{j=1}^{g} \sum_{i=1}^{n_j} (y_{ij} - \bar{y}_{\cdot j})^2$$

and SSLF is the sum of squares due to lack of fit

$$\text{SSLF} = \sum_{j=1}^{g} \sum_{i=1}^{n_j} (\hat{y}_{ij} - \bar{y}_{\cdot j})^2.$$

Use the fact that $\hat{y}_{ij} = \hat{y}_j$ since the x_i are fixed and $\sum_{i=1}^{n_j}(y_{ij} - \bar{y}_{\cdot j}) = 0$ to show that

$$SSE = SSPE + SSLF.$$

14. Denote the ordinary least squares estimator of the regression of Y on X based on the sample $(x_j, y_j), \ j = 1, 2, \ldots, n$ by

$$b_1 = \frac{\sum_{i=1}^{n}(x_i - \bar{x})(y_i - \bar{y})}{\sum_{i=1}^{n}(x_i - \bar{x})^2}, \qquad b_0 = \bar{y} - b_1\bar{x}.$$

(a) Show that $\hat{y}_j = \sum_{i=1}^{n} h_{ij} y_i$ where

$$h_{ij} = \frac{1}{n} + \frac{(x_i - \bar{x})(x_j - \bar{x})}{\sum_{i=1}^{n}(x_i - \bar{x})^2}.$$

(b) Show that $\sum_{i=1}^{n} h_{ij} = 1$; and

$$\sum_{i=1}^{n} h_{ij} x_i = \sum_{i=1}^{n} h_{ij}(x_i - \bar{x}) + \bar{x}.$$

(c) Use the results in (b) and the fact that $E[y_i] = \beta_0 + \beta_1 x_i$ to show that $E[\hat{y}_j] = E[y_j]$.

15. Assume that the random variables X, Y and Z are mutually independent and define the ratio variables $W = \frac{X}{Z}$, $V = \frac{Y}{Z}$ and $U = \frac{Y}{Z}$. Denote the means and variances by μ_x, μ_y, μ_z, μ_w, μ_v, μ_u, σ_x^2, σ_y^2, σ_z^2, σ_w^2, σ_v^2 and σ_u^2. Use the result that the expectation of the product of independent variables is the product of the expectations of the individual variables (i.e., $E[XY] = E[X]E[Y]$) AND use the result that if X and Y are mutually independent then $g(X)$ and $f(Y)$ are also mutually independent for any functions g and f in the following questions.

(a) Show that $\mathrm{Cov}(W, V) = \mu_x \mu_y \sigma_u^2$, $\mathrm{Cov}(U, W) = \mu_x \sigma_u^2$, $\mathrm{Cov}(U, V) = \mu_y \sigma_u^2$.

(b) Show that

$$\rho_{wv} = \frac{\mu_x \mu_y \sigma_u^2}{\sigma_v \sigma_w}, \qquad \rho_{uw} = \frac{\mu_x \sigma_u}{\sigma_w}, \qquad \rho_{vu} = \frac{\mu_y \sigma_u}{\sigma_v}$$

and the partial correlation

$$\rho_{wv \cdot u} = \frac{\rho_{wv} - \rho_{wu}\rho_{vu}}{\sqrt{1 - \rho_{wu}^2}\sqrt{1 - \rho_{vu}^2}} = 0.$$

(c) Use the results in (b) to show that if μ_x and μ_y are positive then ρ_{wv} is positive and discuss the importance of this result to the study of the relationship between ratio type variables.

(d) What does the result $\rho_{wv \cdot u} = 0$ mean?

16. (a) Given three random variables X, Y and Z assume that $\rho_{xz} = \rho_{yz} = 0.60$. Use the expression for $\rho_{xy \cdot z}$ and the fact that $-1 \le \rho_{xy \cdot z} \le 1$ to obtain a range for ρ_{xy}.

(b) Given three random variables X, Y and Z assume $\rho_{xy} = 0.20$, $\rho_{xz} = 0.90$ and $\rho_{yz} = 0.50$ and determine the proportion of the variation in Y explained by both X and Z. Is this proportion greater than $(\rho_{xy}^2 + \rho_{zy}^2)$? Explain the importance of this result.

17. Two samples of observations (x_{i1}, y_{i1}), $i = 1, 2, \ldots, n_1$, and (x_{i2}, y_{i2}), $i = 1, \ldots, n_2$, are selected from two different population groups. Denote the sample means on X and Y by

$$\bar{x}_1 = \frac{1}{n_1}\sum_{i=1}^{n_1} x_{i1}, \qquad \bar{x}_2 = \frac{1}{n_2}\sum_{i=1}^{n_2} x_{i2}$$

$$\bar{y}_1 = \frac{1}{n_1}\sum_{i=1}^{n_1} y_{i1}, \qquad \bar{y}_2 = \frac{1}{n_2}\sum_{i=1}^{n_2} y_{i2}.$$

(a) Show that if $\bar{x}_1 = a\bar{x}_2$ and $\bar{y}_1 = b\bar{y}_2$ and if the sample covariances in each sample are zero then the sample covariance for the combined sample is given by

$$\bar{x}_2\bar{y}_2\frac{n_1 n_2}{n}(1 - a)(1 - b)\left(\frac{1}{n-1}\right).$$

Thus combining two uncorrelated samples with different means yields a combined sample with X and Y correlated!

(b) Discuss the impact of the magnitudes of (a) and (b) on this covariance.

(c) Plot a scatter plot to show this result and hence obtain the "dumbell" effect.

18. Given a sample of observations (x_i, y_i) $i = 1, 2, \ldots, n$, the sample covariance for all n observations is given by $s_{xy} = \sum_{i=1}^{n}(x_i - \bar{x})(y_i -$

$\bar{y})/(n-1)$. If observation (x_i, y_i) is omitted then the sample covariance is denoted by $s_{xy}(i)$. Show that

$$s_{xy}(i) = \frac{(n-1)}{(n-2)} s_{xy} - \frac{n}{(n-1)(n-2)} (x_i - \bar{x})(y_i - \bar{y})$$

and $\quad s_{xy} = \frac{(n-2)}{(n-1)} s_{xy}(i) + \frac{1}{n}(x_i - \bar{x}(i))(y_i - \bar{y}(i))$

where $\bar{x}(i)$ and $\bar{y}(i)$ denote the sample means with (x_i, y_i) omitted. Discuss the impact on the covariance of an outlier (x_i, y_i).

19. The relationship between two random variables Y and X is given by

$$Y = \beta_0 + \beta_1 X \qquad \text{if } X \leq C$$

$$\text{and} \qquad Y = \gamma_0 + \gamma_1 X \qquad \text{if } X \geq C.$$

Assume that at $X = C$ the two equations for Y are equal. Show that the relationship between Y and X can be written

$$Y = \beta_0 + \beta_1 X + \lambda(X - C)Z$$

where

$$Z = 0 \qquad \text{if } X \leq C$$

$$Z = 1 \qquad \text{if } X > C \quad \text{and} \quad \lambda = (\gamma_1 - \beta_1).$$

20. The conditional density function for Y given X if X and Y are bivariate normal is given by

$$f_y(y \mid X = x) = \frac{1}{\sqrt{2\pi}\,\sigma_u} \exp\left[-\frac{1}{2}(y - \beta_0 - \beta_1 x)^2/\sigma_u^2\right].$$

(a) Show that the likelihood of the sample (x_i, y_i) is given by

$$L = \frac{1}{(2\pi)^{n/2}[\sigma_u^2]^{n/2}} \exp\left[-\frac{1}{2}\sum_{i=1}^{n}(y_i - \beta_0 - \beta_1 x_i)^2/\sigma_u^2\right]$$

and that the natural logarithm of the likelihood is given by

$$\ln L = -\frac{n}{2}\ln 2\pi - \frac{n}{2}\ln \sigma_u^2 - \frac{1}{2}\sum_{i=1}^{n}(y_i - \beta_0 - \beta_1 x_i)^2/\sigma_u^2.$$

(b) Maximizing L with respect to $(\beta_0, \beta_1, \sigma_u^2)$ is equivalent to maximizing $\ln L$ with respect to $(\beta_0, \beta_1, \sigma_u^2)$. Show that the values of β_0, β_1 and σ_u^2 that maximize $\ln L$ are given by

$$\tilde{b}_1 = \left[n \sum_{i=1}^{n} x_i y_i - \sum_{i=1}^{n} x_i \sum_{i=1}^{n} y_i \right] / \left[n \sum_{i=1}^{n} x_i^2 - \left(\sum_{i=1}^{n} x_i \right)^2 \right]$$

$$\tilde{b}_0 = \bar{y} - \tilde{\beta}_1 \bar{x}$$

$$\tilde{s}^2 = \frac{1}{n} \sum_{i=1}^{n} (y_i - \tilde{\beta}_0 - \tilde{\beta}_1 x_i)^2.$$

21. Let $\mathbf{e}' = (e_1, e_2, \ldots, e_n)$ denote the vector of residuals from the ordinary least squares regression of $\mathbf{y}' = (y_1, y_2, \ldots, y_n)$ on $\mathbf{x}' = (x_1, x_2, \ldots, x_n)$.

(a) Show that $\sum_{i=1}^{n} e_i = 0$ and $\sum_{i=1}^{n} x_i e_i = 0$.

(b) Determine the angle between the vectors $\mathbf{e}'$ and the vectors $\mathbf{x}'$ and $\mathbf{i}'$ where $\mathbf{i}' = (1, 1, \ldots, 1)$.

(c) What does the result in (b) say about the geometric relationship between $\mathbf{e}$ and the vectors $\mathbf{x}$ and $\mathbf{i}$?

(d) Show that the vector $\hat{\mathbf{y}}' = (\hat{y}_1, \hat{y}_2, \ldots, \hat{y}_n)$ is a linear combination of the vectors $\mathbf{i}$ and $\mathbf{x}$. What does this imply about the geometric relationship between $\hat{\mathbf{y}}$ and $\mathbf{e}$?

(e) Show that $\mathbf{y}$ can be written as a linear combination of $\hat{\mathbf{y}}$ and $\mathbf{e}$. What does this result say about the geometric relationship between $\mathbf{y}, \hat{\mathbf{y}}$ and $\mathbf{e}$?

22. Given a sample of observations (x_i, y_i) $i = 1, 2, \ldots, n$ the means, variances and covariance are denoted by $\bar{x}, \bar{y}, s_x^2, s_y^2$ and s_{xy}. If observation (x_i, y_i) is omitted the means, variances and covariance are denoted by $\bar{x}(i), \bar{y}(i), s_x^2(i), s_y^2(i)$, and $s_{xy}(i)$. The relationships given below have been demonstrated in other questions.

$$s_x^2 = \frac{(n-2)}{(n-1)} s_x^2(i) + \frac{1}{n} [\bar{x}(i) - x_i]^2$$

$$s_{xy} = \frac{(n-2)}{(n-1)} s_{xy}(i) + \frac{1}{n} (\bar{x}(i) - x_i)(\bar{y}(i) - y_i).$$

Assume that

$$(\bar{x}(i) - x_i)^2 = k_x^2 s_x^2(i)$$

$$(\bar{y}(i) - y_i)^2 = k_y^2 s_y^2(i)$$

and show that the Pearson correlation coefficient $r = s_{xy}/s_x s_y$ can be written as

$$r = \frac{r(i) \pm \frac{(n-1)}{(n-2)} \left[\frac{k_x k_y}{n} \right]}{\sqrt{\left[1 + \frac{(n-1)}{(n-2)} \frac{k_x^2}{n} \right] \left[1 + \frac{(n-1)}{(n-2)} \frac{k_y^2}{n} \right]}}$$

where the "$-$" sign occurs when $(\bar{x}(i) - x_i)(\bar{y}(i) - y_i)$ is negative. Use this expression to determine the impact on r of an outlier (x_i, y_i).

4
Multiple Linear Regression

The *multiple linear regression model* is the most commonly applied statistical technique for relating a set of two or more variables. In Chapter 3 the concept of a regression model was introduced to study the relationship between two quantitative variables X and Y. In the latter part of Chapter 3, the impact of another explanatory variable Z on the regression relationship between X and Y was also studied. It was shown that by extending the regression to include the explanatory variable Z, the relationship between Y and X can be studied while controlling or taking into account Z. In a multivariate setting, the regression model can be extended so that Y can be related to a set of p explanatory variables $X_1, X_2, \ldots, X_p$. In this chapter, an extensive outline of the multiple linear regression model and its applications will be presented. A data set to be used as a multiple regression example is described next.

Example

Financial accounting information for the year 1983 was collected from the DATASTREAM database for a sample of 40 UK listed companies. The variables to be employed in the example are listed in Table 4.1. Table 4.2 below shows the observations obtained for the 13 financial variables. This data will be used to estimate a linear relationship between the return on capital employed (RETCAP) and the remaining 12 variables.

The linear model to be estimated is given by

$$\text{RETCAP} = \beta_0 + \beta_1 \text{ WCFTDT} + \beta_2 \text{ LOGSALE} + \beta_3 \text{ LOGASST}$$

$$+ \beta_4 \text{ CURRAT} + \beta_5 \text{ QUIKRAT} + \beta_6 \text{ NFATAST}$$

$$+ \beta_7 \text{ FATTOT} + \beta_8 \text{ PAYOUT} + \beta_9 \text{ WCFTCL}$$

$$+ \beta_{10} \text{ GEARRAT} + \beta_{11} \text{ CAPINT} + \beta_{12} \text{ INVTAST} + U,$$

where U is an error term representing all other factors which influence RETCAP. As we shall see below the assumptions made about the properties of U will be crucial.

Table 4.1. Financial Accounting Variables for 40 UK Companies

Variable Abbr.	Variable Definition
RETCAP	Return on capital employed
WCFTDT	Ratio of working capital flow to total debt
LOGSALE	Log to base 10 of total sales
LOGASST	Log to base 10 of total assets
CURRAT	Current ratio
QUIKRAT	Quick ratio
NFATAST	Ratio of net fixed assets to total assets
FATTOT	Gross fixed assets to total assets
PAYOUT	Payout ratio
WCFTCL	Ratio of working capital flow to total current liabilities
GEARRAT	Gearing ratio (debt-equity ratio)
CAPINT	Capital intensity (ratio of total sales to total assets)
INVTAST	Ratio of total inventories to total assets

4.1 The Multiple Linear Regression Model

4.1.1 The Model, Assumptions and Least Squares Estimation

The Model and Assumptions

In this section we extend the simple linear regression model outlined in Chapter 3. We assume that there are now p *explanatory variables* $X_1, X_2, \ldots, X_p$ which are to be related to a dependent variable Y. The data matrix is assumed to be derived from a random sample of n observations $(x_{i1}, x_{i2}, \ldots, x_{ip}, y_i)$, $i = 1, 2, \ldots, n$, or equivalently, an $n \times (p + 1)$ data matrix as described in Chapter 1.

The $(p + 1)$ random variables are assumed to satisfy the linear model

$$y_i = \beta_0 + \beta_1 x_{i1} + \beta_2 x_{i2} + \cdots + \beta_p x_{ip} + u_i \qquad i = 1, 2, \ldots, n$$

where:

(1) The u_i, $i = 1, 2, \ldots, n$ are values of an unobserved error term U and are mutually independent and identically distributed; $E[u_i] = 0$; $V[u_i] = \sigma_u^2$.

(2) The distribution of the error term U is independent of the joint distribution of $X_1, X_2, \ldots, X_p$ and hence the regression function $E[Y \mid X_1, X_2, \ldots, X_p] = \beta_0 + \beta_1 X_1 + \cdots + \beta_p X_p$; and $V[Y \mid X_1, X_2, \ldots, X_p] = \sigma_{Y \cdot X_1, X_2, \ldots, X_p}^2 = \sigma_u^2$.

(3) The unknown parameters $\beta_0, \beta_1, \beta_2, \ldots, \beta_p$ are constants.

Since the observations $y_1, y_2, \ldots, y_n$ are a random sample they are mutually independent and hence the error terms are also mutually independent. Because of assumption (2) the results are conditional on the observed values of $X_1, X_2, \ldots, X_p$.

Table 4.2. Financial Accounting Data for a Sample of 40 UK Companies 1983

RETCAP	GEARRAT	CAPINT	WCFTDT	LOGSALE	LOGASST	CURRAT
0.26	0.46	0.64	0.25	4.11	4.30	1.53
0.57	0.00	1.79	0.33	4.25	4.00	1.73
0.09	0.24	0.36	0.20	4.44	4.88	0.44
0.32	0.45	1.86	0.21	4.71	4.44	1.23
0.17	0.91	1.26	0.12	4.85	4.75	1.76
0.24	0.26	1.54	0.25	5.61	5.42	1.44
0.53	0.52	3.34	0.40	4.83	4.30	0.83
0.26	0.24	1.38	0.37	4.49	4.35	1.45
0.13	0.19	0.91	0.21	4.13	4.17	2.89
0.16	0.29	1.70	0.18	4.40	4.17	2.13
0.06	0.85	1.60	0.01	4.30	4.09	1.31
0.07	0.02	0.15	0.70	3.62	4.45	4.57
−0.18	0.76	0.60	−0.32	4.13	4.35	0.47
0.12	0.39	2.34	0.11	4.11	3.74	0.85
0.15	0.06	1.19	0.65	4.63	4.55	1.81
0.03	0.00	0.00	1.47	0.00	4.18	12.98
0.08	0.39	1.12	0.08	4.06	4.01	1.43
0.09	0.26	1.42	0.13	4.21	4.06	1.75
0.25	0.15	2.33	0.23	3.99	3.62	1.49
−0.03	0.67	1.62	−0.07	4.51	4.30	1.35
0.03	0.15	0.00	0.04	1.74	4.24	0.29
0.04	0.34	1.65	0.03	4.24	4.02	1.42
0.17	0.38	1.29	0.14	3.52	3.41	1.12
0.07	0.18	1.20	0.05	4.03	3.96	1.50
0.11	0.45	2.40	0.09	4.35	3.97	1.30
0.04	0.54	3.46	−0.01	4.72	4.18	1.08
0.04	0.09	2.11	0.46	4.26	3.93	1.32
0.11	0.17	1.16	0.17	4.67	4.60	1.34
0.14	0.35	2.75	0.08	4.82	4.38	1.32
0.29	0.13	1.88	0.34	5.14	4.86	1.71
0.02	0.13	1.53	0.15	4.14	3.96	1.97
0.10	0.59	3.42	0.12	5.06	4.53	0.91
0.14	0.16	1.11	0.26	6.29	6.25	1.09
0.11	0.17	2.03	0.23	4.77	4.46	1.45
0.29	0.43	1.10	0.26	6.23	6.18	1.83
0.04	0.37	0.66	0.05	4.30	4.47	1.23
0.17	0.04	1.68	0.23	4.27	4.05	2.35
0.16	0.17	1.71	0.25	5.24	5.01	1.88
0.14	0.02	1.38	0.39	3.99	3.85	1.37
0.13	0.23	1.49	0.20	4.54	4.37	2.49

Table 4.2. Financial Accounting Data for a Sample of 40 UK Companies 1983 (continued)

QUIKRAT	NFATAST	INVTAST	FATTOT	PAYOUT	WCFTCL
0.18	0.10	0.74	0.12	0.07	0.25
1.26	0.12	0.27	0.15	0.30	0.33
0.39	0.94	0.01	0.97	0.57	0.50
0.69	0.29	0.29	0.52	0.00	0.23
0.90	0.26	0.33	0.54	0.31	0.21
1.23	0.42	0.06	0.57	0.15	0.37
0.83	0.14	0.00	0.21	0.21	0.59
0.58	0.40	0.36	1.04	0.16	0.44
1.95	0.06	0.29	0.11	0.39	0.21
0.56	0.21	0.58	0.40	0.46	0.21
0.73	0.23	0.34	0.38	0.00	0.01
4.51	0.54	0.00	0.63	0.00	0.70
0.47	0.54	0.00	0.84	0.00	−0.58
0.14	0.41	0.49	0.97	0.00	0.11
1.25	0.65	0.10	0.77	0.26	0.81
12.98	0.05	0.00	0.06	0.00	1.47
0.59	0.36	0.36	0.44	0.00	0.09
0.92	0.31	0.33	0.41	0.60	0.13
0.79	0.21	0.37	0.49	0.23	0.23
0.57	0.20	0.43	0.41	0.00	−0.07
0.29	0.00	0.00	0.00	0.00	0.04
0.71	0.28	0.36	0.43	4.21	0.03
0.61	0.27	0.33	0.37	0.16	0.15
0.90	0.21	0.31	0.31	1.66	0.05
0.86	0.36	0.22	0.60	0.35	0.10
0.66	0.49	0.19	0.51	0.00	−0.20
0.78	0.64	0.15	1.16	0.97	0.55
0.76	0.32	0.23	0.55	0.71	0.17
0.30	0.21	0.61	0.25	0.56	0.09
0.96	0.29	0.30	0.50	0.43	0.37
1.33	0.43	0.18	0.85	0.00	0.19
0.58	0.41	0.20	0.57	0.65	0.14
0.59	0.67	0.11	0.81	0.47	0.39
0.75	0.45	0.27	0.84	0.67	0.26
1.17	0.28	0.20	0.43	0.52	0.42
0.33	0.47	0.39	0.53	1.83	0.05
1.33	0.11	0.39	0.26	0.71	0.24
1.05	0.29	0.31	0.65	0.31	0.32
0.66	0.52	0.24	0.82	0.78	0.39
1.47	0.32	0.28	0.46	0.58	0.30

A sample of n observations on Y and $X_1, X_2, \ldots, X_p$ forms a $[n \times (p+1)]$ data matrix. Equations relating the n observations are given by the n equations

$$y_1 = \beta_0 + \beta_1 x_{11} + \beta_2 x_{12} + \cdots + \beta_p x_{1p} + u_1$$

$$y_2 = \beta_0 + \beta_1 x_{21} + \beta_2 x_{22} + \cdots + \beta_p x_{2p} + u_2$$

$$\vdots$$

$$y_n = \beta_0 + \beta_1 x_{n1} + \beta_2 x_{n2} + \cdots + \beta_p x_{np} + u_n.$$

Using matrix notation this system of n equations can be written as

$$\mathbf{y} = \mathbf{X}\boldsymbol{\beta} + \mathbf{u},$$

where:

$$\mathbf{y} = \begin{bmatrix} y_1 \\ y_2 \\ \vdots \\ y_n \end{bmatrix}_{(n \times 1)} \quad ; \quad \mathbf{X} = \begin{bmatrix} 1 & x_{11} & x_{12} & \cdots & x_{1p} \\ 1 & x_{21} & x_{22} & \cdots & x_{2p} \\ \vdots & & & & \\ 1 & x_{n1} & x_{n2} & \cdots & x_{np} \end{bmatrix}_{n \times (p+1)} ;$$

$$\boldsymbol{\beta} = \begin{bmatrix} \beta_0 \\ \beta_1 \\ \beta_2 \\ \vdots \\ \beta_p \end{bmatrix}_{(p+1) \times 1} \quad ; \quad \text{and} \quad \mathbf{u} = \begin{bmatrix} u_1 \\ u_2 \\ \vdots \\ u_n \end{bmatrix}_{(n \times 1)}.$$

Least Squares Estimation

As in simple linear regression, least squares can be used to estimate the parameter vector $\boldsymbol{\beta}$ given the observations $\mathbf{y}$ and $\mathbf{X}$. The *least squares procedure* determines the $\boldsymbol{\beta}$ vector that minimizes the sum of squares given by

$$\zeta^2 = \sum_{i=1}^{n} (y_i - \beta_0 - \beta_1 x_{i1} - \beta_2 x_{i2} - \cdots - \beta_p x_{ip})^2 = (\mathbf{y} - \mathbf{X}\boldsymbol{\beta})'(\mathbf{y} - \mathbf{X}\boldsymbol{\beta}).$$

Using calculus the resulting equation for $\boldsymbol{\beta}$ is given by the *normal equations*

$$\sum_{i=1}^{n} y_i = n\beta_0 + \beta_1 \sum_{i=1}^{n} x_{i1} + \beta_2 \sum_{i=1}^{n} x_{i2} + \cdots + \beta_p \sum_{i=1}^{n} x_{ip}$$

$$\sum_{i=1}^{n} x_{i1} y_i = \beta_0 \sum_{i=1}^{n} x_{i1} + \beta_1 \sum_{i=1}^{n} x_{i1}^2 + \beta_2 \sum_{i=1}^{n} x_{i1}x_{i2} + \cdots + \beta_p \sum_{i=1}^{n} x_{i1}x_{ip}$$

$$\sum_{i=1}^{n} x_{i2} y_i = \beta_0 \sum_{i=1}^{n} x_{i2} + \beta_1 \sum_{i=1}^{n} x_{i1}x_{i2} + \beta_2 \sum_{i=1}^{n} x_{i2}^2 + \cdots + \beta_p \sum_{i=1}^{n} x_{i2}x_{ip}$$

$$\vdots$$

$$\sum_{i=1}^{n} x_{ip} y_i = \beta_0 \sum_{i=1}^{n} x_{ip} + \beta_1 \sum_{i=1}^{n} x_{i1}x_{ip} + \beta_2 \sum_{i=1}^{n} x_{i2}x_{ip} + \cdots + \beta_p \sum_{i=1}^{n} x_{ip}^2.$$

The normal equations can also be written in matrix notation as

$$\mathbf{X}'\mathbf{y} = \mathbf{X}'\mathbf{X}\boldsymbol{\beta},$$

where:

$$\mathbf{X}'\mathbf{y} = \begin{bmatrix} \sum_{i=1}^{n} y_i \\ \sum_{i=1}^{n} x_{i1} y_i \\ \sum_{i=1}^{n} x_{i2} y_i \\ \vdots \\ \sum_{i=1}^{n} x_{ip} y_i \end{bmatrix}_{(p+1)\times 1} \quad \text{and}$$

$$\mathbf{X}'\mathbf{X} = \begin{bmatrix} n & \sum_{i=1}^{n} x_{i1} & \sum_{i=1}^{n} x_{i2} & \cdots & \sum_{i=1}^{n} x_{ip} \\ \sum_{i=1}^{n} x_{i1} & \sum_{i=1}^{n} x_{i1}^2 & \sum_{i=1}^{n} x_{i1}x_{i2} & & \sum_{i=1}^{n} x_{i1}x_{ip} \\ \sum_{i=1}^{n} x_{i2} & \sum_{i=1}^{n} x_{i1}x_{i2} & \sum_{i=1}^{n} x_{i2}^2 & & \sum_{i=1}^{n} x_{i2}x_{ip} \\ \vdots & \vdots & \vdots & & \vdots \\ \sum_{i=1}^{n} x_{ip} & \sum_{i=1}^{n} x_{i1}x_{ip} & \sum_{i=1}^{n} x_{i2}x_{ip} & \cdots & \sum_{i=1}^{n} x_{ip}^2 \end{bmatrix}_{(p+1)\times(p+1)}$$

The j-th element $\sum_{i=1}^{n} x_{ij} y_i$ of $\mathbf{X'y}$ denotes the product of the vectors $\mathbf{x}_j$ and $\mathbf{y}$. The vector $\mathbf{X'y}$ denotes a vector of $(p+1)$ products between the matrix $\mathbf{X}$ and the vector $\mathbf{y}$. The matrix $\mathbf{X'X}$ is an example of a *sums of squares and cross products matrix*. The diagonal elements are the sums of squares of the elements in each column of $\mathbf{X}$ and the off diagonal elements are the cross products between the various columns of $\mathbf{X}$. Such matrices will play an important role throughout this text. The matrix $\mathbf{X'X}$ is sometimes referred to as the (raw) sum of squares and cross products matrix.

The solution vector for the normal equations is denoted by

$$\mathbf{b} = (\mathbf{X'X})^{-1}\mathbf{X'y}.$$

For the solution vector $\mathbf{b}$ to exist the matrix $(\mathbf{X'X})^{-1}$ must exist. The matrix $(\mathbf{X'X})$ therefore must have full rank $(p+1)$. A necessary and sufficient condition is that the columns of $\mathbf{X}$ must be *linearly independent*, or equivalently, it must not be possible to express any one of the $\mathbf{X}$ columns as a linear combination of the remaining $\mathbf{X}$ columns. If it is possible to express one column of $\mathbf{X}$ in terms of the other columns the condition is called *multicollinearity*. An example of such a violation would be if $\mathbf{x}_3 = \mathbf{x}_1 + \mathbf{x}_2$. Recalling that correlation measures the degree of linear association between two variables, a correlation coefficient of ± 1 between two columns of $\mathbf{X}$ would result in a violation of this full rank condition. If the correlation coefficient between two columns of $\mathbf{X}$ is very close to 1, eg. 0.99, the inverse will exist but $(\mathbf{X'X})^{-1}$ will contain some very large elements. As we shall see later, such occurrences of near multicollinearity cause difficulties for model fitting. The general problem of multicollinearity, its impact, its detection and solution will be discussed later in Section 4.3.

Example

For the UK financial accounting data the least squares model is given by

$$\text{RETCAP} = 0.196 - 0.035 \text{ GEARRAT} - 0.014 \text{ CAPINT}$$

$$+ 0.317 \text{ WCFTDT} + 0.118 \text{ LOGSALE}$$

$$- 0.079 \text{ LOGASST} - 0.209 \text{ CURRAT}$$

$$- 0.162 \text{ QUIKRAT} - 0.372 \text{ NFATAST}$$

$$+ 0.214 \text{ INVTAST} - 0.103 \text{ FATTOT}$$

$$- 0.019 \text{ PAYOUT} + 0.209 \text{ WCFTCL}. \qquad (4.1)$$

Each coefficient of an explanatory variable measures the impact of that variable on the dependent variable RETCAP, holding the other variables fixed. The impact of an increase in WCFTDT of one unit is a change in RETCAP of 0.317 units, assuming that the other variables are held constant. Similarly an increase in CURRAT of one unit will bring about a decrease in RETCAP of 0.209 units, if all the other explanatory variables are held constant. In order to discuss the properties of this estimated linear relationship we must outline additional theory.

The SAS procedure PROC REG was used throughout this chapter to determine the analyses for this example.

Properties of Least Squares Estimators

As in the case of simple linear regression, the least squares estimators $b_0, b_1, \ldots, b_p$ contained in $\mathbf{b}$ are unbiased, $E[b_j] = \beta_j$, or equivalently, $E[\mathbf{b}] = \boldsymbol{\beta}$. The estimators are also *best linear unbiased* in the sense that among all estimators, which are both unbiased and linear functions of the elements of $\mathbf{y}$, the least squares estimators have the smallest variance.

The variances of the elements of $\mathbf{b}$, and the covariances among all possible pairs of elements in $\mathbf{b}$, can be summarized in a matrix called a covariance matrix. Each of the elements of the covariance matrix can be written as $\text{Cov}(b_j, b_k) = E[(b_j - E[b_j])(b_k - E[b_k])]$. The complete matrix containing all possible pairs is given by $E[(\mathbf{b} - E[\mathbf{b}])(\mathbf{b} - E[\mathbf{b}])']$ and can be written as

$$E[(\mathbf{b} - \boldsymbol{\beta})(\mathbf{b} - \boldsymbol{\beta})'] = \begin{bmatrix} V(b_0) & \text{Cov}(b_0, b_1) & \cdots & \text{Cov}(b_0, b_p) \\ \text{Cov}(b_0, b_1) & V(b_1) & & \text{Cov}(b_1, b_p) \\ \vdots & & \ddots & \vdots \\ \text{Cov}(b_0, b_p) & \text{Cov}(b_1, b_p) & \cdots & V(b_p) \end{bmatrix}.$$

The covariance matrix above can be seen to be symmetric.

The covariance matrix is given by $\sigma_u^2 (\mathbf{X'X})^{-1} = \sigma_u^2 \mathbf{C}^2$ where $\mathbf{C}^2 = (\mathbf{X'X})^{-1}$. The elements of $\mathbf{C}^2 = (\mathbf{X'X})^{-1}$ will be denoted by c_{ij}^2 where $i, j = 0, 1, 2, \ldots, p$ to correspond to the subscripts on $\mathbf{b}$. For example, c_{0j}^2 refers to the element of $(\mathbf{X'X})^{-1}$ corresponding to $\text{Cov}(b_0, b_j)$ and hence $\text{Cov}(b_0, b_j) = \sigma_u^2 c_{0j}^2$.

An unbiased estimator, s^2, of σ_u^2 is provided by $(\mathbf{y} - \mathbf{Xb})'(\mathbf{y} - \mathbf{Xb})/(n - p - 1)$. The numerator of this expression is the minimum value of $(\mathbf{y} - \mathbf{X}\boldsymbol{\beta})'(\mathbf{y} - \mathbf{X}\boldsymbol{\beta})$ determined above when deriving the least squares estimator $\mathbf{b}$.

Sums of Squares

The fitted model obtained using **b** is given by

$$\hat{\mathbf{y}} = \mathbf{Xb} = \mathbf{X}(\mathbf{X'X})^{-1}\mathbf{X'y} = \mathbf{Hy},$$

where $\mathbf{H} = \mathbf{X}(\mathbf{X'X})^{-1}\mathbf{X'}$ is equivalent to the hat matrix introduced in Chapter 3. The residuals from the fitted model are given by

$$\mathbf{e} = \mathbf{y} - \hat{\mathbf{y}} = \mathbf{y} - \mathbf{Xb} = (\mathbf{I} - \mathbf{H})\mathbf{y}.$$

The sum of squared residuals can be written

$$\sum_{i=1}^{n}(y_i - \hat{y}_i)^2 = (\mathbf{y} - \hat{\mathbf{y}})'(\mathbf{y} - \hat{\mathbf{y}}) = (n - p - 1)s^2$$

$$= (\mathbf{y} - \mathbf{Xb})'(\mathbf{y} - \mathbf{Xb}) = (\mathbf{y'y} - \mathbf{b'X'y}) = \mathbf{y'}(\mathbf{I} - \mathbf{H})\mathbf{y}.$$

This quantity is referred to as the error sum of squares and is denoted by SSE.

The total sum of squares SST $= \sum_{i=1}^{n}(y_i - \bar{y})^2 = \mathbf{y'y} - n\bar{y}^2$, is equivalent to the quantity defined in simple linear regression. The SST represents the total variation in Y that is to be explained by one or more X variables.

The regression sum of squares, SSR, is also defined as in simple linear regression

$$\text{SSR} = \sum_{i=1}^{n}(\hat{y}_i - \bar{y})^2 = \mathbf{b'X'y} - n\bar{y}^2 = \hat{\mathbf{y}}'\hat{\mathbf{y}} - n\bar{y}^2 = \mathbf{y'Hy} - n\bar{y}^2.$$

Coefficient of Multiple Determination

The ratio of SSR to SST represents the proportion of the total variation in Y explained by the regression model. As in simple linear regression, this ratio is denoted by R^2 and is referred to as the *coefficient of multiple determination*. In small samples, if p is large relative to n there is a tendency for the model to fit the data very well. R^2 therefore is sensitive to the magnitudes of n and p in small samples. In the extreme case if $n = (p+1)$ the model will fit the data exactly.

In order to give a better measure of goodness of fit a penalty function can be employed to reflect the number of explanatory variables used. Using this penalty function the *adjusted* R^2 is given by

$$R'^2 = 1 - \left(\frac{n-1}{n-p-1}\right)(1 - R^2) = 1 - \frac{\text{SSE}/(n-p-1)}{\text{SST}/(n-1)}.$$

Note that the adjusted R^2, R'^2, changes the ratio SSE/SST, to the ratio of SSE/$(n-p-1)$ an unbiased estimator of σ_u^2, and SST/$(n-1)$ an unbiased estimator of σ_y^2.

Example

For the estimated financial accounting model given by (4.1), the sums of squares are SSR = 0.556, SSE = 0.152 and SST = 0.708. The ratio of SSR to SST yields an R^2 value of 0.785 and hence 78.5% of the variation in RETCAP is explained by the 12 variables. The adjusted R^2 value is given by 0.689. An unbiased estimator of σ_u^2 is given by $s^2 = \text{SSE}/(n - p - 1) = 0.152/27 = 0.0056$.

Coefficient of Multiple Correlation

The square root of R^2, denoted by R, is defined to be the *coefficient of multiple correlation*. This correlation is equivalent to the Pearson correlation between the original observations y_i and the predicted or fitted observations $\hat{y}_i$, $i = 1, 2, \ldots, n$. For the estimated financial accounting model, the coefficient of multiple correlation is 0.886 which indicates a strong correlation between the observed and fitted y observations.

Some Population Parameters

The dependent variable Y and the p explanatory variables $X_1, X_2, \ldots, X_p$ together may be viewed as a *multivariate random variable* $\begin{bmatrix} y \\ x \end{bmatrix}$. The mean vector and covariance matrix for this random vector are given by

$$\boldsymbol{\mu} = \begin{bmatrix} \mu_y \\ \boldsymbol{\mu_x} \end{bmatrix} \quad \text{and} \quad \boldsymbol{\Sigma} = \begin{bmatrix} \sigma_y^2 & \boldsymbol{\sigma'_{xy}} \\ \boldsymbol{\sigma_{xy}} & \boldsymbol{\Sigma_{xx}} \end{bmatrix},$$

where (μ_y, σ_y^2) denote the mean and variance of y; $\boldsymbol{\sigma_{xy}}$ denotes the $(p \times 1)$ vector of covariances between y and $X_1, X_2, \ldots, X_p$; and $(\boldsymbol{\mu_x}, \boldsymbol{\Sigma_{xx}})$ denote the mean vector and covariance matrix for the X variables.

The $(p \times 1)$ vector of regression coefficients excluding the intercept β_0 is denoted by $\boldsymbol{\beta^*}$, hence $\boldsymbol{\beta} = \begin{bmatrix} \beta_0 \\ \boldsymbol{\beta^*} \end{bmatrix}$. The vector $\boldsymbol{\beta^*}$ is given by $\boldsymbol{\Sigma_{xx}^{-1}\sigma_{xy}}$ and the total variance in y explained by the p X variables is $\boldsymbol{\beta^{*'}\Sigma_{xx}\beta^*}$. The residual variance in the regression is therefore $\sigma_u^2 = \sigma_{yx}^2 = \sigma_y^2 - \boldsymbol{\beta^{*'}\Sigma_{xx}\beta^*}$. The population coefficient of determination is given by $\rho^2 = \boldsymbol{\beta^{*'}\Sigma_{xx}\beta^*}/\sigma_y^2$. If $\rho^2 = 0$, then R^2 defined above is an unbiased estimator of ρ^2.

4.1.2 Statistical Inference in Multiple Linear Regression

To make inferences about the multiple regression model using the results from the least squares estimation, we require a fourth assumption regarding the distribution of the elements of the error vector **u**.

An Additional Assumption

(4) The elements of the error vector **u** are normally distributed. Since the error terms u_i, $i = 1, 2, \ldots, n$ have mean 0 and common variance σ_u^2 and are mutually independent, the u_i are independent and identically distributed normal random variables.

Inferences for Regression Coefficients

Inferences can be made for the individual regression coefficient β_j using the statistic

$$(b_j - \beta_j)/(s^2 c_{jj}^2)^{1/2},$$

which has a t distribution with $(n-p-1)$ degrees of freedom. This statistic can be used to test the null hypothesis $H_0 : \beta_j = b^*$ by computing the test statistic $(b_j - b^*)/(s^2 c_{jj}^2)^{1/2}$. If H_0 is true, this statistic has a t distribution with $(n-p-1)$ degrees of freedom. A $100(1-\alpha)\%$ confidence interval for β_j can be obtained from the expression

$$b_j \pm t_{\alpha/2;(n-p-1)}(s^2 c_{jj}^2)^{1/2}.$$

It is very important to keep in mind that the inferences made above regarding β_j are conditional on the other X variables being included in the model. Recall from Chapter 3 that the addition of an explanatory variable to the model usually causes the regression coefficients to change.

Example

For the financial accounting data of Table 4.2, the estimated regression coefficients and their standard errors are given in Table 4.3 below. The ratio of each estimate to its standard error determines the t-statistic for testing the null hypothesis that the true regression coefficient is zero. The t-values and their corresponding p-values are also given in the table. Using a 0.10 p-value cutoff, only five of the twelve coefficients are significantly different from zero. For the five significant variables, it would appear that marginal increases in either of the variables LOGSALE and QUIKRAT bring about increases in RETCAP. For the variables LOGASST, CURRAT and NFATAST marginal increases bring about decreases in RETCAP.

Table 4.3. Estimated Regression Coefficients and Significance Levels for UK Financial Accounting Example

Variable	Regression Coefficient Estimate	Standard Error	t value	p value
INTERCEP	0.196	0.1345	1.460	0.155
WCFTDT	0.317	0.2998	1.059	0.299
LOGSALE	0.118	0.0366	3.236	0.003
LOGASST	−0.078	0.0455	−1.725	0.095
CURRAT	−0.209	0.0899	−2.329	0.027
QUIKRAT	0.162	0.0938	1.727	0.095
NFATAST	−0.372	0.1387	−2.685	0.012
INVTAST	0.213	0.1894	1.129	0.268
FATTOT	−0.103	0.0884	−1.169	0.252
PAYOUT	−0.018	0.0178	−1.038	0.308
WCFTCL	0.208	0.1995	1.045	0.305
GEARRAT	−0.036	0.0780	−0.457	0.651
CAPINT	−0.014	0.0237	−0.584	0.564

Inferences for the Model

The overall *goodness of fit* of the model can be evaluated using an F-test in an analysis of variance format. Under the null hypothesis $H_0 : \beta_1 = \beta_2 = \ldots \beta_p = 0$ the statistic

$$\frac{[SSR/p]}{[SSE/(n-p-1)]} = \frac{MSR}{MSE}, \qquad (4.2)$$

has an F distribution with p and $(n-p-1)$ degrees of freedom. The statistic for this test is often summarized in an analysis of variance table as in Table 4.4 below.

Table 4.4. Analysis of Variance Table for Multiple Regression

Source	d.f.	Sum of Squares	Mean Square	F
Regression	p	SSR	MSR $= $ SSR$/p$	MSR/MSE
Error	$n-p-1$	SSE	MSE $= $ SSE$/(n-p-1)$	
Total	$n-1$	SST		

Example

The analysis of variance table for the financial accounting data is given below. The p-value of 0.0001 for the F-statistic indicates that the overall regression is highly significant.

Table 4.5. Analysis of Variance for Financial Accounting Model

Source	d.f.	Sum of Squares	Mean Square	F-value	p-value
Model	12	0.556	0.046	8.204	0.0001
Error	27	0.152	0.006		
Total	39	0.708			

Inference for Reduced Models

Generally there will be at least some X variables that are related to Y, and hence the F-test of goodness of fit usually results in rejection of H_0. A more useful test is one that permits the evaluation of a subset of the explanatory variables relative to a larger set. For notational purposes we assume without loss of generality that we wish to test the null hypothesis $H_0 : \beta_1 = \beta_2 = \cdots = \beta_q = 0$ where $q < p$. The model with all p explanatory variables is called the *full model*, while the model that holds under H_0 with $(p - q)$ explanatory variables is called the *reduced model*. The reduced model is therefore given by

$$y_i = \beta_0 + \beta_{q+1} x_{i\ q+1} + \cdots + \beta_p x_{ip} + u_i \qquad i = 1, 2, \ldots, n.$$

The sums of squares for the reduced model will be denoted by SSR_R and SSE_R. The statistic

$$[(\text{SSR} - \text{SSR}_R)/q]/[\text{SSE}/(n - p - 1)] \qquad (4.3)$$

has an F distribution with q and $(n - p - 1)$ degrees of freedom if the hypothesis H_0 is true. This *partial F-statistic* can also be written conveniently in terms of R_R^2 and R^2, where R_R^2 denotes the value of R^2 for the reduced model. The F-statistic therefore is also given by

$$[(R^2 - R_R^2)/q]/[(1 - R^2)/(n - p - 1)].$$

Example

A second regression model which employs only six of the twelve explanatory variables was estimated for the UK financial accounting data. Table 4.6 below summarizes the results of this regression. Comparing the full model fitted previously to this reduced model yields the F-statistic

$$F = \frac{(0.556 - 0.525)/6}{0.152/27} = 0.92.$$

Comparing this computed F-value to the critical value of $F_{.10;6,27} = 2.00$ allows us to conclude that, when the other six variables are present, the coefficients for GEARRAT, CAPINT, WCFTDT, FATTOT, PAYOUT and INVTAST are not significant at the 10% level.

Table 4.6.   Estimated Regression Coefficients and Significance Levels for Financial Accounting Example Reduced Model

Variable	Regression Coefficient Estimate	Standard Error	t-value	p-value
INTERCEP	0.211	0.102	2.053	0.048
WCFTCL	0.425	0.058	7.248	0.000
NFATAST	−0.505	0.074	−6.764	0.000
QUIKRAT	0.083	0.041	2.019	0.051
LOGSALE	0.097	0.025	3.896	0.000
LOGASST	−0.071	0.032	−2.197	0.035
CURRAT	−0.122	0.040	−3.006	0.005

The R^2 values for the two models are 0.785 and 0.741 which can be used to yield the same F-statistic

$$F = \frac{(0.785 - 0.741)/6}{(1 - 0.785)/27} = 0.92.$$

Dropping the six explanatory variables reduces the proportion of variance explained from 78.5% to 74.1%. This difference is not significant.

Relation to t-Test

In the special case that $q = 1$ the test statistic (4.3) is equivalent to the t-test for a single variable $H_0 : \beta_j = 0$. The t-statistic in this case is the square root of the F-statistic. The relationship is given by

$$t^2 = b_j^2/s^2 c_{jj}^2 = F = [\text{SSR} - \text{SSR}_R]/[\text{SSE}/(n - p - 1)].$$

Inferences for $\boldsymbol{\beta}$ Other than $\mathbf{0}$

To test the hypothesis $H_0 : \beta_1 = \beta_1^*, \quad \beta_2 = \beta_2^*, \ldots, \beta_p = \beta_p^*$ the F-test statistic developed above for $H_0 : \beta_1 = \beta_2 = \cdots = \beta_p = 0$ can be used. The model

$$y_i = \beta_0 + \beta_1 x_{i1} + \beta_2 x_{i2} + \cdots + \beta_p x_{ip} + u_i,$$

can be rewritten as

$$y_i - \beta_1^* x_{i1} - \beta_2^* x_{i2} - \cdots - \beta_p^* x_{ip} = \beta_0 + (\beta_1 - \beta_1^*)x_{i1} + (\beta_2 - \beta_2^*)x_{i2}$$

$$+ \cdots + (\beta_p - \beta_p^*)x_{ip} + u_i,$$

or equivalently,

$$z_i = \beta_0 + \gamma_1 x_{i1} + \gamma_2 x_{i2} + \cdots + \gamma_p x_{ip} + u_i,$$

where $z_i = y_i - \beta_1^* x_{i1} - \beta_2^* x_{i2} - \cdots - \beta_p^* x_{ip}$ and $\gamma_j = (\beta_j - \beta_j^*)$, $j = 1, 2, \ldots p$. Using the z_i in place of the y_i a test of $H_0 : \gamma_1 = \gamma_2 = \cdots = \gamma_p = 0$ can be carried out using the F-statistic (4.2) above.

Inferences for Linear Functions

Inferences about any linear combination of the coefficients $\boldsymbol{\lambda}'\boldsymbol{\beta}$ where $\boldsymbol{\lambda}$ is a $(p+1) \times 1$ vector of constants, can also be made using the t distribution. Under $H_0 : \boldsymbol{\lambda}'\boldsymbol{\beta} = B$, the statistic $(\boldsymbol{\lambda}'\mathbf{b} - B)/(s^2\boldsymbol{\lambda}'C^2\boldsymbol{\lambda})^{1/2}$ has a t distribution with $(n - p - 1)$ degrees of freedom.

As shown above, confidence intervals for individual regression coefficients can be obtained from the least squares fit using the t distribution. If a confidence region is desired for the entire $(p+1) \times 1$ vector $\boldsymbol{\beta}$ of regression coefficients, it would not be appropriate to use the $(p+1)$ individual confidence intervals. Instead a $100(1 - \alpha)\%$ *confidence ellipsoid* for $\boldsymbol{\beta}$ is given by

$$(\mathbf{b} - \boldsymbol{\beta})'\mathbf{X}'\mathbf{X}(\mathbf{b} - \boldsymbol{\beta}) \leq (p+1)s^2 F_{\alpha;p,(n-p-1)}.$$

This confidence region takes into account the correlation among the elements of $\mathbf{b}$.

Similarly a $100(1 - \alpha)\%$ confidence ellipsoid for the $(k \times 1)$ vector of linear combinations of the regression coefficients, $\mathbf{T}\boldsymbol{\beta}$, is given by

$$(\mathbf{Tb} - \mathbf{T}\boldsymbol{\beta})'[\mathbf{T}(\mathbf{X}'\mathbf{X})^{-1}\mathbf{T}']^{-1}(\mathbf{Tb} - \mathbf{T}\boldsymbol{\beta}) \leq (p+1)s^2 F_{\alpha;k,(n-p-1)},$$

where $\mathbf{T}$ is a $(k \times (p+1))$ matrix of known constants.

An estimator of $\boldsymbol{\beta}$ can be obtained subject to the condition $\mathbf{T}\boldsymbol{\beta} = \mathbf{a}$, where $\mathbf{T}(k \times (p+1))$ and $\mathbf{a}(k \times 1)$. The estimator is given by $\hat{\boldsymbol{\beta}}_a = \hat{\boldsymbol{\beta}} + (\mathbf{X}'\mathbf{X})^{-1}\mathbf{T}'[\mathbf{T}(\mathbf{X}'\mathbf{X})^{-1}\mathbf{T}']^{-1}(\mathbf{a} - \mathbf{T}\hat{\boldsymbol{\beta}})$.

Given $\mathbf{T}$ $(k \times (p+1))$, $\boldsymbol{\beta}$ $(p \times 1)$ and $\mathbf{a}$ $(k \times 1)$, a test of $H_0 : \mathbf{T}\boldsymbol{\beta} = \mathbf{a}$ can be carried out by employing the test statistics

$$(\mathbf{T}\hat{\boldsymbol{\beta}} - \mathbf{a})'[\mathbf{T}(\mathbf{X}'\mathbf{X})^{-1}\mathbf{T}']^{-1}(\mathbf{T}\hat{\boldsymbol{\beta}} - \mathbf{a})/ks^2,$$

which has an F distribution with k and $(n - p - 1)$ degrees of freedom if H_0 is true.

An example which uses these techniques to make inferences about regression coefficients will be presented in the example in the next section.

Prediction For New Observations

Given a new set of observations on the X variables $\mathbf{x}_0' = (\mathbf{x}_{01}, \mathbf{x}_{02}, \ldots, \mathbf{x}_{0p})$, a *prediction*, say $\hat{y}_0$, for both the true value of Y and $E[Y]$ can be made using the fitted model. The estimator $\hat{y}_0$, where

$$\hat{y}_0 = b_0 + b_1 x_{01} + b_2 x_{02} + \cdots + b_p x_{0p}, \tag{4.4}$$

provides an estimate of Y and $E[Y]$ at $\mathbf{x}_0$. The $100(1 - \alpha)\%$ confidence intervals for Y and $E[Y]$ are given by

$$\hat{y}_0 \pm t_{\alpha/2;(n-p-1)} s \left(1 + \mathbf{x}_0'(\mathbf{X}'\mathbf{X})^{-1}\mathbf{x}_0\right)^{1/2}$$

$$\text{and} \quad \hat{y}_0 \pm t_{\alpha/2;(n-p-1)} s \left(\mathbf{x}_0'(\mathbf{X}'\mathbf{X})^{-1}\mathbf{x}_0\right)^{1/2}$$

respectively. If confidence intervals are required at r distinct values of $\mathbf{x}$ a Bonferroni approximation to the joint interval can be obtained using (α/r) in place of α.

A $100(1 - \alpha)\%$ confidence interval for the entire regression surface is given by

$$\hat{y}_0 \pm [(p+1)F_{\alpha;(p+1),(n-p-1)}]^{1/2} s (\mathbf{x}_0'(\mathbf{X}'\mathbf{X})^{-1}\mathbf{x}_0)^{1/2}.$$

This expression is useful when confidence intervals are required for several $\mathbf{x}$ values and can be used instead of the Bonferroni intervals.

Predictions Using Statistical Software

A prediction for Y, given observations on $X_1, X_2, \ldots, X_p$, can be made conveniently using the same statistical software that is used to compute the least squares estimators. The original $n \times 1$ vector $\mathbf{y}$ is augmented by adding an $(n+1)$-th element 0 to correspond to the case to be predicted. The $\mathbf{X}$ matrix is augmented by an $(n+1)$-th row with the particular X values, $\mathbf{x}_0$, corresponding to the case to be predicted. In addition, a $(p+1)$-th variable (a new column) X_{p+1} is added to the $\mathbf{X}$ matrix with the element 0 in the first n rows and the element -1 in the last row corresponding to the case to be predicted. The augmented least squares estimator yields the same least squares solution that would be obtained without the augmented case. In addition the coefficient b_{p+1} of the new variable X_{p+1} is the required prediction $\hat{y}_0$, and the standard error of this regression coefficient is the standard error for the prediction.

Example

Using the reduced model given in Table 4.6 predictions were determined for a sample of ten companies from the same data file. Table 4.7 below shows the observed and predicted values of RETCAP as well as the values of the explanatory variables used. The standard errors of the predictions and the *t*-ratio formed from the ratio of the prediction errors to the standard error are also shown in the table. The predicted values of RETCAP for the new sample of ten companies appear to be very close to the actual values. The prediction errors are determined by subtracting column (2) from column (1) in Table 4.7. The sum of squares of the prediction errors divided by 10 yields an average squared error of 0.00129. This average prediction error compares quite favorably with $s^2 = 0.152/27 = 0.00563$.

4.1.3 A Second Regression Example – Air Pollution Data

Estimated Full Model

A second example of multiple regression is provided by a sample of 1960 air pollution data studied by Gibbons, McDonald and Gunst (1987). A random sample of 40 cities was selected from their list of 117. The data are shown in Table 4.8 below. The results of a least squares fit of the model relating TMR to the remaining variables are shown in Table 4.9. From this table it would appear that the only variables that are significant at a *p*-value of 0.10 are GE65 and PERWH. The value of R^2 for this model is 0.877. These calculations were performed using the SAS procedure PROC REG.

Some Reduced Models

A second model relating TMR to the subset of variables PMEAN, GE65, PERWH and NONPOOR was also estimated using this data. The regression results are shown in Table 4.10 below. In this case we can conclude that all four variables are significant at the 0.03 level. Thus, by deleting 7 of the 11 explanatory variables it is possible to obtain a model with four significant variables and an R^2 of 0.866.

By regressing TMR on the variables SMEAN, GE65 and NONPOOR a third model with $R^2 = 0.835$ is obtained. The results from this regression are shown in Table 4.11 below. All three variables are significant at the 0.05 level.

A fourth model relating TMR to the variables can be obtained (see Table 4.12) from the regression on the variables SMIN, PMAX, GE65,

Table 4.7. Predictions for a Sample of Ten Companies Using Fitted UK Financial Accounting Model

(1) Observed RETCAP	(2) Predicted RETCAP	(3) Std. Error of Prediction	t-ratio [(1) − (2)/(3)]	LOGSALE	LOGASST	CURRAT	QUIKRAT	NFATAST	WCFTCL
0.10	0.093	0.0762	0.09	4.4198	4.2682	1.78	0.96	0.41	0.23
0.05	0.011	0.0794	0.49	3.8954	3.7993	1.08	0.77	0.59	0.13
0.23	0.246	0.0803	−0.20	5.6682	5.5590	1.60	1.12	0.51	0.56
0.20	0.187	0.1078	0.12	4.4936	4.3313	5.02	2.42	0.16	0.79
0.13	0.136	0.0773	−0.08	4.7721	4.6358	2.48	1.47	0.29	0.27
0.15	0.090	0.0825	0.73	4.9294	4.9795	1.41	0.07	0.29	0.15
0.12	0.109	0.0857	0.13	4.2554	3.8927	3.20	1.47	0.09	0.17
0.16	0.148	0.0810	0.15	3.7142	3.5922	1.48	1.33	0.52	0.38
0.12	0.189	0.0825	−0.84	4.3035	4.4419	1.38	0.02	0.01	0.10
0.10	0.147	0.0764	−0.62	5.0200	4.4842	1.30	0.50	0.32	0.10

Table[a] 4.8. Air Pollution Data for a Sample of United States Cities, 1960

City	TMR	SMIN	SMEAN	SMAX	PMIN	PMEAN	PMAX
San Jose	664	21	37	105	50	108	3-2
Roanoke	929	29	80	313	42	112	343
Albuquerque	621	2	50	91	61	244	646
Charleston	825	37	56	152	35	78	233
Harrisburg	1008	73	119	220	50	92	189
Greenville	829	51	69	212	39	125	285
Hartford	899	28	128	344	53	114	241
Columbus	721	34	60	145	28	99	160
Orlando	828	21	47	106	23	76	164
Sacramento	810	22	41	147	22	81	149
Philadelphia	1029	87	229	620	70	160	342
Washington	780	51	124	210	76	135	242
Minneapolis	876	23	69	202	43	100	231
Los Angeles	869	25	123	280	50	156	344
Greensboro	747	38	60	71	28	60	94
Jacksonville	863	46	66	133	23	106	193
Madison	734	39	65	166	30	83	215
Wilmington	910	57	228	445	99	221	403
Tacoma	943	36	126	264	46	143	347
Scranton	1400	112	153	365	75	215	537
Canton	964	189	273	399	81	175	323
Atlanta	823	46	91	139	46	112	2326
Baltimore	978	24	165	414	48	148	495
Portland	1037	28	75	212	21	70	185
Springfield MA	996	37	162	396	24	77	182
Salt Lake City	682	27	79	260	47	121	309
Wichita	690	23	54	139	22	102	174
Lorain	821	59	81	351	37	144	417
Hamilton	776	82	100	225	42	86	163
Montgomery	908	30	42	70	26	72	157
San Diego	737	28	77	149	34	81	166
Duluth	1117	46	72	251	28	74	135
Wilkes Barre	1282	61	81	203	39	121	260
Wheeling	1210	152	194	437	74	198	444
San Antonio	734	16	68	233	50	124	296
Cincinnati	1039	45	125	194	63	145	316
Saginaw	854	33	51	107	42	101	202
Baton Rouge	706	27	58	113	57	125	352
New York	1046	60	228	531	79	162	270
Springfield OH	978	58	73	212	39	111	255

[a]TMR: total mortality rate
SMIN: smallest biweekly sulfate reading ($\mu_g/m^3 \times 10$)
SMEAN: arithmetic mean of biweekly sulfate readings ($\mu_g/m^3 \times 10$)
SMAX: largest biweekly sulfate reading ($\mu_g/m^3 \times 10$)
PMIN: smallest biweekly suspended particulate reading ($\mu/m^3 \times 10$)
PMEAN: arithmetic mean of biweekly suspended particulate readings
          ($\mu_g/m^3 \times 10$)
PMAX: largest biweekly suspended particulate reading ($\mu_g/m^3 \times 10$)

Table[a] 4.8. Air Pollution Data for a Sample of United States Cities, 1960 (continued)

City	PM2	GE65	PERWH	NONPOOR	LPOP
San Jose	49.3	70	96.8	89.8	58.0775
Roanoke	52.4	89	87.3	78.0	52.0086
Albuquerque	22.5	47	96.7	84.7	54.1863
Charleston	22.5	48	63.5	67.3	53.3522
Harrisburg	32.1	98	93.3	86.0	55.3791
Greenville	26.6	62	82.4	73.6	53.2176
Hartford	93.2	91	95.3	91.3	58.3857
Columbus	19.6	49	70.7	69.0	53.3843
Orlando	25.7	92	83.4	76.1	55.0309
Sacramento	51.1	69	92.5	89.7	57.0138
Philadelphia	122.4	91	84.3	87.0	66.3778
Washington	134.8	62	75.1	89.5	63.0144
Minneapolis	70.2	92	98.2	89.9	61.7086
Los Angeles	139.2	89	91.2	87.6	68.2883
Greensboro	37.9	61	79.1	79.4	53.9185
Jacksonville	58.6	62	76.6	77.6	56.5840
Madison	18.6	82	98.8	87.2	53.4654
Wilmington	46.5	77	87.6	87.9	55.6367
Tacoma	19.2	94	94.9	83.3	55.0730
Scranton	51.7	119	99.6	78.3	53.7020
Canton	59.4	95	94.6	86.8	55.3192
Atlanta	59.0	65	77.2	79.2	60.0740
Baltimore	95.6	76	77.8	85.5	62.3730
Portland	22.5	113	97.0	85.2	59.1482
Springfield, MA	46.4	105	97.0	87.7	57.2639
Salt Lake City	50.1	70	98.6	88.2	55.8324
Wichita	34.4	67	93.6	86.9	55.3559
Lorain	43.9	72	94.3	88.0	53.3746
Hamilton	42.3	71	94.7	87.0	52.9902
Montgomery	21.4	71	61.7	68.9	52.2843
San Diego	24.3	73	94.5	84.9	60.1410
Duluth	3.6	114	99.2	81.9	54.4185
Wilkes Barre	38.9	111	99.7	75.2	55.4029
Wheeling	20.1	122	97.6	77.8	52.7953
San Antonio	55.1	68	93.1	72.8	58.3705
Cincinnati	146.8	96	87.9	85.1	60.3004
Saginaw	23.5	79	90.0	85.7	52.8047
Baton Rouge	49.8	52	68.2	78.1	53.6184
New York	497.7	97	88.0	86.8	70.2917
Springfield, OH	32.7	101	90.6	84.0	51.1873

[a]PM2:   population density per square mile ×0.1
GE65:   percent of population at least 65 × 10
PERWH: percent of whites in population
NONPOOR: percent of families with income above poverty level
LPOP: logarithm (base 10) of population ×10
Source: Gibbons, Dianne I.; McDonald, Gary C. and Gunst, Richard F, "The Complementary Use of Regression Diagnostics and Robust Estimators," *Naval Research Logistics* 34, 1, February, 1987.

Table 4.9. Results from Air Pollution Multiple Regression – Full Model

Variable	Regression Estimates Regression Coefficient	Standard Error	t-Value	p-Value
INTERCEPT	818.74772	261.01955	3.137	0.0040
SMIN	0.72662852	0.59741131	1.216	0.2340
SMEAN	−0.52761789	0.70205223	−0.752	0.4586
SMAX	0.18247332	0.22518826	0.810	0.4246
PMIN	−0.76311091	1.40306283	−0.544	0.5908
PMEAN	1.07158845	0.95496651	1.122	0.2713
PMAX	−0.01678561	0.27340210	−0.061	0.9515
PM2	0.06337831	0.23867796	0.266	0.7925
GE65	8.25576156	0.93343922	8.844	0.0001
PERWH	−5.14026262	2.1403996	−2.402	0.0232
NONPOOR	−3.81529373	2.9225505	−1.305	0.2024
LPOP	0.906809187	4.441999522	0.204	0.8397

Analysis of Variance

Source	d.f.	Sum of Squares	Mean Squares	F-Value	Prob> F
MODEL	11	984781.29	89525.57137	18.114	0.0001
ERROR	28	138384.49	4942.30321		
C TOTAL	39	1123165.78			

Table 4.10. Results from a Reduced Multiple Regression Model for Air Pollution Data

Variable	Regression Estimates Regression Coefficient	Standard Error	t-Value	p-Value
INTERCEPT	888.39438	136.87954	6.490	0.0001
PMEAN	0.87629068	0.24312950	3.604	0.0010
GE65	8.63763826	0.65656075	13.156	0.0001
PERWH	−0.46869068	1.53845342	−3.047	0.0044
NONPOOR	−0.47450993	2.04901919	−2.316	0.0266

Analysis of Variance

Source	d.f.	Sum of Squares	Mean Squares	F-Value	Prob> F
REGRESSION	4	973222.75031296	243305.68757824	56.79	0.0001
ERROR	35	149943.02468705	4284.08641963		
TOTAL	39	1123165.77500000			

PERWH, and NONPOOR. This model has an R^2 of 0.867. This model differs from the first reduced model by the addition of the variable SMIN.

A comparison of the four reduced models seems to demonstrate that the variables GE65, NONPOOR, PERWH plus one of (SMIN, SMEAN and SMAX) and one of (PMIN, PMEAN and PMAX) should provide a suitable fit. The fact that there are a variety of models that can be used to explain TMR is due to the correlations among the explanatory variables, as well as the correlations between the explanatory variables and TMR.

Table 4.11. Results From a Reduced Multiple Regression Model for Air Pollution Data

Regression Estimates

Variable	Regression Coefficient	Standard Error	t-Value	p-Value
INTERCEPT	1007.48032000	154.02762000	6.541	0.0001
SMEAN	0.70143463	0.22082448	3.176	0.0031
GE65	6.80637269	0.63921794	10.648	0.0001
NONPOOR	−8.97571770	1.89375070	−4.740	0.0001

Analysis of Variance

	d.f.	Sum of Squares	Mean Square	F	Prob > F
REGRESSION	3	38266.36538181	312755.45512727	60.89	0.0001
ERROR	36	184899.40961819	5136.09471162		
TOTAL	39	1123165.77500000			

Table 4.12. Results From a Reduced Multiple Model for Air Pollution Data

Regression Estimates

Variable	Regression Coefficient	Standard Error	t-Value	p-Value
INTERCEPT	868.74689000	139.14899000	6.243	0.0001
SMIN	0.56605908	0.34402843	1.645	0.1091
PMAX	0.25481684	0.09423609	2.704	0.0106
GE65	8.24587315	0.74419128	11.080	0.0001
PERWH	−4.59913300	1.58223070	−2.907	0.0064
NONPOOR	−4.10150100	2.07613540	−1.976	0.0564

Analysis of Variance

	d.f.	Sum of Squares	Mean Square	F	Prob > F
REGRESSION	5	973389.46511945	194677.89302389	44.19	0.0001
ERROR	34	149776.30988055	4405.18558472		
TOTAL	39	1123165.77500000			

The correlation matrix for this data will be studied later in this section.

The use of a fitted air pollution model based on 1960 data, to predict mortality for a sample of ten cities based on 1969 data, will now be illustrated. The model to be used to make the predictions will be the model fitted in Table 4.10. A sample of ten cities was selected randomly from a table of twenty-six cities for the year 1969 in Gibbons, McDonald and Gunst (1987). The predicted values of TMR and the actual values are presented in Table 4.13 below. With the exception of San Bernadino the predicted values are quite close. The total mean square error for the ten predictions was 5365.96 which is larger than the MSE of 4284.09 in Table 4.13. The contribution to this MSE by San Bernadino was 3,384.50 (63.1% of 5365.96).

Table 4.13. Prediction of Mortality Rate for Ten U.S. Cities in 1969 Using 1960 Fitted Model

	City	(1) TMR	(2) Pred TMR	(3) Std. Error	t-ratio [(1) − (2)]/(3)	PMEAN	PERWH	NONPOOR	GE65
1	Buffalo	1012	987.22	70.184	0.35	131	91.5	93.2	99
2	San Bernadino	901	1084.97	69.093	−2.66	140	94.3	89.7	109
3	Spokane	1045	1044.06	69.745	0.01	101	97.7	91.4	111
4	Fort Wayne	812	812.64	70.583	−0.01	87	93.0	94.9	85
5	Peoria	914	959.24	69.761	−0.65	144	95.1	94.3	97
6	Evansville	957	1058.05	70.410	−1.44	95	94.1	90.9	111
7	Trenton	967	958.69	74.574	0.11	77	83.4	93.6	97
8	Utica	1077	1043.44	71.097	0.47	86	97.5	92.6	113
9	Rochester	872	942.64	71.064	−0.99	104	92.5	94.8	98
10	Huntsville	582	552.59	70.619	0.42	64	84.5	86.5	48

Inferences for Linear Functions

From the regression results above in the prediction model for TMR, it is of interest to determine if a model which retains the six pollution variables would be useful. An estimate of the model was carried out subject to the two restrictions:

1. that the coefficients of PMAX, PMEAN and PMIN are equal, and

2. that the coefficients of SMAX, SMEAN and SMIN are equal.

The resulting regression model is given by

$$\text{TMR} = \underset{(0.000)}{980.728} + \underset{(0.317)}{(0.084)} \ (\text{SMIN} + \text{SMEAN} + \text{SMAX})$$

$$+ \underset{(0.049)}{0.161} \ (\text{PMIN} + \text{PMEAN} + \text{PMAX})$$

$$+ \underset{(0.726)}{0.077} \ \text{PM2} + \underset{(0.000)}{8.098} \ \text{GE65} - \underset{(0.039)}{3.932} \ \text{PERWH}$$

$$- \underset{(0.042)}{5.450} \ \text{NONPOOR} - \underset{(0.500)}{1.043} \ \text{LPOP}$$

with $R^2 = 0.8675$. The equation fitted without any restriction above showed an R^2 of 0.8768. Thus the restriction on the coefficients did not change the goodness of fit of the entire model. It is interesting to observe that the coefficients of the non-pollution variables were not affected greatly by the two restrictions.

The **T** matrix required to carry out the restricted least squares is given by the (4×12) matrix below.

$$\mathbf{T} = \begin{bmatrix} 1 & -1 & 0 & 0 & 0 & 0 & 0 & 0 & 0 & 0 & 0 & 0 \\ 1 & 0 & -1 & 0 & 0 & 0 & 0 & 0 & 0 & 0 & 0 & 0 \\ 0 & 0 & 0 & 1 & -1 & 0 & 0 & 0 & 0 & 0 & 0 & 0 \\ 0 & 0 & 0 & 1 & 0 & -1 & 0 & 0 & 0 & 0 & 0 & 0 \end{bmatrix}.$$

The first three columns correspond to the S variables and the second three columns correspond to the P variables. The last six columns correspond to the remaining six variables which are not being restricted.

An F-test for testing the hypothesis $H_0 : \mathbf{T\beta} = \mathbf{0}$ was carried out yielding an F-value of 0.5305 and a p-value of 0.7143. Thus as we have seen from the comparison of the two R^2 values, the restriction has virtually no impact on the fitted relationship.

4.1.4 Standardized Variables, Standardized Regression Coefficients and Partial Correlation

When interpreting the results of an estimated multiple regression model it is sometimes of interest to compare the regression coefficients. The magnitudes of the regression coefficients, however, depend on the scales of measurement used for Y and the explanatory variables included in the model. It is therefore necessary to standardize the variables in order to make meaningful comparisons. In this section an expression is introduced for the least squares regression coefficients in terms of standardized variables. Expressions will also be introduced for variables that have been transformed to have mean zero by subtracting the sample means. We begin by introducing some notation.

Some Notation

The p columns of $\mathbf{X}$ excluding the first denote n observations on each of p variables $X_1, X_2, \ldots, X_p$. The sample means based on the n observations are denoted by $\bar{x}_1, \bar{x}_2, \ldots, \bar{x}_p$ where

$$\bar{x}_j = \sum_{i=1}^{n} x_{ij}/n \qquad j = 1, 2, \ldots, p.$$

The sample variances are denoted by

$$s_j^2 = \sum_{i=1}^{n} (x_{ij} - \bar{x}_j)^2/(n-1) \qquad j = 1, 2, \ldots, p.$$

Similarly, for the n observations on Y, the sample mean and variance are given by

$$\bar{y} = \sum_{i=1}^{n} y_i/n$$

$$\text{and} \qquad s_y^2 = \sum_{i=1}^{n} (y_i - \bar{y})^2/(n-1).$$

The sample covariances among the X variables are denoted by

$$s_{jk} = \sum_{i=1}^{n} (x_{ij} - \bar{x}_j)(x_{ik} - \bar{x}_k)/(n-1)$$

and the sample Pearson correlations by

$$r_{jk} = s_{jk}/\sqrt{s_j^2 s_k^2} = s_{jk}/s_j s_k \qquad j, k = 1, 2, \ldots, p.$$

Non-intercept Portion of Parameter Vector

In the multiple regression model $\mathbf{y} = \mathbf{X}\boldsymbol{\beta} + \mathbf{u}$ it is sometimes convenient to distinguish the $(p \times 1)$ sub-vector of $\boldsymbol{\beta}$, say $\boldsymbol{\beta}^*$, that does not contain the intercept parameter β_0, and hence $\boldsymbol{\beta}' = (\beta_0 \ \boldsymbol{\beta}^{*\prime})$. The least squares estimator of β_0, b_0, can be expressed in terms of the remaining components $\mathbf{b}^*$ of $\mathbf{b}$ as

$$b_0 = \bar{y} - b_1 \bar{x}_{\cdot 1} - b_2 \bar{x}_{\cdot 2} - \cdots - b_p \bar{x}_{\cdot p}$$

where $\mathbf{b}' = (b_0 \ \mathbf{b}^{*\prime})$. The estimated model $\hat{y}_i = b_0 + b_1 x_{i1} + b_2 X_{i2} + \cdots + b_p x_{ip}$ can be written in the form

$$\left(\frac{\hat{y}_i - \bar{y}}{s_y}\right) = \frac{b_1 s_1}{s_y}\left(\frac{x_{i1} - \bar{x}_{\cdot 1}}{s_1}\right) + \frac{b_2 s_2}{s_y}\left(\frac{x_{i2} - \bar{x}_{\cdot 2}}{s_2}\right)$$

$$+ \cdots + \frac{b_p s_p}{s_y}\left(\frac{x_{ip} - \bar{x}_{\cdot p}}{s_p}\right). \tag{4.5}$$

Sample Standardized Values and Mean Corrected Values

The variable values $(x_{ij} - \bar{x}_{\cdot j})/s_j$ are called the *sample standardized values* and are denoted by $\tilde{x}_{ij}$, $j = 1, 2, \ldots, p$; $i = 1, 2, \ldots, n$. The numerator of the standardized value is the *mean corrected value*, $(x_{ij} - \bar{x}_{\cdot j})$, which will be denoted by x_{ij}^*, $j = 1, 2, \ldots, p$; $i = 1, 2, \ldots, n$. Similarly for y we have $\tilde{y}_i = (y_i - \bar{y})/s_y$ and $y_i^* = (y_i - \bar{y})$, $i = 1, 2, \ldots, n$ which are the standardized values and mean corrected values, respectively.

Standardized Regression Coefficients

The coefficients $b_j s_j/s_y$, $j = 1, 2, \ldots, p$ in (4.5) are called the *standardized regression coefficients* and are denoted by d_j. These coefficients are useful for comparing the regression coefficients with respect to their impact on y. The coefficient d_j measures the impact on a standardized unit of y of a one unit change in the standardized value of x_j. To express the standardized regression coefficients in vector form some additional notation must be introduced. Before introducing the notation below, an example is discussed next.

Example

For the financial accounting example the standardized regression coefficients are shown in Table 4.14 for both the full and reduced models. The magnitudes of the standardized regression coefficients can be compared to determine which variables have the greatest impact on RETCAP. For the full model the most important variables appear to be CURRAT and QUIKRAT. For the reduced model the same two variables are the most important. A comparison of the full model and reduced model standardized regression coefficients shows that the coefficient of WCFTCL increases by a factor of 2.

Table 4.14. Standardized Regression Coefficients and Partial Correlation Coefficients for Full and Reduced Financial Accounting Models

	Full Model			Reduced Model		
Variable	Std. Regr. Coeff.	Partial Corr.	p value	Std. Regr. Coeff.	Partial Corr.	p value
GEARRAT	−0.061	−0.008	0.6510	—	—	—
CAPINT	−0.085	−0.012	0.5638	—	—	—
WCFTDT	0.638	0.040	0.2992	—	—	—
LOGSALE	0.890	0.279	0.0032	0.723	0.315	0.0005
LOGASST	−0.335	−0.099	0.0959	−0.303	−0.128	0.0351
CURRAT	−3.027	−0.167	0.0276	−1.765	−0.215	0.0050
QUIKRAT	2.447	0.099	0.0956	1.259	0.110	0.0517
NFATAST	−0.533	−0.211	0.0122	−0.723	−0.581	0.0001
INVTAST	0.274	0.045	0.2689	—	—	—
FATTOT	−0.213	−0.048	0.2524	—	—	—
PAYOUT	−0.101	−0.038	0.3083	—	—	—
WCFTCL	0.482	0.039	0.3051	0.983	0.614	0.0001

Matrix Notation for Standardized Coefficients

When the first column of the $\mathbf{X}$ matrix contains a column of ones the matrix $(\mathbf{X'X})^{-1}$ can be written as

$$(\mathbf{X'X})^{-1} = \begin{bmatrix} \frac{1}{n} + \bar{\mathbf{x}}'(\mathbf{X^{*\prime}X^*})^{-1}\bar{\mathbf{x}} & -\bar{\mathbf{x}}'(\mathbf{X^{*\prime}X^*})^{-1} \\ -(\mathbf{X^{*\prime}X^*})^{-1}\bar{\mathbf{x}} & (\mathbf{X^{*\prime}X^*})^{-1} \end{bmatrix}.$$

The $(p \times 1)$ vector $\bar{\mathbf{x}}$ is the vector of means $\bar{x}_{.1}, \bar{x}_{.2}, \ldots, \bar{x}_{.p}$. The matrix $\mathbf{X}^*$ is the $(n \times p)$ matrix of *mean corrected* X values, which are obtained by subtracting the column means $\bar{x}_{.1}, \bar{x}_{.2}, \ldots, \bar{x}_{.p}$ from their respective column values as shown below.

$$
\mathbf{X}^* = \begin{bmatrix}
x_{11} - \bar{x}_{.1} & x_{12} - \bar{x}_{.2} & \cdots & x_{1p} - \bar{x}_{.p} \\
x_{21} - \bar{x}_{.1} & x_{22} - \bar{x}_{.2} & & x_{2p} - \bar{x}_{.p} \\
x_{31} - \bar{x}_{.1} & & & \\
\vdots & \vdots & & \vdots \\
x_{n1} - \bar{x}_{.1} & x_{n2} - \bar{x}_{.2} & \cdots & x_{np} - \bar{x}_{.p}
\end{bmatrix}.
$$

Corrected Sums of Squares and Cross Products Matrix and Covariance Matrix

The matrix $\mathbf{X}^{*\prime}\mathbf{X}^*$ is a matrix of *sums of squares and cross products of mean deviations* given by

$$
\begin{bmatrix}
\sum_{i=1}^{n}(x_{i1} - \bar{x}_{.1})^2 & \sum_{i=1}^{n}(x_{i1} - \bar{x}_{.1})(x_{i2} - \bar{x}_{.2}) & \cdots & \sum_{i=1}^{n}(x_{i1} - \bar{x}_{.1})(x_{ip} - \bar{x}_{.p}) \\
\sum_{i=1}^{n}(x_{i1} - \bar{x}_{.1})(x_{i2} - x_{.2}) & \sum_{i=1}^{n}(x_{i2} - \bar{x}_{.2})^2 & & \sum_{i=1}^{n}(x_{i2} - \bar{x}_{.2})(x_{ip} - \bar{x}_{.p}) \\
\vdots & \vdots & & \vdots \\
\sum_{i=1}^{n}(x_{i1} - \bar{x}_{.1})(x_{ip} - \bar{x}_{.p}) & & \cdots & \sum_{i=1}^{n}(x_{ip} - \bar{x}_{.p})^2
\end{bmatrix}.
$$

The above matrix is also called a *corrected sums of squares and cross products matrix*. Dividing this matrix by $(n-1)$ yields the *sample covariance matrix* for the X variables denoted by $\mathbf{S}$.

Correlation Matrix

The matrix of standardized X values, excluding the column of unities, is given by

$$
\tilde{\mathbf{X}} = \begin{bmatrix}
\dfrac{x_{11} - \bar{x}_{.1}}{s_1} & \dfrac{x_{12} - \bar{x}_{.2}}{s_2} & \cdots & \dfrac{x_{1p} - \bar{x}_{.p}}{s_p} \\
\dfrac{x_{21} - \bar{x}_{.1}}{s_1} & \dfrac{x_{22} - \bar{x}_{.2}}{s_2} & \cdots & \dfrac{x_{2p} - \bar{x}_{.p}}{s_p} \\
\vdots & \vdots & & \vdots \\
\dfrac{x_{n1} - \bar{x}_{.1}}{s_1} & \dfrac{x_{n2} - \bar{x}_{.2}}{s_2} & \cdots & \dfrac{x_{np} - \bar{x}_{.p}}{s_p}
\end{bmatrix}.
$$

The corresponding matrix of sums of squares and cross products derived from $\widetilde{\mathbf{X}}$ is the *correlation matrix*

$$\mathbf{R} = \widetilde{\mathbf{X}}'\widetilde{\mathbf{X}} = \begin{bmatrix} 1 & r_{12} & r_{13} & \cdots & r_{1p} \\ r_{12} & 1 & r_{23} & \cdots & r_{2p} \\ \vdots & \vdots & \vdots & & \vdots \\ r_{1p} & r_{2p} & \cdots & \cdots & 1 \end{bmatrix}.$$

This correlation matrix can be expressed in terms of the sample covariance matrix $\mathbf{S}$ using the $(p \times p)$ diagonal matrix $\mathbf{D}$ of sample variances $s_1^2, \ldots, s_p^2$.

$$\mathbf{R} = \mathbf{D}^{-1/2}\mathbf{S}\mathbf{D}^{-1/2}.$$

Mean Corrected Form of Linear Model

The vector of mean corrected y values $(y_i - \bar{y})$ is denoted by $\mathbf{y}^*$. Using the mean corrected data matrix $\mathbf{X}^*$ and the mean corrected vector $\mathbf{y}^*$ the least squares estimator of $\boldsymbol{\beta}^*$ is given by

$$\mathbf{b}^* = (\mathbf{X}^{*\prime}\mathbf{X}^*)^{-1}\mathbf{X}^{*\prime}\mathbf{y}^*,$$

$$\text{where} \quad \mathbf{b}' = (b_0, \mathbf{b}^{*\prime}).$$

In terms of the covariance matrix $\mathbf{S}$, $\mathbf{b}^*$ is given by

$$\mathbf{b}^* = \mathbf{S}^{-1}\mathbf{s}_{\mathbf{xy}},$$

where $\mathbf{s}_{\mathbf{xy}}$ denotes the $(p \times 1)$ vector of sample covariances

$$\sum_{i=1}^{n}(x_{ij} - \bar{x}_{\cdot j})(y_i - \bar{y})/(n-1); \quad j = 1, 2, \ldots, p.$$

The expression for $\mathbf{b}^*$ is therefore given by

$$\mathbf{b}^* = s_y\mathbf{D}^{-1/2}\mathbf{R}^{-1}\mathbf{r}_{\mathbf{xy}},$$

where $\mathbf{r}_{\mathbf{xy}}$ is the $p \times 1$ vector of correlations between y and the $\mathbf{x}$ variables. The standardized regression coefficients are given by

$$\mathbf{d} = \mathbf{R}^{-1}\mathbf{r}_{\mathbf{xy}}$$

$$\text{and} \quad \mathbf{b}^* = s_y\mathbf{D}^{-1/2}\mathbf{d}.$$

Partial Correlation Coefficients

Recall from Chapter 3 that the partial correlation coefficient relating Y to X_j controlling for X_k is given by

$$r_{yx_j \cdot x_k} = \frac{r_{yx_j} - r_{yx_k} r_{jk}}{\sqrt{(1 - r_{yx_k}^2)(1 - r_{jk}^2)}}$$

and that the regression coefficient for X_j in the regression of Y on X_j and X_k is given by

$$b_j = \frac{r_{yx_j} - r_{yx_k} r_{jk}}{(1 - r_{jk}^2)} \cdot \frac{s_y}{s_j} = r_{yx_j \cdot x_k} \frac{\sqrt{1 - r_{yx_k}^2}}{\sqrt{1 - r_{jk}^2}} \frac{s_y}{s_j}$$

$$= r_{yx_j \cdot x_k} \frac{s_{y \cdot x_k}}{s_{x_j \cdot x_k}}.$$

The terms $s_{y \cdot x_k}$ and $s_{x_j \cdot x_k}$ are the standard errors of the residuals in the regressions of Y on X_k and X_j on X_k respectively. The partial correlation coefficient can be related to the standardized regression coefficient by writing the standardized regression coefficient as

$$d_j = b_j \frac{s_j}{s_y} = r_{yx_j \cdot x_k} \frac{\sqrt{1 - r_{yx_k}^2}}{\sqrt{1 - r_{jk}^2}}.$$

The standardized regression coefficient is therefore equal to the product of the partial correlation coefficient and the ratio of two square roots. Under the root signs the numerator is the unexplained variance in the regression of Y on X_k, while the denominator is the unexplained variance in the regression of X_j on X_k.

Higher Order Partial Correlation Coefficients

The expression for the first order partial correlation can be extended to *higher order partial correlations* which control for additional variables. As in the first order case partial correlation coefficients can be related to multiple regression coefficients. We must first introduce some notation. The partial correlation between Y and X_j controlling for $(X_1, X_2, \ldots, X_p)_j$ is denoted by $r_{YX_j \cdot (X_1 X_2 \ldots X_p)_j}$ where $(X_1, X_2, \ldots, X_p)_j$ denotes the set of all X variables $X_1, X_2, \ldots, X_p$ excluding X_j. The proportion of unexplained variance in the regression of Y on $(X_1, X_2, \ldots, X_p)_j$ and X_j on

$(X_1, X_2, \ldots, X_p)_j$ are denoted by $(1 - R_y^2)$ and $(1 - R_j^2)$ respectively. The standard errors of the residuals in these two regressions are given by

$$s_{y \cdot (x_1, x_2, \ldots, x_p)_j} = s_y \sqrt{1 - R_y^2}$$

and $\qquad s_{x_j \cdot (x_1, x_2, \ldots, x_p)_j} = s_j \sqrt{1 - R_j^2} \ .$

The relationship between the multiple regression coefficient b_j in the regression of Y on $X_1, X_2, \ldots, X_p$ and the partial correlation coefficient $r_{yx_j \cdot (x_1, x_2, \ldots, x_p)_j}$ is given by

$$b_j = r_{yx_j \cdot (x_1, x_2, \ldots, x_p)_j} \frac{s_y}{s_j} \frac{\sqrt{1 - R_y^2}}{\sqrt{1 - R_j^2}}$$

$$= r_{yx_j \cdot (x_1, x_2, \ldots, x_p)_j} \frac{s_{y \cdot (x_1, x_2, \ldots, x_p)_j}}{s_{x_j \cdot (x_1, x_2, \ldots, x_p)_j}} .$$

The partial correlation coefficient can be related to the standardized regression coefficient $d_j = b_j s_y / s_j$ by the relation

$$r_{yx_j \cdot (x_1, x_2, \ldots, x_p)_j} = d_j \frac{\sqrt{1 - R_j^2}}{\sqrt{1 - R_y^2}} .$$

Example

The partial correlation coefficients for the financial accounting example were shown in Table 4.14 above. For the full model the largest partial correlations are for LOGSALE and NFATAST at 0.279 and –0.211 respectively. Thus controlling for the other explanatory variables the correlation between LOGSALE and RETCAP is positive and between NFATAST and RETCAP the correlation is negative. The partial correlations for the reduced model are somewhat larger than those of the full model. The most dramatic differences are for the variables WCTFCL and NFATAST. For the WCTFCL variable the partial correlation in the full model is 0.039 while for the reduced model this partial correlation is 0.614. For the NFATAST variable the negative partial correlation increases in magnitude from 0.211 to 0.581. These increases in partial correlation reflect the fact that there are some strong correlations among the explanatory variables and that including them in the model results in major changes in the partial relationships. In this case the more variables that are being controlled the weaker the partial relationships become.

Example

Examination of the correlation matrix for the air pollution data given in Table 4.15 below reveals that the strongest correlations with TMR are provided by GE65 (0.824) and PERWH (0.476). A moderately weak group of correlations with TMR are SMEAN (0.313), PMEAN (0.275), SMIN (0.237), PMAX (0.242), SMAX (0.182), and PMIN (0.184).

It is important to note that the correlation between TMR and NON-POOR is 0.0165 indicating nonsignificant zero order correlation ($p=0.7762$) between these two variables. From the multiple regression results discussed above we can conclude that the partial correlation between TMR and NON-POOR after controlling for other variables such as GE65 and SMEAN is significant at the 0.0001 level (see Table 4.11; similarly in Tables 4.9, 4.10 and 4.12). A potential explanation of this result is the correlation between NONPOOR and both GE65 and PERWH (0.317 and 0.347 respectively). Using the formula for calculating the first order partial correlation coefficient presented in Chapter 3, the partial correlations between TMR and NONPOOR after controlling for each of GE65 and PERWH respectively, are given by

$$\frac{0.0165 - (0.8240)(0.3173)}{\sqrt{[1 - (0.8240)^2][1 - (0.3173)^2]}} = -0.456$$

$$\text{and} \quad \frac{0.0165 - (0.3472)(0.3173)}{\sqrt{[1 - (0.3472)^2][1 - (0.3173)^2]}} = -0.143.$$

Thus we would conclude that the partial correlation between TMR and the variable NONPOOR is much stronger after separately controlling for GE65 and PERWH.

It is also interesting to note that although the zero order correlation between TMR and PERWH is positive (0.476) the first order correlation is negative after controlling for the variable GE65. This calculation is left for the reader. The interpretation of this sign change is also left for the reader.

The correlation matrix also reveals the correlations among the various pollution measures. The correlations among the three S variables (SMIN, SMEAN and SMAX) and the correlations among the three P variables (PMIN, PMEAN and PMAX) suggest that perhaps only one of each threesome are required as explanatory variables. Procedures for the selection of a subset of variables to explain TMR will be outlined in the next section.

Table 4.15. Correlation Matrix Relating Air Pollution Variables

	SMIN	SMEAN	SMAX	PMIN	PMEAN	PMAX	PM2	GE65	PERWH	NON-POOR	LPOP
TMR	0.23679	0.31273	0.18181	0.18409	0.27470	0.24162	0.08975	0.82401	0.47622	0.01651	0.13867
SMIN	1.00000	0.55514	0.21960	0.26569	0.23612	0.12680	0.41488	0.03892	0.03367	-0.07177	0.17197
SMEAN	0.55514	1.00000	0.74348	0.29427	0.45665	0.26093	0.33094	0.23185	0.25897	0.26964	0.35314
SMAX	0.21960	0.74348	1.00000	0.10248	0.31851	0.21393	0.13009	0.22203	0.31584	0.32366	0.14577
PMIN	0.26569	0.29427	0.10248	1.00000	0.70853	0.08810	0.23434	-0.03526	-0.03591	-0.27107	0.23352
PMEAN	0.23612	0.45665	0.31851	0.70853	1.00000	0.56750	0.07828	0.09489	0.24957	-0.15939	0.33542
PMAX	0.12680	0.26093	0.21393	0.08810	0.56750	1.00000	-0.04232	0.12366	0.20179	0.02641	0.28395
PM2	0.41488	0.33094	0.13009	0.23434	0.07828	-0.04232	1.00000	-0.01651	-0.09037	0.11445	0.38477
GE65	0.03892	0.23185	0.22203	-0.03526	0.09489	0.12366	-0.01651	1.00000	0.64152	0.31726	0.06663
PERWH	0.03367	0.25897	0.31584	-0.03591	0.24957	0.20179	-0.09037	0.64152	1.00000	0.34717	-0.02073
NON-POOR	-0.07177	0.26964	0.32366	-0.27107	-0.15939	0.02641	0.11445	0.31726	0.34717	1.00000	0.10322
LPOP	0.17197	0.35314	0.14577	0.23352	0.33542	0.28395	0.38477	0.06663	-0.02073	0.10322	1.00000

Orthogonal Columns of **X**

In some specialized applications the columns of **X** are mutually orthogonal $\left(\sum_{i=1}^{n} x_{ij} x_{ik} = 0, \text{ if } j \neq k \right)$ and hence $(\mathbf{X}'\mathbf{X})$ is a diagonal matrix. In this case each element of **b** is given by

$$b_j = \sum_{i=1}^{n} x_{ij} y_k \Big/ \sum_{i=1}^{n} x_{ij}^2 \qquad j = 1, 2, \ldots, p$$

and $b_0 = \bar{y}$.

Another special case arises when the columns of $\mathbf{X}^*$ are mutually uncorrelated and hence $\mathbf{R} = \mathbf{I}$. In this case the elements of $\mathbf{b}^*$ are given by

$$b_j^* = \frac{s_y}{s_{x_j}} r_{x_j y} \qquad j = 1, 2, \ldots, p$$

and $\qquad b_0 = \bar{y} - \sum_{j=1}^{p} b_j^* \bar{x}_j.$

The standardized regression coefficients in this case are given by $d_j = r_{x_j y}$.

4.1.5 Omitted Variables, Irrelevant Variables and Incorrect Specification

A critical assumption made for the multiple linear regression model is that a variable Y can be expressed as a linear function of p explanatory variables $X_1, X_2, \ldots, X_p$ plus an unobserved error term U. The error term is assumed to be independent of the p X variables. Equivalently, we are assuming that <u>ALL</u> other variables not in the model that can influence Y are independent of the explanatory variables $X_1, \ldots, X_p$. These variables combine to make up the error term U. In practice it is not always possible to include all relevant variables, and hence the model could be incorrectly specified. It may be that the analyst is ignorant of some of the variables required or it may not be possible to measure some of the variables known to be required. In some cases variables may be included that are irrelevant. It is important to determine the impact such incorrect specifications have on the results outlined above.

Example

In the financial accounting example discussed above, models of 12 explanatory variables and a subset of 6 explanatory variables, were used to explain RETCAP. A comparison of the estimated regression coefficients in Tables 4.3 and 4.6 for the 6 variables in both models, shows how the regression coefficients and their standard errors can change when some variables are excluded. The greatest difference between the two models is the coefficient of WCFTCL which increases from 0.208 to 0.425. The standard error of this coefficient drops from 0.1995 to 0.058. As a result of these changes the *p*-value of the regression coefficient for WCFTCL decreases from 0.3051 to 0.000.

Notation for Omitted Variable Models

Beginning with the full model outlined in 4.1.1, we partition the explanatory variables into the two sets, $X_1, \ldots, X_q$ which are to be omitted and the remaining variables $X_{q+1}, \ldots, X_p$ which are to be retained. The intercept term will usually be included as part of the reduced model. The full model can be written as

$$\mathbf{y} = \mathbf{X}\boldsymbol{\beta} = \mathbf{X}_1\boldsymbol{\beta}_1 + \mathbf{X}_2\boldsymbol{\beta}_2 + \mathbf{u},$$

where $\mathbf{X}_1(n \times q)$ and $\mathbf{X}_2(n \times (p - q + 1))$ denote the partition of $\mathbf{X}$, and $\boldsymbol{\beta}_1(q \times 1)$ and $\boldsymbol{\beta}_2((p-q+1) \times 1)$ are the corresponding partitions of $\boldsymbol{\beta}$. The reduced model after omitting $\mathbf{X}_1$ is given by

$$\mathbf{y} = \mathbf{X}_2\boldsymbol{\beta}_2^* + \mathbf{u},$$

where $\boldsymbol{\beta}_2^*$ reflects the revised coefficient of $\mathbf{X}_2$ in the absence of $\mathbf{X}_1$.

Least Squares Estimators

The ordinary least squares estimators of the regression coefficients and the residual mean squares are given by

$$\mathbf{b} = (\mathbf{X}'\mathbf{X})^{-1}\mathbf{X}'y \quad \text{and} \quad s^2 = \mathbf{y}'(\mathbf{I} - \mathbf{H})\mathbf{y}/(n - p - 1) = \text{MSE}$$

for the full model and

$$\mathbf{b}_2^* = (\mathbf{X}_2'\mathbf{X}_2)^{-1}\mathbf{X}_2'\mathbf{y} \quad \text{and} \quad s_R^2 = \mathbf{y}'(\mathbf{I} - \mathbf{H}_2)\mathbf{y}/(n - p - 1 + q) = \text{MSE}_R$$

for the reduced model. The matrix $\mathbf{H}_2 = \mathbf{X}_2(\mathbf{X}_2'\mathbf{X}_2)^{-1}\mathbf{X}_2'$ is the hat matrix for the reduced model. If the two subsets of variables $\mathbf{X}_1$ and $\mathbf{X}_2$ are orthogonal to each other then

$$\mathbf{b} = \begin{bmatrix} \mathbf{b}_1 \\ \mathbf{b}_2 \end{bmatrix} = \begin{bmatrix} (\mathbf{X}_1'\mathbf{X}_1)^{-1} \mathbf{X}_1'\mathbf{y} \\ (\mathbf{X}_2'\mathbf{X}_2)^{-1} \mathbf{X}_2'\mathbf{y} \end{bmatrix}$$

and hence $\mathbf{b}_2 = \mathbf{b}_2^*$. In general the explanatory variables in the two sets are not orthogonal and hence $\mathbf{b}_2$ and $\mathbf{b}_2^*$ will not be equal.

Properties of Estimators

The mean vector and covariance matrix for the above estimators are

$$E[\mathbf{b}] = \boldsymbol{\beta} \quad \text{and} \quad \text{Cov}[\mathbf{b}] = \sigma_u^2(\mathbf{X}'\mathbf{X})^{-1}$$

for the full model and

$$E[\mathbf{b}_2^*] = \boldsymbol{\beta}_2 + (\mathbf{X}_2'\mathbf{X}_2)^{-1}\mathbf{X}_2'\mathbf{X}_1\boldsymbol{\beta}_1 = \boldsymbol{\beta}_2 + \text{BIAS}(\mathbf{b}_2^*)$$

$$\text{and} \quad \text{Cov}[\mathbf{b}_2^*] = \sigma_u^2(\mathbf{X}_2'\mathbf{X}_2)^{-1}$$

for the reduced model, where

$$\text{BIAS}(\mathbf{b}_2^*) = (\mathbf{X}_2'\mathbf{X}_2)^{-1}\mathbf{X}_2'\mathbf{X}_1\boldsymbol{\beta}_1.$$

Therefore $\mathbf{b}_2^*$ is a biased estimator of $\boldsymbol{\beta}_2$ unless $\mathbf{X}_1$ and $\mathbf{X}_2$ are orthogonal, or unless $\boldsymbol{\beta}_1 = 0$. If $\mathbf{b}_2^*$ is a biased estimator of $\boldsymbol{\beta}_2$, the *mean square error* for $\mathbf{b}_2^*$ is not equal to $\text{Cov}(\mathbf{b}_2^*)$ but is given by

$$\text{MSE}(\mathbf{b}_2^*) = \sigma_u^2(\mathbf{X}_2'\mathbf{X}_2)^{-1} + (\mathbf{X}_2'\mathbf{X}_2)^{-1}\mathbf{X}_2'\mathbf{X}_1\boldsymbol{\beta}_1\boldsymbol{\beta}_1'\mathbf{X}_1'\mathbf{X}_2(\mathbf{X}_2'\mathbf{X}_2)^{-1}$$

$$= \text{Cov}(\mathbf{b}_2^*) + [\text{BIAS}(\mathbf{b}_2^*)]^2.$$

The mean square error s_R^2 is also a biased estimator of σ_u^2 since

$$E[s_R^2] = \sigma_u^2 + \boldsymbol{\beta}_1'\mathbf{X}_1'(\mathbf{I} - \mathbf{H}_2)\mathbf{X}_1\boldsymbol{\beta}_1/(n - p - 1 + q).$$

Properties for Predictors

For predictions let $\hat{y}_0$ denote the prediction for y at $\mathbf{x} = \mathbf{x}_0$ using the full model, and let $\hat{y}_{0R}$ denote the prediction using the reduced model at $\mathbf{x}_2 = \mathbf{x}_{02}$. The variances for the predictions are given by

$$V(\hat{y}_0) = \sigma_u^2\left(1 + \mathbf{x}_0'(\mathbf{X}'\mathbf{X})^{-1}\mathbf{x}_0\right) \quad \text{and}$$

$$V(\hat{y}_{0R}) = \sigma_u^2\left(1 + \mathbf{x}_{02}'(\mathbf{X}_2'\mathbf{X}_2)^{-1}\mathbf{x}_{02}\right).$$

The mean square error for the prediction $\hat{y}_{0R}$ from the reduced model is given by

$$\text{MSE}(\hat{y}_{0R}) = V(\hat{y}_{0R}) + (\mathbf{x}_{02}'(\mathbf{X}_2'\mathbf{X}_2)^{-1}\mathbf{X}_2'\mathbf{X}_1\boldsymbol{\beta}_1 - \mathbf{x}_{02}'\boldsymbol{\beta}_1)^2.$$

The variances of the elements of $\mathbf{b}_2$ are generally larger than the variances of the elements of $\mathbf{b}_2^*$, and for predictions the variance of $\hat{y}_0$, $V(\hat{y}_0)$, is at least as large as the variance of $\hat{y}_{0R}$, $V(\hat{y}_{0R})$. Thus omitting a subset of the variables reduces the variances of the estimated coefficients and increases the bias. This was seen in the example discussed at the beginning of the section. Comparing the variance of $\hat{y}_0$ and the variances of the elements of $\mathbf{b}_2$, to the variance of $\hat{y}_{0R}$ and the elements of $\mathbf{b}_2^*$, will indicate whether the loss of bias is outweighed by the gain in lower variance.

Irrelevant Variables

In the event that the variables $X_1, X_2, \ldots, X_q$ are irrelevant $(\boldsymbol{\beta}_1 = \mathbf{0})$ the incorrect inclusion of these variables affects the variance of the estimator of $\boldsymbol{\beta}$. Fitting the full model, the covariance matrix for $\mathbf{b}$ is $(\mathbf{X'X})^{-1}\sigma_u^2$ while for the reduced model $\mathbf{b}_2^*$, the covariance matrix is $(\mathbf{X}_2'\mathbf{X}_2)^{-1}\sigma_u^2$. It can be shown that the diagonal elements of $(\mathbf{X'X})^{-1}$ are in general larger than the diagonal elements of $(\mathbf{X}_2'\mathbf{X}_2)^{-1}$. The estimator $\mathbf{b}$ is an unbiased estimator of $\boldsymbol{\beta}$ when $\boldsymbol{\beta}_1 = \mathbf{0}$. The cost of including *irrelevant variables* therefore is a larger variance for the coefficient estimators than would occur if the superfluous variables were omitted.

4.1.6 Geometric Representations

For the simple linear regression model the observations (x_i, y_i), $i = 1, 2, \ldots, n$ are commonly shown geometrically as a two-dimensional scatterplot. The least squares regression is then a line in this X–Y plane. For multiple linear regression the observations $(x_{i1}, x_{i2}, \ldots, x_{ip}, y_i)$ can be represented as n points in a $(p+1)$-dimensional space. The least squares regression relationship $\hat{\mathbf{y}} = \mathbf{Xb}$ can be represented as a p-*dimensional plane* in this $(p+1)$-dimensional space. The n observations represent a scatter of points around the plane. The closer the observations to the plane the higher the value of R^2. The least squares regression plane minimizes the perpendicular distances from the n points to the fitted plane.

As discussed in Chapter 3, the n observations on each variable can also be represented as the tip of a vector in n-dimensional space originating at the origin. The p vectors corresponding to the n observations on each of the X variables generate a p-*dimensional subspace*. The least squares regression given by $\hat{\mathbf{y}}$ is the projection of the vector $\mathbf{y}$ on the p-dimensional space generated by the p vectors $\mathbf{x}_1, \mathbf{x}_2, \ldots, \mathbf{x}_p$. If the origin is moved to the sample means $(\bar{x}_{.1}, \bar{x}_{.2}, \ldots, \bar{x}_{.p}, \bar{y})$, the vectors originating at the mean are given by mean corrected vectors $(\mathbf{y} - \bar{y}\mathbf{i})$, $(\mathbf{x}_1 - \bar{x}_{.1}\mathbf{i})$, $\ldots$, $(\mathbf{x}_p - \bar{x}_{.p}\mathbf{i})$, where $\mathbf{i}$ is a vector of unities. The angle θ between $(\mathbf{y} - \bar{y}\mathbf{i})$ and the plane generated by the mean corrected X vectors is given by $\sqrt{R^2} = \mathrm{Cos}\ \theta$.

4.2 Variable Selection

In practice the regression analyst often has a large number of "candidate" explanatory variables which are believed to be related to the dependent variable Y. Because of the large number of variables and/or because of ignorance of the subset of relevant variables, there is a need to select the appropriate subset. The purpose of *variable selection* techniques is to choose a suitable subset. The manner in which the selection is made will depend somewhat on the purpose of the model. The various purposes of a regression model can be classified into one of the three categories; *description*, *control* or *prediction*. In the case of description, the magnitudes of the individual regression coefficients are of interest, and in addition, the influence of all variables related to the dependent variable are of interest. In the case of description, therefore, it is important to include all relevant variables as well as a variety of possible linear models. If the linear model is to be used for control, it is important that the regression coefficients accurately reflect the impact of the explanatory variables over the entire range of the variables. Variables which are going to be fixed or controlled in the application of the model should be similarly controlled for the sample data. Neither the descriptive nor control applications are generally suited to variable selection methods.

The third category of application, prediction, can usually be handled using variable selection methods. In this case the user is primarily concerned with predicting the behavior of the dependent variable in an inexpensive but effective manner. The objective therefore is usually to optimize the goodness of fit and to minimize the number of variables used. It is the prediction problem that will be the primary focus of the variable selection method. An important concern for the analyst, however, is to ensure that the range of values for the variables used for model estimation are comparable to the range of values for the variables when the model is used for prediction. Even variables which do not appear in the final model, but are related to the dependent variable, must be in the same range for both the estimation phase and the prediction phase.

In Section 4.1 two examples were used to illustrate the techniques of multiple regression. Although twelve explanatory variables in the financial accounting example showed some correlation with the dependent variable RETCAP, only a subset of six were significant in the full model regression. For the air pollution data a variety of models using subsets of 3 to 5 variables from the complete set of eleven variables were observed to be sufficient to maximize R^2.

For each example the correlations between the dependent variable and most of the explanatory variables were significantly different from zero; however, not all the explanatory variables were significant when all were included in a multiple regression model. The reason for the lack of importance of some of the variables in the multiple regression is the intercorrelation among the explanatory variables. It was demonstrated in Chapter 3 for two explanatory variables, that the impact of one explanatory variable on the dependent variable depends on whether the other variable is also included in the model, and also depends on the correlations among the two explanatory variables and the dependent variable. In the case of more than two explanatory variables the relationships are more complex. Given that all explanatory variables are not significant in a multiple regression, it is sometimes desirable to select a subset of the variables that is equally good for prediction purposes. As outlined below, procedures for variable selection have been developed to select the best subsets.

The selection of a subset of variables to be used in a regression model requires that we have some criterion for variable and/or model selection. For a set of p potential explanatory variables there are 2^p possible models. If p is large, therefore, there are many possible models. The fitting of a very large number of models is expensive although continuing advances in computing methodology continue to reduce this cost. More importantly, how do we compare a large set of models and what criteria do we use to select a model. Before discussing variable selection methods we shall discuss various criteria that are commonly used for model selection.

The development of techniques for variable selection, historically, have been closely related to the development of inexpensive computing hardware and the development of variable selection algorithms. Increased computing efficiency has also permitted more evaluation and comparison of the various techniques. The discussion in this section will begin with the *stepwise methods* which were developed first and which are more commonly available in computing software packages. The more recent and more complex procedures will be outlined at the end of this section.

All of the example calculations in this section have been performed using the SAS prodecures PROC STEPWISE and PROC RSQUARE.

4.2.1 Stepwise Regression Methods

Stepwise regression is a sequential process for fitting least squares models, where at each step a single explanatory variable is either added to or deleted from the model in the next fit. The three common methods are usually referred to as *forward selection, backward elimination* and stepwise regression.

Selection and Deletion Criteria

The most commonly used criterion for the addition and deletion of variables in stepwise regression is based on the partial F-statistic introduced in Section 4.1. The partial F-statistic is given by

$$\frac{(\text{SSR} - \text{SSR}_R)/q}{\text{SSE}/(n-p-1)} = \frac{(R^2 - R_R^2)}{(1 - R^2)} \frac{(n-p-1)}{q},$$

where SSR, SSE and R^2 refer to the larger model with p explanatory variables plus an intercept, and SSR_R and R_R^2 refer to the reduced model with $(p - q)$ explanatory variables and an intercept. From this statistic we can see that if the change in R^2, $(R^2 - R_R^2)$ is sufficiently large then the addition of the q variables is considered worthwhile. The partial F-statistic therefore can be used to compare pairs of models in which one model contains all the explanatory variables of the other plus some additional variables. The comparison criterion can be rewritten in the form

$$(R^2 - R_R^2) > \frac{q}{(n-p-1)}(1 - R^2)F_{\alpha;q,(n-p-1)},$$

indicating when the reduced model is inferior to the larger model.

Forward Selection

The *forward selection procedure* begins with no explanatory variables in the model and sequentially adds one variable according to some optimality criterion. Once a variable is included in the model it will remain throughout the process. A common criterion used in forward selection is to add the variable at each step whose partial F-statistic yields the smallest p-value. Variables are entered as long as the partial F-statistic p-value remains below a specified maximum, say PIN. The procedure terminates when the addition of any of the remaining variables would yield a partial F p-value $>$ PIN, or when all variables have been entered. An important disadvantage of this process is that some variables never get in and hence their importance is never determined.

Example

An example of the use of the forward selection procedure is provided by the following analysis of the financial accounting data discussed in Section 4.1. Table 4.16 gives the correlation matrix with the corresponding p-values given underneath each correlation. Table 4.17 summarizes the results of the forward process to obtain a model relating RETCAP to a subset of the twelve explanatory variables. The second step of the forward procedure provides an illustration of the suppression phenomenon discussed in Chapter 3. By entering the second variable QUIKRAT after WCFTCL, the p-value of the coefficient of WCFTCL decreases considerably and R^2 triples in value. It is interesting to examine the correlation matrix and to determine the correlations with the variable RETCAP. While the largest correlation with RETCAP is provided by the variable WCFTCL (0.32486), the second variable to enter was QUIKRAT which has a very weak correlation with RETCAP (−0.10985). The variable QUIKRAT has a correlation of 0.70703 with WCFTCL, and hence the partial correlation between QUIKRAT and RETCAP after controlling for WCFTCL is given by

$$\frac{-0.10985 - (0.70703)(0.32486)}{\sqrt{(1 - (0.70703)^2)}\sqrt{(1 - (0.32486)^2)}} = -0.508.$$

Thus the variation in QUIKRAT that is not related to WCFTCL, explains 26% of the variation in RETCAP that remains to be explained after WCFTCL. The R^2-value of 0.336 at this point is composed of 0.106 from WCFTCL, plus $(0.26)(0.894) = 0.230$ from the addition of QUIKRAT after WCFTCL. This numerical example demonstrates clearly that the value of an added variable in a regression is measured by its contribution after taking into account the contribution of the other variables present. The residual variation in the added variable, after removing the variation related to the other explanatory variables, must be strongly related to the residual variation in the dependent variable, after removing the variation related to the other explanatory variables.

A Useful Analogy

A useful analogy to explain the marginal importance of variables in a regression is given by the employer who wishes to hire college graduates to solve various problems. The first graduate hired should be the one that can solve the most problems. The second graduate hired is the one who can solve the largest number of the problems that still remain to be solved after hiring the first graduate. The second graduate hired is not necessarily the one who can solve the second largest number of problems.

Example (continued)

The phenomenon discussed above repeats itself in the next step when NFA-TAST is entered. R^2 increases again by a large amount and the p-values of the regression coefficients are further reduced. With step 4 the increases in R^2 become less dramatic and eventually the increases in R^2 become very small. It is also of interest to look at the behavior of the p-values of the estimated coefficients. In step 5 the variable CURRAT is entered and the p-value of QUIKRAT is 0.233. In step 6 however after LOGASST is entered the p-value of QUIKRAT drops to 0.052. The stopping rule employed in this example is to end the process if no variables can be added at a p-value of less than 0.50. It is important to carry the forward process far beyond the point of conventional coefficient p-values of 0.05 or 0.10. This permits the analyst to view the impact on the coefficients of the addition of other variables and may suggest alternative models.

Table 4.16. Financial Accounting Correlation Coefficients Among Variables

	GEARRAT	CAPINT	WCFTDT	LOGSALE	LOGASST	CURRAT
RETCAP	−0.16792	0.30964	0.23329	0.29455	0.14200	−0.09715
	0.3003	0.0519	0.1474	0.0650	0.3821	0.5509
GEARRAT	1.00000	0.25324	−0.56113	0.24950	0.03901	−0.33094
	0.0000	0.1149	0.0002	0.1205	0.8111	0.0370
CAPINT	0.25324	1.00000	−0.25158	0.43699	−0.16428	−0.35298
	0.1149	0.0000	0.1173	0.0048	0.3111	0.0255
WCFTDT	−0.56113	−0.25158	1.00000	−0.45363	0.06424	0.82053
	0.0002	0.1173	0.0000	0.0033	0.6937	0.0001
LOGSALE	0.24950	0.43699	−0.45363	1.00000	0.56866	−0.64080
	0.1205	0.0048	0.0033	0.0000	0.0001	0.0001
LOGASST	0.03901	−0.16428	0.06424	0.56866	1.00000	−0.04627
	0.8111	0.3111	0.6937	0.0001	0.0000	0.7768
CURRAT	−0.33094	−0.35298	0.82053	−0.64080	−0.04627	1.00000
	0.0370	0.0255	0.0001	0.0001	0.7768	0.0000
QUIKRAT	−0.32016	−0.35821	0.82503	−0.66261	−0.02505	0.98477
	0.0440	0.0232	0.0001	0.0001	0.8780	0.0001
NFATAST	−0.06681	−0.02862	−0.04179	0.35876	0.28112	−0.26983
	0.6821	0.8609	0.7979	0.0230	0.0789	0.0922
INVTAST	0.19912	0.21274	−0.32033	0.16963	−0.27155	−0.20205
	0.2180	0.1875	0.0439	0.2954	0.0901	0.2112
FATTOT	−0.04816	0.10288	−0.07290	0.36288	0.13897	−0.29597
	0.7679	0.5276	0.6548	0.0214	0.3924	0.0637
PAYOUT	−0.10929	0.01840	−0.15240	0.09174	−0.04931	−0.10833
	0.5020	0.9103	0.3478	0.5734	0.7625	0.5058
WCFTCL	−0.55195	−0.23758	0.96200	−0.31033	0.18319	0.70115
	0.0002	0.1399	0.0001	0.0513	0.2579	0.0001

	QUIKRAT	NFATAST	INVTAST	FATTOT	PAYOUT	WCFTCL
RETCAP	−0.10985	−0.29743	0.10348	−0.25567	−0.14049	0.32486
	0.4998	0.0623	0.5252	0.1113	0.3872	0.0408
GEARRAT	−0.32016	−0.06681	0.19912	−0.04816	−0.10929	−0.55195
	0.0440	0.6821	0.2180	0.7679	0.5020	0.0002
CAPINT	−0.35821	−0.02862	0.21274	0.10288	0.01840	−0.23758
	0.0232	0.8609	0.1875	0.5276	0.9103	0.1399
WCFTDT	0.82503	−0.04179	−0.32033	−0.07290	−0.15240	0.96200
	0.0001	0.7979	0.0439	0.6548	0.3478	0.0001
LOGSALE	−0.66261	0.35876	0.16963	0.36288	0.09174	−0.31033
	0.0001	0.0230	0.2954	0.0214	0.5734	0.0513
LOGASST	−0.02505	0.28112	−0.27155	0.13897	−0.04931	0.18319
	0.8780	0.0789	0.0901	0.3924	0.7625	0.2579
CURRAT	0.98477	−0.26983	−0.20205	−0.29597	−0.10833	0.70115
	0.0001	0.0922	0.2112	0.0637	0.5058	0.0001
QUIKRAT	1.00000	−0.21162	−0.34911	−0.26661	−0.13336	0.70703
	0.0000	0.1899	0.0272	0.0963	0.4120	0.0001
NFATAST	−0.21162	1.00000	−0.36942	0.84441	0.05184	0.03827
	0.1899	0.0000	0.0190	0.0001	0.7507	0.8146
INVTAST	−0.34911	−0.36942	1.00000	−0.19402	0.14671	−0.32889
	0.0272	0.0190	0.0000	0.2303	0.3663	0.0382
FATTOT	−0.26661	0.84441	−0.19402	1.00000	0.00403	−0.01254
	0.0963	0.0001	0.2303	0.0000	0.9803	0.9388
PAYOUT	−0.13336	0.05184	0.14671	0.00403	1.00000	−0.12173
	0.4120	0.7507	0.3663	0.9803	0.0000	0.4543
WCFTCL	0.70703	0.03827	−0.32889	−0.01254	−0.12173	1.00000
	0.0001	0.8146	0.0382	0.9388	0.4543	0.0000

Table 4.17. Results of Forward Selection Procedure for Dependent Variable RETCAP

Step No.	Variable Entered	R^2-Value	Variables in Model	Estimated Coefficient	p-Value
1	WCFTCL	0.106	WCFTCL	0.140	0.041
2	QUIKRAT	0.336	WCFTCL	0.348	0.000
			QUIKRAT	−0.045	0.001
3	NFATAST	0.588	WCFTCL	0.435	0.000
			QUIKRAT	−0.062	0.000
			NFATAST	−0.373	0.000
4	LOGSALE	0.654	WCFTCL	0.395	0.000
			QUIKRAT	−0.042	0.002
			NFATAST	−0.419	0.000
			LOGSALE	0.050	0.014
5	CURRAT	0.703	WCFTCL	0.410	0.000
			QUIKRAT	0.049	0.233
			NFATAST	−0.497	0.000
			LOGSALE	0.058	0.003
			CURRAT	−0.097	0.024
6	LOGASST	0.741	WCFTCL	0.425	0.000
			QUIKRAT	0.083	0.052
			NFATAST	−0.505	0.000
			LOGSALE	0.098	0.001
			CURRAT	0.122	0.005
			LOGASST	−0.071	0.035
7	WCFTDT	0.755	WCFTCL	0.186	0.330
			QUIKRAT	0.074	0.084
			NFATAST	−0.511	0.001
			LOGSALE	0.095	0.103
			CURRAT	−0.127	0.004
			LOGAST	−0.057	0.103
			WCFTDT	0.353	0.191
8	FATTOT	0.766	WCFTCL	0.162	0.397
			QUIKRAT	0.067	0.119
			NFATAST	-0.385	0.005
			LOGSALE	0.100	0.000
			CURRAT	−0.121	0.005
			LOGASST	−0.065	0.068
			WCFTDT	0.396	0.145
			FATTOT	−0.102	0.240
9	PAYOUT	0.773	WCFTCL	0.187	0.333
			QUIKRAT	0.061	0.154
			NFATAST	−0.364	0.009
			LOGSALE	0.101	0.000
			CURRAT	−0.114	0.009
			LOGASST	−0.069	0.055
			WCFTDT	0.354	0.197
			FATTOT	−0.113	0.199
			PAYOUT	−0.017	0.333

Table 4.17. Results of Forward Selection Procedure for Dependent Variable RETCAP (continued)

Step No.	Variable Entered	R^2-Value	Variables in Model	Estimated Coefficient	p-Value
10	INVTAST	0.780	WCFTCL	0.199	0.306
			QUIKRAT	0.124	0.125
			NFATAST	−0.350	0.012
			LOGSALE	0.100	0.000
			CURRAT	−0.176	0.031
			LOGASST	−0.062	0.090
			WCFTDT	0.361	0.203
			FATTOT	−0.109	0.214
			PAYOUT	−0.016	0.344
			INVTAST	0.161	0.352

Backward Elimination

The *backward elimination* method in a sense is the opposite of the forward selection procedure. The backward procedure begins with all variables in the model. At each step of the backward process a variable is removed. The variable removed each time is the one yielding the largest p-value. Variables are removed as long as the p-value > POUT, where POUT is some prescribed p-value limit. The process continues until all variables in the model if removed would yield a p-value < POUT or until there are no variables remaining.

Example

The results of a backward process applied to the RETCAP model are illustrated in Table 4.18. The process continued as long as at least one of the p-values was greater than 0.100. The results of the backward process are similar to the results from the forward procedure except that the procedure is reversed. The order of entry in the forward process was (WCFTCL, QUIKRAT, NFATAST, LOGSALE, CURRAT, LOGASST, WCFTDT, FATTOT, PAYOUT and INVTAST) while the order of exit in the backward process was (GEARRAT, CAPINT, INVTAST, WCFTCL, PAYOUT, FATTOT, LOGASST and QUIKRAT). For either process the variables GEARRAT, CAPINT, INVTAST, PAYOUT and FATTOT seem to be less important. Using the backward process only one model was obtained in which the p-value was less than 0.01 for all coefficients. The model at the end of step 8 provides an excellent fit ($R^2 = 0.721$) with only 4 highly significant explanatory variables.

Table 4.18. Results of Backward Elimination Procedure for Dependent Variable
RETCAP

Step No.	Variable Removed	R^2-Value	Variables in Model	Estimated Coefficients	p-Value
0		0.785	GEARRAT	−0.035	0.651
			CAPINT	−0.013	0.563
			WCFTDT	0.317	0.299
			LOGSALE	0.118	0.003
			LOGASST	−0.078	0.027
			CURRAT	−0.209	0.027
			QUIKRAT	0.162	0.095
			NFATAST	−0.372	0.012
			INVTAST	0.213	0.268
			FATTOT	−0.103	0.252
			PAYOUT	−0.018	0.308
			WCFTCL	0.208	0.305
			CURRAT	-0.209	0.027
1	GEARRAT	0.783	CAPINT	−0.015	0.513
			WCFTDT	0.369	0.186
			LOGSALE	0.117	0.002
			LOGASST	−0.079	0.085
			CURRAT	−0.197	0.027
			QUIKRAT	0.146	0.100
			NFATAST	−0.372	0.011
			INVTAST	0.184	0.302
			FATTOT	−0.104	0.239
			PAYOUT	−0.016	0.343
			WCFTCL	0.190	0.331
2	CAPINT	0.780	WCFTDT	0.351	0.202
			LOGSALE	0.100	0.000
			LOGASST	−0.061	0.090
			CURRAT	−0.176	0.031
			QUIKRAT	0.124	0.125
			NFATAST	−0.349	0.012
			INVTAST	0.161	0.352
			FATTOT	−0.109	0.214
			PAYOUT	−0.016	0.343
			WCFTCL	0.198	0.305
3	INVTAST	0.773	WCFTDT	0.354	0.196
			LOGSALE	0.100	0.000
			LOTAST	−0.068	−0.055
			CURRAT	−0.113	0.009
			QUIKRAT	0.061	0.153
			NFATAS	−0.363	0.008
			FATTOT	−0.112	0.199
			PAYOUT	−0.016	0.332
			WCFTCL	0.186	0.333
4	WCFTCL	0.766	WCFTDT	0.605	0.000
			LOGSALE	0.099	0.000
			LOGASST	−0.057	0.085
			CURRAT	−0.115	0.008
			QUIKRAT	0.055	0.190
			NFATAS	−0.353	0.010
			FATTOT	−0.120	0.168
			PAYOUT	−0.014	0.395

Table 4.18. Results of Backward Elimination Procedure for Dependent Variable RETCAP (continued)

Step No.	Variable Removed	R^2-Value	Variables in Model	Estimated Coefficients	p-Value
5	PAYOUT	0.760	WCFTDT	0.612	0.000
			LOGSALE	0.099	0.000
			LOGASST	−0.055	0.095
			CURRAT	−0.121	0.004
			QUIKRAT	0.060	0.145
			NFATAST	−0.373	0.006
			FATTOT	−0.109	0.200
6	FATTOT	0.747	WCFTDT	0.601	0.000
			LOGSALE	0.094	0.000
			LOGASST	−0.045	0.163
			CURRAT	−0.128	0.003
			QUIKRAT	0.067	0.107
			NFATAST	−0.509	0.000
7	LOGASST	0.731	WCFTDT	0.601	0.000
			LOGSALE	0.068	0.000
			CURRAT	−0.112	0.007
			QUIKRAT	0.045	0.246
			NFATAST	−0.506	0.000
8	QUIKRAT	0.721	WCFTDT	0.611	0.000
			LOGSALE	0.060	0.000
			CURRAT	−0.068	0.000
			NFATAST	−0.474	0.000

The advantage of the backward elimination method over the forward selection method is that the backward method permits the user to see the impact of all the variables simultaneously. A disadvantage when there are a large number of candidate explanatory variables is that the sample size, n, may not be large relative to the number of explanatory variables, p.

Stepwise Method

The forward selection procedure described above has the disadvantage that it does not eliminate variables which can become nonsignificant after other variables have been added. The *stepwise procedure* is a modified forward method which, later in the process, permits the elimination of variables that have become nonsignificant. At each step of the process the p-values are determined for all variables in the model. If the largest of these p-values is > POUT then that variable is removed. After the included variables have been examined for exclusion the excluded variables are examined for inclusion. At each step of the process there can be at most one exclusion followed by at most one inclusion. To prevent infinite cycles PIN must be at least as small as POUT.

Example

For the financial accounting example a stepwise process was applied (see Table 4.19). The conditions for entry and to stay were $p = 0.15$. The first five steps of the stepwise process were identical to the forward process. At the end of step 5, however, the variable QUIKRAT has a p-value of 0.233 and hence in step 6 QUIKRAT is removed. In step 7 the variable WCFTDT is entered with a p-value of 0.057. As a result of the entry of WCFTDT the p-value of WCFTCL becomes 0.676 and hence in step 8 the variable WCFTCL is removed. At the end of step 8 $R^2 = 0.721$ and the four variables (LOGSALE, WCFTDT, CURRAT and NFATAST) are all significant at the 0.000 level.

Table 4.19. Results of Stepwise Procedure for Dependent Variable RETCAP

Step No.	Variable Entered or Removed	R^2-Value	Variables in Model	Estimated Coefficients	p-Value
1,2,3,4,5	SAME AS FORWARD PROCEDURE IN TABLE 4.				
6	QUIKRAT Removed	0.690	LOGSALE	0.0498	0.006
			CURRAT	−0.0493	0.000
			NFATAST	−0.4619	0.000
			WCFTCL	0.4179	0.000
7	WCFTDT Entered	0.722	WCFTDT	0.509	0.056
			LOGSALE	0.058	0.001
			CURRAT	−0.066	0.000
			NFATAST	−0.474	0.000
			WCFTCL	0.076	0.676
8	WCFTCL Removed	0.721	WCFTDT	0.611	0.000
			LOGSALE	0.060	0.000
			CURRAT	−0.068	0.000
			NFATAST	−0.474	0.000

The final models at the end of the backward and stepwise processes in this case are identical although this is not always the case. If the forward process had been terminated at a p-value of 0.10 or 0.15, the final model would have been step 4 in Table 4.17 with $R^2 = 0.654$ and with the four variables WCFTCL, LOGSALE, NFATAST and QUIKRAT. If step 6 of the forward process is used as a final model, $R^2 = 0.741$, and the six variables are WCFTCL, LOGSALE, NFATAST, QUIKRAT, CURRAT and LOGASST. A comparison of the final model obtained from the backward and stepwise processes, to the step 6 model from the forward process, suggests that the variable WCFTDT can be substituted for the variable WCFTCL. In step 5 of both the forward and stepwise processes, the entry of the variable CURRAT resulted in a p-value of 0.233 for QUIKRAT. With the entry of LOGASST on step 6 of the forward process, however, the p-value of QUIKRAT decreased again to 0.052. A comparison of the coefficients of QUIKRAT in steps 4 and 5 reveals a change in sign from negative to positive. Additional model variations will be studied later in this section.

Selecting a Model

The PIN and POUT values should be chosen so that a large variety of models can be studied. After examining the results from all three procedures it is often possible to choose a good model or to suggest other models that should be tried. Various models should also be determined using other criteria which will be discussed below. The three stepwise methods described above typically fit only a small number of the possible models when the total number of explanatory variables is large. It is therefore possible to miss the best subset. For this reason more complex methods have been developed to allow the user to see more models.

4.2.2 Other Selection Criteria and Methods

Before introducing additional methods for model selection it is necessary to introduce additional *selection criteria*. We begin with a discussion of R^2. If the set of explanatory variables in one model is not nested or contained in the set of explanatory variables of the other model being compared, the F-statistic employed in the stepwise methods cannot be used. Some other criterion is required for comparing models in this case. It is sometimes possible for the model containing the fewer explanatory variables to yield a larger R^2 value.

The R^2 Measure

The R^2 measure has already been introduced as a measure of goodness of fit. This measure has the property that it cannot decrease with the addition of a variable to a model. One approach to model selection is to use R^2 as the comparison criterion. Given a set of p potential explanatory variables, determine the maximum R^2 value R_j^2, for each possible model size j, $j = 1, 2, \ldots, p$ where j denotes the number of explanatory variables. The sequence of R_j^2 values will be monotonic with decreasing differences $(R_j^2 - R_{j-1}^2)$ and typically will behave as in Figure 4.1.

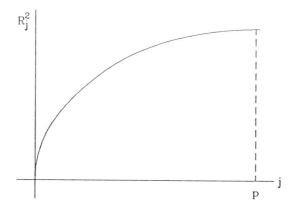

Figure 4.1. Behavior of R^2, as a Function of j.

For the financial accounting example discussed above the R^2 values obtained from the forward procedure showed a pattern similar to the graph in Figure 4.1.

A possible model selection criterion is to select the smallest model j such that the difference $(R_p^2 - R_j^2) < \Delta$ where Δ is a small number. While such a criterion is simple to use, it does not take into account the number of variables used in the models and hence tends to overfit or use too many variables. A good criterion must also have a penalty function which takes into account the number of variables used.

Adjusted R^2 and Mean Square Error

The adjusted R^2, R'^2, introduced earlier, was designed to provide a measure of goodness of fit which takes into account the number of explanatory variables. The R'^2 measure does not always increase as additional variables are included. One possible criterion is to continue to choose the larger model as long as $R_j'^2$ increases. Thus we would select the model j with the largest value of $R_j'^2$.

Denoting the adjusted R^2 for the full model by R'^2 we would reject the model with j explanatory variables as inferior if $(R'^2 > R_j'^2)$. Using the definitions of adjusted R^2 this criterion becomes

$$(R'^2 - R_j'^2) > \frac{(p-j)}{(n-p-1)}(1 - R^2),$$

which is equivalent to the criterion based on the partial F-statistic with critical value $F = 1$.

The adjusted R^2 criterion may also be written in terms of a mean square error comparison between the two models. The adjusted R^2 is given by

$$R'^2 = 1 - \frac{(n-1)}{(n-p-1)}(1 - R^2)$$

$$= 1 - \frac{(n-1)\text{MSE}}{\text{SST}}.$$

Therefore $R'^2 > R'^2_j$ is equivalent to $\text{MSE} < \text{MSE}_j$ where MSE and MSE_j are the mean square errors for the complete model and the model with j explanatory variables respectively. A maximum adjusted R^2 criterion is therefore equivalent to a minimum mean square error criterion.

Mallow's C_p Statistic

In addition to mean square error, an important consideration in variable selection models is bias. Recall from 4.1.5 that the omission of relevant explanatory variables from a model can result in biased estimators of the regression coefficients. Recall also from Chapter 1 that the mean square error can be written as the sum of the variance and the square of the bias. It is therefore possible in variable selection to optimize the mean square error and yet have a large bias. It would be helpful therefore to have a criterion which also took into account the impact on the bias. *Mallow's C_p statistic* is designed with such a goal in mind.

The mean square error for the prediction of y_i using a particular model, can be written

$$\text{MSE}(\hat{y}_i) = V[\hat{y}_i] + [\text{BIAS}(\hat{y}_i)]^2.$$

The total mean square error over all n observations is given by

$$\sum_{i=1}^{n} \text{MSE}(\hat{y}_i) = \sum_{i=1}^{n} V[\hat{y}_i] + \sum_{i=1}^{n} [\text{BIAS}(\hat{y}_i)]^2.$$

Dividing through the equation by σ_u^2 we have an expression for the total standardized mean square error given by

$$\frac{\sum_{i=1}^{n} \text{MSE}(\hat{y}_i)}{\sigma_u^2} = \sum_{i=1}^{n} \frac{V[\hat{y}_i]}{\sigma_u^2} + \sum_{i=1}^{n} \frac{[\text{BIAS}(\hat{y}_i)]^2}{\sigma_u^2}.$$

Assuming that estimators obtained from the full model are unbiased, and using estimators of $V[\hat{y}_i]$ and σ_u^2 from the full model, the expression for total standardized mean square error can be written as

$$C_j = (n-j-1)\frac{\text{MSE}_j}{\text{MSE}} - [n - 2(j+1)],$$

where j denotes the number of explanatory variables in the fitted model excluding the intercept. We shall use C_j to denote the value of Mallow's C_p statistic when there are j explanatory variables.

If the bias is negligible it can be shown that $C_j \approx (j + 1)$. To compare models with respect to bias it is useful to plot C_j vs. $(j + 1)$ and compare the plotted points to a 45° line through the origin. It is therefore desirable to select models which both minimize C_j and at the same time maintain C_j at or below the value $(j+1)$. If the sample size n is large, the multiplier $(n-j-1)$ is also large, and hence small changes in MSE_j can produce large changes in C_j. If $\mathrm{MSE}_j < \mathrm{MSE}$ then it can be shown that $C_j < (j + 1)$.

Expected Mean Square Error of Prediction

Sometimes the data $(y_i, x_{i1}, x_{i2}, \ldots, x_{ij})$ $i = 1, 2, \ldots, n$ can be viewed as a sample from a $(j + 1)$-dimensional multivariate normal. In this case a selection criterion based on the *expected mean square error of prediction* can be used. Denoting $\hat{y}$ as the prediction for y based on the sample, we want to minimize $E[(y-\hat{y})^2]$ where the expectation is taken over all possible samples from the $(j+1)$-dimensional multivarite normal. The prediction $\hat{y}$ is assumed to be based on the regression of Y on the X's. This criterion is estimated by $S_j = \mathrm{SSE}_j / (n-j-1)(n-j-3)$ where SSE_j is the error sum of squares for the j variable regression. The selection of variables is made by minimizing S_j. Since SSE_j necessarily is nonincreasing as j increases, and since $1/(n-j-1)(n-j-3)$ increases as j increases, the criteria S_j will be minimized for some subset of explanatory variables.

Press Statistic

We have already noted in Chapter 3 that the MSE can be greatly influenced by a few observations that are distant from the remaining observations. The PRESS statistic introduced in Chapter 3 to detect outliers was based on this principle. In the absence of outliers we would prefer to select a model with a relatively low value of

$$\mathrm{PRESS} = \sum_{i=1}^{n} [y_i - \hat{y}_{(i)}]^2,$$

where $\hat{y}_{(i)}$ denotes the predicted value of y_i obtained after omitting the i-th observation from the fitting process. In comparing models the one with the lower value of PRESS is a better model for prediction. The PRESS statistic is sometimes used as a model selection criterion.

Maximum R^2 Method

Two recent modifications of the stepwise method are the *maximum R^2 improvement* and the *minimum R^2 improvement* methods. These two methods move sequentially from level to level where the level is the number of explanatory variables in the model. At each level j the methods attempt to switch variables that are included with variables that have been excluded. The maximum R^2 improvement tries to switch an included variable with an excluded variable which provides the largest increase in R^2. This process continues until no included variable can be switched that would increase R^2. The resulting model is said to be the best model for that level of explanatory variables. This model, however, is not necessarily the best j variable model since not all j variable models have been compared. The method then moves to the next level by adding the variable yielding the largest increase in R^2.

Example

For the financial accounting example the maximum R^2 method provided results which help to tie together the results obtained from the three stepwise methods discussed above. Step 1 of the maximum R^2 procedure selects WCFTCL as the best one-variable model ($R^2 = 0.106$). In step 2 beginning with WCFTCL, the variable that increases R^2 by the most is the variable QUIKRAT yielding $R^2 = 0.336$. At this stage WCFTCL is replaced by WCFTDT which yields an R^2 of 0.341. At the end of step 2 the best two-variable model contains WCFTDT and QUIKRAT. Step 3 of the procedure first brings in the variable NFATAST to increase R^2 to 0.571 and then replaces QUIKRAT by CURRAT to increase R^2 to 0.606. The variable WCFTDT is now replaced by WCFTCL which yields the best three variable model ($R^2 = 0.615$). In step 4 the variable LOGSALE is added to get $R^2 = 0.690$ and then WCFTCL is replaced by WCFTDT to yield $R^2 = 0.721$. At this stage the best four variable model has been obtained. In step 5 the variable QUIKRAT is entered which yields the best five variable model ($R^2 = 0.732$), and in step 6 the variable LOGASST is entered which yields the best six variable model with $R^2 = 0.747$. The best models in steps 6 and 7 are obtained by simply adding another variable in the order FATTOT ($R^2 = 0.760$) and INVTAST ($R^2 = 0.766$). The final R^2 for all twelve variables is 0.785.

Minimum R^2 Method

The minimum R^2 improvement procedure is similar to the maximum R^2 method except at each step within each level the variable chosen for switching is the one providing the smallest increase in R^2. The starting point at each level is the maximum R^2 model from the previous level. The reason for the choice of the model with minimum R^2 improvement at each step within any level is to show the user a large number of reasonable models at each level. Typically for each switch the results of the regression are provided. Both the R^2 improvement procedures are designed to produce a large number of possible regressions at each level. These two methods will tend to yield more possible solutions than the stepwise methods. In addition the minimum R^2 method usually produces more solutions than the maximum R^2 method. As suggested earlier it is important that other criteria such as R'^2 and C_j also be employed to evaluate and compare the fitted models.

Example

The minimum R^2 method was applied to the financial accounting example. The first level results consist of twelve steps corresponding to the twelve possible one-variable models. The second level begins with the best one-variable model WCFTCL ($R^2 = 0.106$) and adds the one variable yielding the smallest increase in R^2. The first eleven steps at this level consist of all eleven possible selections of one variable from the remaining eleven. The one yielding the largest increase in R^2 is QUIKRAT ($R^2 = 0.336$). At this step the substitution of WCFTDT for WCFTCL increases R^2 to 0.341. The resulting model (WCFTDT, QUIKRAT) is the best two-variable model. The third level begins with the best two level model and adds the variable CURRAT. The sequence of changes in level three are given by (CURRAT, LOGASST), (LOGASST, WCFTCL), (WCFTCL, INVTAST), (INVTAST, PAYOUT), (WCFTDT, WCFTCL), (PAYOUT, INVTAST), (QUIKRAT, CURRAT), (INVTAST, CAPINT), (CURRAT, QUIKRAT), (CAPINT, FATTOT), (WCFTCL, WCFTDT), (FATTOT, NFATAST), (WCFTDT, WCFTCL), (QUIKRAT, CURRAT). The final level three model contains the variables WCFTCL, NFATAST and CURRAT with $R^2 = 0.615$. The fourth level begins by entering the variable QUIKRAT. The sequence of changes in level four are given by (QUIKRAT, INVTAST), (INVTAST, GEARRAT), (GEARRAT, FATTOT), (NFATAST, CAPINT), (CURRAT, QUIKRAT), (FATTOT, NFATAST), (CAPINT, LOGSALE), (WCFTCL, WCFTDT), (QUIKRAT, CURRAT). The final level four model contains the variables WCFTDT, LOGSALE, CURRAT and NFATAST with $R^2 = 0.721$. The remaining levels of the minimum R^2 process will not be described here.

As suggested above, the advantage of the minimum R^2 procedure is that it shows more possible models. Table 4.21 illustrates 6 four-variable models in which all coefficients are significant at the 0.05 level. The R^2-values vary from a low of 0.619 to a high of 0.721. The total number of variables employed over the six models is eight. Model 1 with $R^2 = 0.619$ and model 5 with $R^2 = 0.667$ have no variables in common. A comparison of the models reveals that LOGSALE and CAPINT are close substitutes as are WCFTDT and WCFTCL; FATTOT and NFATAST, and CURRAT and QUIKRAT.

The other criteria for judging the quality of multiple regression models can also be applied to the financial accounting examples. Table 4.20 shows the values of R'^2, C_j and S_j for the maximum R^2 model at each level. With the objectives that R'^2 should be maximized, S_j should be minimized, and C_j should be less than or equal to $(j + 1)$, the MAX R^2 models corresponding to levels 5 through 7 appear to be adequate.

Table 4.20. Values of Various Selection Criteria for Max R^2 Models

Level	Variables in Model	R^2	R'^2	C_j	s_j
1	WCFTCL	0.106	0.082	76.2	0.0005
2	WCFTDT, QUIKRAT	0.341	0.305	48.7	0.0004
3	WCFTCL, NFATAST, CURRAT	0.615	0.583	16.2	0.0002
4	WCFTDT, CURRAT, NFATAST, LOGSALE	0.721	0.689	5.1	0.0002
5	WCFTDT, QUIKRAT, CURRAT, LOGSALE, NFATAST	0.732	0.692	5.7	0.0002
6	WCFTDT, QUIKRAT, CURRAT, LOGASST, LOGSALE, NFATAST	0.747	0.701	5.7	0.0002
7	WCFTDT, QUIKRAT, CURRAT, FATTOT, LOGASST, LOGSALE, NFATAST	0.760	0.708	6.1	0.0002
8	WCFTDT, QUIKRAT, CURRAT, FATTOT, LOGASST, INVTAST, NFATAST, LOGSALE	0.766	0.706	7.3	0.0002
9	WCFTDT, QUIKRAT, CURRAT, FATTOT, PAYOUT, WCFTCL, LOGASST, LOGSALE, NFATAST	0.773	0.705	8.5	0.0002
10	WCFTDT, QUIKRAT, CURRAT, FATTOT, PAYOUT, WCFTCL, LOGASST, LOGSALE, NFATAST, INVTAST	0.780	0.704	9.6	0.0002
11	WCFTDT, QUIKRAT, CURRAT, FATTOT, PAYOUT, WCFTCL, LOGASST, LOGSALE, NFATAST, INVTAST, CAPINT	0.783	0.698	11.2	0.0002
12	All twelve variables	0.785	0.689	13	0.0002

Table 4.21. Comparison of Several Four Variable Models for RETCAP

Variables	Model 1 Coeff	p-value	Model 2 Coeff	p-value	Model 3 Coeff	p-value	Model 4 Coeff	p-value	Model 5 Coeff	p-value	Model 6 Coeff	p-value
CAPINT	0.049	0.010	0.048	0.012	0.038	0.043						
CURRAT	-0.057	0.000									-0.069	0.000
FATTOT	-0.251	0.000	-0.239	0.000								
WCFTCL	0.418	0.000	0.419	0.000	0.427	0.000	0.395	0.000				
QUIKRAT			-0.054	0.000	-0.055	0.000	-0.042	0.002	-0.058	0.000		
NFATAST					-0.352	0.000	-0.419	0.000	-0.415	0.000	-0.474	0.000
WCFTDT									0.564	0.000	0.612	0.000
LOGSALE							0.050	0.014	0.058	0.003	0.061	0.001
R^2	0.619		0.623		0.634		0.654		0.667		0.721	

All Possible Regressions

As outlined above, stepwise methods do not necessarily show the best model or the best subset at each level of the number of explanatory variables. With recent advances in computer hardware and software it is now possible to examine all possible regressions. With faster computers the necessary calculations are performed more quickly and cheaply, and with more sophisticated algorithms the necessary calculations are performed more efficiently. The major difficulty remaining for the analyst is to digest the large amount of information produced and use it to make the best selection.

A number of algorithms have been developed to determine the best subsets without computing all possible regressions. A technique commonly available in statistical software was developed by Furnival and Wilson (1974). This algorithm determines which regressions are necessarily inferior without computing them. As a result many of the regressions do not need to be computed. Typically the output produced from such computer software gives the results for only the best subsets at each level of the number of explanatory variables. Usually best is defined as the largest value of R^2 or R'^2 and the smallest value of C_j or S_j.

Survey articles discussing variable selection procedures are available in Hocking (1976) and Thompson (1978).

4.2.3 Variable Selection Caution

Statistical procedures which allow the data to determine the model usually yield significance levels which are artificially inflated. In a study of this behavior Freedman (1983) stated:

> If the number of variables is comparable to the number of data points, and if the variables are only imperfectly correlated among themselves, then a very modest search procedure will produce an equation with a relatively small number of explanatory variables, most of which come in with significant coefficients, and a highly significant R^2. This will be so even if Y is totally unrelated to the X's.

In the Freedman study 100 observations on 51 variables were generated independently from a standard normal distribution. The 51st variable was selected as the Y variable. The Y variable was therefore independent of the 50 X variables. The regression of Y on all 50 X variables yielded $R^2 = 0.50$ with $p = 0.53$ and 1 coefficient out of the 50 significant at the 5% level. A total of 15 coefficients were significant at the 25% level. A second regression was estimated using only the 15 variables significant at the 25% level. Six of these fifteen coefficients were significant at the 5% level and $R^2 = 0.36$ (significant at 0.0005). Thus after the second pass we might wish to conclude that the model consisting of these six variables represents a good fitting model. Remember the 51 variables are mutually independent. This experiment was carried out ten times resulting in 9 models significant at the 0.05 level and with the number of .05 significant regressors varying from 1 to 9. Freedman presents theoretical expressions that can be used to approximate these quantities which are functions of the number of regressors, p, and the sample size n. This article should be read often by users of variable selection methods. A useful defence against this problem is to use model validation methods to be outlined later in this chapter. If possible the sample should be split into subsamples and the variable selection procedure should be applied to each subsample.

The article by Lovell (1983) also discusses the "data mining" aspects of variable selection.

4.3 Multicollinearity and Biased Regression

The presence of multicollinearity among the columns of $\mathbf{X}$ can have significant impacts on the quality and stability of the fitted model. In this section a discussion of the effects of multicollinearity and its detection will be presented. A common approach to the multicollinearity problem is to omit explanatory variables. As outlined in 4.1.5 however omitted variables lead to biased coefficients. This section will also outline techniques of *biased regression*.

4.3.1 Multicollinearity

Multicollinearity and its Importance

Recall from Section 4.1 that the term *multicollinearity* was used to describe a situation where the columns of the matrix $\mathbf{X}$ are either linearly dependent or very close to being linearly dependent. When multicollinearity is present, the $\mathbf{X'X}$ matrix is singular or near singular and hence $(\mathbf{X'X})^{-1}$ is either undefined or contains some elements which are very large. As can be seen from the expression for $\mathbf{b}$ and $\text{Cov}(\mathbf{b})$, $(\mathbf{X'X})^{-1}$ is an important determinant of the properties of $\mathbf{b}$.

In Chapter 3 the effects of adding a second explanatory variable Z to a simple linear regression model were studied. It was shown that if the correlation between X and Z is relatively large, the variance of the regression coefficient for X would increase, and the magnitude of this coefficient could also change. In some cases the sign of the regression coefficient also changes. It is possible in some circumstances for the individual regression coefficients to be insignificant due to the large variances, even though the value of R^2 for the regression is relatively large. Thus it is possible to reject the null hypothesis that all regression coefficients are zero even though the individual coefficients are not significant. In the case of suppression, it is possible for highly correlated variables which are weakly correlated with Y to yield a relatively large R^2.

Geometric View of Multicollinearity

Using the two explanatory variable case discussed in Chapter 3 a geometric view of the multicollinearity problem can easily be obtained. In Figures 4.2 and 4.3 three-dimensional views of samples of observations on Y, X and Z are shown. In both figures the scatter of points in the $X - Z$ plane (the dots at the bottom of the vertical lines) is also shown. The points denoted by a cross "x" show the fitted regression plane of Y on X and Z.

The dots "•" near the crosses are the actual observations. In Figure 4.2 the scatter of points in the $X - Z$ plane is broad indicating a low correlation between X and Z. As a result the plane generated by the fitted points is well defined. In comparison Figure 4.3 shows an example of high correlation between X and Z. In the second figure the fitted plane is not as well defined in that there are many planes that could be drawn through the points with little variation in the fitted crosses. In fact the points can almost be fitted as well by a line.

Example

To provide a numerical example for multicollinearity we employ the Canadian financial market data set given in Table 4.22. This data set represents the values of Bank of Canada assets (BANKCAN), monthly return on 91 day Canadian treasury bills (TRSBILL), Canadian Consumer Price Index (CPI) , Canada/U.S.A. spot exchange rate (USSPOT) and Canada/U.S.A. one month forward exchange rate (USFORW). The data are monthly observations for the period September 1977 to December 1980.

The correlation matrix relating the variables in Table 4.22 is shown in Table 4.23. The results of the multiple regression of TRSBILL on BANKCAN, CPI, USSPOT and USFORW are shown in Table 4.24. From the correlation matrix it is clear that the magnitudes of the correlations are relatively high. The correlation between USSPOT and USFORW is extremely high at 0.99865. From the regression results, it is interesting to note that the coefficient of BANKCAN is not significantly different from zero although the correlation between TRSBILL and BANKCAN is 0.75912. In addition, the

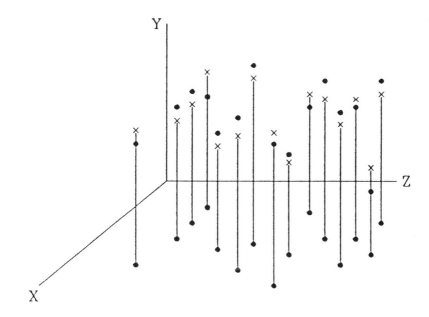

Figure 4.2. Fitted Regression Plane with Low Multicollinearity

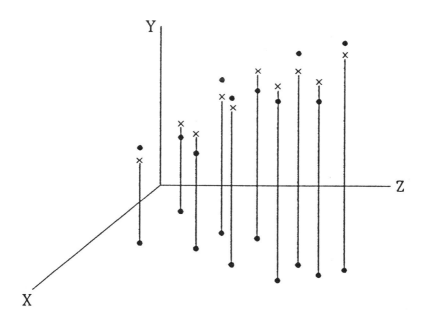

Figure 4.3. Fitted Regression Plane with High Multicollinearity

Table 4.22. Canadian Financial Market Data September 1977 to December 1980

Year	Month	BANKCAN	TRSBILL	CPI	USSPOT	USFORW
1977	S	12107	7.09	69.0	1.0735	1.0750
	O	12503	7.19	69.6	1.0774	1.0777
	N	12303	7.25	70.1	1.1036	1.1039
	D	13416	7.18	70.6	1.1085	1.1093
1978	J	12471	7.14	70.8	1.0915	1.0914
	F	12948	7.24	71.3	1.1085	1.1086
	M	13439	7.62	72.1	1.1170	1.1171
	A	13783	8.18	72.3	1.1373	1.1378
	M	13119	8.13	73.3	1.1283	1.1285
	J	13804	8.24	73.9	1.1190	1.1196
	J	13541	8.43	75.0	1.1222	1.1219
	A	13380	8.77	75.1	1.1380	1.1384
	S	14464	9.02	74.9	1.1511	1.1516
	O	16200	9.52	75.7	1.1892	1.1892
	N	13889	10.29	76.3	1.1658	1.1648
	D	15106	10.43	76.5	1.1687	1.1677
1979	J	14012	10.80	77.1	1.1880	1.1871
	F	14281	10.78	77.8	1.2010	1.2016
	M	14992	10.90	78.8	1.1863	1.1871
	A	14841	10.84	79.3	1.1573	1.1580
	M	14513	10.84	80.1	1.1424	1.1433
	J	15338	10.82	80.5	1.1664	1.1673
	J	14989	10.91	81.1	1.1654	1.1658
	A	15636	11.32	81.4	1.1724	1.1731
	S	16381	11.57	82.1	1.1662	1.1662
	O	15133	12.86	82.7	1.1610	1.1599
	N	15566	13.61	83.5	1.1835	1.1828
	D	15746	13.63	84.0	1.1666	1.1655
1980	J	14821	13.54	84.5	1.1659	1.1654
	F	15010	13.56	85.2	1.1561	1.1557
	M	17245	14.35	86.1	1.1429	1.1401
	A	15705	15.64	86.6	1.1969	1.1927
	M	16516	12.54	87.6	1.1891	1.1919
	J	15956	11.15	88.6	1.1587	1.1604
	J	16035	10.10	89.3	1.1468	1.1514
	A	17476	10.21	90.1	1.1583	1.1586
	S	16091	10.63	90.9	1.1573	1.1570
	O	17044	11.57	91.7	1.1707	1.1679
	N	17791	12.87	92.9	1.1790	1.1758
	D	17313	16.31	93.4	1.1928	1.1872

regression coefficient for USFORW is negative even though the correlation between TRSBILL and USFORW is 0.71802.

All of the calculations for this example throughout this section were performed using SAS PROC REG.

Table 4.23. Pearson Correlation Coefficients for Financial Market Data

	TRSBILL	BANKCAN	USSPOT	USFORW	CPI
TRSBILL	1.00000	0.75912	0.72534	0.71802	0.90886
BANKCAN	0.75912	1.00000	0.76236	0.74400	0.80568
USSPOT	0.72534	0.76236	1.00000	0.99865	0.66166
USFORW	0.71802	0.74400	0.99865	1.00000	0.65389
CPI	0.90886	0.80568	0.66166	0.65389	1.00000

Table 4.24. Results for TRSBILL Multiple Regression

Analysis of Variance

Source	DF	Sum of Squares	Mean Square	F-value	Prob $>$ F
Model	4	191.47145	47.86786218	39.744	0.0001
Error	35	42.15440585	1.20441160		
Total	39	233.62585			

$R^2 = 0.8196$

Variable	Parameter Estimate	Standard Error	t for H_0: Parameter$= 0$	P value
INTERCEP	−36.75495422	7.60007104	−4.836	0.0001
BANKCAN	−0.000323553	0.000301726	−1.072	0.2909
CPI	0.21314334	0.05880803	3.624	0.0009
USSPOT	434.23241	106.74402	4.068	0.0003
USFORW	−403.82760	107.22821	−3.766	0.0006

Impact of Multicollinearity

It was shown in Section 4.1.4 that the vector of estimated regression coefficients for the explanatory variables excluding the intercepts, can be written in terms of the correlation matrix $\mathbf{R}$ as

$$\mathbf{b}^* = s_y \mathbf{D}^{-1/2} \mathbf{R}^{-1} \mathbf{r}_{\mathbf{X}y},$$

and the covariance matrix for $\mathbf{b}^*$ is given by

$$\sigma_u^2 \mathbf{S}^{-1} \frac{1}{(n-1)} = \sigma_u^2 \mathbf{D}^{-1/2} \mathbf{R}^{-1} \mathbf{D}^{-1/2} \left(\frac{1}{n-1} \right). \tag{4.6}$$

Variance Inflation Factor

It can be shown that the diagonal elements of $\mathbf{R}^{-1}$ are given by $f_j = 1/(1 - R_j^2)$, $j = 1, 2, \ldots, p$. The quantity R_j^2 denotes the multiple coefficient of determination in the regression of X_j on the remaining X's in the explanatory set. If the variable X_j has a strong linear relationship with at least one of the remaining X's, then R_j^2 will be close to 1, and hence $f_j = 1/(1 - R_j^2)$ will be relatively large. The variance of b_j^* will be $\left(\frac{\sigma_u^2}{n-1}\right)/(1 - R_j^2)s_{x_j}^2$ which will be large if R_j^2 is close to 1. For this reason the factor f_j is commonly called the *variance inflation factor* (VIF). If the variable X_j is orthogonal to the other X's, R_j^2 has the value 0 and the variance of b_j is minimized.

The expected value of the squared distance between $\mathbf{b}^*$ and the true $\boldsymbol{\beta}^*$ is given by $\sigma_u^2 \sum_{j=1}^{p} f_j s_{x_j}^2$. Thus when the variance inflation factors, f_j, are large, the elements of $\mathbf{b}^*$ tend to be large relative to $\boldsymbol{\beta}^*$. An additional impact of multicollinearity therefore is that the regression coefficients will tend to be too large in absolute value.

Example

For the financial data, the variance inflation factors for the four variables in the multiple regression are BANKCAN (6.84), CPI (5.76), USSPOT (391.62) and USFORW (383.05). The variance inflation factors for USSPOT and USFORW are therefore quite large. Including both USSPOT and US-FORW in the model results in an extremely large increase in the standard errors of the two regression coefficients.

Multicollinearity Diagnostics

The simplest device for detecting multicollinearity is the correlation matrix, which can be used to show large correlations between pairs of explanatory variables. It is possible however for a strong linear relationship to exist among several explanatory variables even though the underlying simple correlations are not large. The squared multiple correlation measure R_j^2, which relates variable X_j to the remaining explanatory variables, is useful for detecting multicollinearity involving more than two variables. The value of R_j^2 should be determined for each j, $j = 1, 2, \ldots, p$ and perhaps for a variety of subsets of the explanatory variables. Some statistics software for stepwise regression will determine the R_j^2 values relating each excluded variable to all the included variables. The diagonal elements of the matrix $\mathbf{R}^{-1}$, $1/(1 - R_j^2)$, $j = 1, 2, \ldots, p$; provide the same information about multicollinearity as the R_j^2 measure. The measure $(1 - R_j^2)$ is sometimes called the *tolerance* and is also used to detect multicollinearity.

Using the Eigenvalue Structure

The reader who is unfamiliar with the concepts of *eigenvalues* and *eigenvectors* for symmetric matrices is advised to review this topic in the appendix before continuing with this section.

The eigenvalue structure of $\mathbf{R}$ can also be used to provide information about multicollinearity. The correlation matrix $\mathbf{R}$ is a positive definite symmetric matrix and therefore can be written as the product of three matrices, $\mathbf{R} = \mathbf{V\Lambda V'}$. The matrix $\mathbf{\Lambda}$ is a diagonal matrix of nonnegative unique elements called the eigenvalues of $\mathbf{R}$. The diagonal elements of $\mathbf{\Lambda}$ are denoted by λ_j, $j = 1, 2, \ldots, p$ and are usually ordered from largest to smallest. The $p \times p$ matrix $\mathbf{V}$ is an orthogonal matrix and hence $\mathbf{VV'} = \mathbf{V'V} = \mathbf{I}$ and $\mathbf{V'} = \mathbf{V^{-1}}$. The columns of $\mathbf{V}$ are unique and are called the eigenvectors of $\mathbf{R}$. The eigenvectors share a one to one relationship with the eigenvalues. The inverse of the matrix $\mathbf{R}$ is given by $\mathbf{R^{-1}} = \mathbf{V\Lambda^{-1}V'}$ and hence the eigenvalues of $\mathbf{R^{-1}}$ are the inverse elements of the eigenvalues of $\mathbf{R}$. The eigenvectors of $\mathbf{R}$ and $\mathbf{R^{-1}}$ are the same.

The eigenvalue structure of $\mathbf{R}$ given by $(\lambda_1, \lambda_2, \ldots, \lambda_p)$ can be used as an indicator of the correlation structure of $\mathbf{R}$. The rank of $\mathbf{R}$ is equal to the number of nonzero eigenvalues. A zero eigenvalue indicates a linear dependency and an eigenvalue close to zero indicates a near linear dependency. The eigenvector $\mathbf{v}_j$ corresponding to a very small eigenvector λ_j can be used to determine which variables are contributing to the linear dependencies. The elements of $\mathbf{v}_j$ which are relatively large in magnitude indicate which variables are the major contributors to the collinearity.

Multicollinearity Condition Number

A commonly used *collinearity diagnostic* is the ratio $\kappa = \lambda_1/\lambda_p$ where λ_1 and λ_p are the largest and smallest eigenvalues respectively. The ratio κ is commonly called the *multicollinearity condition number*. If the condition number is less than 100 multicollinearity is not considered to be a problem, while if κ exceeds 1000 the multicollinearity is considered to be severe.

The diagonal elements of $\mathbf{R^{-1}}$ were earlier labelled as the variance inflation factors and denoted by $f_1, f_2, \ldots, f_p$. Large values of f_j indicated that the variable X_j was strongly correlated to the other X's. The condition number κ defined above can be shown to be related to the variance inflation factors by

$$f_m \leq \kappa < p \sum_{j=1}^{p} f_j,$$

where $f_m = \max_j f_j.$

The largest variance inflation factor f_m is therefore a lower bound for the condition number κ. The maximum f_m is sometimes used as a tolerance test for stepwise regression. A more conservative tolerance test is provided by $\sum\limits_{j=1}^{p} f_j$ which is necessarily contained in the interval for κ. Additional discussion on the multicollinearity condition number can be found in Berk (1977).

Information from Eigenvectors

The matrix of eigenvectors $\mathbf{V}$ derived from $\mathbf{R}^{-1}$ can also be employed to detect the sources of multicollinearity by examining the variances of the estimated regression coefficients. Using (4.6) the covariance matrix of estimated regression coefficients is given by

$$\frac{\sigma_u^2}{n-1}\,\mathbf{D}^{-1/2}\mathbf{R}^{-1}\mathbf{D}^{-1/2} = \frac{\sigma_u^2}{n-1}\,\mathbf{D}^{-1/2}\mathbf{V}\mathbf{\Lambda}^{-1}\mathbf{V}'\mathbf{D}^{-1/2}.$$

For an individual regression coefficient b_j^* this expression is given by

$$\frac{\sigma_u^2}{(n-1)s_{x_j}^2}\sum_{i=1}^{p} v_{ij}/\lambda_i \qquad j=1,2,\ldots p$$

where v_{ij} and λ_i denote elements of $\mathbf{V}$ and Λ respectively. The ratio $m_{ji} = [v_{ij}/\lambda_i]/\left[\sum\limits_{i=1}^{p} v_{ij}/\lambda_j\right]$ represents the proportion of the variance of b_j^* that comes from the term v_{ij}/λ_i. A relatively small value of λ_i (near linear dependency) results in a large value of v_{ij}/λ_i. For the variable j, if the value of m_{ji} is relatively large then the i-th component is an important contributor to the variance of b_j^*. If for a given i, the ratio m_{ji} is relatively large for several coefficients say b_j^* and b_ℓ^*, then the linear dependency represented by λ_i is related to the variables j and ℓ.

Inclusion of the Intercept

While the most common approach to multicollinearity detection employs the correlation matrix $\mathbf{R}$, it is believed by some that this *mean centered* approach masks the important role played by the constant term (see Belsey (1984)). Recall that collinearity is a function of the relation among the columns of the $\mathbf{X}$ matrix, and that the $\mathbf{X}$ matrix normally includes a column of unities corresponding to the constant term. The correlation matrix $\mathbf{R}$ however, is based on the mean centered data matrix $\mathbf{X}^*$ defined in 4.1.4. In situations where it is natural to have the model pass through the origin mean centering is justified. In other situations it may be of practical importance to use centering about some natural origin other than zero such as $-273°C$ or a base index value of 100. Such centering is said to be model dependent rather than the data dependent mean centering which occurs in the calculation of $\mathbf{R}$.

If the column of unities is to be included, the matrix $\mathbf{R}$ in the above discussion is replaced by the matrix $\mathbf{T}$ where $\mathbf{T} = \mathbf{E}^{-1/2}(\mathbf{X'X})\mathbf{E}^{-1/2}$ and where $\mathbf{E}$ is the diagonal matrix of diagonal elements of $\mathbf{X'X}$. The $\mathbf{T}$ matrix has the diagonal elements of unity. The eigenvalue structure of $\mathbf{T}$ is used to determine the multicollinearity diagnostics based on eigenvalues. In the case of the $\mathbf{T}$ matrix, there will be one additional eigenvalue since the dimension of $\mathbf{T}$ is one unit larger than the dimension of $\mathbf{R}$.

Example

The collinearity diagnostics for the TRSBILL multiple regression are given in Table 4.25 for both cases, with and without the column of 1's in $\mathbf{X}$. If the intercept is included the condition number for the smallest eigenvalue is 5030152 which exceeds the benchmark value of 1000, indicating severe multicollinearity. Without the intercept the largest condition number drops to 2579. In the intercept case the condition number for the second and third smallest eigenvalues are also relatively large at 16646 and 5941 respectively.

Table 4.25. Multicollinearity Diagnostics for TRSBILL Regression

			Without Intercept			
No.	Eigenvalue	Condition Number	Var Prop BANKCAN	Var Prop CPI	Var Prop USSPOT	Var Prop USFORW
1	3.330	1.00	0.011	0.012	0.000	0.000
2	0.570	5.80	0.049	0.097	0.001	0.001
3	0.080	38.56	0.934	0.887	0.000	0.000
4	0.0010	2579	0.005	0.002	0.998	0.998

			With Intercept				
No.	Eigenvalue	Condition Number	Var Prop Intercept	Var Prop BANKCAN	Var Prop CPI	Var Prop USSPOT	Var Prop USFORW
1	4.991	1.0	0.000	0.000	0.000	0.000	0.000
2	0.008	624	0.015	0.069	0.000	0.000	0.000
3	0.001	5941	0.010	0.656	0.922	0.000	0.000
4	3.0×10^{-4}	16646	0.954	0.270	0.038	0.001	0.001
5	9.9×10^{-7}	5030152	0.021	0.005	0.003	0.999	0.999

In the intercept case, we can see from the proportion of variance columns in the table, that the regression coefficients for USSPOT and USFORW are strongly related to the smallest eigenvalue. The second smallest eigenvalue contributes a large proportion of the variance in the estimate of the intercept, while the third largest eigenvalue contributes a large proportion of the variance in the regression coefficients for CPI and BANKCAN. From these results we can see the impact of the correlations between USSPOT and USFORW and between CPI and BANKCAN. Similar results can be observed by examining the without-intercept diagnostics.

The variation in the estimated regression coefficients over 14 possible models of the four variables is shown in Table 4.26. The four models with only one explanatory variable indicate that individually all four variables have a significant positive relationship with TRSBILL. The six two-variable models indicate that when CPI and BANKCAN are together BANKCAN is not significant. When USSPOT and USFORW are together the coefficients are both significant, are both larger by a factor of 15 and are opposite in sign. When BANKCAN combines with one of USSPOT or USFORW its coefficient decreases by factor of 2. In the three variable models whenever CPI and BANKCAN are included, BANKCAN is not significant. Whenever USSPOT and USFORW are both included, their coefficient signs are opposite, and their magnitudes are 15 times larger than in the corresponding one variable models.

Table 4.26. Summary of Estimated Regression Coefficients and p-values for Various Models with Dependent Variable TRSBILL

	Variables in Equation			
Equation	BANKCAN	CPI	USSPOT	USFORW
1				56.77(0.000)
2			57.27(0.000)	
3	0.0012(0.000)			
4		0.27(0.000)		
5	0.0002(0.51)	0.23(0.007)		
6	0.0007(0.002)			31.33(0.005)
7	0.0007(0.003)		33.57(0.002)	
8			538.03(0.000)	−488.98(0.001)
9		0.19(0.000)		28.95(0.002)
10		0.18(0.000)	30.64(0.001)	
11	−0.0003(0.54)	0.22(0.002)	33.31(0.001)	
12	0.0006(0.006)		454.87(0.001)	−424.37(0.002)
13		0.16(0.000)	424.48(0.000)	−397.35(0.001)
14	−.0002(0.55)	0.23(0.002)		31.20(0.002)

Multicollinearity Remedies

A fitted regression model with multicollinearity can be a good predictive model, provided that the range of values of the explanatory variables used for prediction is comparable to the range used for estimation. It is important to keep in mind that the multicollinearity can sometimes be caused by artificially restricting the sample in some way. Multicollinearity may also be natural to the population and can only be eliminated by model respecification. Multicollinearity can also be a result of too few observations for too many variables.

When multicollinearity is present, additional data is often the best way of reducing its effects. In some circumstances it may be possible to eliminate variables on a theoretical basis that would also reduce collinearity. As outlined in 4.1.5 however, the omission of relevant variables can yield biased estimators of the coefficients for the included variables. Thus if the model is to be used for some other purpose than prediction, other methods of estimation may be preferable. Rather than eliminating variables it is sometimes preferable to combine the collinear variables into a single variable or index. If prediction is the primary purpose of the model the variable selection methods of the previous section can be used. In the next section various techniques of biased estimation will be introduced which are useful when multicollinearity is present.

4.3.2 Biased Estimation

An alternative approach to estimation in the presence of multicollinearity is to use a *biased estimation* method. One method, *ridge regression*, will retain the variables in their original form. Two other methods, *principal component regression* and *latent root regression*, will involve linear transformations of the variables. The advantage of the principal component and latent root methods is that they retain all of the relevant variables. The purpose of biased estimation methods is to reduce the mean square error of the estimated coefficients at the expense of a little bias. In some applications substantial reductions in mean square error can be obtained with only small amounts of bias.

Ridge Regression

We have seen that severe multicollinearity can increase the variance of the least squares estimator. If however the requirement that the estimator be unbiased is removed, it is usually possible to obtain an estimator with much less variance and only a very small bias. One method of finding such an estimator is called ridge regression. The term ridge was borrowed from the term ridge analysis in response surface methodology because of the similarity of the mathematics.

Using the notation introduced in 4.1 the ridge regression estimator is given by

$$\mathbf{b}_R = [\mathbf{X}'\mathbf{X} + k\mathbf{I}]^{-1}\mathbf{X}'\mathbf{y}$$

where k is a specified positive constant called the biasing parameter. The mean vector is given by

$$E[\mathbf{b}_R] = [\mathbf{X}'\mathbf{X} + k\mathbf{I}]^{-1}\mathbf{X}'\mathbf{X}\boldsymbol{\beta} = \boldsymbol{\beta} - k(\mathbf{X}'\mathbf{X} + k\mathbf{I})^{-1}\boldsymbol{\beta},$$

which is a linear function of $\boldsymbol{\beta}$. The covariance matrix is given by

$$\mathrm{Cov}(\mathbf{b}_R) = [\mathbf{X}'\mathbf{X} + k\mathbf{I}]^{-1}\mathbf{X}'\mathbf{X}[\mathbf{X}'\mathbf{X} + k\mathbf{I}]^{-1}\sigma_u^2.$$

The sum of the mean square elements of $\mathbf{b}_R$ can be written as

$$\mathrm{MSE}(\mathbf{b}_R) = \sigma_u^2 \sum_{j=1}^{p} \frac{\lambda_j}{(\lambda_j + k)^2} + k^2\boldsymbol{\beta}'(\mathbf{X}'\mathbf{X} + k\mathbf{I})\boldsymbol{\beta}, \tag{4.7}$$

where $\lambda_1, \lambda_2, \ldots, \lambda_p$ are the eigenvalues of $\mathbf{X}'\mathbf{X}$.

The first term of (4.7) represents the sum of the variances of the elements of $\mathbf{b}_R$, while the second term denotes the sum of the squares of the bias terms. The impact of k on each of the two components of the MSE can be determined from this expression. As k increases the variance term decreases while the bias term increases. The objective in ridge regression is to choose k in such a way that the decrease in the variance achieved through the introduction of k is larger than the increase in bias. The objective is to have $\mathrm{MSE}(\mathbf{b}_R)$ less than the variance of the least squares estimator.

Determination of k

The determination of the "ideal" value of k is usually a trial and error process. A range of values of k, usually $(0 \le k \le 1)$, are tried and the elements of $\mathbf{b}_R$ are plotted to observe their behavior over the range of k. The plots are called the ridge trace and are used to find a value of k for which the estimates appear to be stable. There is no ideal method for choosing an appropriate k. The simplest suggestion is to choose

$$k = \frac{ps^2}{\mathbf{b}'\mathbf{b}},$$

where $\mathbf{b}$ and s^2 are based on the least squares solution. Ideally a small value of k can be determined in a region of stable values of $\mathbf{b}_R$ such that MSE is less than the variance.

Example

The ridge regression technique was applied to the Financial Market Data of Table 4.22. The results of the analysis are summarized in Table 4.27. It would appear that by the time $k = 0.100$ the regression coefficients are stable. At this value of k the estimated model is

$$\text{TRSBILL} = -37.505 + 0.0001045 \text{ BANKCAN} + 0.15046 \text{ CPI}$$

$$+ 20.251 \text{ USSPOT} + 9.69 \text{ USFORW}.$$

Comparing this equation to the ordinary least squares regression estimate developed earlier ($k = 0$ in Table 4.27), we can observe considerable changes in the coefficients of USSPOT and USFORW. The MSE increased from 1.204 at $k = 0.000$ to 1.779 at $k = 0.100$. Using the above equation for TRSBILL, the correlation between the estimates and the true values is 0.8571, and the mean of the estimates is 11.03. In comparison, the least squares estimate shows a multiple correlation of 0.9053 and a mean of 10.58.

The variance inflation factors and coefficient standard deviations for the four variables, have also changed as a result of the ridge parameter k. The variance inflation factors for $k = 0.100$ are BANKCAN (3.326), CPI (2.920), USSPOT (5.423), USFORW (5.334) which can be compared to the factors for $k = 0.000$, BANKCAN (6.842), CPI (5.765), USSPOT (391.620) and USFORW (383.050). Thus the VIF values were reduced by a factor of 2 for BANKCAN and CPI while for USSPOT and USFORW the reduction is by a factor of 60. The ordinary least squares coefficient standard errors are estimated to be BANKCAN (0.00030), CPI (0.05881), USSPOT (106.744) and USFORW (107.230). For the ridge parameter $k = 0.100$, the estimated standard errors are given by BANKCAN (0.00018), CPI (0.03736), USSPOT (4.328), USFORW (4.472). Although the latter estimates of the coefficient standard errors are biased, they certainly indicate a considerable reduction in standard error for the ridge regression coefficient estimators.

The SAS procedure PROC RIDGEREG was used to perform the calculations for this example.

Principal Component Regression

Earlier in this section we have seen how the correlation matrix for the explanatory variables $\mathbf{R}$, can be expressed in the form $\mathbf{R} = \mathbf{V}\boldsymbol{\Lambda}\mathbf{V}' = \sum_{j=1}^{p} \lambda_j \mathbf{v}_j \mathbf{v}_j'$, where $\boldsymbol{\Lambda}$ is the diagonal matrix of eigenvalues $\lambda_1, \lambda_2, \ldots, \lambda_p$ and $\mathbf{V}$ is the corresponding matrix of eigenvectors $\mathbf{v}_1, \mathbf{v}_2, \ldots, \mathbf{v}_p$. The ordinary least squares estimator of $\boldsymbol{\beta}^*$ (excludes β_0) is given by $\mathbf{b}^* = s_y \mathbf{D}^{-1/2} \mathbf{R}^{-1} \mathbf{r}_{xy}$ and its covariance matrix by $\sigma_u^2 \mathbf{D}^{-1/2} \mathbf{R}^{-1} \mathbf{D}^{-1/2}/(n-1)$.

Table 4.27. Summary of Ridge Regression Estimates

Estimated Coefficients

Value of k	Intercept	BANKCAN	CPI	USSPOT	USFORW	MSE
0.000	−36.755	−0.00032355	0.213143	434.232	−403.83	1.20441
0.005	−39.872	−0.00022040	0.215474	101.822	−70.16	1.53891
0.010	−39.974	−0.00018042	0.209945	63.687	−32.06	1.62123
0.015	−39.860	−0.00014844	0.204363	48.907	−17.40	1.65695
0.020	−39.697	−0.00012066	0.199121	41.034	−9.67	1.67790
0.025	−39.521	−0.00009593	0.194262	36.132	−4.90	1.69240
0.030	−39.346	−0.00007364	0.189767	32.779	−1.67	1.70350
0.035	−39.176	−0.00005339	0.185605	30.336	0.65	1.71261
0.040	−39.012	−0.00003490	0.181741	28.475	2.40	1.72042
0.045	−38.856	−0.00001794	0.178148	27.007	3.76	1.72734
0.050	−38.706	−0.00000233	0.174796	25.819	4.85	1.73361
0.055	−38.563	0.00001209	0.171662	24.836	5.74	1.73937
0.060	−38.426	0.00002544	0.168724	24.009	6.48	1.74474
0.065	−38.295	0.00003784	0.165965	23.302	7.11	1.74978
0.070	−38.169	0.00004938	0.163367	22.692	7.64	1.75455
0.075	−38.049	0.00006014	0.160915	22.158	8.10	1.75908
0.080	−37.932	0.00007019	0.158598	21.687	8.50	1.76341
0.085	−37.820	0.00007960	0.156403	21.269	8.85	1.76755
0.090	−37.712	0.00008843	0.154321	20.894	9.17	1.77154
0.095	−37.607	0.00009672	0.152343	20.557	9.44	1.77537
0.100	−37.505	0.00010451	0.150459	20.251	9.69	1.77907
0.105	−37.406	0.00011186	0.148664	19.972	9.91	1.78265
0.110	−37.310	0.00011878	0.146950	19.717	10.12	1.78612
0.115	−37.217	0.00012532	0.145311	19.482	10.30	1.78949
0.120	−37.126	0.00013150	0.143743	19.266	10.46	1.79276
0.125	−37.037	0.00013736	0.142240	19.066	10.62	1.79594
0.130	−36.950	0.00014290	0.140798	18.880	10.76	1.79904
0.135	−36.866	0.00014816	0.139413	18.706	10.88	1.80207
0.140	−36.782	0.00015315	0.138081	18.544	11.00	1.80502
0.145	−36.701	0.00015789	0.136799	18.392	11.11	1.80791
0.150	−36.621	0.00016240	0.135564	18.249	11.21	1.81074
0.155	−36.543	0.00016669	0.134373	18.114	11.31	1.81351
0.160	−36.465	0.00017078	0.133223	17.987	11.39	1.81622
0.165	−36.389	0.00017467	0.132112	17.867	11.48	1.81888
0.170	−36.315	0.00017839	0.131039	17.753	11.55	1.82150
0.175	−36.241	0.00018193	0.129999	17.645	11.62	1.82407
0.180	−36.168	0.00018531	0.128993	17.542	11.69	1.82660
0.185	−36.097	0.00018855	0.128018	17.444	11.75	1.82909
0.190	−36.026	0.00019164	0.127073	17.351	11.81	1.83154
0.195	−36.956	0.00019460	0.126155	17.261	11.86	1.83396

Table 4.27. Summary of Ridge Regression Estimates (continued)

Estimated Coefficients

Value of k	Intercept	BANKCAN	CPI	USSPOT	USFORW	MSE
0.200	−35.887	0.00019743	0.125265	17.176	11.91	1.83634
0.205	−35.819	0.00020013	0.124399	17.094	11.96	1.83869
0.210	−35.752	0.00020273	0.123557	17.015	12.01	1.84102
0.215	−35.685	0.00020522	0.122739	16.939	12.05	1.84331
0.220	−35.619	0.00020760	0.121942	16.866	12.09	1.84559
0.225	−35.554	0.00020989	0.121166	16.796	12.13	1.84783
0.230	−35.489	0.00021208	0.120410	16.729	12.16	1.85005
0.235	−35.425	0.00021419	0.119673	16.663	12.20	1.85226
0.240	−35.361	0.00021621	0.118954	16.600	12.23	1.85444
0.245	−35.298	0.00021815	0.118253	16.539	12.26	1.85660
0.250	−35.235	0.00022001	0.117568	16.480	12.29	1.85874

Since $\mathbf{R}^{-1} = \mathbf{V}\boldsymbol{\Lambda}^{-1}\mathbf{V}' = \sum_{j=1}^{p} \frac{1}{\lambda_j}\mathbf{v}_j\mathbf{v}_j'$, we can see that for values of λ_j close to zero, the contribution to $\mathbf{R}$ from $\lambda_j\mathbf{v}_j\mathbf{v}_j'$ will be minimal while the contribution to $\mathbf{R}^{-1}$ from $\frac{1}{\lambda_j}\mathbf{v}_j\mathbf{v}_j'$ will be large. Thus if λ_j is small, removal of the term $\lambda_j\mathbf{v}_j\mathbf{v}_j'$ from $\mathbf{R}$ will have a negligible impact, while removal of $\frac{1}{\lambda_j}\mathbf{v}_j\mathbf{v}_j'$ from $\mathbf{R}^{-1}$ will have a large impact.

If the corresponding element of $\mathbf{V}'\mathbf{r}_{xy}$ is also small there is value in removing λ_j and $\mathbf{v}_j$ from $\mathbf{R}$ and $\mathbf{R}^{-1}$. At the cost of a small bias in the estimation of $\boldsymbol{\beta}^*$, a considerable reduction could be achieved in the magnitude of the variances of the regression coefficients. The necessary conditions are that both λ_j and the corresponding element of $\mathbf{V}'\mathbf{r}_{xy}$ be negligible. It is important to keep in mind that in practice a negligible value of λ_j does not always correspond to a negligible value of $\mathbf{V}'\mathbf{r}_{xy}$.

If the k smallest values of λ_j are removed, the new estimator of $\boldsymbol{\beta}^*$ is given by $\mathbf{b}_{k/p}^* = s_y\mathbf{D}^{-1/2}\mathbf{R}_1^{-1/2}\mathbf{r}_{xy}$ where $\mathbf{R}_1 = \mathbf{V}_1\boldsymbol{\Lambda}_1\mathbf{V}_1'$; $\boldsymbol{\Lambda}_1$ is the $(p-k) \times (p-k)$ diagonal matrix remaining after removal of the lowest k values $\lambda_p, \lambda_{p-1}, \ldots, \lambda_{p-k+1}$ from $\boldsymbol{\Lambda}$; and $\mathbf{V}_1$ is the $p \times (p-k)$ matrix of corresponding eigenvectors for $\lambda_1, \lambda_2, \ldots, \lambda_{p-k}$. The variance of this new estimator is given by $\sigma_u^2\mathbf{V}_1\boldsymbol{\Lambda}_1^{-1}\mathbf{V}_1'$ which should be considerably less than $\sigma_u^2\mathbf{V}\boldsymbol{\Lambda}^{-1}\mathbf{V}'$. The bias introduced by omitting the k terms is given by

$$\sum_{j=p-k+1}^{p} \mathbf{v}_j\mathbf{v}_j'\boldsymbol{\beta}^*.$$

The term principal component regression is derived from the fact that the variables $\mathbf{Z} = \widetilde{\mathbf{X}}\mathbf{V}$ are called the principal components of $\widetilde{\mathbf{X}}'\widetilde{\mathbf{X}}$, and that regressing $\tilde{y}$ on $\widetilde{\mathbf{X}}$ is equivalent to regressing $\tilde{y}$ on $\mathbf{Z}$. Deleting the k eigenvalues and eigenvectors is equivalent to deleting the last k principal components $\mathbf{z}_p, \mathbf{z}_{p-1}, \ldots, \mathbf{z}_{p-k+1}$ as explanatory variables. Since the variance of $\mathbf{z}_j$ is λ_j the last k principal components have negligible variance. The theory of principal components will be outlined in Chapter 9.

Latent Root Regression

The technique of latent root regression is very similar to principal component regression except that an augmented correlation matrix $\mathbf{R}^*$ is used. This augmented matrix also contains the correlations for the dependent variable $\mathbf{y}$,

$$\mathbf{R}^* = \begin{bmatrix} \mathbf{R} & \mathbf{r}_{xy} \\ \mathbf{r}'_{xy} & 1 \end{bmatrix}.$$

The eigenvalues are then determined for $\mathbf{R}^*$. Eigenvalues for $\mathbf{R}^*$ that are negligible are also negligible relative to $\mathbf{R}$. The negligible values also correspond to variation unrelated to $\mathbf{y}$, and hence deletion of these elements should not result in a loss of explanatory power for $\mathbf{y}$. The disadvantage of the latent root approach is that the estimator of $\boldsymbol{\beta}^*$ is not obtained directly from the eigenvalue, eigenvector results as in principal component regression.

4.4 Residuals, Influence, Outliers and Model Validation

The fitting of a multiple linear regression model to a sample of observations $(y_i, x_{i1}, x_{i2}, \ldots, x_{ip})$ $i = 1, 2, \ldots, n$, is not complete without an assessment of the sensitivity of the results to the observed data. As outlined in Chapter 3, some observations can have a large influence on the magnitude of the estimated coefficients. These influential observations may be or not be outliers. The techniques introduced in Chapter 3 will be extended here to multiple regression. It is important to emphasize at the outset that the sample data must already have been studied using univariate and bivariate analysis techniques designed to identify outliers and influential observations. The methods outlined in this section should be viewed as additions to the previously introduced procedures.

A multiple regression model is normally used to characterize or predict a relationship among variables for some target population. The sample data used to estimate the relationship is assumed to be representative of this target population. Our concern in this section is whether the fitted model will apply to the target population. Equivalently, if repeated samples were obtained from this population, would the model still hold for the new data. In many research applications the sample values of the explanatory variables cannot be controlled by the researcher. In addition, it is often difficult to define the population in sufficient detail to guarantee that all the relevant factors are taken into account. Obvious dangers, such as changes in population characteristics over time and extrapolation beyond the range of the sample, are important concerns. The techniques of *model validation* are designed to measure model stability across data samples, consistency with theory and knowledge, and performance under data extrapolation.

4.4.1 Residuals and Influence

In Chapter 3 the discussion of simple linear regression included a discussion of residuals and influence. For the most part this section will simply generalize these techniques to the multiple regression model.

Leverage

In Section 4.1.1 the vector of predictions $\hat{\mathbf{y}} = \mathbf{H}\mathbf{y}$ was introduced where $\mathbf{H} = \mathbf{X}(\mathbf{X}'\mathbf{X})^{-1}\mathbf{X}'$. The diagonal elements of $\mathbf{H}$, h_{ii}, $i = 1, 2, \ldots, n$ are $\mathbf{x}'_i(\mathbf{X}'\mathbf{X})^{-1}\mathbf{x}_i$ where $\mathbf{x}'_i$ denotes the i-th row of $\mathbf{X}$. It can be shown that $\sum_{i=1}^{n} h_{ii} = (p+1)$ and $1 \geq h_{ii} \geq 1/n$, $i = 1, 2, \ldots, n$. As in Chapter 3, h_{ii} is commonly called the *leverage* of observation i.

Using the mean deviation from $\mathbf{X}$, $\mathbf{X}^*$, the *hat matrix* can be written as

$$\mathbf{H} = (1/n)\mathbf{d}\mathbf{d}' + \mathbf{X}^*(\mathbf{X}^{*'}\mathbf{X}^*)^{-1}\mathbf{X}^{*'}$$

where $\mathbf{d}$ is an $(n \times 1)$ vector of unities. The diagonal elements of $\mathbf{H}$ are therefore given by

$$h_{ii} = (1/n) + (\mathbf{x}_i - \bar{\mathbf{x}})'(\mathbf{X}^{*'}\mathbf{X}^*)^{-1}(\mathbf{x}_i - \bar{\mathbf{x}}).$$

The second term of this expression describes an ellipsoid in n-dimensional space with centre at $\bar{\mathbf{x}}$. All points $\mathbf{x}_i$ that are on the ellipsoid are said to have the same Mahalanobis distance from $\bar{\mathbf{x}}$. Thus, h_{ii} is large or small depending on how far $\mathbf{x}_i$ is away from the centre $\bar{\mathbf{x}}$. As we shall see later, this distance takes into account the covariances among the columns of $\mathbf{X}$.

If $\lambda_1, \lambda_2, \ldots, \lambda_p$ denote the eigenvalues of $\mathbf{X}^{*'}\mathbf{X}^*$ ranked from largest to smallest, and if $\mathbf{v}_1, \mathbf{v}_2, \ldots, \mathbf{v}_p$ denote the corresponding eigenvectors, the expression for h_{ii} can be written

$$h_{ii} = 1/n + \sum_{\ell=1}^{p}(\mathbf{v}'_\ell\mathbf{x}^*_i/\sqrt{\lambda_\ell})^2.$$

From this expression, we can conclude that h_{ii} tends to be large if $\mathbf{x}_i$ is primarily in the direction of an eigenvector, $\mathbf{v}_\ell$, with a relatively small eigenvalue λ_ℓ. If n is small relative to p it is difficult to identify outliers using this measure.

Range of h

If the multiple regression contains an intercept and if no row of the **X** matrix is repeated, the range of h_{ii} is bounded below by $1/n$ and above by 1. If the model does not contain an intercept the lower bound becomes 0. The sum $\sum_{i=1}^{n} h_{ii}$ is equal to $(p+1)$ and hence the average is $(p+1)/n$. Values of h_{ii} which exceed $2(p+1)/n$ are usually considered to be large.

Leverage and Multicollinearity

Points with high leverage can sometimes produce a high correlation between two or more columns of the **X** matrix. In the case of two explanatory variables, say X and Z, if two points say (x_1, z_1) and (x_2, z_2) have high leverage relative to the remainder of the data then a high correlation between X and Z may exist primarily because of these two observations. This result is shown in Figure 4.4. In this figure the two extreme points would produce a high correlation between X and Z even though there is little correlation among the remaining values of X and Z. In this case the two extreme points may not show as high leverage points because they are close together. If the source of the multicollinearity is points with high leverage, different adjustments for multicollinearity should be considered than in the case of more genuine correlation.

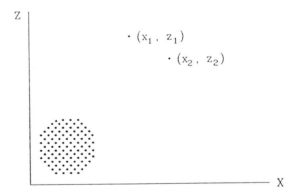

Figure 4.4. Leverage and Multicollinearity

Residuals and Deleted Residuals

The vector of residuals was defined in Section 4.1 as $\mathbf{e} = (\mathbf{I} - \mathbf{H})y$. The covariance matrix for this residual vector is given by $\sigma_u^2(\mathbf{I} - \mathbf{H})$ and hence the variance of the residual e_i is given by $\sigma_u^2(1 - h_{ii})$. Using s^2 to estimate σ_u^2 the standardized or *internally studentized residuals* are given by $r_i = e_i/s(1 - h_{ii})^{1/2}$ as in Chapter 3.

If observation i (i-th row of $\mathbf{X}$) is omitted from the ordinary least squares fit the predicted values of $\mathbf{y}$ are given by

$$\hat{\mathbf{y}}_{(i)} = \mathbf{X}(\mathbf{X}'_{(i)}\mathbf{X}_{(i)})^{-1}\mathbf{X}'_{(i)}\mathbf{y}_{(i)}$$

where the subscript (i) indicates that observation i has been omitted from $\mathbf{X}$ and $\mathbf{y}$. The corresponding residuals are given by $\mathbf{e}_{(i)} = (\mathbf{y} - \hat{\mathbf{y}}_{(i)})$ and the estimate of σ_u^2 by $s_{(i)}^2 = \mathbf{e}'_{(i)}\mathbf{e}_{(i)}/(n - p - 2)$. The *studentized or externally studentized residual* is given by $t_i = e_i/s_{(i)}(1 - h_{ii})^{1/2}$ as in Chapter 3.

The *PRESS* statistic also has the same form as in simple linear regression

$$\text{PRESS} = \sum_{i=1}^{n} e_{(i)}^2 = \sum_{i=1}^{n} e_i^2/(1 - h_{ii})^2.$$

Measures of Influence

Cook's D. While large values of h_{ii} indicate observations which have a large impact on $\hat{\mathbf{y}}$ and on $(\hat{\mathbf{y}} - \hat{\mathbf{y}}_{(i)})$, a preferred measure of influence is Cook's D_i statistic given by

$$D_i = (\hat{\mathbf{y}} - \hat{\mathbf{y}}_{(i)})'(\hat{\mathbf{y}} - \hat{\mathbf{y}}_{(i)})/(p + 1)s^2$$

$$= \left[\frac{h_{ii}}{1 - h_{ii}}\right]\frac{e_i^2}{(p+1)s^2(1 - h_{ii})} = \left[\frac{h_{ii}}{1 - h_{ii}}\right]\frac{r_i^2}{(p+1)}$$

$$= (\mathbf{b}_{(i)} - \mathbf{b})'(\mathbf{X}'\mathbf{X})(\mathbf{b}_{(i)} - \mathbf{b})/(p + 1)s^2.$$

Replacing p by 1 in these expressions gives the definitions used in Chapter 3. A useful criterion for Cook's D is to conclude that the influence is relatively large if D_i exceeds 1. Since Cook's D_i statistic contains two important components, r_i^2 and $[h_{ii}/(1 - h_{ii})(p + 1)]$, it is sometimes useful to look at the two parts separately.

Taking the logarithm of D_i we have the expression

$$\log D_i = \log r_i^2 + \log\{[h_{ii}/(1 - h_{ii})(p + 1)]\}.$$

A plot of $\log r_i^2$ vs. $\log\{[h_{ii}/(1 - h_{ii})(p + 1)]\}$ can then be used to show the components. Constant $\log D_i$ contours are provided by lines with slope minus 1 as in the figure below.

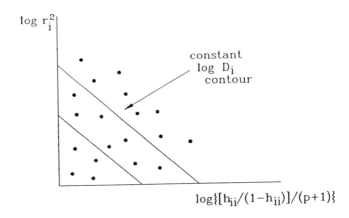

Figure 4.5. Cook's D Decomposition

The DFFIT Family

The generalizations of the DFFIT family of measures of influence introduced in Chapter 3 are summarized below.

$$\text{DFFITS}_i = (\hat{y}_i - \hat{y}_{(i)i})/s_{(i)}\sqrt{h_{ii}}$$

$$= \left[\frac{h_{ii}}{1 - h_{ii}}\right]^{1/2} t_i,$$

and $$\text{DFBETA}_{ji} = (b_j - b_{(i)j})/s_{(i)}c_{jj},$$

where c_{jj}^2 denotes the diagonal element of $(\mathbf{X'X})^{-1}$ corresponding to β_j. The recommended upper bounds for these statistics are $2\sqrt{p/n}$ and $2/\sqrt{n}$ respectively (see Belsey, Kuh and Welsch (1980)).

$$\text{COVRATIO} = \frac{|s_{(i)}^2(\mathbf{X'}_{(i)}\mathbf{X}_{(i)})^{-1}|}{|s^2(\mathbf{X'X})^{-1}|}$$

$$= \frac{(n - p)^p}{[(n - p - 1) + t_i]^p(1 - h_{ii})}.$$

The expression for COVRATIO in Chapter 3 must be adjusted by p, the number of $\mathbf{X}$ variables. Approximate bounds for COVRATIO given by Belsey, Kuh and Welsch (1980) are $|\text{COVRATIO} - 1| \leq 3p/n$. Values of COVRATIO outside this region reflect observations that result in large changes in the covariance matrix for $\mathbf{b}$.

Andrews–Pregibon

An alternative statistic designed to measure influence was introduced by
Andrews and Pregibon (1978). Their statistic is given by

$$R_i = \frac{(n-p-2)s_{(i)}^2 |\mathbf{X}_{(i)}'\mathbf{X}_{(i)}|}{(n-p-1)s^2|\mathbf{X}'\mathbf{X}|},$$

which is the product of two ratios. The ratio $(n-p-2)s_{(i)}^2/(n-p-1)s^2$
denotes a ratio of the error sum of squares without and with the i-th
observation vector. The ratio $|\mathbf{X}_{(i)}'\mathbf{X}_{(i)}|/|\mathbf{X}'\mathbf{X}|$ is a ratio of the measures
of generalized variance among the columns of $\mathbf{X}$ without and with the i-th
observation vector. Small values of R_i are indications that deleting the
i-th observation has brought about a marked reduction in one or both of
$s^2(n-p-1)$ and $|\mathbf{X}'\mathbf{X}|$.

Draper and John (1981) have shown that R_i can be written as

$$R_i = (1-h_{ii}) - e_i^2 / \sum_{i=1}^{n} e_i^2$$

$$= (1-h_{ii})\left[1 - \frac{r_i^2}{(n-p-1)}\right].$$

As in the case of D_i the value of R_i is governed by relative magnitudes of
both h_{ii} and e_i^2. Relatively large values of e_i and/or h_{ii} lead to reductions
in R_i. Thus R_i and D_i tend to move in opposite directions. Large values
of D_i and small values of R_i are indications of a large influence for the i-th
observation.

Given the variety of measures of influence outlined above, what measure or measures should be used? Draper and John (1981) suggest the
combination of D_i, $(1-h_{ii})$ and $e_i^2/(1-h_{ii})$. The rationale given by
Draper and John is that D_i measures the influence on the fitted equation,
$e_i^2/(1-h_{ii})$ is a measure of the observation i as an outlier, and $(1-h_{ii})$ is
a measure of the distance of observation i from the centre of the data. The
statistics commonly produced by computer software such as r_i, t_i, h_{ii} and
D_i will also satisfy these requirements. A useful summary of the concepts
discussed in this section can be found in Cook and Weisberg (1982).

Example

For the financial accounting data discussed in sections 4.1, 4.2 and 4.3, a
multiple regression model relating RETCAP to the variables WCFTCL,
NFATAST, QUIKRAT, LOGSALE, LOGASST and CURRAT was esti-
mated using ordinary least squares. The results of this regression are displayed in Table 4.6 in section 4.1. A summary of the residuals and influence

diagnostics is presented in Table 4.28. The studentized residuals column contains two observations with values greater than 2 in absolute value. Corresponding to observation number 2 the residual is 4.659 and for observation number 21 the residual is –2.071. The H hat column shows two values 0.903 and 0.853, that are close to the upper bound of 1. These values correspond to observations 16 and 21 respectively. The average value of the H hat values is $(p + 1)/n = 7/40 = 0.175$. Observations 16 and 21 appear to have a very high leverage effect on the estimated relationship.

The values of Cook's D and DFFITS indicate that observations 16 and 21 also have a large influence on the estimated relationship. From the sign on the DFFITS measures, we can conclude that adding observation 16 increases the magnitude of $\hat{y}$ while observation 21 decreases the magnitude of $\hat{y}$. From the individual DFFITS values for each variable, we can conclude that adding observation 16 increases the magnitude of the regression coefficient for CURRAT, LOGASST and QUIKRAT and decreases the coefficient of LOGSALE. After adding observation 21, the regression coefficients increase for CURRAT, LOGSALE and NFATAST and decrease for LOGASST, WCFTCL and the intercept. Observation 21 seems to have a substantial influence on the regression coefficients. In addition to observations 16 and 21, observation 2 also has some measure of influence on the fit. For this observation, Cook's D and DFITS are smaller but some of the individual regression coefficients are changed by a larger amount than for observation 16.

An examination of observations 2, 16 and 20 in Table 4.2 permits an explanation of the above diagnostic results. For observation 2 the value of RETCAP is the largest in the sample and the value of QUIKRAT is also relatively large. For observation 16 the value of RETCAP is very small while WCFTCL, QUIKRAT and CURRAT are quite large. In addition, the value of LOGSALE is 0.0. For observation 21 the values of most variables are quite small except for LOGASST. Thus, observation 21 is some distance from the centre of the data. Examination of other rows of the data table reveal that there are other observations which yield some unusual values although the residuals and influence diagnostics do not indicate that these observations have a substantial impact on the fitted model. It could be that "masking" has hidden the impact of these observations. This topic will be addressed in the next section.

The COVRATIO column indicates that omitting observations 16 or 21 brings about substantial increases in the magnitudes of the covariance matrix elements. For observation 2 the opposite is true. Observation 7, when omitted, also results in a large increase in the elements of the covariance matrix. Observation 7 has a very large value of RETCAP which seems to increase s^2 when omitted.

In Table 4.28 the largest value of Cook's D, 3.248 corresponds to observation 21. This value of Cook's D is more than 3 times as large as the next

Table 4.28. Influence Diagnostics for Financial Accounting Data

Obs No.	Standard Residual	Student Residual	H Hat	Cook's D	COV RATIO	DFFITS	DFBETA INTERCEP
1	1.036	1.037	0.209	0.041	1.244	0.533	−0.039
2	3.652	4.659	0.131	0.287	0.038	1.809	0.966
3	1.274	1.286	0.375	0.140	1.395	0.998	−0.017
4	1.442	1.467	0.043	0.013	0.821	0.311	0.057
5	0.058	0.057	0.056	0.000	1.314	0.014	−0.007
6	−0.091	−0.089	0.164	0.000	1.481	−0.039	0.017
7	0.084	0.083	0.455	0.001	2.272	0.076	0.030
8	0.873	0.869	0.078	0.009	1.142	0.253	0.023
9	−0.805	−0.801	0.096	0.010	1.194	−0.261	−0.005
10	0.690	0.684	0.230	0.020	1.455	0.374	−0.064
11	−0.960	−0.958	0.059	0.008	1.081	−0.241	−0.087
12	−0.325	−0.320	0.166	0.003	1.454	−0.143	−0.008
13	0.956	0.954	0.476	0.119	1.944	0.910	−0.020
14	0.353	0.349	0.074	0.001	1.305	0.099	0.063
15	−1.361	−1.379	0.223	0.076	1.065	−0.739	−0.063
16	0.822	0.818	0.903	0.900	11.056	2.497	−0.083
17	0.358	0.353	0.059	0.001	1.283	0.089	0.022
18	−0.086	−0.084	0.044	0.000	1.296	−0.018	−0.004
19	0.429	0.424	0.091	0.003	1.312	0.134	0.104
20	−1.792	−1.857	0.075	0.037	0.654	−0.529	0.024
21	−1.975	−2.071	0.853	3.248	3.507	−5.000	−1.242
22	−0.783	−0.779	0.048	0.005	1.143	−0.176	−0.063
23	0.221	0.217	0.115	0.001	1.387	0.078	0.073
24	−0.850	−0.846	0.057	0.006	1.126	−0.208	−0.109
25	−0.248	−0.245	0.070	0.001	1.317	−0.067	−0.036
26	1.094	1.097	0.179	0.037	1.167	0.513	0.023
27	−1.817	−1.886	0.185	0.107	0.727	−0.899	−0.382
28	−0.556	−0.550	0.035	0.002	1.203	−0.104	0.018
29	−0.378	−0.373	0.067	0.001	1.289	−0.100	0.016
30	0.553	0.548	0.064	0.003	1.242	0.144	−0.057
31	−0.676	−0.670	0.060	0.004	1.197	−0.170	−0.063
32	−1.036	−1.037	0.083	0.014	1.072	−0.312	−0.028
33	0.228	0.225	0.324	0.004	1.814	0.155	−0.126
34	−0.260	−0.256	0.038	0.000	1.272	−0.051	0.005
35	−0.027	−0.026	0.325	0.000	1.838	−0.018	0.014
36	0.940	0.938	0.126	0.018	1.174	0.356	−0.091
37	−0.585	−0.579	0.085	0.005	1.261	−0.177	−0.029
38	−0.772	−0.767	0.077	0.007	1.183	−0.222	0.137
39	0.307	0.302	0.117	0.002	1.378	0.110	0.052
40	.009716	0.009	0.069	0.000	1.333	0.002	−0.000

Table 4.28. Influence Diagnostics for Financial Accounting Data (continued)

Obs No.	DFBETA WCFTCL	DFBETA NFATAST	DFBETA QUIKRAT	DFBETA LOGSALE	DFBETA LOGASST	DFBETA CURRAT
1	0.123	−0.093	−0.415	−0.270	0.217	0.352
2	0.657	−1.313	1.008	0.839	−0.915	−1.067
3	0.240	0.673	−0.040	−0.419	0.265	−0.074
4	0.051	−0.151	0.140	0.118	−0.063	−0.150
5	−0.002	−0.003	−0.004	−0.000	0.005	0.005
6	−0.001	0.010	−0.021	−0.014	−0.007	0.021
7	0.043	−0.050	0.052	0.032	−0.029	−0.059
8	0.154	0.053	−0.137	−0.065	0.013	0.104
9	0.049	0.137	0.036	−0.063	0.032	−0.063
10	−0.003	0.056	−0.331	−0.092	0.060	0.335
11	0.104	0.084	−0.090	−0.069	0.071	0.074
12	−0.006	−0.050	−0.061	−0.032	0.034	0.044
13	−0.768	0.279	0.298	−0.042	0.099	−0.203
14	0.001	0.035	−0.013	−0.003	−0.045	0.010
15	−0.541	−0.286	0.130	0.069	0.060	−0.048
16	−0.073	0.022	0.165	−0.305	0.202	0.144
17	−0.025	0.037	−0.045	−0.028	−0.001	0.045
18	0.005	−0.003	0.006	−0.000	0.004	−0.007
19	0.038	−0.045	0.024	0.041	−0.092	−0.028
20	0.337	0.140	0.002	−0.070	−0.022	−0.054
21	−0.553	1.401	−0.026	3.898	−2.904	1.059
22	0.083	0.005	−0.005	−0.028	0.053	−0.009
23	0.014	−0.009	0.019	0.002	−0.045	−0.025
24	0.066	0.074	−0.061	−0.035	0.068	0.054
25	0.015	0.001	−0.037	−0.033	0.043	0.032
26	−0.393	0.173	0.175	0.188	−0.152	−0.084
27	−0.484	−0.459	0.074	0.005	0.355	−0.004
28	0.020	0.023	−0.018	−0.004	−0.027	0.019
29	0.020	0.026	0.041	−0.013	−0.003	−0.045
30	0.041	−0.054	−0.005	0.041	0.022	0.004
31	0.045	−0.083	0.002	−0.037	0.085	−0.021
32	0.053	0.044	−0.222	−0.187	0.110	0.200
33	−0.010	0.041	−0.016	−0.015	0.098	0.016
34	−0.000	−0.023	0.010	−0.007	0.005	−0.012
35	−0.000	0.006	−0.000	−0.002	−0.010	0.000
36	−0.132	0.218	−0.240	−0.205	0.174	0.235
37	−0.007	0.079	0.048	−0.044	0.047	−0.058
38	−0.007	0.057	0.041	−0.046	−0.071	−0.050
39	0.045	0.059	−0.035	−0.023	−0.031	0.025
40	−0.000	0.000	−0.001	0.000	−0.000	0.001

largest value, 0.900, corresponding to observation 16. If observation 21 is omitted the regression equation becomes

$$\text{RETCAP} = 0.333 + 0.456 \text{ WCFTCL} - 0.605 \text{ NFATAST}$$

$$+ 0.085 \text{ QUIKRAT} + 0.004 \text{ LOGSALE}$$

$$+ 0.019 \text{ LOGASST} - 0.163 \text{ CURRAT.}$$

The large changes in the coefficients of LOGSALE and LOGASST result in these coefficients becoming nonsignificant ($p = 0.93$ and $p = 0.73$ respectively). Since the omission of this one observation suggests that the two variables LOGSALE and LOGASST are superfluous, it is of interest to examine the values of Cook's D for the full model regression before carrying out the variable selection procedure discussed in 4.2.

For the full model regression given in Table 4.3 there are two very large values of Cook's D, 3.757 and 5.379 (not shown) corresponding to observations 16 and 21 respectively. After omitting these two observations a variety of variable selection procedures were carried out. The best five variable model found has an R^2 of 0.783 and is given by

$$\text{RETCAP} = 0.244 + 0.658 \text{ WCFTDT} + 0.047 \text{ LOGASST}$$

$$- 0.176 \text{ CURRAT}$$

$$+ 0.067 \text{ QUIKRAT} - 0.620 \text{ NFATAST,}$$

with all coefficients significant at the 0.10 level. There are no models with more than five variables in which all variables are significant at the 0.10 level. This model reaches the maximum adjusted R^2 value of 0.75 and has a C_p value of 3.6.

If observations 16 and 21 are not excluded, the estimated model for the same variables is given by

$$\text{RETCAP} = 0.117 - 0.445 \text{ NFATAST} - 0.011 \text{ QUIKRAT}$$

$$+ 0.626 \text{ WCFTDT}$$

$$+ 0.043 \text{ LOGASST} - 0.078 \text{ CURRAT,}$$

which has an R^2 value of 0.637. For this model the intercept and the coefficient of QUIKRAT have p-values of 0.32 and 0.80 respectively. For observation 21 the Cook's D value is 1.212 while all remaining Cook's D values are below 0.300.

If observation 21 is omitted, the estimated model is given by

$$\text{RETCAP} = 0.241 - 0.620 \text{ NFATAST} + 0.071 \text{ QUIKRAT}$$

$$+ 0.655 \text{ WCFTDT}$$

$$+ 0.047 \text{ LOGASST} - 0.175 \text{ CURRAT,}$$

which has an R^2 value of 0.787. All coefficients are significant at the 0.05 level and all Cook's D values are below 0.365. It would appear that the estimated model with observation 21 omitted is a suitable model for the data.

Influential Subsets

The measures of influence outlined above can be used to identify influential individual cases. The determination of influential cases in this manner
may omit blocks of influential observations that, because of masking, do
not show up when examined individually. Determination of the influence
of subsets of observations on the overall fit requires a procedure for subset
selection. When selecting a subset of say k observations for examination
and comparison to the rest of the data, there can be a sampling problem in
that what is unusual is determined by comparing the subset to the remainder. Depending on the choice of subset, the two groups being compared
could turn out to be similar when compared to each other.

Given a subset of k observations the influence of this subset can be measured by generalizing the single case measures discussed above. We shall
use the subscript (I) with brackets to denote the vectors and matrices corresponding to the complete set with the subset of k removed. The subscript
I without brackets will refer to the subset of k observations removed from
the complete set.

The matrix $\mathbf{X}$ is partitioned $\mathbf{X} = \begin{bmatrix} \mathbf{X}_{(I)} \\ \mathbf{X}_I \end{bmatrix}$ and the corresponding $\mathbf{y}$ vec-

tor is given by $\mathbf{y} = \begin{bmatrix} \mathbf{y}_{(I)} \\ \mathbf{y}_I \end{bmatrix}$ where $\mathbf{X}_I$ and $\mathbf{y}_I$ denote the observations to

be excluded. The least squares estimator of $\boldsymbol{\beta}$ based on the observations
excluding the subset $(\mathbf{y}_I \ \mathbf{X}_I)$ is given by

$$\mathbf{b}_{(I)} = (\mathbf{X}'_{(I)}\mathbf{X}_{(I)})^{-1}(\mathbf{X}'_{(I)}\mathbf{y}_{(I)})$$

$$= [\mathbf{X}'\mathbf{X} - \mathbf{X}'_I\mathbf{X}_I]^{-1}[\mathbf{X}'\mathbf{y} - \mathbf{X}'_I\mathbf{y}_I]$$

$$= \mathbf{b} - (\mathbf{X}'\mathbf{X})^{-1}\mathbf{X}'_I(\mathbf{I}_k - \mathbf{H}_I)^{-1}\mathbf{e}_I,$$

where $\mathbf{H}_I = \mathbf{X}_I(\mathbf{X}'\mathbf{X})^{-1}\mathbf{X}'_I$, $\mathbf{e}_I = (\mathbf{I}_K - \mathbf{H}_I)\mathbf{y}_I$ and $\mathbf{I}_k$ denotes a $k \times k$
identity matrix.

Cook's D and Influential Subsets

The generalization of Cook's D_i statistic is given by

$$D_I = (\mathbf{b}_{(I)} - \mathbf{b})'[\mathbf{X}'\mathbf{X}](\mathbf{b}_{(I)} - \mathbf{b})/(p+1)s^2$$

$$= \mathbf{e}'_I(\mathbf{I}_k - \mathbf{H}_I)^{-1}\mathbf{H}_I(\mathbf{I}_k - \mathbf{H}_I)^{-1}\mathbf{e}_I/(p+1)s^2.$$

A computation of D_I over all all possible subsets $\binom{n}{k}$ would permit a comparison and hence show *influential subsets* if present.

Andrews–Pregibon

The generalization of the Andrews and Pregibon (1978) statistic is given by

$$R_I = \frac{(n-p-k-1)s^2_{(I)}|\mathbf{X}'_{(I)}\mathbf{X}_{(I)}|}{(n-p-1)s^2|\mathbf{X}'\mathbf{X}|}$$

$$= [1 - Q_k/(n-p-1)s^2][|\mathbf{I}_k - \mathbf{H}_I|],$$

where $Q_k = \mathbf{e}'_I[\mathbf{I}_k - \mathbf{H}_I]^{-1}\mathbf{e}_I$.

DFFITS

The extension of the DFFITS measure given by Welsch (1982) is

$$\text{DFFITS}_{(I)} = \{\mathbf{e}'_I\mathbf{H}_I[\mathbf{I}_k - \mathbf{H}_I]^{-3}\mathbf{e}_I/s^2_{(I)}k^2(n-k)\}^{1/2}.$$

Application of Multiple Case Measures

Using one or more of the above measures and appropriate bounds, a straight-forward but lengthy process would be to compute these measures over all possible subsets $\binom{n}{k}$. If n is large however such a process may not be practical. One shortcut to this process is to eliminate from consideration all subsets of size k that contain smaller subsets already declared influential; eg., if $\mathbf{x}_i$, $\mathbf{x}_j$ and $\mathbf{x}_\ell$ have been declared singly influential then we would not consider the subsets $(\mathbf{x}_i, \mathbf{x}_j)$, $(\mathbf{x}_i, \mathbf{x}_j)$ and $(\mathbf{x}_j, \mathbf{x}_\ell)$ among the sets for $k = 2$ and $(\mathbf{x}_i, \mathbf{x}_j, \mathbf{x}_\ell)$ would not be considered among the sets for $k = 3$. As suggested by Welsch (1982) this means we are looking for influential subsets which do not contain any previously declared influential subsets.

Using Cluster Analysis

A procedure developed by Gray and Ling (1984) uses *cluster analysis* (see Chapter 10) to identify influential subsets. The $\mathbf{y}$ vector is appended to the data matrix $\mathbf{X}$ to yield a new data matrix $\mathbf{Z} = [\mathbf{X}\ \mathbf{y}]$, and a new hat matrix is defined by $\mathbf{H}^*$ where

$$\mathbf{H}^* = \mathbf{Z}(\mathbf{Z}'\mathbf{Z})^{-1}\mathbf{Z}' = \mathbf{H} + \mathbf{e}\mathbf{e}'/\text{SSE}$$

The matrix $\mathbf{H}^*$ contains the information in $\mathbf{H}$ and the information in the residual vector $\mathbf{e}$. Gray and Ling demonstrate that jointly influential cases tend to result in submatrices of $\mathbf{H}^*$ containing elements of relatively large absolute value. It is then possible to rearrange the elements of $\mathbf{H}^*$ to obtain a block diagonal form. Each block represents a cluster of influential elements. Cluster analysis (see Chapter 10) can also be used to determine the clusters of elements of $\mathbf{H}^*$. This approach demonstrates the importance of the off diagonal elements of $\mathbf{H}$ in the determination of joint influence. Relatively large off-diagonal elements of $\mathbf{H}^*$ say h_{ij}^* are indicators of joint influence. The previous measures of influence have ignored these off diagonal elements.

4.4.2 Outliers

As discussed in Chapter 3, there is a relationship between the measurement of influence and the detection of outliers. Not all observations that are outliers have a strong influence on the fitted model, and by the same token not all observations that have a major influence on the fitted model are outliers. In Chapter 3 two test procedures were outlined for the detection of outliers. One procedure used the internally studentized residual while the other used the externally studentized residual. In this section these test procedures will be generalized to the multiple regression model. In addition a discussion of procedures for multivariate outliers will also be presented.

Mean Shift Model

The most common model for outliers in multiple linear regression is called the *mean shift model*. A model for k outliers is given by

$$\mathbf{y} = \mathbf{X}\boldsymbol{\beta} + \mathbf{D}\boldsymbol{\gamma} + \mathbf{u},$$

where the additional term $\mathbf{D}\boldsymbol{\gamma}$ consists of a $k \times 1$ unknown parameter vector $\boldsymbol{\gamma}$ and an $n \times k$ matrix $\mathbf{D}$. The matrix $\mathbf{D}$ consists of k, $(n \times 1)$ column vectors $\mathbf{d}_1, \mathbf{d}_2, \ldots, \mathbf{d}_k$ whose elements are zero except for a 1 in one of the rows corresponding to outlier observations. Thus only k rows of $\mathbf{D}$ contain a single '1' element, all remaining elements of $\mathbf{D}$ are zero.

If the k observations believed to contain outliers can be specified "a priori" then a test of that hypothesis can be carried out by testing $H_0 : \boldsymbol{\gamma} = \mathbf{0}$. The F-statistic for the test is given by

$$F = \left(\frac{(n - p - 1 - k)}{k} \right) \frac{Q_k}{[\mathrm{SSE} - Q_k]},$$

where SSE $= [\mathbf{y'y} - \boldsymbol{\beta}'\mathbf{X'y}]$ denotes the error sum of squares for the model with no outliers, and Q_k is the additional sum of squares accounted for by the parameter $\boldsymbol{\gamma}$. Q_k is given by

$$Q_k = \mathbf{e}_I'[\mathbf{I}_k - \mathbf{H}_I]\mathbf{e}_I, \tag{4.8}$$

where

$$\mathbf{e}_I = (\mathbf{I}_k - \mathbf{H}_I)\mathbf{y}_I \quad \text{and} \quad \mathbf{H}_I = \mathbf{X}_I(\mathbf{X'X})^{-1}\mathbf{X}_I'.$$

Under the assumption of normality, if H_0 is true, the statistic has an F distribution with k and $(n - p - 1 - k)$ degrees of freedom.

In the special case that $k = 1$ the F-statistic becomes

$$(n - p - 2)r_i^2/(n - p - 1 - r_i^2) = t_i^2,$$

where $r_i = e_i/s(1 - h_{ii})^{1/2}$ and t_i^2 is the square of the t-statistic for externally studentized residuals.

In general the outlier observations cannot be prespecified and hence a diagnostic tool is required to identify them. The two statistical tests presented in Chapter 3 for a single outlier in simple linear regression can be extended to multiple linear regression. The two test statistics are given by t_i defined above and $R_n = \max_i |r_i|$.

Multiple Outliers

By computing the single outlier statistics for each of the n observations $(k = 1)$ it is possible to identify more than one outlier. This procedure however has been found to have low power in the presence of multiple outliers. There is no guarantee that such a process would determine all outliers present. It is possible, when several outliers are close together, that *masking* of one outlier by a neighbor will result in neither observation being declared an outlier. At the opposite extreme, it is also possible that, if k is too large, outliers can be missed due to *swamping*. In the case of swamping the influence of an outlier is hidden by $(k - 1)$ more normal observations. The methods currently available for detecting multiple outliers in large data sets are subject to the risk of both masking and swamping.

The reduction in the residual sum of squares, Q_k, resulting from the deletion of k observations given by (4.8), can also be written as $\sum_{j=1}^{k} d_i^2$, where d_i^2 is the square of the statistic $d_i = e_i/\sqrt{1 - h_{ii}}$, $i = 1, 2, \ldots, n$. Given the $\binom{n}{k}$ subsets of size k, the maximum value of Q_k over these subsets can be used to detect the k most likely outliers. Approximations to the critical F-statistic can then be obtained using the Bonferroni inequality. If n is large, the determination of $\max Q_k$ for all $\binom{n}{k}$ subsets, for each level of k can be a lengthy computation.

One approach to the identification of outliers is to define a separate dummy variable for each observation as in the mean shift model discussed above. Variable selection procedures can then be used to identify significant dummy variables and hence identify influential outliers. This approach could also be carried out simultaneously with a variable selection procedure being used to select a subset of variables. Since outliers can have an impact on the outcome of a variable selection process, this combined approach could eliminate this problem.

Another approach suggested by Marasingle (1985) is also based on d_i^2. Beginning with the complete set of n observations , the observation for which $|d_i|$ is maximum is determined (equivalently one may use $|r_k|$). After deleting this observation from the set, the maximum $|d_i|$ is once again determined. This observation yields a second outlier. This process is repeated until a subset of k outliers is identified. Having identified the k outliers the reduction in residual sum of squares Q_k^* is determined. The statistic Q_k^* can be obtained from $Q_k^* = \sum_{i=1}^{k} d_i^2$, where the d_i, $i = 1, 2, \ldots, k$ correspond to the k outliers identified by the above stepwise process.

This procedure does not necessarily determine the maximum value of Q_k over all $\binom{n}{k}$ sets. Marasingle claims that in general $\max Q_k$ is close to Q_k^*. The stepwise process outlined above does however provide the maximum increase in Q_k^* at each stage.

The test statistic is given by $F_k = (S - Q_k^*)/S$ where $S = (n-p)s^2$. The no outlier hypothesis is rejected when F_k is below a critical value. Tables for the critical value of F_k are presented in Marasingle for $k = 2, 3, 4, 5$ in the simple linear regression case.

Since the above process is susceptible to the masking problem, Marasingle also suggests an alternative procedure. This procedure can be applied when the user can identify a subset of k observations, which contains as a subset, the observations suspected of being outliers. The statistic F_k is computed and compared to a critical value. If the hypothesis of no outliers is rejected, the observation with the largest value of $|d_i|$ (or $|r_k|$) is deleted and F_{k-1} is then determined. The process is repeated until the null hypothesis of no outliers is accepted.

Another approach is to delete alternately the observation with the largest value of t_i, followed by the largest value of the leverage h_{ii}, etc., until the values of t_i and h_{ii} become sufficiently small.

4.4.3 Missing Variable Plots

In Section 3.3, the use of missing variable plots, after fitting a simple linear regression model, was introduced to identify other potential explanatory variables. In this section these concepts will be extended to multiple regression. The two most common approaches, the added variable plot, and the partial residual plot, were introduced earlier for the simple linear regression model.

Added Variable Plot

The *added variable plot* requires residuals from two different regressions. The residual vector $\mathbf{e}$, obtained from the regression of Y on $X_1, X_2, \ldots, X_p$, is to be related to the residual vector $\mathbf{v}$, obtained from the regression of a potential variable Z on $X_1, X_2, \ldots, X_p$. The residual vectors $\mathbf{e}$ and $\mathbf{v}$ are given by

$$\mathbf{e} = (\mathbf{I} - \mathbf{H})\mathbf{y} = \mathbf{y} - \mathbf{X}\mathbf{b} \quad \text{where} \quad \mathbf{b} = (\mathbf{X}'\mathbf{X})^{-1}\mathbf{X}'\mathbf{y}$$

$$\text{and} \quad \mathbf{v} = (\mathbf{I} - \mathbf{H})\mathbf{z} = \mathbf{z} - \mathbf{X}\mathbf{g} \quad \text{where} \quad \mathbf{g} = (\mathbf{X}'\mathbf{X})^{-1}\mathbf{X}'\mathbf{z}.$$

The added variable plot displays the points (v_i, e_i), $i = 1, 2, \ldots, n$. If the plot shows a relationship (not a random scatter) between $\mathbf{v}$ and $\mathbf{e}$, then the variable Z will provide additional information (after $X_1, X_2, \ldots, X_p$) regarding the variation in Y. It is worth noting that Z can be a function of the X variables already included. (eg. $Z = X_1 X_2$.)

In the regression of $\mathbf{e}$ on $\mathbf{v}$, the intercept will necessarily be zero and the slope is given by

$$a = \mathbf{z}'(\mathbf{I} - \mathbf{H})\mathbf{y}/\mathbf{z}'(\mathbf{I} - \mathbf{H})\mathbf{z}.$$

The estimator a is equivalent to the ordinary least squares estimator of θ obtained from fitting the model

$$\mathbf{y} = \mathbf{X}\beta + \theta\mathbf{z} + \mathbf{u}.$$

The residuals from this regression, say $\mathbf{e}^* = \mathbf{y} - \mathbf{X}\mathbf{b} - a\mathbf{z}$, are identical to the residuals obtained from regressing $\mathbf{e}$ on $\mathbf{v}$.

Given a set of p potential explanatory variables $X_1, X_2, \ldots, X_p$ it is sometimes useful to obtain an added variable plot for each of these X variables. Thus the variable Z is now one of the X variables. Denoting by $\mathbf{X}(k)$ the $\mathbf{X}$ matrix with the k-th column removed, the two vectors of residuals are given by $\mathbf{e}(k) = \mathbf{y} - \mathbf{X}(k)\mathbf{b}(k)$ and $\mathbf{v}(k) = \mathbf{z} - \mathbf{X}(k)\mathbf{g}(k)$, where $\mathbf{b}(k)$ and $\mathbf{g}(k)$ denote $\mathbf{b}$ and $\mathbf{g}$ with elements b_k and g_k removed, and where $\mathbf{z}$ is the k-th column of $\mathbf{X}$. The added variable plot is determined from plotting the elements of $\mathbf{e}(k)$ against $\mathbf{v}(k)$. The set of $k = 1, 2, \ldots, p$ plots have also been called the partial leverage regression plots.

Partial Residual Plot

Another approach to the determination of the importance of a potential explanatory variable is the *partial residual plot*. The advantage of this plot over the added variable plot is that it can be carried out after only one multiple regression on all p explanatory variables. The residuals from this one regression are used to determine the partial residual plot for each of the p variables. Denoting the residuals from the multiple regression by $\mathbf{e} = \mathbf{y} - \mathbf{Xb} = (\mathbf{I} - \mathbf{H})\mathbf{y}$, the partial residual plot for variable X_k is determined by plotting the *augmented residual*, $\mathbf{e} + b_k \mathbf{x}_k$, against $\mathbf{x}_k$, where b_k denotes the element of $\mathbf{b}$ corresponding to X_k.

As in the case of the added variable plot, the slope estimator obtained from the ordinary least squares fit for the partial residual plot yields the same estimator b_k as the multiple regression. In addition, the residuals around this linear fit are identical to the residuals from the multiple regression. Thus the two plots yield similar information. The major difference between the two plots is that the horizontal axes are $(\mathbf{I} - \mathbf{H}(k))\mathbf{x}_k$ for the added variable plot and $\mathbf{x}_k$ for the partial residual plot. As a result, the variation in the two plots will be different with the partial residual plot showing more variation than the added variable plot.

Nonlinear Relationships

We can generalize the missing variable problem to the case where the missing variable is given by $\mathbf{f}$ which is a nonlinear function of some variable X_k not in $\mathbf{X}$. The variable X_k can, however, be a function of the variables in X. The model is given by $\mathbf{y} = \mathbf{X\beta} + \mathbf{f} + \mathbf{u}$. In place of $\mathbf{f}$ the term $\mathbf{x}_k \beta_k$ is added to the model as if $\mathbf{f} = \mathbf{x}_k \beta_k$. If the multiple regression $\mathbf{y} = \mathbf{X\beta} + \mathbf{x}_k \beta_k + \mathbf{u}$ is estimated, the residuals are $\mathbf{e} = (\mathbf{I} - \mathbf{H})\mathbf{y}$, and the expectation of the residuals is $(\mathbf{I} - \mathbf{H})\mathbf{f}$. If the residuals $\mathbf{e}$ are plotted against $\mathbf{x}_k$, the display of points will present a scatter around the expected value $(\mathbf{I} - \mathbf{H})\mathbf{f}$. If the true function $\mathbf{f}$ is linear $\mathbf{f} = \mathbf{x}_k \beta_k$, then the display should be a random scatter around a horizontal line at 0. If the scatter shows a particular functional form, say $\mathbf{f}$, then a new set of residuals based on $\mathbf{f}$ in place of $\mathbf{x}_k \beta_k$ should be determined. If $\mathbf{f}$ is the correct functional form the plot of the residuals against $\mathbf{f}$ should appear as a random scatter around a horizontal line. Several examples of this approach are presented in Mansfield and Conerly (1987).

The partial plot can also be used to detect nonlinearity. One approach would be to fit a number of nonlinear functions of X_k, i.e., $X_k^2, X_k^3, X_k^{1/2}$, $\log X_k$ perhaps in a backward stepwise fashion. The augmented residuals from the final model are then plotted against X_k. The augmented residuals would be given by $\mathbf{e} + \hat{\mathbf{f}}$ where $\hat{\mathbf{f}}$ denotes the fitted part of the model corresponding to the various functions of X_k. For many applications a simple quadratic term can be used as a preliminary step (see Mallows (1986)).

4.4.4 Model Validation

If additional data can be obtained from the same population through additional samples, then the ideal approach to model validation is to compare fitted models across samples. If additional samples are not available, large data sets can usually be split so that comparisons can be made between subsamples. Such data splitting techniques are usually called *cross validation*. For the remainder of this discussion we assume that there are two samples available to be compared. The data for the two samples is denoted by $\mathbf{y}^{(1)}$, $\mathbf{y}^{(2)}$, $\mathbf{X}^{(1)}$ and $\mathbf{X}^{(2)}$ respectively. The resulting least squares estimators, based on two separate regressions, are given by $\mathbf{b}^{(1)} = (\mathbf{X}^{(1)\prime}\mathbf{X}^{(1)})^{-1}\mathbf{X}^{(1)\prime}\mathbf{y}^{(1)}$ and $\mathbf{b}^{(2)} = (\mathbf{X}^{(2)\prime}\mathbf{X}^{(2)})^{-1}\mathbf{X}^{(2)\prime}\mathbf{y}^{(2)}$. The error sums of squares and R^2 values are given by $\text{SSE}^{(1)}$, $\text{SSE}^{(2)}$, $R^{(1)^2}$ and $R^{(2)^2}$ respectively. The most important characteristic to be studied is the stability of the estimated regression coefficients over the two samples.

Chow Test

A comparison of the two fitted models can be carried out using the *Chow test*, which tests the null hypothesis that the samples are drawn from a common population, and hence the same true regression coefficients.

The Chow test statistic is given by

$$\frac{(\text{SSE} - \text{SSE}^{(1)} - \text{SSE}^{(2)})/(p+1)}{(\text{SSE}^{(1)} + \text{SSE}^{(2)})/(n_1 + n_2 - 2(p+1))} \;,$$

where SSE denotes the error sum of squares derived from the combined sample regression under the assumption of a common parameter $\boldsymbol{\beta}$. The two samples are assumed to contain n_1 and n_2 observations respectively. If the null hypothesis is true this statistic follows an F distribution with $(p+1)$ and $(n_1 + n_2 - 2p - 2)$ degrees of freedom.

Example

To provide a cross validation example a second sample of 40 observations was selected from the population of Financial Accounting Data introduced in Section 4.1. Table 4.30 summarizes the data. A multiple regression analysis relating RETCAP to the twelve explanatory variables showed that two observations (21 and 36) produced very large values of Cook's D. After omitting these two observations the multiple regression yielded the equation shown in Table 4.29. For comparison the equations from the first sample and also the combined sample are shown in the table. The equation for the first sample omits observation 21 which yielded a large value of Cook's D.

Table 4.29.   Comparison of Multiple Regression Results From Two Separate Samples and Combined Sample For Financial Accounting Data

Variable	Second Sample		First Sample		Combined Sample	
	Coefficient	p-value	Coefficient	p-value	Coefficient	p-value
GEARRAT	−0.020	0.853	−0.099	0.209	0.016	0.771
CAPINT	−0.080	0.040	0.025	0.371	0.006	0.724
WCFTDT	−0.214	0.592	0.351	0.217	0.321	0.099
LOGSALE	0.500	0.012	−0.079	0.386	0.036	0.573
LOGASST	−0.487	0.010	0.116	0.218	−0.028	0.658
CURRAT	0.189	0.051	−0.149	0.099	−0.088	0.125
QUIKRAT	0.080	0.466	0.029	0.778	0.013	0.852
NFATAST	−0.304	0.060	−0.560	0.001	−0.347	0.002
INVTASS	0.230	0.338	−0.061	0.772	−0.030	0.837
FATTOT	−0.111	0.236	−0.080	0.339	−0.083	0.186
PAYOUT	0.002	0.923	−0.023	0.173	−0.015	0.197
WCFTCL	1.115	0.006	0.178	0.344	0.264	0.059
INTERCEPT	0.249	0.096	0.354	0.018	0.273	0.006
SSE	0.0762		0.1251		0.2845	
R^2	0.696		0.820		0.703	

The equation for the combined sample omits observations 21 and 36 from the second sample and observation 21 from the first sample.

A comparison of the regression coefficients for the three samples in Table 4.29 shows that few of the variables have significant p-values, and that the estimated coefficients vary over the three samples. In addition to the INTERCEPT only two variables NFATAST and CURRAT seem to be important in all three models. In the case of CURRAT the coefficient does not retain the same sign in all cases. The value of the F-statistic for Chow's test is given by

$$F = \frac{(0.2845 - 0.1251 - 0.0762)/13}{(0.1251 + 0.0762)/(39 + 38 - 26)} = 1.621$$

which has a p-value of approximately 0.10. There is therefore only very weak evidence that the models are different for the two samples.

Earlier in this chapter variable selection procedures were used to examine a variety of models for the first sample. Using the combined sample, variable selection procedures were used to obtain a four variable model which contains the variables NFATAST, CURRAT, WCFTDT and WCFTCL. Table 4.31 shows the results of fitting this four variable model to the two samples and the combined sample. For the variables NFATAST and CURRAT the coefficients remain stable and significant. For the variables WCFTDT and WCFTCL the coefficients are not always significant and show larger differences between samples. Given that the correlation between WCFTDT and WCFTCL is quite high this instability might be

Table 4.30. Second Sample of Financial Accounting Data

OBS	RETCAP	GEARRAT	CAPINT	WCFTDT	LOGSALE	LOGASST
1	0.19	0.15	2.47	0.16	5.2297	4.8375
2	0.22	0.54	0.64	0.16	4.1495	4.3402
3	0.17	0.49	3.18	0.20	5.3831	4.8811
4	0.12	0.39	1.55	0.08	4.1225	3.9333
5	0.21	0.11	1.56	0.34	4.7795	4.5877
6	0.12	0.19	1.74	0.25	4.1503	3.9086
7	0.15	0.35	1.39	0.16	5.6998	5.5577
8	0.10	0.39	1.60	0.09	4.4162	4.2128
9	0.08	0.50	1.58	0.04	4.7108	4.5126
10	0.31	0.41	1.88	0.11	4.4678	4.1928
11	0.21	0.08	1.43	0.33	4.3899	4.2336
12	0.22	0.16	1.55	0.37	4.0253	3.8344
13	0.20	0.13	0.96	0.48	3.8573	3.8764
14	0.11	0.23	1.09	0.15	3.9068	3.8685
15	0.38	0.27	3.13	0.20	5.1631	4.6669
16	0.23	0.00	5.44	0.24	5.7130	4.9772
17	0.32	0.11	2.51	0.09	4.7114	4.3123
18	0.13	0.55	1.51	0.05	4.6763	4.4972
19	0.29	0.00	0.45	0.60	4.5233	4.8709
20	0.09	0.28	3.82	0.09	4.9876	4.4058
21	−2.22	1.78	3.21	−1.28	4.0554	3.5485
22	0.17	0.28	2.07	0.11	4.2837	3.9679
23	−0.04	0.46	2.79	−0.04	4.7616	4.3153
24	0.26	0.00	2.34	0.23	4.2468	3.8779
25	0.21	0.20	1.07	0.30	4.4106	4.3829
26	0.15	0.66	1.08	0.21	4.3984	4.3634
27	0.23	0.11	2.46	0.07	4.8314	4.4399
28	0.20	0.33	1.47	0.28	4.2050	4.0364
29	0.19	0.30	1.38	0.14	4.3139	4.1727
30	0.08	0.35	1.21	0.10	4.9510	4.8675
31	0.19	0.19	1.36	0.14	5.5754	5.4405
32	0.20	0.21	0.81	0.35	4.7722	4.8638
33	0.14	0.30	1.48	0.20	4.9993	4.8282
34	0.04	0.18	1.83	0.07	4.1786	3.9151
35	0.10	0.13	0.96	0.12	5.7613	5.7801
36	−0.09	0.68	0.77	−0.22	3.9671	4.0802
37	0.10	0.23	12.14	0.14	5.6884	5.6334
38	0.20	0.05	2.35	0.12	4.7908	4.4200
39	0.13	0.22	1.37	0.13	5.4876	5.3501
40	0.08	0.19	1.64	0.14	4.0891	3.8737

Table 4.30. Second Sample of Financial Accounting Data (continued)

OBS	CURRAT	QUIK-RAT	NFATAST	INVTASS	FAT-TOT	PAY-OUT	WCFTCL
1	1.33	0.54	0.28	0.42	0.36	0.31	0.16
2	0.93	0.83	0.13	0.04	0.16	0.45	0.26
3	1.09	0.84	0.43	0.13	0.74	0.50	0.26
4	1.09	0.50	0.23	0.37	0.50	0.65	0.08
5	1.74	1.10	0.30	0.20	0.50	0.25	0.34
6	1.89	1.00	0.34	0.31	0.38	0.80	0.25
7	1.38	0.73	0.48	0.22	0.62	0.46	0.25
8	1.57	0.94	0.26	0.30	0.42	1.03	0.12
9	1.28	0.74	0.25	0.31	0.33	0.00	0.04
10	1.10	0.66	0.17	0.31	0.25	0.25	0.12
11	1.49	1.06	0.40	0.17	0.71	0.61	0.36
12	1.38	0.97	0.42	0.17	0.62	0.25	0.37
13	1.00	0.61	0.68	0.13	0.97	0.60	0.48
14	1.23	0.92	0.40	0.15	0.64	0.80	0.18
15	1.39	0.33	0.21	0.38	0.32	0.39	0.25
16	1.29	0.24	0.27	0.50	0.38	0.36	0.24
17	1.34	0.86	0.09	0.31	0.13	0.53	0.09
18	1.14	0.44	0.24	0.42	0.40	0.00	0.06
19	1.21	1.18	0.57	0.01	0.58	0.21	0.60
20	1.28	0.34	0.34	0.46	0.50	1.52	0.10
21	1.06	0.50	0.16	0.37	0.30	0.00	−1.28
22	1.36	0.67	0.26	0.37	0.32	0.22	0.12
23	1.11	0.72	0.19	0.28	0.32	0.00	−0.04
24	1.83	1.20	0.21	0.27	0.26	0.53	0.23
25	2.72	1.77	0.24	0.24	0.36	0.42	0.40
26	0.58	0.29	0.70	0.15	1.07	0.00	0.30
27	0.88	0.88	0.17	0.00	0.22	0.67	0.07
28	0.91	0.77	0.53	0.07	1.16	0.21	0.33
29	1.28	0.49	0.25	0.42	0.33	0.52	0.16
30	2.36	1.44	0.31	0.27	0.51	1.08	0.18
31	1.35	0.96	0.22	0.22	0.36	0.40	0.15
32	3.98	2.63	0.21	0.26	0.34	0.51	0.63
33	0.54	0.26	0.72	0.09	0.74	0.53	0.27
34	1.57	1.08	0.28	0.23	0.54	4.21	0.07
35	1.40	0.57	0.12	0.28	0.21	0.43	0.15
36	1.45	0.60	0.62	0.19	0.71	0.00	−0.46
37	1.56	0.83	0.33	0.23	0.52	0.12	0.18
38	1.42	0.80	0.04	0.37	0.07	0.33	0.13
39	1.73	0.75	0.26	0.41	0.52	0.53	0.17
40	1.57	0.74	0.17	0.34	0.27	0.91	0.14

expected. The F-statistic for Chow's test is given by

$$F = \frac{(0.3154 - 0.1802 - 0.1161)/5}{(0.1802 + 0.1161)/(39 + 38 - 10)} = 0.864.$$

Using a different model for each of the two samples does not produce a significant increase in explained variation.

Table 4.31. Comparison of Multiple Regression Results For Reduced Model For Two Separate Samples and Combined Sample

	Second Sample		First Sample		Combined Sample	
Variable	Coefficient	p-value	Coefficient	p-value	Coefficient	p-value
NFATAST	−0.418	0.000	−0.523	0.000	−0.465	0.000
CURRAT	−0.093	0.001	−0.091	0.000	−0.089	0.000
WCFTDT	0.098	0.704	0.353	0.149	0.310	0.035
WCFTCL	0.544	0.042	0.238	0.162	0.288	0.011
INTERCEPT	0.296	0.000	0.350	0.000	0.322	.000
SSE	0.1161		0.1802		0.3154	
R^2	0.537		0.741		0.674	

Validation by Prediction

While the Chow test is useful for comparing samples with respect to a given model it does not allow for the possibility of variable selection. In practice we may have used a variable selection procedure to select a subset of explanatory variables. Our interest in cross validation may include an interest in whether the selected subset is adequate across samples. Since variation between samples can lead to different selections of variables for the two samples, we should also study the prediction ability of the fitted model in the other sample.

Using the parameter estimate $\mathbf{b}^{(1)}$ from the first sample, predictions for the second sample are obtained from $\mathbf{X}^{(2)}$ using $\tilde{\mathbf{y}}_{(1)}^{(2)} = \mathbf{X}^{(2)}\mathbf{b}^{(1)}$. The error sum of squares for these predictions is given by

$$\text{SSE}_{(1)}^{(2)} = \sum_{i=1}^{n_2} (\tilde{y}_{(1)i}^{(2)} - y_i^{(2)})^2,$$

and the mean square by $\text{MSE}_{(1)}^{(2)} = \text{SSE}_{(1)}^{(2)}/n_2$. This error mean square can be compared to the error mean square obtained from the use of $\hat{\mathbf{b}}^{(1)}$ in the first sample, $\text{MSE}^{(1)} = \text{SSE}^{(1)}/n_1$. If the overall variation of the $\mathbf{X}$ variables around their centroid is similar in the two samples the two MSE values should be comparable. When a sample is subdivided for the purpose of cross validation, algorithms are sometimes used to obtain two samples which are similar with respect to the variation of the $\mathbf{X}$ variables around the centroid.

Another useful measure is to compute the correlation between $\mathbf{y}^{(2)}$ and $\tilde{\mathbf{y}}_{(1)}^{(2)}$. Denoting the square of this correlation by $R_{(1)}^{(2)^2}$ it can be compared to the $R^{(1)^2}$ value based on the first sample. If the two are close the model is stable over the two samples.

A *double cross validation* procedure repeats the above process by interchanging the two samples. The model obtained from the second sample is tested on the first sample by computing the predictions $\tilde{y}_{(2)}^{(1)} = \mathbf{X}^{(1)}\mathbf{b}^{(2)}$ for $\mathbf{y}^{(1)}$ and determining $\text{MSE}_{(2)}^{(1)} = \sum_{i=1}^{n_1} \left(\tilde{y}_{(2)i}^{(1)} - y_i^{(1)} \right)^2 / n_1$ and $R_{(2)}^{(1)^2}$.

Example

In Section 4.4.1 the model used to explain RETCAP for the financial accounting example was given by

$$\text{RETCAP} = \underset{(.015)}{0.241} - \underset{(.000)}{0.620}\,\text{NFATAST} + \underset{(.000)}{0.071}\,\text{QUIKRAT}$$

$$+ \underset{(.058)}{0.655}\,\text{WCFTDT}$$

$$+ \underset{(.023)}{0.047}\,\text{LOGASST} - \underset{(.000)}{0.175}\,\text{CURRAT}. \tag{4.9}$$

This regression omitted observation number 21 as explained in 4.4.1. The error sum of squares was 0.1481 and R^2 was 0.787. The value of MSE is $0.1481/39 = 0.0038$. Using this model to predict RETCAP for the second sample an $\text{MSE}_{(1)}^{(2)}$ of 0.0066 was obtained with $R_{(1)}^{(2)^2} = 0.723$.

Using the same variables a model was fitted to the second sample yielding the equation

$$\text{RETCAP} = \underset{(.693)}{0.087} + \underset{(.000)}{1.456}\,\text{WCFTDT} - \underset{(.004)}{0.494}\,\text{NFATAST}$$

$$- \underset{(.104)}{0.188}\,\text{QUIKRAT}$$

$$+ \underset{(.662)}{0.020}\,\text{LOGASST} + \underset{(.773)}{0.025}\,\text{CURRAT}$$

with $R^2 = 0.873$ and $\text{SSE} = 0.7453$.

For the second sample observation 21 has a large residual and has a large influence on the fitted model. If this observation is omitted the fitted model is given by

$$\text{RETCAP} = \underset{(0.027)}{0.218} + \underset{(0.000)}{0.539} \text{ WCFTDT} - \underset{(0.000)}{0.286} \text{ NFATAST}$$

$$- \underset{(0.643)}{0.023} \text{ QUIKRAT}$$

$$+ \underset{(0.919)}{0.002} \text{ LOGASST} + \underset{(0.370)}{0.033} \text{ CURRAT}$$

with $R^2 = 0.576$ and SSE $= 0.1341$. A considerable reduction in SSE results from the omission of this observation. When the estimated model obtained from sample 1 was used to predict RETCAP for sample 2 without observation 21 the $\text{MSE}_{(1)}^{(2)}$ obtained was .00745 and the value of $R_{(1)}^{(2)^2}$ obtained was 0.450. Thus the removal of the outlier observation causes some reduction in the measures of goodness of fit.

To perform a double cross validation a regression model estimated from the second sample was used to predict the first. Using variable selection procedures the following model was selected for the second sample. Observations 21 and 36 were omitted because of very large influence values.

$$\text{RETCAP} = \underset{(0.000)}{0.249} + \underset{(0.000)}{0.664} \text{ WCFTCL} - \underset{(0.000)}{0.405} \text{ NFATAST}$$

$$+ \underset{(0.068)}{0.020} \text{ CAPINT} - \underset{(0.000)}{0.093} \text{ CURRAT}.$$

$$(4.10)$$

The R^2 and SSE values are given by 0.580 and 0.1053 respectively. Using this model to predict RETCAP for the first sample yielded $R_{(2)}^{(1)^2} = 0.651$ and $\text{MSE}_{(2)}^{(1)} = 0.00756$.

Press Statistic

A relatively simple approach to cross validation is to examine the PRESS statistic introduced in Chapter 3. For each observation y_i the model fitted from the remaining $(n-1)$ observations is used to predict the observation y_i resulting in the estimator $\hat{y}_{(i)}$. Repeating this process for y_i, $i = 1, 2, \ldots, n$ gives some evidence of the presence of points with large influence. As shown in 4.4 the PRESS statistic can be written as PRESS $= \sum_{i=1}^{n} \left(\frac{e_i}{1 - h_{ii}} \right)^2$ and hence the leverage h_{ii} and the residual e_i have an impact on PRESS. The PRESS statistic is useful in small samples to detect model instability over the sample.

Example

For the financial accounting example the value of the PRESS statistic for the first sample of forty observations after omitting observation 21 was 0.2166 which compares to the SSE of 0.1481. The model fitted is given by (4.9). For the second set of forty observations the value of the PRESS statistic is 0.1484 which compares to the SSE of 0.1053. The model fitted in this case is given by (4.10). In both samples the statistic indicates stability after removal of the influential observations.

Theoretical Consistency

In addition to the use of cross validation, a fitted model should be examined and compared to what would be expected from theory. The signs and relative magnitudes of the regression coefficients are usually important characteristics. If multicollinearity is present some coefficients may be relatively unstable and should be used with caution.

4.5 Qualitative Explanatory Variables

In Chapter 3 it was demonstrated that a qualitative variable may also be included as an explanatory variable in a regression model. In the example presented, the qualitative variable Z was assigned the values 0 or 1 depending on which of two categories the observation belonged to. This variable Z is commonly referred to as a *dummy variable*. A dummy variable is one example of a class of variables called *indicator variables* which are used to represent qualitative variables. Indicator variables vary depending on the types of *coding* used. The dummy variable is derived from dummy coding. Other forms of coding to be discussed in this section are effect coding and cell parameter coding.

4.5.1 Dummy Variables

Dummy variables are extremely useful not only in multiple regression but also in other multivariate analyses. In this section we shall demonstrate how dummy variables can be used in a variety of ways to measure variation due to one or more classification variables and also to permit the study of interaction between them.

In the examples of Section 3.4 the qualitative variable being studied contained only two categories. We now extend the technique to qualitative variables having more than two categories. Our discussion will be based on an example which will incorporate seasonal effects in a supply function.

The relationship between supply and price is to be studied for an agricultural commodity. The supply function is given by

$$S = \beta_0 + \beta_1 P + U$$

where $S = $ quantity supplied and $P = $ price. In order to permit the study of seasonal variation four dummies are defined below to represent the four quarters of the year.

$$Q_1 = 1 \quad \text{if in first quarter}$$
$$= 0 \quad \text{otherwise,}$$

$$Q_2 = 1 \quad \text{if in second quarter}$$
$$= 0 \quad \text{otherwise,}$$

$$Q_3 = 1 \quad \text{if in third quarter}$$
$$= 0 \quad \text{otherwise,}$$

$$Q_4 = 1 \quad \text{if in fourth quarter}$$
$$= 0 \quad \text{otherwise.}$$

A multiple regression model which includes all four dummies is given by

$$S = \beta_0 + \beta_1 P + \gamma_1 Q_1 + \gamma_2 Q_2 + \gamma_3 Q_3 + \gamma_4 Q_4 + U.$$

This model yields four parallel lines described by the four equations

$$S = (\beta_0 + \gamma_1) + \beta_1 P + U$$
$$S = (\beta_0 + \gamma_2) + \beta_1 P + U$$
$$S = (\beta_0 + \gamma_3) + \beta_1 P + U$$
$$S = (\beta_0 + \gamma_4) + \beta_1 P + U.$$

A perusal of the four intercepts reveals that we are using five parameters to describe four different intercepts, and hence we cannot separate β_0 from the parameters $\gamma_1, \gamma_2, \gamma_3$ and γ_4. The model is therefore over-parameterized.

The $\mathbf{X}$ matrix for this multiple regression is given by Table 4.32. It is assumed that the data in $\mathbf{X}$ are in rows in order of time. A study of the pattern of the columns of $\mathbf{X}$ quickly reveals that the columns are not linearly independent. The column corresponding to β_0 can be expressed as the sum of the columns corresponding to the four quarters, $\gamma_1, \gamma_2, \gamma_3$ and γ_4. It will not be possible therefore to determine $(\mathbf{X}'\mathbf{X})^{-1}$ and hence the

Table 4.32. **X** Matrix For Model With Dummy Variables

$$
\begin{array}{cccccc}
\beta_0 & \beta_1 & \gamma_1 & \gamma_2 & \gamma_3 & \gamma_4 \\
1 & P & 1 & 0 & 0 & 0 \\
1 & P & 0 & 1 & 0 & 0 \\
1 & P & 0 & 0 & 1 & 0 \\
1 & P & 0 & 0 & 0 & 1 \\
1 & P & 1 & 0 & 0 & 0 \\
1 & P & 0 & 1 & 0 & 0 \\
1 & P & 0 & 0 & 1 & 0 \\
1 & P & 0 & 0 & 0 & 1 \\
\vdots & & & & & \\
1 & P & 1 & 0 & 0 & 0 \\
1 & P & 0 & 1 & 0 & 0 \\
1 & P & 0 & 0 & 1 & 0 \\
1 & P & 0 & 0 & 0 & 1 \\
\end{array}
$$

least squares estimators are undefined. Once again we can see that the model has been over-parameterized.

In order to obtain a full rank **X** matrix we must delete one of the columns corresponding to β_0, γ_1, γ_2, γ_3 and γ_4. It is customary to delete one of the four dummy variables. This permits one category to be treated as a base case and the remaining quarters to be measured relative to it. If in this example we delete the first quarter dummy Q_1 we obtain the equation

$$S = \beta_0 + \beta_1 P + \gamma_2 Q_2 + \gamma_3 Q_3 + \gamma_4 Q_4 + U$$

and hence the four quarters are described respectively by the four equations

$$S = \beta_0 + \beta_1 P + U$$

$$S = (\beta_0 + \gamma_2) + \beta_1 P + U$$

$$S = (\beta_0 + \gamma_3) + \beta_1 P + U \tag{4.11}$$

$$S = (\beta_0 + \gamma_4) + \beta_1 P + U.$$

Fitting this model results in four parallel lines with intercepts given by β_0, $(\beta_0+\gamma_2)$, $(\beta_0+\gamma_3)$ and $(\beta_0+\gamma_4)$ respectively. A plot showing the four parallel lines appears in Figure 4.6. The advantage of this particular formulation is that we can easily compare the four intercepts.

To test the null hypothesis that the four intercepts are equal we need only test $H_0 : \gamma_2 = \gamma_3 = \gamma_4 = 0$. A comparison of the model (4.11) to the reduced model

$$S = \beta_0 + \beta_1 P + U$$

using the F-statistic (4.3) from section 4.1 is a test of H_0.

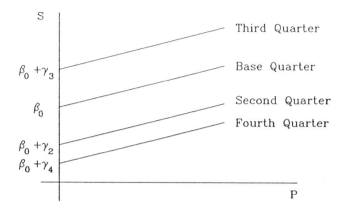

Figure 4.6. Four Parallel Lines Showing Seasonal Variation in Relation Between Supply and Price

The F-statistic is given by

$$\frac{[\text{SSR} - \text{SSR}_R]/3}{\text{SSE}/(n-5)} = \frac{(R^2 - R_R^2)/3}{(1-R^2)/(n-5)},$$

and is compared to $F_{\alpha;3,(n-5)}$ to test H_0. The sums of squares are SSR and SSE for the full model and SSR_R for the reduced model.

The individual coefficient estimates may also be used to compare each of the three quarters to the base case. If the t-statistic for the coefficient has a significant p-value, then the corresponding quarter coefficient is significantly different from zero, and hence that quarter has an intercept which differs from β_0. It is important to note that these t-tests are only useful for comparing each quarter to the base case within the context of the particular model fitted. The particular model fitted assumes that the supply functions are parallel. This assumption may be overly restrictive. To permit different slopes, slope shifter variables are introduced below.

4.5.2 Slope Shifters

A more general model which permits the four quarters to have different intercepts and slopes can be obtained by defining the *slope shifters* PQ_1, PQ_2, PQ_3 and PQ_4. These slope shifters are obtained from the products of Q_1, Q_2, Q_3 and Q_4 with P and are often referred to as *interaction variables*. The $\mathbf{X}$ matrix for the regression after adding the slope shifters is shown in Table 4.33.

Table 4.33. **X** Matrix For Model With Dummy Variables And Slope Shifters

β_0	β_1	γ_1	γ_2	γ_3	γ_4	ρ_1	ρ_2	ρ_3	ρ_4
1	P	1	0	0	0	P	0	0	0
1	P	0	1	0	0	0	P	0	0
1	P	0	0	1	0	0	0	P	0
1	P	0	0	0	1	0	0	0	P
1	P	1	0	0	0	P	0	0	0
1	P	0	1	0	0	0	P	0	0
1	P	0	0	1	0	0	0	P	0
1	P	0	0	0	1	0	0	0	P
$\vdots$	$\vdots$	$\vdots$			$\vdots$		$\vdots$		
1	P	1	0	0	0	P	0	0	0
1	P	0	1	0	0	0	P	0	0
1	P	0	0	1	0	0	0	P	0
1	P	0	0	0	1	0	0	0	P

Once again only three of these four new variables can be used along with the variable P.

Omitting the first quarter dummy variable and corresponding slope shifter yields the model

$$S = \beta_0 + \beta_1 P + \gamma_2 Q_2 + \gamma_3 Q_3 + \gamma_4 Q_4$$
$$+ \rho_2 P Q_2 + \rho_3 P Q_3 + \rho_4 P Q_4 + U.$$

The equations for the four quarters derived from this model are given by

$$S = \beta_0 + \beta_1 P + U$$
$$S = (\beta_0 + \gamma_2) + (\beta_1 + \rho_2)P + U$$
$$S = (\beta_0 + \gamma_3) + (\beta_1 + \rho_3)P + U$$
$$S = (\beta_0 + \gamma_4) + (\beta_1 + \rho_4)P + U.$$

The values of the fitted coefficients for the four models obtained in this fashion are equivalent to the fitted coefficients that would be obtained if linear models were fitted to the data for the four quarters separately. An example for this system of four equations is shown in Figure 4.7.

A variety of hypotheses can be tested by comparing the complete model

$$S = \beta_0 + \beta_1 P + \gamma_2 Q_2 + \gamma_3 Q_3 + \gamma_4 Q_4 + \rho_2 P Q_2 + \rho_3 P Q_3 + \rho_4 P Q_4 + U. \quad (4.12)$$

to one of the reduced models

$$S = \beta_0 + \beta_1 P + U$$

or $\quad S = \beta_0 + \beta_1 P + \gamma_2 Q_2 + \gamma_3 Q_3 + \gamma_4 Q_4 + U$

or $\quad S = \beta_0 + \beta_1 P + \rho_2 P Q_2 + \rho_3 P Q_3 + \rho_4 P Q_4 + U.$

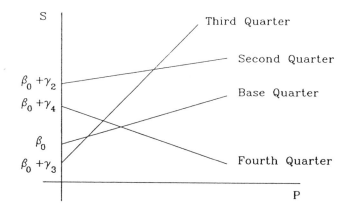

Figure 4.7.   Seasonal Variation in Relationship Between Supply and Price

The null hypotheses being tested in each of these three cases would be

$H_0 : \gamma_2 = \gamma_3 = \gamma_4 = \rho_2 = \rho_3 = \rho_4 = 0$ [The four quarters have identical equations]

$H_0 : \rho_2 = \rho_3 = \rho_4 = 0$ [The four quarters have the same slope but different intercept]

$H_0 : \gamma_2 = \gamma_3 = \gamma_4 = 0$ [The four quarters have different slopes but the same intercept].

These three tests can be carried out using the F-statistic

$$\frac{(\mathrm{SSR} - \mathrm{SSR}_R)/q}{\mathrm{SSE}/(n - p - 1)} = \frac{(R^2 - R_R^2)/q}{(1 - R^2)/(n - p - 1)},$$

where SSR and SSE denote the sums of squares from the complete model (4.12), and SSR_R denotes the sum of squares for the fitted model corresponding to H_0. The numerator degrees of freedom, q, denotes the number of parameters set equal to zero by H_0.

The test of the hypothesis that the four quarters have identical equations employs the null hypothesis $H_0 : \gamma_2 = \gamma_3 = \gamma_4 = \rho_2 = \rho_3 = \rho_4 = 0$ for the model which contains $Q_2, Q_3, Q_4, PQ_2, PQ_3$ and PQ_4. This test is not the same as the test of $H_0 : \gamma_2 = \gamma_3 = \gamma_4$ for the model which contains dummy variables but no slope shifters. It is possible to reject the first hypothesis but accept the second hypothesis. It is important to keep in mind that the second test compares the four lines after they have been "forced" to be parallel. It is possible when the lines are not parallel that omitting the slope shifter variables will introduce bias into the estimates of the intercepts. The figure below illustrates this point for an example of two lines.

The lines A and B in Figure 4.8 are the true lines with samples of points scattered around each. The line C is the line that would result from a least squares fit to the scatter of points if no allowance was made for different slopes. Fitting a pair of lines with the restriction that they be parallel would result in the two lines having essentially the same intercept. The two lines would be almost collinear.

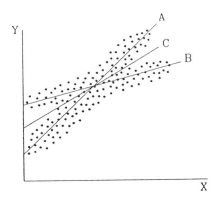

Figure 4.8. Bias Due to Omission of Slope Shifters

When employing both dummy variables and slope shifters it is important to keep in mind that the intercept is merely a way of locating the line geometrically. Figure 4.9 shows a scatter plot illustrating the relationship between Y and X for two different categories A and B. An examination of the fitted lines reveals that the parallel lines 1 and 2, or the lines with common intercept 3 and 4, are sufficient to capture the difference between the two scatter plots. The addition of slope shifters after dummies in the case of 1 and 2, or the addition of dummies after slope shifters in 3 and 4 will probably lead to insignificant increases in R^2.

Figure 4.9 is also useful for displaying the inherent multicollinearity problem that occurs with the simultaneous use of dummy variables and slope shifters. Notice that lines 1 and 3 are reasonably close to the best fitting line for group A and yet the slopes and intercepts are quite different. Similarily for lines 2 and 4 and group B. There is usually a very high correlation between dummy variables and the corresponding slope shifters which can be seen by examining the X matrix in Table 4.33. Notice how the columns for each dummy and the corresponding slope shifter have a very large proportion of zeroes in common. As a result the correlation between say Q_1

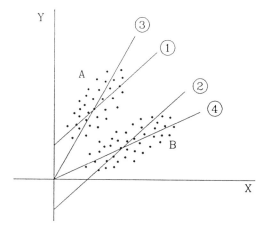

Figure 4.9. Comparison Of No Dummy And No Slope Shifter Regimes For Two Different Groups

and PQ_1 will be close to 1. Thus the standard errors of the coefficients of Q_1 and PQ_1 will tend to be large when both variables are included in a regression model.

4.5.3 An Example From Real Estate Pricing

To provide an example illustrating the use of dummy variables and slope shifter variables, we shall employ a sample of 116 real estate sales records from a multiple listing service. The sales are for a particular year for bungalows between 1000 and 1600 square feet sold in a certain sector of a large city. The variables to be used are SQF (number of square feet), LISTP (list price), BROKER (broker category), $P1$, $P2$ and $P3$ (time period of the year when house sold). The BROKER variable divides the listing real estate brokers into two large firms; BROKER $= 1$; and all other firms: BROKER $= 0$. The time periods are $P1$ (January, February, March or April), $P2$ (July, August, September or October) and $P3$ (May, June, November and December), which were chosen to represent three different levels of sales activity in the market. The observations are displayed in Table 4.34. All of the calculations throughout this section have been performed using SAS PROC REG.

Comparison Of Brokers

To determine the impact of the broker group, on the relationship between LISTP and SQF, several multiple regression models involving the dummy

Table 4.34. Real Estate Sales Data

OBS	LISTP	SQF	BRO-KER	P1	P2	P3	OBS	LISTP	SQF	BRO-KER	P1	P2	P3	OBS	LISTP	SQF	BRO-KER	P1	P2	P3
1	79000	1165	1	1	0	0	40	86000	1080	1	0	0	1	79	95800	1340	0	1	0	0
2	85000	1170	0	1	0	0	41	95000	1060	0	0	0	1	80	83000	1160	0	1	0	0
3	91000	1160	0	1	0	0	42	109800	1232	1	0	0	1	81	89800	1200	0	1	0	0
4	107800	1306	1	1	0	0	43	123600	1200	0	0	0	1	82	91800	1202	0	1	0	0
5	83000	1120	1	1	0	0	44	95800	1080	0	0	1	0	83	97800	1278	0	1	0	0
6	85000	1040	0	1	0	0	45	130000	1460	0	0	1	0	84	106150	1260	0	1	0	0
7	85800	1130	0	1	0	0	46	91200	1100	0	0	1	0	85	119600	1284	0	1	0	0
8	89800	1232	1	1	0	0	47	93800	1140	0	0	1	0	86	87000	1237	1	0	0	1
9	97600	1364	1	1	0	0	48	119600	1248	0	0	1	0	87	91800	1050	0	0	0	1
10	99400	1260	0	1	0	0	49	89800	1050	0	0	1	0	88	101800	1320	0	0	0	1
11	103800	1302	0	1	0	0	50	107000	1286	0	0	1	0	89	115000	1490	1	0	0	1
12	77800	1040	0	1	0	0	51	111800	1250	1	0	0	1	90	128400	1491	0	0	0	1
13	103800	1278	1	0	0	1	52	77800	1056	0	0	0	1	91	89000	1154	0	0	0	1
14	107000	1408	1	0	0	1	53	91800	1135	0	0	0	1	92	93800	1224	0	0	1	0
15	109800	1225	0	0	0	1	54	97800	1280	0	0	0	1	93	97800	1220	0	0	0	1
16	93800	1160	0	0	0	1	55	100600	1100	0	0	1	0	94	103000	1152	0	0	0	1
17	95600	1040	0	0	0	1	56	81000	1060	0	0	0	1	95	105800	1392	0	0	0	1
18	102000	1424	1	0	0	1	57	91800	1080	0	0	0	1	96	129800	1480	0	0	1	0
19	111800	1232	0	0	0	1	58	95000	1080	0	0	0	1	97	95800	1042	0	0	1	0
20	95800	1270	0	0	1	0	59	95800	1140	0	0	0	1	98	96600	1167	0	0	1	0

Table 4.34. Real Estate Sales Data (continued)

OBS	LISTP	SQF	BRO-KER	P1	P2	P3
21	105800	1298	0	0	1	0
22	135000	1340	0	0	1	0
23	139800	1380	0	0	1	0
24	99800	1200	1	0	1	0
25	109400	1180	0	0	1	0
26	109800	1200	1	0	1	0
27	110800	1260	0	0	1	0
28	97200	1219	0	0	1	0
29	101000	1154	0	0	1	0
30	105800	1221	0	0	1	0
31	111800	1205	0	0	1	0
32	91800	1170	0	0	1	0
33	107800	1200	1	0	1	0
34	95600	1120	0	0	0	1
35	97800	1130	0	0	0	1
36	99800	1270	1	0	0	1
37	103800	1120	0	0	0	1
38	103800	1072	0	0	0	1
39	105000	1130	0	0	0	1

OBS	LISTP	SQF	BRO-KER	P1	P2	P3
60	103800	1160	0	0	0	1
61	105000	1089	0	0	0	1
62	93000	1040	0	0	0	1
63	99800	1200	0	0	0	1
64	107500	1257	0	0	0	1
65	95800	1260	1	1	0	0
66	77000	1106	1	1	0	0
67	77800	1120	0	1	0	0
68	79000	1190	0	1	0	0
69	85000	1264	0	1	0	0
70	87000	1232	0	1	0	0
71	87800	1160	0	1	0	0
72	89800	1190	1	1	0	0
73	93800	1392	1	1	0	0
74	97600	1220	1	1	0	0
75	118400	1450	1	1	0	0
76	87200	1206	1	1	0	0
77	87800	1150	0	1	0	0
78	89800	1204	0	1	0	0

OBS	LISTP	SQF	BRO-KER	P1	P2	P3
99	112000	1246	0	0	1	0
100	114000	1440	0	0	1	0
101	135000	1564	0	0	1	0
102	111000	1224	0	0	1	0
103	115000	1325	0	0	1	0
104	127000	1270	0	0	1	0
105	91800	1340	0	0	1	0
106	105800	1270	1	0	1	0
107	119800	1460	0	0	1	0
108	87800	1001	1	0	0	1
109	101800	1227	0	0	0	1
110	77800	1124	1	1	0	0
111	78000	1070	0	1	0	0
112	78000	1120	0	1	0	0
113	95800	1200	0	1	0	0
114	105800	1308	0	1	0	0
115	97400	1113	0	0	0	1
116	105000	1345	1	0	0	1

variable BROKER and slope shifter BRSQF were estimated. The results are shown in Table 4.35. Comparing equations 2 and 1 of the table, adding the dummy variable BROKER increases R^2 from 0.440 to 0.494 which has a p-value of 0.001. With the BROKER variable included in equation 2 the equations for the two broker groups are

$$\text{LISTP} = 5337.49 + 79.01 \text{ SQF} \qquad \text{BROKER} = 0$$

$$\text{LISTP} = -2192.38 + 79.01 \text{ SQF} \qquad \text{BROKER} = 1.$$

The result is therefore two parallel lines with common slope 79.01 and intercepts 5337.49 for all the smaller brokers and –2192.38 for the two major brokers.

Table 4.35. Regression Relationships Between LISTP and SQF Controlling for Broker Effects

Coefficients of Variables in Model

Equation Number	β_0 Intercept	β_1 SQF	γ BROKER	ρ BRSQF	R^2
1	6895.04 (0.485)	76.34 (0.000)			0.440
2	5337.49 (0.572)	79.01 (0.000)	–7529.87 (0.001)		0.494
3	3694.19 (0.696)	80.38 (0.000)		–6.21 (0.001)	0.496
4	1066.33 (0.920)	82.54 (0.000)	12600.62 (0.589)	–16.35 (0.387)	0.498

Comparing equations 1 and 3, adding the slope shifter variable increases R^2 from 0.440 to 0.496 and has a p-value of 0.001. The equations for the two broker groups using equation 3 are given by

$$\text{LISTP} = 3694.19 + 80.38 \text{ SQF} \qquad \text{BROKER} = 0$$

$$\text{LISTP} = 3694.19 + 74.17 \text{ SQF} \qquad \text{BROKER} = 1.$$

In this case the two lines have the common intercept 3694.19 and different slopes 80.38 and 74.17 respectively. Equations 2 and 3 therefore seem to indicate that the two major brokers use lower list prices particularly for larger houses.

A comparison of equations 1 and 4 provides a test of the hypothesis $H_0 : \gamma = \rho = 0$ in the model

$$\text{LISTP} = \beta_0 + \beta_1 \text{ SQF} + \gamma \text{ BROKER} + \rho \text{ BRSQF} + U.$$

A test of H_0 here is a comparison of the relationship between LISTP and SQF for the two groups without any restrictions such as common slope or common intercept in the two previous tests.

The F-statistic for this test is given by

$$F = \frac{(0.498 - 0.440)/2}{(1 - 0.498)/112} = 6.470$$

and is significant at the 0.00 level. The two equations for the two broker groups are given by

$$\text{LISTP} = 1066.33 + 82.54 \text{ SQF} \qquad \text{BROKER} = 0$$

$$\text{LISTP} = 13,666.95 + 66.19 \text{ SQF} \qquad \text{BROKER} = 1.$$

It would appear that in comparison to the remaining brokers the two major brokers use a pricing policy which does not put much emphasis on the size of the house.

It is also of interest to compare the standard errors of the coefficients in equation 4 to the corresponding coefficients in equations 2 and 3. The standard error for the coefficient of BROKER increases from 2162.1 in equation 2 to 23275.2 in equation 4 while for BRSQF the standard error increases from 1.75 in equation 3 to 18.83 in equation 4. In both cases the standard error increases by a factor of 10 with the addition of the third variable. As discussed above increases of this magnitude can be expected in models which include both a dummy variable and the corresponding slope shifter variable.

The four graphs in Figure 4.10 illustrate the fit of the four models to the bungalow data. The observations corresponding to BROKER = 1 are denoted by *, while the observations corresponding to BROKER = 0 are denoted by △. It is interesting to compare the fitted models obtained in the four graphs. In panels (b), (c) and (d) we can see that for the relevant range of SQF the line corresponding to BROKER = 0 is higher than the line corresponding to BROKER = 1. Thus even though the brokers in category 1 have a larger intercept, the LISTP for this group remains below the LISTP for category 0. The observations corresponding to BROKER = 1 are denoted by *, while the observations corresponding to BROKER = 0 are denoted by △.

Seasonal Effects

Table 4.36 contains the estimation results for several regression models relating LISTP to SQF controlling for seasonal effects. Comparing equation 2 to equation 1 permits a test of the hypothesis $H_0 : \gamma_1 = \gamma_2 = 0$ in the model

$$\text{LISTP} = \beta_0 + \beta_1 \text{ SQF} + \gamma_1 P_1 + \gamma_2 P_2 + U.$$

(a)

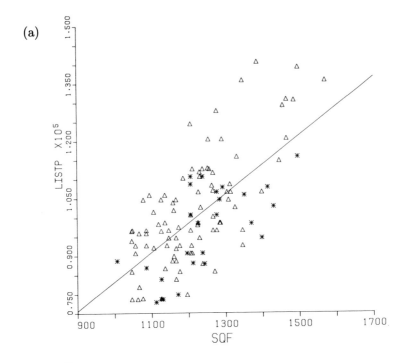

(b)

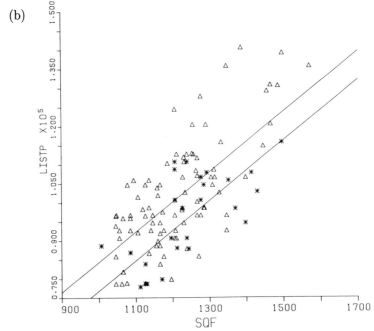

Figure 4.10. LISTP vs. SQF Controlling for Broker Effects

(c)

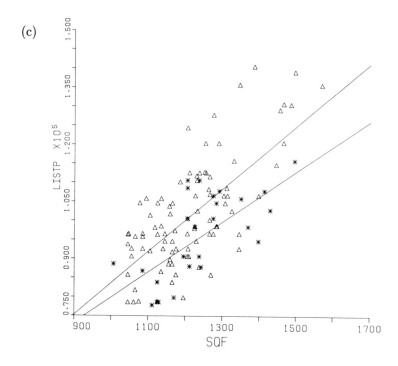

(d)

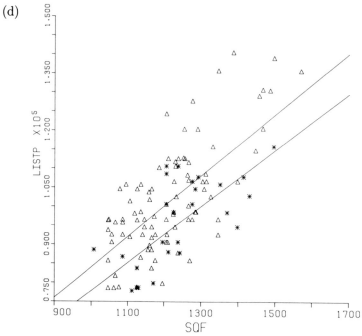

Figure 4.10. LISTP vs. SQF Controlling for Broker Effects (continued)

The F-statistic for the test of H_0 is given by

$$F = \frac{(0.643 - 0.440)/2}{0.357/(112)} = 31.843,$$

and hence the hypothesis H_0 of equal intercepts given equal slopes is rejected. The estimated coefficients for equation 2 result in the three models

$$\text{LISTP} = 15699.45 + 70.75 \text{ SQF} \qquad \text{for } P3$$

$$\text{LISTP} = 5187.74 + 70.75 \text{ SQF} \qquad \text{for } P1$$

$$\text{LISTP} = 20012.86 + 70.75 \text{ SQF} \qquad \text{for } P2.$$

Using the equation 2 model we can conclude that list prices are highest for $P2$ and lowest for $P1$. List prices for $P1$ seem to be quite a bit lower than for $P2$ and $P3$. These coefficients are obtained under the assumption that the slopes must remain the same. From the p-values for the coefficients of $P1$ and $P2$ we can conclude that the intercepts in $P1$ and $P2$ are significantly different from the intercept for $P3$ at the 0.00 and 0.02 levels respectively.

Table 4.36. Regression Relationship Between LISTP and SQF Controlling for Seasonal Effects

Coefficients of Variables in Model

Equation Number	β_0 Intercept	β_1 SQF	γ_1 $P1$	γ_2 $P2$	ρ_1 $P1\text{SQF}$	ρ_2 $P2\text{SQF}$	R^2
1	6895.04 (0.485)	76.34 (0.000)					0.440
2	15699.45 (0.053)	70.75 (0.000)	−10511.71 (0.000)	4313.41 (0.024)			0.643
3	15210.38 (0.065)	70.95 (0.000)			−8.34 (0.000)	3.73 (0.017)	0.639
4	39011.64 (0.001)	51.19 (0.000)	−56765.47 (0.008)	−35500.56 (0.053)	38.56 (0.028)	32.74 (0.028)	0.665

A comparison of equations 3 and 1 in Table 4.36 allows a test of the hypothesis $H_0 : \rho_1 = \rho_2 = 0$ in the model

$$\text{LISTP} = \beta_0 + \beta_1 \text{ SQF} + \rho_1 P1\text{SQF} + \rho_2 P2\text{SQF} + U.$$

The F-statistic for the test of H_0 is given by

$$F = \frac{(0.639 - 0.440)/2}{(0.361)/(112)} = 30.870,$$

and therefore the hypothesis of a common line for the three seasons must be rejected. Under the assumption of a common intercept the three lines have different slopes. The three lines obtained from equation 3 are given by

$$\text{LISTP} = 15210.38 + 70.95 \text{ SQF} \qquad \text{for } P3$$

$$\text{LISTP} = 15210.38 + 62.61 \text{ SQF} \qquad \text{for } P1$$

$$\text{LISTP} = 15210.38 + 74.68 \text{ SQF} \qquad \text{for } P2.$$

Again it would appear that the highest prices are in $P2$ and the lowest prices are in $P1$. In this model the slope differences are estimated under the assumption that the intercepts are equal. The p-values for the slope shifter coefficients indicate that the slope differences from $P3$ are significant at the 0.00 and 0.02 levels for $P1$ and $P2$ respectively.

A comparison of equation 4 to equation 1 in Table 4.36 provides a test of the null hypothesis $H_0 : \gamma_1 = \gamma_2 = \rho_1 = \rho_2 = 0$ in the model

$$\text{LISTP} = \beta_0 + \beta_1 \text{ SQF} + \gamma_1 P1 + \gamma_2 P2 + \rho_1 P1\text{SQF} + \rho_2 P2\text{SQF} + U. \quad (4.13)$$

The F-statistic for this test is given by

$$F = \frac{(0.665 - 0.440)/4}{(0.335)/(110)} = 18.470,$$

which suggests that there are differences among periods with respect to the relationship between LISTP and SQF.

Other hypotheses can also be tested using the equations in Table 4.36. A comparison of equation 4 to equation 2 is a test of the null hypothesis $H_0 : \rho_1 = \rho_2 = 0$ in (4.13). The F-statistic is given by

$$F = \frac{(0.665 - 0.643)/2}{(1 - 0.665)/110} = 3.612,$$

which is significant at the 0.04 level. Similarly a comparison of equation 4 to equation 3 is a test of the null hypothesis $H_0 : \gamma_1 = \gamma_2 = 0$ in (4.13). The F-statistic for this test is given by

$$F = \frac{(0.665 - 0.639)/2}{(1 - 0.665)/110} = 4.269$$

which is significant at the 0.02 level.

Using equation 4 the three lines are obtained as

$$\text{LISTP} = 39011.64 + 51.19 \text{ SQF} \qquad \text{for } P3$$

$$\text{LISTP} = -17753.83 + 89.75 \text{ SQF} \qquad \text{for } P1$$

$$\text{LISTP} = 3511.08 + 83.93 \text{ SQF} \qquad \text{for } P2.$$

Thus in comparison to $P3$, it would appear that in periods 1 and 2 the prices for small bungalows are much less, but for large bungalows the opposite is true. The crossover points are at 1472 square feet and 1084 square feet for $P1$ and $P2$ respectively. Since the bungalows in the sample range from 1000 square feet to 1600 square feet we can conclude that for the sample range the $P1$ line is generally below the $P3$ line and that the $P2$ line is above the $P3$ line after 1084 square feet.

The graphs in Figure 4.11 can be used to show the relationship between LISTP and SQF for the three periods. The observations corresponding to $P3$ are denoted by □, while the observations corresponding to $P1$ and $P2$ are denoted by △ and * respectively. Panel (a) shows a single line for all three periods. Panel (b) shows the three parallel lines assuming only intercept differences, while panel (c) shows the result for a common intercept and different slopes for the three periods. Panel (d) shows the results if both slopes and intercepts are allowed to vary by period. From these three panels we can see that within the relevant range of SQF the line for $P2$ is usually higher than the line for $P3$ and $P1$, and the line for $P3$ is usually higher than the line for $P1$.

Since the above comparisons are of $P1$ and $P2$ to $P3$ it may be of interest to compare the three periods using an alternative base case. Table 4.37 shows the results of the regressions for the two alternative base case solutions. In equation 1 of Table 4.37, we can conclude that the intercept for the period $P2$ is not significantly different from zero, and that the intercept and slope for $P1$ are not significantly different from those of $P2$. For $P3$ the intercept and slope are significantly different from $P2$ at the 0.05 and 0.03 levels respectively. This latter conclusion is also obtainable from equation 4 in Table 4.36. In equation 2 of Table 4.37, we can conclude that the intercept for period $P1$ is not significantly different from zero, and that the intercept and slope for $P2$ are not significantly different from those of $P1$. For $P3$ the intercept and slope are significantly different from $P1$ at the 0.01 and 0.03 levels respectively.

From the above analyses it would appear that the intercepts for $P1$ and $P2$ are not significantly different from zero but that the intercept for $P3$ is positive. Also, the intercept and slope for $P1$ are not significantly different from the intercept and slope for $P2$. It would appear that periods $P1$ and $P2$ might be combined. If periods $P1$ and $P2$ are treated as the base case the resulting equation is given by

$$\text{LISTP} = -22,429.69 + 99.16 \text{ SQF} + 61441.33 \text{ } P3 - 47.97 \text{ } P3SQF$$
$$\quad\quad\quad (0.081) \quad\quad (0.000) \quad\quad\quad (0.002) \quad\quad\quad (0.003)$$

with $R^2 = 0.497$. To compare this model to the full model in Table 4.37) the F-statistic is given by

$$F = \frac{(0.665 - 0.497)/2}{(0.335)/110} = 27.582,$$

(a)

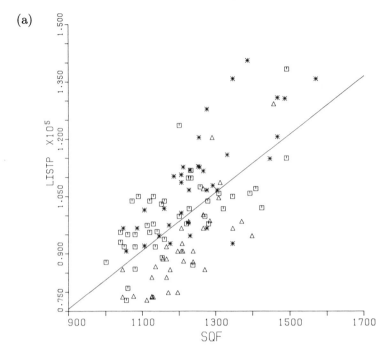

(b)

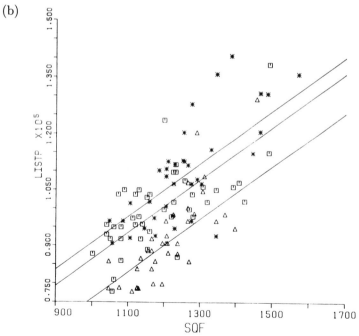

Figure 4.11. LISTP vs. SQF Controlling for Seasonal Effects

(c)

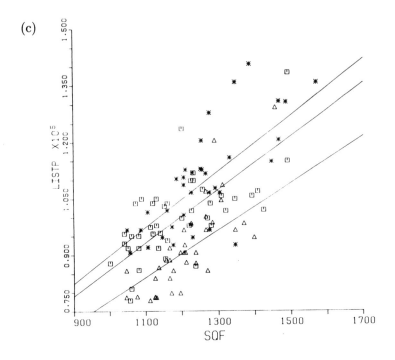

(d)

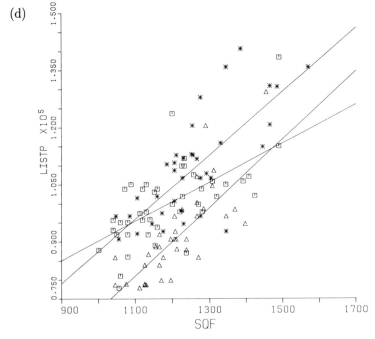

Figure 4.11. LISTP vs. SQF Controlling for Seasonal Effects (continued)

Table 4.37. Regression Relationship Between LISTP and SQF Controlling for Seasonal Effects

Coefficients of Variables in Model

Equation Number	Intercept	SQF	$P1$	$P2$	$P3$
1	3511.09 (0.799)	83.93 (0.000)	−21264.91 (0.338)		35500.56 (0.053)
2	−17753.82 (0.306)	89.75 (0.000)		21264.91 (0.338)	56765.47 (0.008)

Equation Number	$P1SQF$	$P2SQF$	$P3SQF$	R^2
1	5.82 (0.747)		−32.74 (0.028)	0.665
2		−5.82 (0.747)	−38.56 (0.028)	0.665

and hence at conventional significance levels we cannot conclude that periods $P1$ and $P2$ can be combined without loss of information.

If the variable $P2$ is added to the previous model, the equation becomes

$$LISTP = -13,338.30 + 86.09 \ SQF + 14,143.42 \ P2$$
$$\underset{(0.208)}{} \quad \underset{(0.000)}{} \quad \underset{(0.000)}{}$$

$$+ 52,349.95 \ P3 - 34.90 \ P3SQF$$
$$\underset{(0.001)}{} \quad \underset{(0.009)}{}$$

with $R^2 = 0.665$. If instead the variable $P2SQF$ is added to the previous model, the equation becomes

$$LISTP = -4768.74 + 79.06 \ SQF + 43,780.39 \ P3$$
$$\underset{(0.659)}{} \quad \underset{(0.000)}{} \quad \underset{(0.007)}{}$$

$$+ 11.43 \ P2SQF - 27.87 \ P3SQF$$
$$\underset{(0.000)}{} \quad \underset{(0.038)}{}$$

with $R^2 = 0.662$. We can conclude therefore that adding either the dummy variable or the slope shifter is sufficient to maximize R^2. A comparison of the standard errors of the coefficients for the $P2$ dummy and slope shifter $P2SQF$ when both appear, to the case where only one appears, reveals that the standard error increases by a factor of 10. Thus when both the dummy and slope shifter for $P2$ are included the coefficients must be relatively large to be significant.

4.5.4 Other Methods

Effect Coding

An alternative approach to the coding of categorical variables is called *effect coding*. The method is best illustrated using an example. In this method the indicator variables in the supply function of 4.5.1 are coded as follows:

$$
\begin{aligned}
Q_1 &= 1 &&\text{if in first quarter} \\
&= 0 &&\text{if in quarters 2 or 3} \\
&= -1 &&\text{if in quarter 4} \\[1em]
Q_2 &= 1 &&\text{if in second quarter} \\
&= 0 &&\text{if in quarters 1 or 3} \\
&= -1 &&\text{if in quarter 4} \\[1em]
Q_3 &= 1 &&\text{if in third quarter} \\
&= 0 &&\text{if in quarters 1 or 2} \\
&= -1 &&\text{if in quarter 4.}
\end{aligned}
$$

A multiple regression model which includes the dummy variables Q_1, Q_2 and Q_3 is given by

$$ S = \beta_0 + \beta_1 P + \gamma_1 Q_1 + \gamma_2 Q_2 + \gamma_3 Q_3 + U. $$

This model yields four parallel lines for the four quarters given by

$$ S = (\beta_0 + \gamma_1) + \beta_1 P + U $$

$$ S = (\beta_0 + \gamma_2) + \beta_1 P + U $$

$$ S = (\beta_0 + \gamma_3) + \beta_1 P + U $$

$$ S = (\beta_0 - \gamma_1 - \gamma_2 - \gamma_3) + \beta_1 P + U. $$

In this case the four intercepts sum to $4\beta_0$ and hence β_0 may be viewed as the mean of the four intercepts. The parameters γ_1, γ_2 and γ_3 measure respectively the difference between the intercepts in the first three quarters and the average intercept β_0. These parameters are called quarter effects hence the term effect coding. The quarter effect for the fourth quarter is given by $-(\gamma_1 + \gamma_2 + \gamma_3)$. The four quarter effects are measured relative to the average intercept β_0 rather than to a base case as in dummy variable coding.

It is important to note that this average intercept is not necessarily equivalent to the intercept that would be obtained if one line was fitted to all observations. The intercept in this latter case would be a weighted average of the four separate intercepts.

The corresponding slope shifter variables can be obtained by adding the interaction or product variables PQ_1, PQ_2 and PQ_3. The complete multiple regression model is given by

$$S = \beta_0 + \beta_1 P + \gamma_1 Q_1 + \gamma_2 Q_2 + \gamma_3 Q_3 + \rho_1 \, PQ_1 + \rho_2 \, PQ_2 + \rho_3 \, PQ_3 + U.$$

The corresponding $\mathbf{X}$ matrix is given in Table 4.38.

The slopes for the four quarters are respectively $(\beta_1 + \rho_1)$, $(\beta_1 + \rho_2)$, $(\beta_1 + \rho_3)$ and $(\beta_1 - \rho_1 - \rho_2 - \rho_3)$. The parameter β_1 measures the average slope and ρ_1, ρ_2, ρ_3 and $-(\rho_1 + \rho_2 + \rho_3)$ measure the quarterly slope effects which are measured relative to the average slope β_1.

Table 4.38. **X** Matrix For Seasonal Effects Model

β_0	β_1	γ_1	γ_2	γ_3	ρ_1	ρ_2	ρ_3
1	P	1	0	0	P	0	0
1	P	0	1	0	0	P	0
1	P	0	0	1	0	0	P
1	P	-1	-1	-1	$-P$	$-P$	$-P$
$\vdots$	$\vdots$	$\vdots$	$\vdots$	$\vdots$	$\vdots$	$\vdots$	$\vdots$
1	P	1	0	0	P	0	0
1	P	0	1	0	0	P	0
1	P	0	0	1	0	0	P
1	P	-1	-1	-1	$-P$	$-P$	$-P$

Example

The example introduced in 4.5.3 is continued here. The broker effects model can also be estimated using effect coding. The variable BROKER was defined to be +1 if the broker was one of the two major brokers and -1 otherwise. The estimated model for the data in Table 4.34 is given by

$$\text{LISTP} = \underset{(0.633)}{7366.64} + \underset{(0.000)}{74.36} \text{ SQF} + \underset{(0.589)}{6300.31} \text{ BROKER} - \underset{(0.387)}{8.18} \text{ BRSQF}.$$

Using these results we can conclude that the average of the two intercepts is 7366.64 and that the average of the two slopes is 74.36. For the two major brokers the intercept is higher than the average by 6300.31 while for the remaining brokers the intercept is below this average by 6300.31. Similarly for the slopes the major brokers have a slope which is 8.18 below the average while for the other brokers the slope is above the average by this amount. The two relationships are therefore given by

$$\text{LISTP} = 13,666.95 + 66.19 \text{ SQF} \qquad \text{for major brokers}$$

$$\text{LISTP} = 1066.33 + 82.54 \text{ SQF} \qquad \text{for other brokers.}$$

For the seasonal effects model, effect coding was introduced by defining $P1 = 1$ for period 1, 0 for period 2 and -1 for period 3. Similarly $P2 = 1$ for period 2, 0 for period 1 and -1 for period 3. The estimated model for the data in Table 4.34 is given by

$$\text{LISTP} = 8256.30 + 74.96 \text{ SQF} - 26,010.13 \ P1 - 4745.21 \ P2$$

$$+ 14.79 \ P1\text{SQF} + 8.97 \ P2\text{SQF}.$$

For the three seasons the average intercept is therefore 8,256.30 and the average slope is 74.96. The intercepts for the periods $P1$, $P2$ and $P3$ are given by $-17,753.83$, 3,511.09 and 39,011.64 respectively. The corresponding slopes for the three periods are 89.75, 83.93 and 51.20 respectively.

While the slope and intercept parameter estimates give the average values over the three periods these values should not be construed as the values that would be obtained if the $P1$ and $P2$ effects were omitted. The least squares relationship between LISTP and SQF without the $P1$ and $P2$ effects is given by

$$\text{LISTP} = 6,895.04 + 76.34 \text{ SQF}.$$

Cell Parameter Coding

In the discussion of dummy variables at the beginning of this section, it was demonstrated that one of the dummy variables and one of the slope shifters should be omitted from the $\mathbf{X}$ matrix in order to achieve a full rank $\mathbf{X'X}$ matrix. Using this approach, the column of unities and the column for the explanatory variable then represent the cell corresponding to the omitted dummy and slope shifter. This category or cell was referred to as the base case. If instead, the column of unities and the explanatory variable are omitted, the coefficients of the dummy variables and slope shifters correspond to the regression coefficients for the category or cell defined by the dummy. This approach is referred to as *cell parameter coding*. In this approach inferences regarding comparisons of the cell parameters must be carried out using linear functions of the parameters as outlined at the end of Section 4.1.2. Recently the larger statistical software packages have made such procedures available. What is required is a multiple linear regression model estimation procedure which permits the intercept to be omitted, and which permits the estimation to be carried out subject to a set of linear restrictions on the coefficients. If this feature is not present, the user must perform linear transformations of the explanatory variables in order to obtain new regression coefficients which are equivalent to the desired linear combinations.

Example

For the real estate example discussed above, the column of unities and the SQF column can be left out of the $\mathbf{X}$ matrix, while retaining the three dummy variables P1, P2, P3 and the three slope shifters P1SQF, P2SQF and P3SQF. The linear model is given by

$$\text{LISTP} = \beta_1 \text{ P1} + \beta_2 \text{ P2} + \beta_3 \text{ P3} + \alpha_1 \text{ P1SQF} + \alpha_2 \text{ P2SQF} + \alpha_3 \text{ P3SQF} + U.$$

An estimate of this model shows immediately the regression coefficients for the three separate models one for each period. The estimated model is given by

$$\text{LISTP} = \underset{(0.306)}{-17753.82} \text{ P1} + \underset{(0.799)}{3511.09} \text{ P2} + \underset{(0.001)}{39011.64} \text{ P3}$$

$$+ \underset{(0.000)}{89.75} \text{ P1SQF} + \underset{(0.000)}{83.93} \text{ P2SQF} + \underset{(0.001)}{51.19} \text{ P3SQF}$$

with $R^2 = 0.665$ (note: this is the full model R^2 obtained with dummy variables). The significances of the individual coefficients in the model for each of the three categories can be seen immediately from the significances of regression coefficients. The p-values are shown in brackets underneath the estimated coefficients.

To make comparisons among the three cells, estimates of the model can be made subject to various linear restrictions. To estimate the model subject to the restriction of equal slopes, the linear transformation matrix $\mathbf{T}$ is given by

$$\mathbf{T} = \begin{bmatrix} 0 & 0 & 0 & 1 & -1 & 0 \\ 0 & 0 & 0 & 1 & 0 & -1 \end{bmatrix} \quad \text{for the restriction } \mathbf{T}\boldsymbol{\beta} = \mathbf{0}.$$

The coefficient vector $\boldsymbol{\beta}$ is assumed to contain the parameters in the order given by the above equation. The restrictions in this case are $(\alpha_1 - \alpha_2) = 0$ and $(\alpha_1 - \alpha_3) = 0$ which are equivalent to the condition $\alpha_1 = \alpha_2 = \alpha_3$. The estimated model under these restrictions is given by

$$\text{LISTP} = 5187.74 \text{ P1} + 20012.86 \text{ P2} + 15699.45 \text{ P3}$$

$$+ 70.75(\text{P1SQF} + \text{P2SQF} + \text{P3SQF}).$$

A test of the null hypothesis $H_0 : \alpha_1 = \alpha_2 = \alpha_3$ yields an F-statistic of 3.598 which has a p-value of 0.031.

In a similar fashion, to estimate the model subject to the restriction that the intercepts are equal, the transformation matrix $\mathbf{T}$ is given by

$$\mathbf{T} = \begin{bmatrix} 1 & -1 & 0 & 0 & 0 & 0 \\ 1 & 0 & -1 & 0 & 0 & 0 \end{bmatrix} \quad \text{for } \mathbf{T}\boldsymbol{\beta} = \mathbf{0}.$$

The restrictions in this case are $(\beta_1 - \beta_2) = 0$ and $(\beta_1 - \beta_3) = 0$ which are equivalent to the condition $\beta_1 = \beta_2 = \beta_3$. The estimated model under these restrictions is given by

$$\text{LISTP} = 15,210.38(\text{P1} + \text{P2} + \text{P3}) + 62.60 \text{ P1SQF}$$

$$+ 74.67 \text{ P2SQF} + 70.95 \text{ P3SQF}.$$

A test of the null hypothesis $H_0 : \beta_1 = \beta_2 = \beta_3$ is provided by $F = 4.232$ which has a p-value of 0.017.

The above two sets of conditions can be combined to test the hypothesis that the three periods have the same linear model. The $\mathbf{T}$ matrix contains the four rows defined in pairs above and is given by

$$\mathbf{T} = \begin{bmatrix} 1 & -1 & 0 & 0 & 0 & 0 \\ 1 & 0 & -1 & 0 & 0 & 0 \\ 0 & 0 & 0 & 1 & -1 & 0 \\ 0 & 0 & 0 & 1 & 0 & -1 \end{bmatrix}.$$

In this case the estimated model is the reduced model which uses the same intercept and slope for each period given by

$$\text{LISTP} = 6895.04(\text{P1} + \text{P2} + \text{P3}) + 76.34(\text{P1SQF} + \text{P2SQF} + \text{P3SQF}).$$

A test of the null hypothesis that all three periods have the same model is given by $F = 18.478$ which has a p-value of 0.000. The reader should compare all of these test results to the results obtained in 4.5.3 using dummy variable coding.

We can also estimate the model subject to the restriction that periods 1 and 2 have the same model, but period 3 is different. The $\mathbf{T}$ matrix in this case is given by

$$\mathbf{T} = \begin{bmatrix} 1 & -1 & 0 & 0 & 0 & 0 \\ 0 & 0 & 0 & 1 & -1 & 0 \end{bmatrix}.$$

The estimated model under these restrictions is given by

$$\text{LISTP} = -22,429.69(\text{P1} + \text{P2}) + 99.16(\text{P1SQF} + \text{P2SQF})$$

$$+ 39,011.64 \text{ P3} + 51.19 \text{ P3SQF}.$$

The F-statistic for this restriction is 27.59 which has a p-value of 0.000.

Fitting the model subject to only one restriction, either equal intercepts for P1 and P2 or equal slopes for P1 and P2, one of the $\mathbf{T}$ matrices below is required.

$$\mathbf{T} = \begin{bmatrix} 1 & -1 & 0 & 0 & 0 & 0 \end{bmatrix} \quad \text{or} \quad T = \begin{bmatrix} 0 & 0 & 0 & 1 & -1 & 0 \end{bmatrix}.$$

For these two cases the estimated models are

$$\text{LISTP} = -4768.74(\text{P1} + \text{P2}) + 79.06 \text{ P1SQF} + 90.48 \text{ P2SQF}$$
$$+ 39,011.64 \text{ P3} + 51.19 \text{ P3SQF}$$

$$\text{LISTP} = -13,338.30 \text{ P1} + 805.12 \text{ P2} + 86.09(\text{P1SQF} + \text{P2SQF})$$
$$+ 39,011.64 \text{ P3} + 51.19 \text{ P3SQF}.$$

The F-test statistics in these two cases showed p-values of 0.338 and 0.747 respectively. Thus either of these two models can be used to describe the relationship.

4.5.5 Two Qualitative Explanatory Variables and Interaction

When there are two classification variables to be included as explanatory variables, two systems of dummy variables must be defined. As an example we suppose that the relationship between income (Y) and education (X) is to be studied for both males and females in two different occupation groups, say A and B. Defining dummy and slope shifter variables for the two classifications we have

$$S = 1 \qquad \text{for females}$$
$$= 0 \qquad \text{for males}$$

$$\text{and} \qquad G = 1 \qquad \text{for group } B$$
$$= 0 \qquad \text{for group } A.$$

The corresponding slope shifter variables are given by XS and XG respectively.

The model which includes both the dummies and the slope shifters is given by

$$Y = \beta_0 + \beta_1 X + \gamma S + \rho XS + \theta G + \phi XG + U. \tag{4.14}$$

The slopes and intercepts for the four groups can be summarized in tables as shown in Table 4.39.

Table 4.39. Summary of Group Parameters

	INTERCEPTS			SLOPES	
	MALES	FEMALES		MALES	FEMALES
GROUP A	β_0	$\beta_0 + \gamma$	GROUP A	β_1	$\beta_1 + \rho$
GROUP B	$\beta_0 + \theta$	$\beta_0 + \gamma + \theta$	GROUP B	$\beta_1 + \phi$	$\beta_1 + \rho + \phi$

A comparison of the four cells for intercepts reveals a common sex effect γ for both groups and a common group effect θ for both sexes. The same statement can be made about the slope parameters. We have therefore modelled the relationship between income and education without allowing any *interaction* between sex and group.

A useful way of representing the variation of the intercept and slope parameters in the four cells is to use a graph as shown in Figure 4.12. The two graphs are drawn assuming γ and θ are positive and ρ and ϕ are negative.

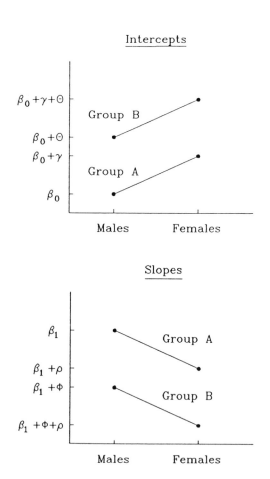

Figure 4.12. Relationships Among Parameters Without Interaction

Interaction

In order to allow interaction we must define an *interaction dummy variable* say SG which has the value 1 if both S and $G = 1$ and has the value 0 otherwise. The corresponding *interaction slope shifter variable* is given by XSG. The model is given by

$$Y = \beta_0 + \beta_1 X + \gamma S + \rho XS + \theta G + \phi XG + \delta SG + \eta XSG + U. \quad (4.15)$$

This model is completely general and fitting it is therefore equivalent to fitting four separate models for the four categories. The intercepts and slopes for the four models are summarized in Table 4.40.

Table 4.40. Summary of Group Parameters

	INTERCEPTS	
	MALES	FEMALES
GROUP A	β_0	$\beta_0 + \gamma$
GROUP B	$\beta_0 + \theta$	$\beta_0 + \gamma + \theta + \delta$

	SLOPES	
	MALES	FEMALES
GROUP A	β_1	$\beta_1 + \rho$
GROUP B	$\beta_1 + \phi$	$\beta_1 + \rho + \phi + \eta$

There is now an interaction between sex and occupation group. The difference between intercepts for males and females is γ for group A and $(\gamma + \delta)$ for group B. The difference between these two differences, δ, is the interaction between sex and occupation group. Similarly for the slopes the interaction term is measured by η. Figure 4.13 illustrates the interaction between group effects and sex effects. The lack of parallelism between the two lines in each case reflects the interaction.

The interaction parameters δ and η can be written as

$$\delta = \{[\beta_0 + \gamma + \theta + \delta] - [\beta_0 + \theta]\} - \{[\beta_0 + \gamma] - \beta_0\}$$
$$= \{\text{Female intercept} - \text{Male intercept}\}_{\text{GROUP B}}$$
$$- \{\text{Female intercept} - \text{Male intercept}\}_{\text{GROUP A}}$$
$$= \{\text{GROUP B intercept} - \text{GROUP A intercept}\}_{\text{Females}}$$
$$- \{\text{GROUP B intercept} - \text{GROUP A intercept}\}_{\text{Males}};$$

Intercepts

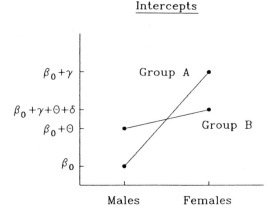

Slopes

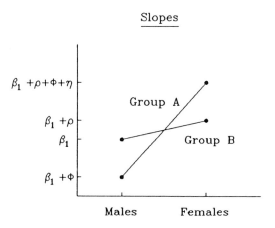

Figure 4.13. Relationships Among Parameters With Interaction

$$\eta = \{[\beta_1 + \rho + \phi + \eta] - [\beta_1 + \phi]\} - \{[\beta_1 + \rho] - \beta_1\}$$
$$= \{\text{Female slope} - \text{Male slope}\}_{\text{GROUP B}}$$
$$- \{\text{Female slope} - \text{Male slope}\}_{\text{GROUP A}}$$
$$= \{\text{GROUP B slope} - \text{GROUP A slope}\}_{\text{Females}}$$
$$- \{\text{GROUP B slope} - \text{GROUP A slope}\}_{\text{Males}}.$$

These expressions are sometimes useful for characterizing the interaction parameters δ and η. If the interaction parameter δ is positive we can conclude that the female intercept exceeds the male intercept by a larger amount in group B than in group A. Equivalently we can say that the group B intercept exceeds the group A intercept by a larger amount for females than for males.

Interaction and Outliers

The presence of significant interaction parameters can sometimes be indicative of outliers and may not represent a population interaction. Interactions should be carefully evaluated.

A variety of hypotheses can be tested by comparing the full model to the various reduced models obtained by omitting some or all of the dummy and slope shifter variables. A test for no interaction could be carried out by testing $H_0 : \delta = \eta = 0$ and hence comparing the models given by (4.14) and (4.15).

4.5.6 An Example from Automobile Fuel Consumption

A sample of 98 automobiles was selected from *Fuel Consumption Guide 1985* published by Transport Canada. Table 4.41 presents the information on fuel consumption, COMBRATE, domestic or foreign manufacturer (FOR $= 1$ if foreign, 0 otherwise), number of cylinders, CYLIND (4, 6 or 8) and WEIGHT. For the variable CYLIND two dummy variables C4 and C6 were created to correspond to CYLIND $= 4$ and 6 respectively. Slope shifter variables and interaction variables were defined by determining various products as follows:

$$FORC4 = FOR*C4$$

$$FORC6 = FOR*C6$$

$$FW = FOR*WEIGHT$$

$$C4W = C4*WEIGHT$$

$$C6W = C6*WEIGHT$$

$$FC4W = FOR*C4*WEIGHT$$

$$FC6W = FOR*C6*WEIGHT.$$

The linear model relating COMBRATE to the complete set of explanatory variables is given by

$$\begin{aligned}
COMBRATE = {} & \beta_0 + \beta_1 WEIGHT + \gamma_1 C4 + \gamma_2 C6 + \gamma_3 FOR \\
& + \rho_1 C4W + \rho_2 C6W + \rho_3 FW \\
& + \phi_1 FORC4 + \phi_2 FORC6 + \theta_1 FC4W \\
& + \theta_2 FC6W + U.
\end{aligned}$$

Table 4.41. Fuel Consumption, Weight, Number of Cylinders for a Sample of Domestic and Foreign Automobiles

COMBRATE	WEIGHT	FOR	CYLIND	TYPE
58	2500	0	4	Plymouth Horizon
84	2500	1	4	Volkswagen Scirocco
112	4000	0	8	Chev Impala/Caprice
91	3000	1	4	Mercedes 190E
101	4000	0	6	Pontiac Parisienne
68	2500	1	4	Nissan Stanza
72	2500	0	4	Pontiac Sunbird
60	1750	1	4	Nissan Micra
67	2750	1	4	Toyota Camry
113	4000	0	8	Buick Regal
76	2750	0	4	Buick Somerset Regal
110	3500	1	6	BMW 635CSI
85	3000	1	4	Volvo 740 GLE
88	3000	0	6	Chev Celebrity
68	2500	1	4	Subaru GL HB/SHA
109	4000	0	8	Pontiac Parisienne
91	2750	0	4	Pontiac Sunbird
77	2500	0	4	Ford EXP
82	2750	0	4	Chrysler Lebaron
83	2750	1	4	Subaru Turbo XT
109	3500	0	8	Pontiac Firebird
128	4000	0	8	Ford LTD Crown Victoria
74	2500	0	4	Chev Cavalier
100	3000	1	4	Volvo 760 Turbo
76	2500	1	4	Hyundai Pony
79	2750	0	4	Mercury Topaz
92	3000	0	4	Dodge Lancer
78	2750	0	4	Pontiac Sunbird
63	2250	0	4	Pontiac 1000
74	2750	0	4	Oldsmobile Firenza
115	4000	0	8	Lincoln Mark VII
113	3500	0	8	Chev Monte Carlo
67	2000	1	4	Honda Civic
68	2500	0	4	Plymouth Reliant
103	3500	0	6	Mercury Marquis
104	3000	0	6	Buick Century
109	3500	0	8	Chev Monte Carlo
58	2500	0	4	Plymouth Turismo
107	3500	0	4	Ford T-Bird
138	4000	1	8	Mercedes 500SEL
123	3500	0	8	Chev Camaro
96	3000	0	6	Ford Mustang
83	2750	1	4	BMW IA
104	3500	0	8	Mercury Capri
88	3000	0	4	Chrysler Lebaron
101	4000	0	6	Chev Impala/Caprice
101	3500	1	6	BMW IA
75	2750	0	4	Pontiac Grand Am
75	2750	0	4	Mercury Lynx EF

Table 4.41. Fuel Consumption, Weight, Number of Cylinders for a Sample of Domestic and Foreign Automobiles (continued)

COMBRATE	WEIGHT	FOR	CYLIND	TYPE
58	2500	0	4	Dodge Omni
100	3000	1	6	Porsche Carrera
113	3500	0	8	Pontiac Firebird
109	4000	0	8	Buick Regal
109	3500	0	8	Chev Camaro
78	3000	0	4	Chev Camaro
132	4000	0	8	Ford LTD/Victoria
71	2500	1	4	Subaru GL
90	3000	0	6	Chev Citation
71	2750	0	4	Dodge Aries
107	3500	0	6	Buick Electra
94	3000	0	4	Chrysler New Yorker
97	3500	0	6	Chev Camaro
78	2500	1	4	Volkswagen Scirocco
74	2500	0	4	Pontiac Sunbird
76	3000	0	4	Chev Celebrity
81	3000	1	4	Volvo 74 GLE
93	3000	0	6	Audi 5000FS
79	2750	0	4	Pontiac Sunbird
90	3000	0	6	Chev Celebrity
84	2750	0	4	Chrysler Laser
78	3000	0	4	Plymouth Caravelle
69	2250	1	4	Toyota Tercel
86	3000	1	4	Saab 900
68	2250	0	4	Mercury Lynx
73	2250	1	4	Nissan Pulsar
70	2000	1	4	Honda Civic
105	3000	1	4	Turb 16
103	3500	0	6	Ford Thunderbird
136	3000	1	6	Porsche Carrera
79	2700	1	4	Mazda 626
113	3500	0	8	Chev Camaro
74	2500	1	4	Subaru XT
79	2500	1	4	Honda Accord
73	2500	1	4	Toyota Corolla Sport
86	2750	0	4	Cadillac Cimarron
74	3000	1	4	Nissan 200XS
78	3000	0	4	Chev Camaro
73	2500	1	4	Volkswagen GTI
90	3000	0	6	Pontiac Grand Am
76	2750	1	4	Toyota Celica
72	3000	0	4	Pontiac Fiero
76	2750	0	4	Olds Calais
73	2500	1	4	Volkswagen GTI
75	2250	0	4	Dodge Colt
101	3500	0	8	Cadillac Fleetwood
111	2750	1	8	Mazda RX7
89	3000	1	4	Porsche 944
97	3500	0	6	Buick Regal

Table 4.42. Summary of Parameters

	INTERCEPTS	
	DOMESTIC	FOREIGN
CYLIND = 4	$\beta_0 + \gamma_1$	$\beta_0 + \gamma_1 + \gamma_3 + \phi_1$
CYLIND = 6	$\beta_0 + \gamma_2$	$\beta_0 + \gamma_2 + \gamma_3 + \phi_2$
CYLIND = 8	β_0	$\beta_0 + \gamma_3$

	SLOPES	
	DOMESTIC	FOREIGN
CYLIND = 4	$\beta_1 + \rho_1$	$\beta_1 + \rho_1 + \rho_3 + \theta_1$
CYLIND = 6	$\beta_1 + \rho_2$	$\beta_1 + \rho_2 + \rho_3 + \theta_2$
CYLIND = 8	β_1	$\beta_1 + \rho_3$

Using the definitions of the explanatory variables the intercepts and slopes are given in Table 4.42. The parameters γ_1 and γ_2 measure cylinder intercept effects relative to CYLIND = 8, and ρ_1 and ρ_2 measure cylinder slope effects relative to CYLIND = 8. The parameters γ_3 and ρ_3 measure foreign vs. domestic effects for intercepts and slopes respectively. The interaction parameters, ϕ_1 and ϕ_2 for intercepts and θ_1 and θ_2 for slopes are explained below.

From Table 4.42 we can conclude that the interaction parameters ϕ_1 and ϕ_2 are equivalent to the expressions below.

$$\phi_1 = \{\text{Foreign Intercept} - \text{Domestic Intercept}\}_{\text{CYLIND} = 4}$$
$$- \{\text{Foreign Intercept} - \text{Domestic Intercept}\}_{\text{CYLIND} = 8};$$

$$\phi_2 = \{\text{Foreign Intercept} - \text{Domestic Intercept}\}_{\text{CYLIND} = 6}$$
$$- \{\text{Foreign Intercept} - \text{Domestic Intercept}\}_{\text{CYLIND} = 8}.$$

A positive interaction parameter ϕ_1 would therefore indicate that the Foreign intercept is greater than the Domestic intercept by a larger amount for CYLIND = 4 than for CYLIND = 8. Similarly, if ϕ_2 is positive the Foreign intercept exceeds the Domestic intercept by a larger amount for CYLIND = 6 than for CYLIND = 8. In each case the interaction is characterized by using four of the six cells. In both cases CYLIND = 8 is involved in the comparison. To compare CYLIND = 4 to CYLIND = 6 the quantity $(\phi_1 - \phi_2)$ can be used since

$$(\phi_1 - \phi_2) = \{\text{Foreign Intercept} - \text{Domestic Intercept}\}_{\text{CYLIND} = 4}$$
$$- \{\text{Foreign Intercept} - \text{Domestic Intercept}\}_{\text{CYLIND} = 6}.$$

These interactions can also be characterized in terms of comparisons of differences between categories of CYLIND. The parameter ϕ_1 is therefore also given by

$$\phi_1 = \{\text{Intercept CYLIND=4} - \text{Intercept CYLIND=8}\}_{\text{Foreign}}$$
$$- \{\text{Intercept CYLIND=4} - \text{Intercept CYLIND=8}\}_{\text{Domestic}}.$$

Similar statements can be made regarding the slope shifter interactions θ_1 and θ_2.

The estimated parameters for a variety of models are shown in Table 4.44. In addition to the variable WEIGHT equation 2 includes the dummy variables FOR, $C4$ and $C6$. Including the three dummy variables results in R^2 increasing from 0.705 to 0.806. The resulting values of the intercepts are given in Table 4.43. A comparison of the two equations is a test of the null hypothesis $H_0 : \gamma_1 = \gamma_2 = \gamma_3 = 0$ in the model

$$\text{COMBRATE} = \beta_0 + \beta_1 \text{WEIGHT} + \gamma_1\ C4 + \gamma_2\ C6 + \gamma_3\ \text{FOR} + U.$$

The F-statistic for this hypothesis is given by

$$F = \frac{(0.806 - 0.705)/3}{(1 - 0.806)/93} = 16.139$$

which is significant at conventional levels.

Table 4.43. Intercept Parameter Estimates For Constant Slope Model Without Interaction

| | INTERCEPTS FOR EQUATION 2 | |
	FOR = 0	FOR = 1
CYLIND = 4	24.485	30.983
CYLIND = 6	37.359	43.857
CYLIND = 8	45.205	51.703

From the equation 2 results we can conclude that foreign automobiles use more fuel than domestic automobiles. In each cylinder class the intercepts for foreign vehicles are approximately six units larger than for domestic vehicles. It is also apparent that automobiles with more cylinders use more fuel than those with fewer cylinders. This conclusion however also requires the assumption of common slope parameter and no interaction between number of cylinders and the foreign parameter.

The results for equation 3 in Table 4.44 illustrate the effect of adding dummy interaction variables. The resulting intercept differences are illustrated in Table 4.45. The differences between the four cylinder intercepts

Table 4.44. Regression Results Showing Relationships Between COMBRATE and FOR, $C4$ and $C6$

Equation Number	R^2	β_0 Intercept	β_1 WEIGHT	γ_3 FOR	γ_1 C4	γ_2 C6	ϕ_1 FORC4	ϕ_2 FORC6	ρ_3 FW	ρ_1 C4W	ρ_2 C6W	θ_1 FC4W	θ_2 FC6W
1	0.705	1.061 (0.857)	0.029 (0.000)										
2	0.806	45.205 (0.000)	0.019 (0.000)	6.498 (0.000)	−20.720 (0.000)	−7.846 (0.009)							
3	0.826	42.870 (0.000)	0.019 (0.000)	17.756 (0.004)	−17.698 (0.000)	−8.588 (0.006)	−14.409 (0.024)	−1.797 (0.810)					
4	0.827	67.365 (0.002)	0.013 (0.030)	−6.093 (0.605)	−53.081 (0.022)	15.651 (0.572)			0.005 (0.267)	0.010 (0.136)	−0.008 (0.318)		
5	0.853	65.556 (0.021)	0.013 (0.091)	−13.956 (0.730)	−64.915 (0.040)	−2.219 (0.947)	33.624 (0.445)	143.619 (0.032)	0.009 (0.439)	0.015 (0.096)	−0.003 (0.777)	−0.015 (0.264)	−0.044 (0.027)
6	0.852	57.775 (0.000)	0.015 (0.000)	16.637 (0.003)	−55.311 (0.000)			118.588 (0.019)		0.012 (0.011)	−0.003 (0.001)	−0.005 (0.027)	−0.037 (0.018)

Table 4.45. Intercept Parameter Estimates For Constant Slope Model With Interaction

INTERCEPTS FOR EQUATION 3

	FOR = 0	FOR = 1
CYLIND = 4	25.172	28.519
CYLIND = 6	34.282	50.241
CYLIND = 8	42.870	60.626

is now much small than the differences among the six and eight cylinder intercepts. The differences for the six and eight cylinder intercepts are almost equal, which is also reflected by the non-significant interaction term FORC6.

A comparison of equations 3 and 2 is a test of the hypothesis $H_0 : \phi_1 = \phi_2 = 0$ in the model

$$\text{COMBRATE} = \beta_0 + \beta_1 \text{ WEIGHT}$$

$$= \gamma_1 C4 + \gamma_2 C6 + \gamma_3 \text{FOR} + \phi_1 \text{FORC4} + \phi_2 \text{FORC6} + U.$$

The F-statistic is given by $\dfrac{(0.826 - 0.806)/2}{(1 - 0.826)/91} = 5.230$ which is significant at the 0.01 level. From the equation 3 results we can conclude that the difference in fuel consumption between foreign automobiles and domestic automobiles depends on the number of cylinders and that this difference is larger for six and eight cylinder automobiles than for four cylinder automobiles.

Equation 4 in Table 4.44 illustrates the estimated model obtained when slope shifter variables for WEIGHT are added to the model. The model does not include parameters to measure interaction between number of cylinders and domestic vs foreign. The resulting intercept and slope parameter estimates are summarized in Table 4.46. By including slope shifters the model permits the impact of WEIGHT on fuel economy to vary by type of automobile. It would appear from these estimates that the impact of WEIGHT on COMBRATE is greater for foreign automobiles than for domestic automobiles. An examination of the cell means in Table 4.46 reveals that foreign automobiles tend to have smaller intercepts and larger slopes than the domestic automobiles. The intercepts now appear to be six units smaller for foreign vehicles, while the slopes for foreign vehicles are 0.005 units larger than for domestic vehicles. Note how the foreign intercepts are now smaller which is the opposite result to Table 4.43 (omitted variable bias).

A comparison of equation 4 to equation 2 is a test of the null hypothesis $H_0 : \rho_1 = \rho_2 = \rho_3 = 0$ in the model

$$\text{COMBRATE} = \beta_0 + \beta_1 \text{ WEIGHT} + \gamma_1 \text{ C4} + \gamma_2 \text{ C6} + \gamma_3 \text{ FOR}$$

$$+ \rho_1 \text{ C4W} + \rho_2 \text{ C6W}$$

$$+ \rho_3 \text{ FW} + U.$$

The F-statistic for this hypothesis is given by

$$\frac{(0.827 - 0.806)/3}{(1 - 0.827)/90} = 3.642$$

which is significant at the 0.04 level. Thus, under the assumption of no interaction between cylinder type and location of manufacturer, there appears to be some variation in the impact of WEIGHT on COMBRATE.

Table 4.46. Parameter Estimates For Model Without Interaction

	INTERCEPTS FOR EQUATION 4	
	FOR = 0	FOR = 1
CYLIND = 4	14.284	8.191
CYLIND = 6	83.016	76.923
CYLIND = 8	67.365	61.272

	SLOPES FOR EQUATION 4	
	FOR = 0	FOR = 1
CYLIND = 4	0.023	0.028
CYLIND = 6	0.005	0.010
CYLIND = 8	0.013	0.018

Equation 5 in Table 4.44 shows the estimated parameters for the full model given by

$$\text{COMBRATE} = \beta_0 + \beta_1 \text{ WEIGHT} + \gamma_1 \text{C4} + \gamma_2 \text{C6} + \gamma_3 \text{FOR}$$

$$+ \rho_1 \text{C4W} + \rho_2 \text{C6W} + \rho_3 \text{FW}$$

$$+ \phi_1 \text{FORC4} + \phi_2 \text{FORC6}$$

$$+ \theta_1 \text{FC4W} + \theta_2 \text{FC6W} + U. \tag{4.16}$$

It would appear from the significance levels for the coefficients that the parameter estimates for $\gamma_3, \gamma_2, \theta_1, \rho_3, \rho_2$ and ϕ_1 are not significantly different from zero. A comparison of this equation to equation 4, which excludes the interaction between CYLIND and FOR, is a test of the null hypothesis $H_0: \phi_1 = \phi_2 = \theta_1 = \theta_2 = 0$ in (4.15). The F-statistic is given by

$$F = \frac{(0.853 - 0.827)/4}{(1 - 0.853)/86} = 3.803$$

which is significant at the 0.01 level. Thus, although there are many non-significant coefficients in equation 5, the additional information provided by equation 5 relative to equation 4 cannot be ignored. The multicollinearity problem created by the presence of the various dummy variables and their corresponding slope shifters once again results in relatively large standard errors for the estimated coefficients.

The estimated coefficients in equation 5 can be used to obtain intercept and slope coefficients as summarized in Table 4.47. The estimates of the intercepts and slopes provided by these two tables are identical to the ordinary least squares regression coefficients that would be obtained if the regression model relating COMBRATE to WEIGHT was determined separately for the six cells. The regression results obtained seem to be particularly unusual for foreign 6-cylinder automobiles. In this case the intercept is relatively large and the slope coefficient is negative.

Table 4.47. Parameter Estimates For Model With Interaction

| | INTERCEPTS FOR EQUATION 5 | |
	FOR = 0	FOR = 1
CYLIND = 4	0.641	20.309
CYLIND = 6	63.337	193.000
CYLIND = 8	65.556	51.600

| | SLOPES FOR EQUATION 5 | |
	FOR = 0	FOR = 1
CYLIND = 4	0.028	0.022
CYLIND = 6	0.010	-0.025
CYLIND = 8	0.013	0.022

Since equation 5 yielded so many nonsignificant coefficients it is of interest to determine if some of the variables can be dropped without reducing the R^2 level of 0.853. Using stepwise regression techniques the estimated model given by equation 6 in Table 4.44 was selected. All of the estimated coefficients in the model are significant at the 0.03 level and $R^2 = 0.852$. The equations for the 6 cells are summarized in Table 4.48. A comparison of the intercepts and slopes for the six cells to the previous estimates obtained from equation 5 reveals only minor differences. The linear relationship for foreign automobiles with 6 cylinders remained downward sloping. The stepwise approach should be limited to the prediction problem environment.

An examination of the intercept interaction parameter estimates in equation 5 (ignoring the p-values) would suggest that the foreign intercept exceeds the domestic intercept by a larger amount for 4-cylinder autos than for 8-cylinder autos. A similar result is also true for the comparison between

6-cylinder autos and 8-cylinder autos. For the slope interaction parameter estimates the opposite conclusion is obtained in that the domestic slope exceeds the foreign slope by a larger amount for 4 and 6-cylinder autos in comparison to 8-cylinder autos.

Table 4.48. Parameter Estimates For Model From Stepwise Selection

| | INTERCEPTS FOR EQUATION 6 | |
	FOR = 0	FOR = 1
CYLIND = 4	2.464	19.101
CYLIND = 6	57.775	193.000
CYLIND = 8	57.775	74.412

| | SLOPES FOR EQUATION 6 | |
	FOR = 0	FOR = 1
CYLIND = 4	0.027	0.022
CYLIND = 6	0.012	-0.025
CYLIND = 8	0.015	0.015

Figure 4.14 shows the six linear relationships fitted by equation 5 plotted on a scatter plot relating COMBRATE and WEIGHT. The symbols $\triangle$, $\square$, $\bigcirc$ are used to denote CYLIND = 4, 6 and 8 respectively. If the symbols are darkened then the point refers to a foreign automobile. It is worth noting that the linear relationship for foreign automobiles with 8 cylinders is based on only two points and hence the line passed through these two points. For foreign automobiles with 6 cylinders we can see that the downward sloping line is due to one outlier (the Porsche Carrera). Since there are only four points in this group the single outlier has a major influence on the direction of the line. We can also see from Table 4.47 that the relationship for 4-cylinder cars is similar for both foreign and domestic automobiles. For 6 and 8-cylinder domestic cars the regression lines are relatively flat with the 8-cylinder line lying above the 6-cylinder line by an almost constant amount of 10 units. Because of the high correlation between number of cylinders and weight it would appear that a quadratic relationship between COMBRATE and WEIGHT, excluding CYLIND, could provide an equally good fit for domestic automobiles. For foreign automobiles there is an insufficient number of points for CYLIND = 6 and CYLIND = 8 to obtain a reliable estimate for these two categories separately.

Since the scatterplot between COMBRATE and WEIGHT shows that there are several outliers, it is of interest to determine the influence diagnostics for the full model fitted in equation 5. For observation 40 (Mercedes

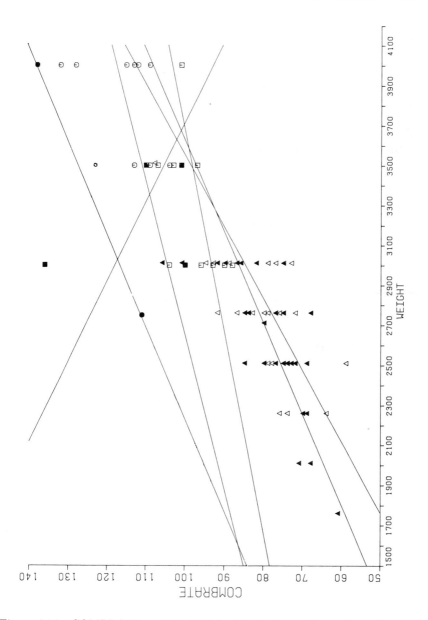

Figure 4.14.   COMBRATE vs WEIGHT for CYLIND and Domestic vs Foreign

500 SEL) the value of Cook's D is 25.781. This observation causes the 8-cylinder foreign group to be different from the 8-cylinder domestic group. The influence is large partially because there are only 2 observations on foreign automobiles with 8 cylinders. There are two additional observations with moderately large values of Cook's D, observations 51 (0.973) and 79 (0.973) which are both Porsche Carrera vehicles with 6 cylinders. The values of COMBRATE for these two observations are 100 and 136 respectively. Observations 51 and 79 have opposite effects on the coefficients of FORC6 and FC6W. This result is also influenced by the fact that there are only two additional observations on foreign automobiles with 6 cylinders both of which have WEIGHT values of 3500. In conclusion it would appear that there are an insufficient number of observations on foreign manufactured automobiles with 6 and 8 cylinders to be able to distinguish them from other automobiles. As suggested above, outliers can have an impact on the interaction parameter estimates and significant interactions may not truthfully represent the population characteristics. The significant interaction between 6-cylinder vehicles and FOR observed in this example illustrates this phenomenon.

4.5.7 Other Approaches

A comparison of the four occupation groups defined in 4.5.5 could have been carried out by using an alternative parameterization. We could have defined dummy variables

$$MA = 1 \quad \text{if male from group A}$$
$$= 0 \quad \text{otherwise}$$

$$MB = 1 \quad \text{if male from group B}$$
$$= 0 \quad \text{otherwise}$$

$$FA = 1 \quad \text{if female from group A}$$
$$= 0 \quad \text{otherwise}$$

$$FB = 1 \quad \text{if female from group B}$$
$$= 0 \quad \text{otherwise.}$$

Using any three of these four dummy variables and their corresponding slope shifters for X we can fit a linear model. The resulting intercepts and slopes are equivalent to fitting separate relationships for the four groups. The disadvantage of this particular parameterization is that it does not allow for an easy test for interaction between sex and group. The model is given by

$$Y = \beta_0 + \beta_1 X + \rho MB + \theta FA + \gamma FB + \phi XMB + \eta XFA + \delta XFB + U.$$

Table 4.49. Summary of Group Parameters

| | INTERCEPTS | |
	MALES	FEMALES
GROUP A	β_0	$\beta_0 + \theta$
GROUP A	$\beta_0 + \rho$	$\beta_0 + \gamma$

| | SLOPES | |
	MALES	FEMALES
GROUP A	β_1	$\beta_1 + \eta$
GROUP A	$\beta_1 + \phi$	$\beta_1 + \delta$

The intercepts and slopes for the four groups are summarized in Table 4.49.

Testing the hypothesis of equal category intercepts and slopes can be carried out by testing $H_0 : \theta = \rho = \gamma = \eta = \phi = \delta = 0$. To test the hypothesis that there is no interaction is more difficult in that we must test $H_0 : \theta = (\gamma - \rho)$ and $\eta = (\delta - \phi)$.

Effect Coding

The effect coding approach introduced in 4.5.4 above may also be applied here. The indicator variables for SEX and OCCUPATION are defined as follows:

$$S = \quad 1 \quad \text{for females}$$
$$= -1 \quad \text{for males}$$

$$G = \quad 1 \quad \text{for group B}$$
$$= -1 \quad \text{for group A.}$$

An interaction indicator is obtained from the product SG. The corresponding slope shifters are given by XS, XG and XSG respectively.

The complete multiple regression model is given by

$$Y = \beta_0 + \beta_1 X + \gamma S + \rho XS + \theta G + \phi XG + \delta SG + \eta XSG + U.$$

The intercepts and slopes for the four sex-occupation groups are summarized in Table 4.50.

Table 4.50. Summary of Group Parameters

	INTERCEPTS	
	MALES	FEMALES
GROUP A	$\beta_0 - \gamma - \theta + \delta$	$\beta_0 + \gamma - \theta - \delta$
GROUP B	$\beta_0 - \gamma + \theta - \delta$	$\beta_0 + \gamma + \theta + \delta$

	SLOPES	
	MALES	FEMALES
GROUP A	$\beta_1 - \rho - \phi + \eta$	$\beta_1 + \rho - \phi - \eta$
GROUP B	$\beta_1 - \rho + \phi - \eta$	$\beta_1 + \rho + \phi + \eta$

An examination of the four intercept parameters in the four cells of Table 4.50 shows that the sum of the four cells is $4\beta_0$ and hence β_0 is the average intercept. The average intercept for males is $(\beta_0 - \gamma)$ and for females the average intercept is $(\beta_0 + \gamma)$. The male effect is $-\gamma$ and the female effect is $+\gamma$. The sex effect is 2γ. The average female is γ units from the overall average. The average difference between males and females is 2γ. Similarly for the occupation groups the group A effect is $-\theta$ and the group B effect is $+\theta$. The occupation effect is 2θ.

The interaction between sex and occupation can be determined by comparing the difference between the female intercept and the average intercept in each group. In group A the difference is $(\gamma - \delta)$, while in group B the difference is $(\gamma + \delta)$. The difference between the two differences is 2δ. The parameter δ therefore measures the interaction between sex and group. The parameter δ can be expressed as

$$\delta = \{\text{Intercept Females} - \text{Average Intercept Males \& Females}\}_{\text{Group B}}$$
$$- \{\text{Intercept Females} - \text{Average Intercept Males \& Females}\}_{\text{Both Groups}},$$

or equivalently as

$$\delta = \{\text{Intercept Group B} - \text{Average Intercept Both Groups}\}_{\text{Females}}$$
$$- \{\text{Intercept Group B} - \text{Average Intercept Both Groups}\}_{\text{Males \& Females}}$$

In a similar fashion the slope parameters also measure sex effects $(\pm\rho)$, group effects $(\pm\phi)$ and interaction effect η.

Example for Automobile Data

The effect coding approach can also be applied to the automobile fuel consumption data. The effect coding variables are defined as follows:

$$
\begin{aligned}
C4 = \ & 1 \text{ if number of cylinders} = 4 \\
= \ & 0 \text{ if number of cylinders} = 6 \\
= \ & {-1} \text{ if number of cylinders} = 8
\end{aligned}
$$

$$
\begin{aligned}
C6 = \ & 0 \text{ if number of cylinders} = 4 \\
= \ & 1 \text{ if number of cylinders} = 6 \\
= \ & {-1} \text{ if number of cylinders} = 8
\end{aligned}
$$

$$
\begin{aligned}
FOR = \ & 1 \text{ if automobile if foreign} \\
= \ & {-1} \text{ if automobile is domestic.}
\end{aligned}
$$

The corresponding interaction variables are defined by the product variables FORC4 = FOR*C4, FORC6 = FOR*C6, and the slope shifter variables by C4W = C4*WEIGHT, C6W = C6*WEIGHT, FW = FOR*WEIGHT, FC4W = FORC4*WEIGHT and FC6W = FORC6*WEIGHT.

The linear model with all explanatory variables is given by

$$
\begin{aligned}
\text{COMBRATE} = \ & \beta_0 + \beta_1 \text{WEIGHT} + \gamma_1 C4 + \gamma_2 C6 + \gamma_3 \text{FOR} \\
& + \rho_1 C4W + \rho_2 C6W + \rho_3 FW \\
& + \phi_1 \text{FORC4} + \phi_2 \text{FORC6} + \theta_1 \text{FC4W} + \theta_2 \text{FC6W} + U.
\end{aligned}
$$

As a result the cell parameters for intercepts and slopes are as given in Table 4.51.

Using this effect coding parameterization the interaction parameters can be expressed as shown below.

$$
\begin{aligned}
\phi_1 = \ & \{\text{Intercept CYLIND=4} - \text{Average Intercept ALL CYLIND}\}_{\text{Foreign}} \\
& -\{\text{Intercept CYLIND=4} - \text{Average Intercept ALL CYLIND}\}_{\text{All Cars}}
\end{aligned}
$$

or

$$
\begin{aligned}
& \{\text{Intercept Foreign} - \text{Average Intercept Foreign \& Domestic}\}_{\text{CYLIND=4}} \\
& -\{\text{Intercept Foreign} - \text{Average Intercept Foreign \& Domestic}\}_{\text{Average for All CYLIND}},
\end{aligned}
$$

$$
\begin{aligned}
\phi_2 = \ & \{\text{Intercept CYLIND=6} - \text{Average Intercept ALL CYLIND}\}_{\text{Foreign}} \\
& -\{\text{Intercept CYLIND=6} - \text{Average Intercept ALL CYLIND}\}_{\text{All Cars}}
\end{aligned}
$$

or

{Intercept Foreign – Average Intercept Foreign & Domestic}$_{\text{CYLIND}=6}$
$-$ {Intercept Foreign – Average Intercept Foreign & Domestic}$_{\text{Average for All CYLIND}}$.

For the category CYLIND $= 8$ the interaction parameters ϕ_1 and ϕ_2 can be used to yield

$- (\phi_1 + \phi_2) =$
{Intercept CYLIND=8 – Average Intercept ALL CYLIND}$_{\text{Foreign}}$
$-$ {Intercept CYLIND=8 – Average Intercept ALL CYLIND}$_{\text{All Cars}}$

or

{Intercept Foreign – Average Intercept Foreign & Domestic}$_{\text{CYLIND}=8}$
$-$ {Intercept Foreign – Average Intercept Foreign & Domestic}$_{\text{Average for All CYLIND}}$.

Table 4.51. Summary of Group Parameters

| | INTERCEPTS | |
	DOMESTIC	FOREIGN
CYLIND $= 4$	$\beta_0 + \gamma_1 - \gamma_3 - \phi_1$	$\beta_0 + \gamma_1 + \gamma_3 + \phi_1$
CYLIND $= 6$	$\beta_0 + \gamma_2 - \gamma_3 - \phi_2$	$\beta_0 + \gamma_2 + \gamma_3 + \phi_2$
CYLIND $= 8$	$\beta_0 - \gamma_1 - \gamma_2 - \gamma_3 + \phi_1 + \phi_2$	$\beta_0 - \gamma_1 - \gamma_2 + \gamma_3 - \phi_1 - \phi_2$

| | SLOPES | |
	DOMESTIC	FOREIGN
CYLIND $= 4$	$\beta_1 + \rho_1 - \rho_3 - \theta_1$	$\beta_1 + \rho_1 + \rho_3 + \theta_1$
CYLIND $= 6$	$\beta_1 + \rho_2 - \rho_3 - \theta_2$	$\beta_1 + \rho_2 + \rho_3 + \theta_2$
CYLIND $= 8$	$\beta_1 - \rho_1 - \rho_2 - \rho_3 + \theta_1 + \theta_2$	$\beta_1 - \rho_1 - \rho_2 + \rho_3 - \theta_1 - \theta_2$

The interaction parameters for the slopes θ_1 and θ_2 can be expressed in a similar fashion. Effect coding therefore makes comparisons to the overall average while for dummy coding the comparisons involve a base case.

The fitted model for the fuel economy data is given by

$$
\begin{aligned}
\text{COMBRATE} = \underset{(0.000)}{74.656} + \underset{(0.022)}{0.009} \text{ WEIGHT} + \underset{(0.032)}{31.479} \text{ FOR} \\
- \underset{(0.159)}{64.182} \text{ C4} + \underset{(0.036)}{53.512} \text{ C6} - \underset{(0.000)}{21.644} \text{ FORC4} \\
+ \underset{(0.189)}{33.353} \text{ FORC6} - \underset{(0.061)}{0.008} \text{ FW} + \underset{(0.006)}{0.016} \text{ C4W} \\
- \underset{(0.022)}{0.017} \text{ C6W} + \underset{(0.288)}{0.005} \text{ FC4W} - \underset{(0.169)}{0.010} \text{ FC6W}.
\end{aligned}
$$

From the estimates of the interaction parameters we can conclude that for 4-cylinder autos the difference between the average intercept and the foreign intercept is greater than the difference between the average intercept and the foreign intercept taken over all autos. For 6-cylinder vehicles the reverse is true. For 8-cylinder autos the intercept interaction is estimated by $21.644 - 33.353 = -11.709$, hence the average intercept exceeds the foreign intercept by a larger amount for 8-cylinder autos than in the case of the overall averages.

Cell Parameter Coding

As outlined previously, the $\mathbf{X}$ matrix can contain dummy variables and slope shifters corresponding to all possible categories, provided that the column of unities and the explanatory variable are omitted from the matrix. In the case of two qualitative variables the total number of categories is determined from the cross-classification of the two variables. For each category of the cross-classification, dummy variables and slope shifters are defined.

Example for Income vs. Education

Earlier in this section an example was used which studied the relationship between income (Y) and education (X) for males and females and for two different occupation groups A and B. At the beginning of this section four dummy variables MA, MB, FA and FB were defined to represent the four categories: males in group A, males in group B, females in group A and females in group B, respectively. The four corresponding slope shifters are defined by MAX, MBX, FAX and FBX respectively. The full model is therefore given by

$$Y = \beta_1 \text{ MA} + \beta_2 \text{ MB} + \beta_3 \text{ FA} + \beta_4 \text{ FB}$$
$$+ \beta_5 \text{ MAX} + \beta_6 \text{ MBX} + \beta_7 \text{ FAX} + \beta_8 \text{ FBX} + U.$$

To test the hypothesis of no interaction between sex and occupation the linear transformation matrix $\mathbf{T}$ is given by

$$\mathbf{T} = \begin{bmatrix} 1 & -1 & -1 & 1 & 0 & 0 & 0 & 0 \\ 0 & 0 & 0 & 0 & 1 & -1 & -1 & 1 \end{bmatrix},$$

which yields the conditions

$$(\beta_1 + \beta_4) - (\beta_2 + \beta_3) = 0 \quad \text{and} \quad (\beta_5 + \beta_8) - (\beta_6 + \beta_7) = 0.$$

To test the hypothesis of no sex effect the $\mathbf{T}$ matrix is given by

$$\mathbf{T} = \begin{bmatrix} 1 & 1 & -1 & -1 & 0 & 0 & 0 & 0 \\ 0 & 0 & 0 & 0 & 1 & 1 & -1 & -1 \end{bmatrix},$$

which yields the conditions

$$(\beta_1 + \beta_2) - (\beta_3 + \beta_4) = 0 \quad \text{and} \quad (\beta_5 + \beta_6) - (\beta_7 + \beta_8) = 0.$$

Finally the test for no group effects is determined using

$$\mathbf{T} = \begin{bmatrix} 1 & -1 & 1 & -1 & 0 & 0 & 0 & 0 \\ 0 & 0 & 0 & 0 & 1 & -1 & 1 & -1 \end{bmatrix},$$

which yields the conditions

$$\beta_1 - \beta_2 + \beta_3 - \beta_4 = 0 \quad \text{and} \quad \beta_5 - \beta_6 + \beta_7 - \beta_8 = 0.$$

Example for Automobile Data

For the automobile fuel consumption data dummy variables D1 to D6 are defined for the 6 categories as follows: D1 (FOR = 0, CYLIND = 4), D2 (FOR = 0, CYLIND = 6), D3 (FOR = 0, CYLIND = 8), D4 (FOR = 1, CYLIND = 4), D5 (FOR = 1, CYLIND = 6), D6 (FOR = 1, CYLIND = 8). In a similar fashion the corresponding slope shifters for the variable WEIGHT are defined by DWEIGHT1 through DWEIGHT6 respectively. The resulting full model is given by

$$\begin{aligned} \text{COMBRATE} = {} & \beta_1 \text{ D1} + \beta_2 \text{ D2} + \beta_3 \text{ D3} + \beta_4 \text{ D4} + \beta_5 \text{ D5} + \beta_6 \text{ D6} \\ & + \alpha_1 \text{ DWEIGHT1} + \alpha_2 \text{ DWEIGHT2} + \alpha_3 \text{ DWEIGHT3} \\ & + \alpha_4 \text{ DWEIGHT4} + \alpha_5 \text{ DWEIGHT5} + \alpha_6 \text{ DWEIGHT6}. \end{aligned}$$

For the data given in Table 4.41 the estimated full model is given by

$$\begin{aligned} \text{COMBRATE} = {} & \underset{(0.962)}{0.64} \text{ D1} + \underset{(0.001)}{63.34} \text{ D2} + \underset{(0.021)}{65.56} \text{ D3} \\ & + \underset{(0.062)}{20.31} \text{ D4} + \underset{(0.000)}{193.00} \text{ D5} + \underset{(0.078)}{51.60} \text{ D6} \\ & + \underset{(0.000)}{0.0279} \text{ DWEIGHT1} + \underset{(0.070)}{0.0102} \text{ DWEIGHT2} \\ & + \underset{(0.091)}{0.0128} \text{ DWEIGHT3} + \underset{(0.000)}{0.0221} \text{ DWEIGHT4} \\ & - \underset{(0.097)}{0.0250} \text{ DWEIGHT5} + \underset{(0.012)}{0.0216} \text{ DWEIGHT6}. \end{aligned}$$

The p-values here are relevant to a test that the particular parameter is zero.

To test the hypothesis of no interaction a restricted model is fitted based on the $\mathbf{T}$ matrix given by

$$
\mathbf{T} = \begin{bmatrix}
2 & -1 & -1 & -2 & 1 & 1 & 0 & 0 & 0 & 0 & 0 & 0 \\
-1 & 2 & -1 & 1 & -2 & 1 & 0 & 0 & 0 & 0 & 0 & 0 \\
0 & 0 & 0 & 0 & 0 & 0 & 2 & -1 & -1 & -2 & 1 & 1 \\
0 & 0 & 0 & 0 & 0 & 0 & -1 & 2 & -1 & 1 & -2 & 1
\end{bmatrix},
$$

where the coefficient vector $\boldsymbol{\beta}$ is assumed to contain the coefficients in the order shown in the above equations. The estimated model is given by

$$
\begin{aligned}
\mathrm{COMBRATE} = {} & 14.28\ \mathrm{D1} + 83.02\ \mathrm{D2} + 67.36\ \mathrm{D3} \\
& + 8.19\ \mathrm{D4} + 76.92\ \mathrm{D5} + 61.27\ \mathrm{D6} \\
& + 0.0226\ \mathrm{DWEIGHT1} + 0.0047\ \mathrm{DWEIGHT2} \\
& + 0.0125\ \mathrm{DWEIGHT3} + 0.0271\ \mathrm{DWEIGHT4} \\
& + 0.0092\ \mathrm{DWEIGHT5} + 0.0171\ \mathrm{DWEIGHT6}.
\end{aligned}
$$

The F-statistic for testing the restriction $H_0 : \mathbf{T}\boldsymbol{\beta} = 0$ has the value 3.424 with 4 degrees of freedom and a p-value of 0.012. The hypothesis $\mathbf{T}\boldsymbol{\beta} = 0$ is equivalent to the four conditions $(2\beta_1 - \beta_2 - \beta_3 - 2\beta_4 + \beta_5 + \beta_6) = 0$, $(2\beta_2 - \beta_1 - \beta_3 - 2\beta_5 + \beta_4 + \beta_6) = 0$, $(2\alpha_1 - \alpha_2 - \alpha_3 - 2\alpha_4 + \alpha_5 + \alpha_6) = 0$ and $(2\alpha_2 - \alpha_1 - \alpha_3 - 2\alpha_5 + \alpha_4 + \alpha_6) = 0$. The reader should verify that these interaction contraints are equivalent to the interaction contraints given by

$$
\begin{aligned}
[\beta_1 - (\beta_1 + \beta_2 + \beta_3)/3] - [\beta_4 - (\beta_4 + \beta_5 + \beta_6)/3] = 0 \\
[\beta_2 - (\beta_1 + \beta_2 + \beta_3)/3] - [\beta_5 - (\beta_4 + \beta_5 + \beta_6)/3] = 0 \\
[\alpha_1 - (\alpha_1 + \alpha_2 + \alpha_3)/3] - [\alpha_4 - (\alpha_4 + \alpha_5 + \alpha_6)/3] = 0 \\
[\alpha_2 - (\alpha_1 + \alpha_2 + \alpha_3)/3] - [\alpha_5 - (\alpha_4 + \alpha_5 + \alpha_6)/3] = 0.
\end{aligned}
$$

To test the hypothesis of no foreign effects for slopes and no interaction, one additional row must be added to the constraint matrix $\mathbf{T}$ given above. The additional row is given by

$$
\begin{bmatrix} 0 & 0 & 0 & 0 & 0 & 0 & 1 & 1 & 1 & -1 & -1 & -1 \end{bmatrix}.
$$

This restriction is equivalent to the condition that $[(\alpha_1 + \alpha_2 + \alpha_3)/3 - (\alpha_4 + \alpha_5 + \alpha_6)/3] = 0$. The estimated model under these conditions is given by

$$
\begin{aligned}
\mathrm{COMBRATE} = {} & 6.04\ \mathrm{D1} + 82.28\ \mathrm{D2} + 63.79\ \mathrm{D3} \\
& + 12.89\ \mathrm{D4} + 89.13\ \mathrm{D5} + 70.64\ \mathrm{D6} \\
& + 0.0255\ (\mathrm{DWEIGHT1} + \mathrm{DWEIGHT4}) \\
& + 0.0050\ (\mathrm{DWEIGHT2} + \mathrm{DWEIGHT5}) \\
& + 0.0136\ (\mathrm{DWEIGHT3} + \mathrm{DWEIGHT6}).
\end{aligned}
$$

The F-statistic for testing the null hypothesis of no interaction plus no foreign effect on the slopes has the value 3.016 with 5 degrees of freedom and a p-value of 0.015.

To test the hypothesis of no CYLIND effects on slopes, as well as no interaction, two rows must be added to the no interaction **T** matrix and are given by

$$\begin{bmatrix} 0 & 0 & 0 & 0 & 0 & 0 & 2 & -1 & -1 & 2 & -1 & -1 \\ 0 & 0 & 0 & 0 & 0 & 0 & -1 & 2 & -1 & -1 & 2 & -1 \end{bmatrix}.$$

The reader should verify that these two conditions are equivalent to the conditions

$$[(\alpha_1 + \alpha_4)/2 - (\alpha_1 + \alpha_2 + \alpha_3 + \alpha_4 + \alpha_5 + \alpha_6)/6] = 0 \quad \text{and}$$
$$[(\alpha_2 + \alpha_5)/2 - (\alpha_1 + \alpha_2 + \alpha_3 + \alpha_4 + \alpha_5 + \alpha_6)/6] = 0.$$

The estimated model under these conditions is given by

$$\begin{aligned} \text{COMBRATE} = {} & 35.34 \text{ D1} + 48.87 \text{ D2} + 58.40 \text{ D3} \\ & + 17.45 \text{ D4} + 30.98 \text{ D5} + 40.51 \text{ D6} \\ & + 0.0149 \text{ (DWEIGHT1} + \text{DWEIGHT2} + \text{DWEIGHT3)} \\ & + 0.0235 \text{ (DWEIGHT4} + \text{DWEIGHT5} + \text{DWEIGHT6)}. \end{aligned}$$

The F-statistic for testing this restriction is given by $F = 3.663$ with 6 degrees of freedom has a p-value of 0.003.

A variety of other tests for restricted models are possible. The above examples are intended only to illustrate the methodology but are not exhaustive.

Analysis of Variance

A special case of a multiple linear regression model occurs when all of the explanatory variables are indicator variables. In this case the model is equivalent to an *analysis of variance model*. The **X** matrix is simply a series of columns of zeroes, ones and in some cases minus ones. In this case the **X** matrix is called a design matrix. A detailed discussion of such models will be provided in Chapter 5.

4.6 Additional Topics in Linear Regression

The first five sections of this chapter have presented a comprehensive survey of procedures for the multiple linear regression model. The purpose of this chapter is to outline several additional, more general models which may also be viewed as linear relationships among a dependent variable Y and a set of explanatory variables $X_1, X_2, \ldots, X_p$.

The first topic of this section introduces several examples of *curvilinear models* which can be viewed as linear models of various functions of the explanatory variables. These models can be estimated using the ordinary least squares method used for the multiple regression model.

The second topic of this section presents special cases of the *general linear model*. The general linear model is less restrictive than the multiple linear regression model in that the error terms are no longer assumed to be mutually uncorrelated and identically distributed. Estimation techniques are more complex in that the covariance matrix of the error vector must also be estimated. These topics are usually discussed in textbooks on Econometrics.

4.6.1 Curvilinear Regression Models

In Chapter 3 a variety of curvilinear models that could be expressed as a linear relationship between known functions of Y and X were introduced. In this section we extend this concept to models which can be expressed as a multiple regression model involving functions of one or more explanatory variables.

Polynomial Models

One example of a *polynomial model* discussed in 3.3 is the nonlinear model $Y = \beta_0 + \beta_1 X^k + U$ where k is a specified numerical constant. This model can be viewed as a special case of the general polynomial model

$$Y = \beta_0 + \beta_1 X + \beta_2 X^2 + \cdots + \beta_p X^p + U,$$

which can also be viewed as a special case of the multiple regression model introduced in 4.1.

The most commonly used example of a polynomial model is the *quadratic model* given by

$$Y = \beta_0 + \beta_1 X + \beta_2 X^2 + U.$$

For many applications the quadratic model is sufficient to capture the nonlinearity that is present.

Example

For the data relating CPT to LOADS introduced in Section 3.2 a quadratic model

$$CPT = \beta_0 + \beta_1 \text{ LOADS} + \beta_2 \text{ LOADS} * \text{LOADS} + U$$

was estimated using ordinary least squares. The fitted model is given by

$$CPT = \underset{(.0001)}{43.770} - \underset{(.0016)}{0.021} \text{ LOADS} + \underset{(.0334)}{0.00000497} \text{ LOADS} * \text{LOADS}$$

with $R^2 = 0.353$. The R^2 value obtained here can be compared to the R^2 values of 0.308 and 0.399 for the fitted linear model and the fitted nonlinear model in Chapter 3.

The use of a polynomial model of higher order than the quadratic is generally discouraged in practice unless its application can be justified by the characteristics of the problem. Some reasons for this are:

(a) A polynomial function can have many changes in direction and hence is not useful for extrapolation. Also, applications seldom require more than two changes in direction.

(b) Polynomials of higher order usually produce multicollinearity problems. This multicollinearity can often be reduced by centering the explanatory variable.

(c) The $\mathbf{X}'\mathbf{X}$ matrix can be ill-conditioned in polynomial models and hence the inverse $(\mathbf{X}'\mathbf{X})^{-1}$ will be unreliable.

Grafted Quadratics

If a model is required to relate Y to X and two or more changes of direction are required an alternative approach is provided by fitting a *grafted quadratic*. A grafted quadratic model is obtained by fitting a series of quadratic models in such a way that the curve is smooth. In Figure 4.15 portions of two quadratics are shown with joint point at $X = X_0$. The two curves are joined so that they have the same value at $X = X_0$ and also have the same slope at this point.

If the joint point $X = X_0$ is known, the two quadratics can be fitted using multiple regression by defining a new variable Z as follows:

$$Z = (X - X_0)^2 \qquad \text{if } X \geq X_0$$
$$= 0 \qquad \text{otherwise.}$$

The model

$$Y = \beta_0 + \beta_1 X + \beta_2 X^2 + \beta_3 Z$$

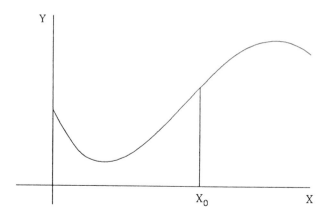

Figure 4.15. Grafted Quadratics with Joint Point X_0

describes the two quadratics. The quadratic to the left of X_0 is given by

$$Y = \beta_0 + \beta_1 X + \beta_2 X^2,$$

and the quadratic on the right is given by

$$Y = (\beta_0 + \beta_3 X_0^2) + (\beta_1 - 2\beta_3 X_0)X + (\beta_2 + \beta_3)X^2.$$

At the point $X = X_0$ both curves have the same Y value $(\beta_0 + \beta_1 X_0 + \beta_2 X_0^2)$ and the same slope $(\beta_1 + 2\beta_2 X_0)$.

Example

To provide an example of an application of a grafted quadratics polynomial fit, the relationship between net income, NETINC and net income rank, RANK, was studied for a sample of Canadian firms. The population of firms ranked in the top 500 was divided into two groups corresponding to whether they were at least 50% owned by Canadian shareholders. A sample of 100 firms was selected from each group. The relationship between LNNETINC = ln(NETINC) and RANK when plotted has the typical shape for a quantile plot of a bell shaped distribution. Such a plot can usually be fitted by two quadratics joined at the median (RANK = 50). The data used for this example is presented in Table 4.52. The fitted curves showing the relationship between LNNETINC and RANK for the two groups are shown in Figure 4.16.

The results of fitting separate grafted quadratics with joint point at RANK = 50 for the two groups are summarized in Table 4.53. The equation for the fitted model is given by

$$\text{LNNETINC} = \beta_0 + \beta_1 \text{ RANK} + \beta_2 \text{ RANK2} + \gamma \text{ XRANK2} + U,$$

Table 4.52. Comparison of NETINC Distributions for Samples of 100 Canadian Companies

At Least 50% Canadian Owned				Less Than 50% Canadian Owned			
RANK	NETINC	RANK	NETINC	RANK	NETINC	RANK	NETINC
1	775	51	15826	1	3781	51	17147
2	978	52	16101	2	3796	52	17155
3	1054	53	16560	3	4059	53	17321
4	1061	54	17703	4	4394	54	17613
5	1102	55	19927	5	4690	55	18766
6	1169	56	20324	6	4821	56	19500
7	1264	57	20761	7	4842	57	19600
8	1429	58	20762	8	4875	58	20000
9	1500	59	21693	9	5071	59	20242
10	1582	60	21749	10	5100	60	20870
11	1595	61	21764	11	5249	61	21185
12	1613	62	22330	12	5400	62	21916
13	1660	63	22576	13	5711	63	22075
14	1687	64	23551	14	5879	64	23381
15	1750	65	23600	15	5904	65	27420
16	2573	66	25118	16	6456	66	28060
17	2734	67	26000	17	6673	67	29421
18	2794	68	26049	18	7558	68	29565
19	2915	69	27227	19	7842	69	31579
20	3076	70	27283	20	8093	70	35123
21	3641	71	27810	21	8717	71	37265
22	4290	72	29064	22	8865	72	37955
23	4364	73	30268	23	9000	73	38587
24	4503	74	32978	24	9090	74	40171
25	4654	75	35670	25	9106	75	40232
26	4929	76	36405	26	9599	76	45602
27	5551	77	38300	27	9743	77	46118
28	5803	78	39076	28	10079	78	47800
29	5981	79	46054	29	10252	79	48764
30	6076	80	46147	30	10587	80	49300
31	6192	81	48569	31	10636	81	52000
32	6560	82	57700	32	10967	82	52688
33	7363	83	58375	33	11290	83	60604
34	7757	84	60300	34	11301	84	62950
35	8037	85	65240	35	12307	85	67599
36	8648	86	65371	36	12458	86	78278
37	8705	87	66222	37	12525	87	89883
38	9515	88	73100	38	12877	88	95092
39	10079	89	73700	39	12898	89	101463
40	14034	90	87845	40	13160	90	103650
41	11469	91	108000	41	14010	91	130000
42	11932	92	110100	42	14700	92	178300
43	12177	93	116944	43	14891	93	193300
44	13423	94	130500	44	14901	94	252100
45	13563	95	150112	45	15336	95	253878
46	13619	96	155074	46	15431	96	261745
47	14015	97	197764	47	15542	97	283000
48	14441	98	200695	48	16096	98	283200
49	14765	99	373200	49	16435	99	406631
50	14956	100	404000	50	16598	100	418435

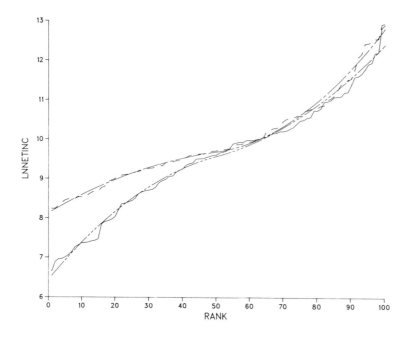

Figure 4.16. Relationship Between LNNETINC and RANK

where the variables RANK2 and XRANK2 are defined by

$$\text{RANK2} = (\text{RANK})^2$$

$$\text{XRANK2} = (\text{RANK} - 50)^2 \quad \text{if RANK} > 50$$
$$= 0 \qquad\qquad\quad \text{otherwise.}$$

All regression coefficients were significant at the 0.000 level.

The calculations were performed using SAS PROC REG.

From the information in Table 4.52 we can conclude that the equations for the two groups are as given below.

At least 50% Canadian owned

$$\text{LNNETINC} = 6.44453 + 0.10153 \, \text{RANK}$$

$$- 0.00079 \, \text{RANK2} + 0.00145 \, \text{XRANK2}.$$

Less than 50% Canadian owned

$$\text{LNNETINC} = 8.12764 + 0.04998 \, \text{RANK}$$

$$- 0.00038 \, \text{RANK2} + 0.00138 \, \text{XRANK2}.$$

Table 4.53. Results of Fitting Grafted Quadratics to LNNETINC vs NETINCR

	At least 50% held in Canada	Less than 50% held in Canada	Combined Sample
n	100	100	200
SST	214.744	134.152	358.466
SSR	212.965	133.328	355.862
SSE	1.779	0.824	2.604
R^2	0.9917	0.9939	0.9927
INTERCEPT	6.44453	8.12764	8.12764
RANK	0.10153	0.04998	0.04998
RANK2	−0.00079	−0.00038	−0.00038
XRANK2	0.00145	0.00138	0.00138
H			0.05155
HRANK			−0.00041
HRANK2			0.00075
HXRANK2			−1.68311

For comparison purposes a grafted quadratic was also fitted to the combined sample with dummy variables and slope shifter variables included. The results for this regression are also shown in Table 4.53. The dummy variable H was defined as $H = 1$ if at least 50% Canadian and $H = 0$ otherwise. The slope shifter variables were denoted by HRANK, HRANK2 and HXRANK2 respectively. All of the regression coefficients in Table 4.53 were significant at the 0.000 level except the coefficient of HXRANK2 which had a p-value of 0.523.

It is easy to extend the grafted quadratic technique to include additional joint points to fit more complex functions. In some applications it may be preferable to estimate the joint points. This topic will not be discussed here. A discussion of joint point estimation can be found in Gallant and Fuller (1973).

Grafted quadratics are a special case from the family of spline functions which are piecewise polynomials of order k. In such models the joint points are called knots. The model usually requries that the first $(k-1)$ derivatives be equal at each joint point. An overview of the methodology of *spline functions* is provided in Smith (1979).

Response Surface Models

Up to this point our discussion of polynomial models has involved only a single explanatory variable X. If there are two explanatory variables X_1 and X_2 a general second order model is given by

$$Y = \beta_0 + \beta_1 X_1 + \beta_2 X_2 + \beta_3 X_1^2 + \beta_4 X_2^2 + \beta_5 X_1 X_2 + U,$$

which is often referred to as a two-dimensional or quadratic *response surface*. In addition to the two squared terms an interaction term derived from the product of X_1 and X_2 is also included in the model. Response surface methodology is particularly useful in experimental design in industrial settings and is summarized extensively in Myers (1976), Box, Hunter and Hunter (1978) and Montgomery (1984).

Example

To provide an example for a response surface model the bank employee salary data in Table 4.54 was analyzed. The dependent variable is LSALARY = ln(SALARY) and the independent variables are AGE and education (EDUC). The ordinary least squares fit of the response surface was obtained using SAS PROC REG and is given by

$$\text{LSALARY} = \underset{(0.000)}{8.7712} - \underset{(0.045)}{0.2025} \text{ EDUC} + \underset{(0.002)}{0.0731} \text{ AGE}$$

$$+ \underset{(0.000)}{0.0133} \text{ EDUC} * \text{EDUC}$$

$$- \underset{(0.003)}{0.0007} \text{ AGE} * \text{AGE} - \underset{(0.085)}{0.0017} \text{ AGE} * \text{EDUC}$$

with $R^2 = 0.672$. The linear portion of the model yields an $R^2 = 0.479$. The addition of the two quadratic terms increases R^2 to 0.661. The crossproduct terms then bring R^2 up to 0.672. The fitted equation indicates that if AGE is held constant, a one unit increase in EDUC yields an increase in LSALARY of $-0.2025 + 0.0266$ EDUC $- 0.0017$ AGE, and hence the marginal impact of EDUC increases with the level of EDUC and decreases with the level of AGE. Similarly, a one unit increase in AGE holding EDUC constant yields an increase in LSALARY of

$$0.0731 - 0.0014 \text{ AGE} - 0.0017 \text{ EDUC},$$

and hence the marginal impact of AGE decreases with increases in EDUC.

Figure 4.17 shows the relationship between LSALARY and AGE at various levels of EDUC and Figure 4.18 shows the relationship between LSALARY and EDUC at various levels of AGE. The two sets of equations are summarized below.

EDUC	EQUATION
8	LSALARY = 7.4224 + .0595 AGE − .0007 AGE*AGE
11	LSALARY = 8.1530 + .0544 AGE − .0007 AGE*AGE
14	LSALARY = 8.5430 + .0493 AGE − .0007 AGE*AGE
17	LSALARY = 9.1724 + .0442 AGE − .0007 AGE*AGE
20	LSALARY = 10.0412 + .0391 AGE − .0007 AGE*AGE

Table 4.54. Bank Employee Salary Data

SALARY	AGE	EDUC	SALARY	AGE	EDUC
10980	37.50	12	27250	29.50	18
8040	26.25	12	13320	27.42	15
10020	26.08	8	10080	33.50	12
20400	32.50	16	7260	62.00	12
11640	55.58	17	9000	44.50	12
10140	28.42	15	10020	63.00	12
8280	59.50	15	8520	24.00	12
26000	28.83	19	11100	24.17	12
10740	32.00	15	6480	60.00	12
9000	24.83	12	11040	25.58	12
7860	55.92	8	7260	25.00	12
10620	30.67	15	15960	30.33	14
10080	51.92	8	8940	24.33	12
12300	56.00	12	14460	29.67	12
10500	51.42	8	11640	39.50	12
6540	60.50	12	11940	40.50	15
7860	25.25	12	9240	24.08	12
27250	30.08	19	16080	35.67	16
26500	38.00	19	14100	31.00	15
14400	37.17	15	9900	32.67	12
24250	33.75	16	12300	27.08	15
10440	24.00	12	41500	37.08	16
31250	31.67	19	11100	24.42	12
9720	25.08	12	13020	28.75	15
19500	32.67	16	22620	33.42	16
15720	33.50	15	9420	49.08	15
8340	23.75	12	16920	39.67	16
16920	31.75	12	8640	23.25	12
11580	43.42	15	11400	31.50	12
12060	29.75	15	9360	45.50	12
7860	24.50	12	8640	50.25	8
7860	44.58	12	22700	31.92	16
17200	28.42	16	12660	27.58	15
21250	30.67	16	10320	39.33	12
7980	24.83	12	16320	26.58	12
11100	33.50	15	14100	59.83	12
12660	32.50	15	13560	32.08	15
8760	52.00	12	29000	31.33	19
14220	27.67	15	17950	29.67	16
20220	33.33	15	8700	59.83	8
10380	32.67	15	9360	25.00	15
12660	25.58	15	14400	43.33	12
11700	30.83	8	14280	39.83	16
12300	60.67	8	10920	34.75	15
15360	28.33	15	33000	46.00	17
9420	32.08	12	10860	55.17	15
12780	30.50	15	10680	26.67	12
9180	23.33	12	11400	59.42	8
8700	53.08	12	9060	24.83	12
8280	23.67	12	23250	27.50	16

AGE		EQUATION
20	LSALARY =	9.9532 − .2365 EDUC + .0133 EDUC*EDUC
30	LSALARY =	10.3342 − .2535 EDUC + .0133 EDUC*EDUC
40	LSALARY =	10.5752 − .2705 EDUC + .0133 EDUC*EDUC
50	LSALARY =	10.6762 − .2875 EDUC + .0133 EDUC*EDUC
60	LSALARY =	10.6372 − .3045 EDUC + .0133 EDUC*EDUC

In Figure 4.17 we can see quite clearly that the relationship between LSALARY and AGE depends on the level of EDUC. For the five levels of EDUC the curves are almost parallel and are well separated. Thus we can conclude that the EDUC level places a person on a certain grid level and that at a given point in time the LSALARY decreases with AGE. (Could this be the impact of inflation on starting salaries?) In contrast to Figure 4.17, Figure 4.18 contains curves which are not well separated. The curves generally have the same shape with LSALARY decreasing as a function of EDUC until EDUC = 11 at all AGE levels and then increasing after EDUC = 11. The greatest response to changes in EDUC comes for the youngest AGE category and the least response comes from the highest AGE category. Because the curves cross we cannot say that at every level of EDUC the order of the AGE relationships is maintained.

4.6.2 Generalized Least Squares

To this point throughout Chapter 4 the unobserved error vector $\mathbf{u}$ in the multiple regression model was assumed to have a distribution with mean vector $\mathbf{0}$ and covariance matrix $E[\mathbf{uu'}] = \sigma_u^2 \mathbf{I}$. This linear model is a special case of a more general class of linear models where the covariance matrix is assumed to have the form $E[\mathbf{uu'}] = \sigma_u^2 \mathbf{\Omega}$. In this general class of linear models the ordinary least squares estimator derived in 4.1 is no longer minimum variance unbiased. An estimator called the *generalized least squares estimator* is now the minimum variance unbiased estimator. This section will introduce the general form for this estimator. Various special cases of this estimator will be applied in the case of *heteroscedasticity, error components models* and *autocorrelation*.

The Generalized Least Squares Estimator

For the linear model $\mathbf{y} = \mathbf{X}\boldsymbol{\beta} + \mathbf{u}$ under the assumption that $E[\mathbf{uu'}] = \sigma_u^2 \mathbf{\Omega}$ the ordinary least squares estimator given by $\mathbf{b} = (\mathbf{X'X})^{-1}\mathbf{X'y}$ has the covariance matrix

$$\sigma_u^2 (\mathbf{X'X})^{-1}(\mathbf{X'\Omega X})(\mathbf{X'X})^{-1}.$$

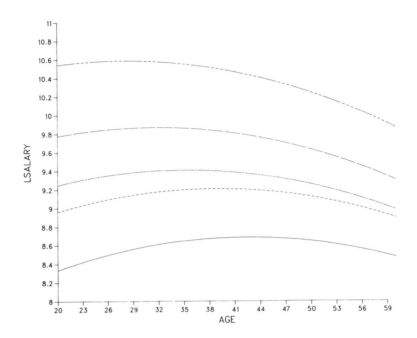

Figure 4.17. Relationship Between LSALARY and AGE Controlling For EDUC

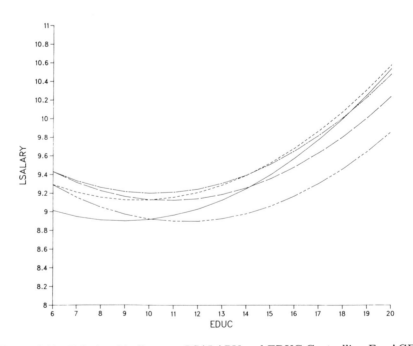

Figure 4.18. Relationship Between LSALARY and EDUC Controlling For AGE

Although $\mathbf{b}$ is still an unbiased estimator it is no longer the minimum variance unbiased estimator. The most efficient or best linear unbiased estimator is given by the *generalized least squares estimator*

$$\tilde{\mathbf{b}} = (\mathbf{X}'\mathbf{\Omega}^{-1}\mathbf{X})^{-1}(\mathbf{X}'\mathbf{\Omega}^{-1}\mathbf{y}),$$

where $\mathbf{\Omega}^{-1}$ is assumed to exist and to be known.

If $\mathbf{\Omega}^{-1}$ is known it is possible to write $\tilde{\mathbf{b}}$ in the form

$$\tilde{\mathbf{b}} = (\mathbf{X}^{*\prime}\mathbf{X}^{*})^{-1}\mathbf{X}^{*\prime}\mathbf{y}^{*}$$

where $\mathbf{X}^{*} = \mathbf{PX}$, $\mathbf{y}^{*} = \mathbf{Py}$ and $\mathbf{P}'\mathbf{P} = \mathbf{\Omega}^{-1}$. The generalized least squares estimator is therefore equivalent to an ordinary least squares estimator for the transformed model

$$\mathbf{y}^{*} = \mathbf{X}^{*}\boldsymbol{\beta} + \mathbf{u}^{*}$$

where $\mathbf{u}^{*} = \mathbf{Pu}$.

The covariance matrix for $\tilde{\mathbf{b}}$ is given by $\sigma_u^2(\mathbf{X}^{*\prime}\mathbf{X}^{*})^{-1} = \sigma_u^2(\mathbf{X}'\mathbf{\Omega}^{-1}\mathbf{X})^{-1}$, and is estimated by the unbiased estimator $s^2(\mathbf{X}'\mathbf{\Omega}^{-1}\mathbf{X})^{-1}$ where

$$s^2 = (\mathbf{y} - \mathbf{X}\tilde{\mathbf{b}})'\mathbf{\Omega}^{-1}(\mathbf{y} - \mathbf{X}\tilde{\mathbf{b}})/(n - p - 1).$$

If the ordinary least squares estimator, $\mathbf{b}$, is incorrectly used when $\mathbf{\Omega} \neq \mathbf{I}$, the estimated covariance matrix for $\mathbf{b}$ is biased, and hence inferences regarding $\boldsymbol{\beta}$ are inconsistent.

Unknown Covariance Matrix

In practice the covariance matrix for the error vector $\mathbf{u}$ is usually unknown. In general without further information on the structure of $\mathbf{\Omega}$, the matrix $\mathbf{\Omega}^{-1}$ required for the generalized least squares estimator cannot be estimated from the sample data. If the ordinary least squares estimator $\mathbf{b}$ is used, the matrix

$$\mathbf{W} = (\mathbf{y} - \mathbf{Xb})(\mathbf{y} - \mathbf{Xb})'$$

has rank 1, and hence $\mathbf{W}^{-1}$ does not exist. In addition, the number of unknown parameters in the matrix $\mathbf{\Omega}$ is $n(n+1)/2$ which exceeds the number of observations, and hence there is an insufficient number of degrees freedom. Since there are in general $(p + 1)$ parameters in the unknown vector $\boldsymbol{\beta}$, there are at most only $(n - p - 1)$ degrees of freedom available for estimating parameters in $\mathbf{\Omega}$.

Feasible Generalized Least Squares

If a consistent estimator $\mathbf{W}$ of $\boldsymbol{\Omega}$ does exist then the estimator

$$\tilde{\mathbf{b}} = (\mathbf{X}'\mathbf{W}^{-1}\mathbf{X})^{-1}\mathbf{X}'\mathbf{W}^{-1}\mathbf{y} \qquad (4.17)$$

is called a *feasible generalized least squares estimator* of $\boldsymbol{\beta}$. The estimator, $\tilde{\mathbf{b}}$, may be viewed as an ordinary least squares estimator for the linear model relating the transformed variables $\hat{\mathbf{y}}^* = \widehat{\mathbf{P}}'\mathbf{y}$ to $\widehat{\mathbf{X}}^* = \widehat{\mathbf{P}}'\mathbf{X}$, and $\tilde{\mathbf{b}} = (\widehat{\mathbf{X}}^{*\prime}\widehat{\mathbf{X}}^*)^{-1}\widehat{\mathbf{X}}^{*\prime}\hat{\mathbf{y}}^*$, where $\widehat{\mathbf{P}}\widehat{\mathbf{P}}' = \mathbf{W}^{-1}$. In large samples, with the normality assumption and very general conditions, the inference techniques generally used for the multiple regression model can be applied for the transformed model $\hat{\mathbf{y}}^* = \widehat{\mathbf{X}}^*\boldsymbol{\beta} + \hat{\mathbf{u}}^*$ where $\hat{\mathbf{u}}^* = \widehat{\mathbf{P}}'\mathbf{u}$, $\hat{\mathbf{y}}^* = \widehat{\mathbf{P}}'\mathbf{y}$ and $\widehat{\mathbf{X}}^* = \widehat{\mathbf{P}}'\mathbf{X}$.

Heteroscedasticity

The simplest example of a covariance matrix, $\sigma_u^2\boldsymbol{\Omega}$, which is not $\sigma_u^2\mathbf{I}$ is the diagonal matrix $\sigma_u^2\mathbf{D}$ where $\mathbf{D}$ has at least two distinct diagonal elements. If all n diagonal elements are mutually distinct from one another, then the number of unknown parameters in $\boldsymbol{\Omega}$ is n which exceeds the available degrees of freedom $(n-p-1)$. The diagonal matrix must therefore contain sufficient structure that will eliminate some of the unknown parameters. Usually it is possible to group the data into a small number of subsets each containing a distinct value of σ_u^2. Alternatively the unknown variances can be expressed as a simple function of a small number of unknown parameters. Examples will be outlined below.

In Section 3.3 in the case of a simple linear regression model, two simple models for heteroscedasticity were introduced. These models were special cases of the more general model

$$\sigma_u^2 = \alpha_0 X^k$$

where k is a specified numerical constant. In this case the covariance matrix of the simple linear regression model, $\sigma_u^2\boldsymbol{\Omega}$, has the diagonal elements $\alpha_0 X^k$ and off diagonal elements zero. The generalized least squares estimator is given by

$$\mathbf{b} = (\mathbf{X}'\boldsymbol{\Omega}^{-1}\mathbf{X})^{-1}(\mathbf{X}'\boldsymbol{\Omega}^{-1}\mathbf{y}) = (\mathbf{X}^{*\prime}\mathbf{X}^*)^{-1}\mathbf{X}^{*\prime}\mathbf{y}^*,$$

where $\mathbf{X}^*$ and $\mathbf{y}^*$ denote the transformed values of $\mathbf{X}$ and $\mathbf{y}$ given by $\mathbf{X}^* = \mathbf{D}\mathbf{X}$ and $\mathbf{y}^* = \mathbf{D}\mathbf{y}$, and where $\mathbf{D}$ is the diagonal matrix with diagonal elements $X^{-k/2}$. In this case it is not necessary to estimate $\boldsymbol{\Omega}$ since it has been specified completely.

Models for Heteroscedasticity

A more general model of heteroscedasticity for a multiple regression model assumes that the diagonal elements of $\boldsymbol{\Omega}$ are given by

$$\alpha_0 + \sum_{j=1}^{p} \alpha_j X_j^2,$$

where the parameter vector $\boldsymbol{\alpha}' = [\alpha_0, \alpha_1, \ldots, \alpha_p]$ must be estimated from the data. A simple approach to obtaining a consistent estimator for $\boldsymbol{\Omega}$ is to use the squared residuals, e_i^2, $i = 1, 2, \ldots, n$ obtained from a least squares fit. The squared residuals are regressed on the explanatory variables $X_1^2, X_2^2, \ldots, X_p^2$ to obtain an estimator, $\mathbf{a}$, of $\boldsymbol{\alpha}$. The estimator $\mathbf{W}$ is then constructed using $\mathbf{a}$. The feasible generalized least squares estimator (4.17) is then determined. Although the estimator $\mathbf{a}$ is consistent it is not the most efficient estimator of $\boldsymbol{\alpha}$.

The above example of heteroscedasticity can be derived from a *random coefficient model* where the regression coefficients β_j, $j = 0, 1, 2, \ldots, p$ are assumed to be mutually independent random variables with mean $\bar{\beta}_j$ and variances α_j. The linear model expressed in terms of $\bar{\beta}_j$ is given by

$$Y = \bar{\beta}_0 + \bar{\beta}_1 X_1 + \cdots + \bar{\beta}_p X_p + U \qquad \text{where} \quad U = \sum_{j=1}^{p} (\beta_j - \bar{\beta}_j) X_j$$

and $V(U) = \alpha_0 + \alpha_1 X_1^2 + \cdots + \alpha_p X_p^2$.

A more general model of heteroscedasticity assumes that the diagonal elements are functions of the $\boldsymbol{\beta}$ parameters as well as the $\mathbf{X}$ matrix and additional parameters $\boldsymbol{\alpha}$. A discussion of estimation techniques for this class of models is outlined in Jobson and Fuller (1980).

In some applications it may be possible to divide the observations into subgroups so that the error variance can be assumed constant within the group. Assuming g groups we have g linear models

$$Y_\ell = \beta_0 + \beta_1 X_{1\ell} + \cdots + \beta_p X_{p\ell} + U_\ell, \qquad \ell = 1, 2, \ldots, g;$$

where $E[U_\ell^2] = \sigma_\ell^2$. Assume a random sample of observations $n = \sum_{j=1}^{g} n_\ell$, where $n_\ell = $ no. of observations from group ℓ. The feasible generalized least squares estimator is given by

$$\tilde{\mathbf{b}} = (\mathbf{X}'\mathbf{W}^{-1}\mathbf{X})^{-1}\mathbf{X}'\mathbf{W}^{-1}\mathbf{y} = (\mathbf{X}^{*\prime}\mathbf{X}^*)^{-1}\mathbf{X}^{*\prime}\mathbf{y}^*,$$

where $\mathbf{X}^* = \mathbf{D}\mathbf{X}$, $\mathbf{y}^* = \mathbf{D}\mathbf{y}$ and $\mathbf{D}$ is the diagonal matrix with diagonal elements $1/s_\ell$. Estimators s_ℓ^2 for the variances σ_ℓ^2 are obtained from the ordinary least squares residuals $e_{i\ell}$ with

$$s_\ell^2 = \sum_{i=1}^{n_\ell} e_{i\ell}^2 / (n_\ell - p - 1). \tag{4.18}$$

Tests For Heteroscedasticity

Statistical tests for the presence of heteroscedasticity can be divided into two groups, specific and nonspecific. In the specific case, the nature of the heteroscedasticity is specified and then estimated under this specification, as in the examples discussed above. In comparison, nonspecific tests of heteroscedasticity are designed to detect a general lack of homogeneity of variance.

For models where heteroscedasticity is assumed to be characterized by a model involving the X variables and unknown parameters, the tests can usually be carried out at the same time as the estimation of parameters. If a linear regression model is used to estimate the covariance matrix parameters, conventional techniques for making inferences about regression coefficients can usually be applied. The problem with such tests is that the error terms in the linear regression model also have heteroscedasticity, and hence the error variances for the parameters will be inconsistent. A summary of some more efficient procedures is available in Fomby, Hill and Johnson (1984).

Under a broad general assumption that

$$\sigma_u^2 = f(\boldsymbol{\alpha}'\mathbf{X}).$$

Breusch and Pagan (1979) have developed a test for heteroscedasticity based on the test statistic $V = 1/2 \sum_{i=1}^{n} (\hat{v}_i - \bar{v})^2$ where $v_i = ne_i^2/(n-p-1)s^2$. $\hat{v}_i$ denotes the ordinary least squares estimator of v_i in the model

$$v_i = \boldsymbol{\alpha}'\mathbf{x}_i + \varepsilon_i$$

and e_i is the corresponding residual. The term s^2 is given by $s^2 = \sum_{i=1}^{n} e_i^2/(n-p-1)$. Under normality in the linear model and under the homoscedasticity assumption, the statistic V has a χ^2 distribution with p degrees of freedom where it is assumed that the first column of $\mathbf{X}$ is a column of unities.

For the linear model in which the variances are assumed to be homogeneous across subsets a test of equal variances can be carried out using the statistic

$$2n \ \ln s - 2 \sum_{\ell=1}^{q} n_\ell \ \ln s_\ell$$

which has a χ^2 distribution with g degrees of freedom if the variances are homogeneous. The estimators s_ℓ^2 of the variances σ_ℓ^2 can be obtained from the ordinary least squares estimators as in (4.18). The estimator s^2 is the conventional estimator of σ_u^2 under the assumption of homoscedasticity. This test will be more powerful if the feasible generalized least squares estimator $\tilde{\mathbf{b}}$ of $\boldsymbol{\beta}$ is determined and the variance estimators determined from the residuals from this second model fit.

If the variances are assumed to be monotonically related to some other variable Z a test developed by Goldfield and Quandt (1965) can be used to test for heteroscedasticity. The observations are ranked according to Z and divided into 3 groups representing the $(n-k)/2$ smallest Z values, the k middle Z values and the $(n-k)/2$ largest Z values. Separate ordinary least squares estimation is carried out for the lowest and highest groups yielding two error sums of squares SSE_L and SSE_H. Under the assumption of homoscedasticity the statistic $\mathrm{SSE}_H/\mathrm{SSE}_L$ has an approximate F distribution with $(n-k-2p)/2$ degrees of freedom for both the numerator and denominator.

Alternative tests for heteroscedasticity in the case of an assumed monotonic relationship can be developed using a measure of correlation between the variable Z and the values $|e_i|$ or e_i^2. The Spearman rank correlation coefficient or Kendall's coefficient of concordance can be used to test for heteroscedasticity.

Although the consistent estimation of $\boldsymbol{\Omega}$ requires information about the structure of $\boldsymbol{\Omega}$, it is possible to estimate the matrix $\mathbf{X}'\boldsymbol{\Omega}\mathbf{X}$ and hence to estimate the covariance matrix of $\hat{\boldsymbol{\beta}}$ when $\boldsymbol{\Omega}$ is a diagonal matrix. A consistent estimator of $\mathbf{X}'\boldsymbol{\Omega}\mathbf{X}$ is provided by $\mathbf{X}'\mathbf{W}\mathbf{X}$ where $\mathbf{W}=\mathbf{dd}'$ and $\mathbf{d}$ is the $n \times 1$ vector of squared ordinary least squares residuals $(e_1^2, e_2^2, \ldots, e_n^2)$. A consistent estimator of the covariance matrix of $\mathbf{b}$ is given by $(\mathbf{X}'\mathbf{X})^{-1}(\mathbf{X}'\mathbf{W}\mathbf{X})(\mathbf{X}'\mathbf{X})^{-1}$. Inferences regarding $\boldsymbol{\beta}$ can be made using this estimator of the covariance matrix of $\mathbf{b}$.

A test for heteroscedasticity can be based on the above calculations. The statistic $\mathbf{f}'\mathbf{B}^{-1}\mathbf{f}$ in large samples has a χ^2 distribution with $p(p+1)/2$ degrees of freedom under homoscedasticity. The $(0.5p(p+1) \times 1)$ vector $\mathbf{f}$ consists of the differences between $\mathbf{X}'\mathbf{W}\mathbf{X}$ and $s^2(\mathbf{X}'\mathbf{X})$ excluding the upper triangle above the diagonal. The matrix $\mathbf{B}$ denotes the matrix

$$\sum_{i=1}^{n}(e_i^2 - s^2)^2(\mathbf{Q}_i - \mathbf{Q})(\mathbf{Q}_i - \mathbf{Q})'$$

where $\mathbf{Q}$ denotes the $(0.5p(p+1) \times 1)$ vector of elements of $\mathbf{X}'\mathbf{X}/n$ excluding the upper triangle above the diagonal and $\mathbf{Q}_i$ denotes the $(0.5p(p+1) \times 1)$ vector of elements of $\mathbf{X}_i\mathbf{X}_i'$ excluding the upper triangle above the main diagonal.

Error Components Models

A useful class of models which have a non-diagonal covariance matrix are the *error components models*. These models are usually derived from data which combines cross section and time series observations such as n individuals observed in T time periods. The linear model for individual i in time period t is given by

$$y_{it} = \sum_{j=1}^{p+1} \beta_j x_{itj} + u_{it} \qquad i = 1, 2, \ldots, n; \quad t = 1, 2, \ldots, T;$$

where $u_{it} = \mu_i + \lambda_t + v_{it}$

and where the variables μ_i, λ_t and v_{it} are mutually independent and are each independently and identically distributed with zero mean and variances σ_μ^2, σ_λ^2, σ_v^2.

For the i-th individual the model is given by

$$\mathbf{y}_i = \mathbf{X}_i \boldsymbol{\beta} + \mu_i \mathbf{E}_i + \mathbf{I}_T \boldsymbol{\lambda} + \mathbf{V}_i$$

where $\mathbf{y}_i(T \times 1)$ and $\mathbf{X}_i(T \times (p+1))$ are observations, $\mathbf{E}_i(T \times 1)$ is a vector of unities, $\mathbf{V}_i(T \times 1)$ a vector of unobserved random distributions, $\boldsymbol{\beta}((p+1) \times 1)$ a vector of unknown parameters, $\boldsymbol{\lambda}(T \times 1)$ a vector of unobserved time shocks, and $\mathbf{I}_T(T \times T)$ an identity matrix.

The $T \times T$ covariance matrix for the vector $\mathbf{y}_i$ is given by

$$\boldsymbol{\Omega}_{ii} = \begin{bmatrix} \sigma_\mu^2 & \sigma_\mu^2 & \cdots & \sigma_\mu^2 \\ \sigma_\mu^2 & & & \vdots \\ \vdots & \vdots & & \vdots \\ \sigma_\mu^2 & \sigma_\mu^2 & \cdots & \sigma_\mu^2 \end{bmatrix} + \begin{bmatrix} \sigma_\lambda^2 + \sigma_v^2 & 0 & \cdots & 0 \\ 0 & \sigma_\lambda^2 + \sigma_v^2 & & \\ \vdots & & \ddots & \\ 0 & & \cdots & 0 \quad \sigma_\lambda^2 + \sigma_v^2 \end{bmatrix}$$

$$= \sigma_\mu^2 \mathbf{J}_T + \sigma_\lambda^2 \mathbf{I}_T + \sigma_v^2 \mathbf{I}_T$$

where $\mathbf{J}_T$ is a $T \times T$ matrix of unities, $\mathbf{J}_T = \mathbf{E}_i \mathbf{E}_i'$. The $T \times T$ matrix $\boldsymbol{\Omega}_{ij}$ which summarizes the covariances between the vectors $\mathbf{y}_i$ and $\mathbf{y}_j$ is given by

$$\boldsymbol{\Omega}_{ij} = \begin{bmatrix} \sigma_\lambda^2 & 0 & \cdots & 0 \\ 0 & \sigma_\lambda^2 & & \vdots \\ \vdots & & \ddots & 0 \\ 0 & \cdots & 0 & \sigma_\lambda^2 \end{bmatrix} = \sigma_\lambda^2 \mathbf{I}_T.$$

The complete $Tn \times Tn$ covariance matrix is given by

$$\Omega = \begin{bmatrix} \Omega_{11} & \Omega_{12} & \cdots & \Omega_{1n} \\ \Omega_{12} & \cdots & \cdots & \vdots \\ \vdots & & & \vdots \\ \Omega_{1n} & & & \Omega_{nn} \end{bmatrix}.$$

The complete model in matrix notation is given by

$$\mathbf{y} = \mathbf{X}\boldsymbol{\beta} + \mathbf{E}\boldsymbol{\mu} + \mathbf{F}\boldsymbol{\lambda} + \mathbf{V}$$

where $\mathbf{y}_{Tn \times 1}$ and $\mathbf{X}_{Tn \times (p+1)}$ summarize the observations, $\boldsymbol{\mu}(N \times 1)$, $\boldsymbol{\lambda}(T \times 1)$ and $\mathbf{V}(Tn \times 1)$ are vectors of unobserved random variables and

$$\mathbf{E}_{(Tn \times n)} = \begin{bmatrix} \mathbf{E}_1 & 0 & \cdots & 0 \\ 0 & \mathbf{E}_2 & & \vdots \\ \vdots & & & \vdots \\ 0 & 0 & \cdots 0 & \mathbf{E}_n \end{bmatrix}$$

$$\mathbf{F}_{(Tn \times T)} = \begin{bmatrix} \mathbf{I}_T \\ \mathbf{I}_T \\ \vdots \\ \mathbf{I}_T \end{bmatrix}.$$

The covariance matrix Ω is given by

$$\Omega = \sigma_\mu^2 \mathbf{A} + \sigma_\lambda^2 \mathbf{B} + \sigma_v^2 \mathbf{I}$$

where $\mathbf{A} = \mathbf{E}\mathbf{E}'$, $\mathbf{B} = \mathbf{F}\mathbf{F}'$ and $\mathbf{I}$ is a $(Tn \times Tn)$ identity.

The generalized least squares estimator of $\boldsymbol{\beta}$ is therefore given by

$$\tilde{\mathbf{b}} = (\mathbf{X}'\Omega^{-1}\mathbf{X})^{-1}(\mathbf{X}'\Omega^{-1}\mathbf{y}).$$

If σ_μ^2, σ_λ^2 and σ_v^2 are known the Fuller and Battese (1974) transformation matrix $\mathbf{P}$ can be used to obtain the estimator $\tilde{\mathbf{b}}$. The estimator is given by

$$\tilde{\mathbf{b}} = (\mathbf{X}^{*\prime}\mathbf{X}^*)^{-1}\mathbf{X}^{*\prime}\mathbf{y}^*$$

where $\quad \mathbf{X}^* = \mathbf{P}\mathbf{X} \quad$ and $\quad \mathbf{y}^* = \mathbf{P}\mathbf{y}.$

The elements of $\mathbf{y}^*$ and $\mathbf{X}^*$ are given by

$$y_{it}^* = y_{it} - \alpha_1 \bar{y}_{i\cdot} - \alpha_2 \bar{y}_{\cdot t} + \alpha_3 \bar{y}_{\cdot\cdot}$$

$$x_{kit}^* = x_{kit} - \alpha_1 \bar{x}_{ki\cdot} - \alpha_2 \bar{x}_{k\cdot t} + \alpha_3 \bar{x}_{k\cdot\cdot}.$$

where $\alpha_1 = 1 - \sigma_e/(\sigma_e^2 + T\sigma_\mu^2)^{1/2}$, $\alpha_2 = 1 - \sigma_e/(\sigma_e^2 + n\sigma_\lambda^2)^{1/2}$ and $\alpha_3 = (\alpha_1 + \alpha_2 - 1) + \sigma_e/(\sigma_e^2 + T\sigma_\mu^2 + n\sigma_\lambda^2)^{1/2}$. The dot notation used here is common to analysis of variance and is defined by

$$\bar{y}_{i\cdot} = \sum_{t=1}^{T} y_{it}/T, \qquad \bar{y}_{\cdot t} = \sum_{i=1}^{n} y_{it}/n, \qquad \bar{y}_{\cdot\cdot} = \sum_{i=1}^{n}\sum_{t=1}^{T} y_{it}/nT,$$

$$\bar{x}_{ki\cdot} = \sum_{t=1}^{T} x_{kit}/T, \qquad \bar{x}_{k\cdot t} = \sum_{i=1}^{n} x_{kit}/n, \qquad \bar{x}_{k\cdot\cdot} = \sum_{i=1}^{n}\sum_{t=1}^{T} x_{kit}/nT.$$

Since the variances σ_μ^2, σ_λ^2 and σ_v^2 are usually unknown they must also be estimated from the data. Using consistent estimators of these parameters a feasible generalized least squares estimator of $\boldsymbol{\beta}$ can be obtained. Consistent estimators can be obtained from the residuals from three distinct regressions. The residual sum of squares $\mathbf{v}'\mathbf{v}$ is obtained from the residual vector $\mathbf{v}$ from the regression of y deviations $(y_{ij} - \bar{y}_{i\cdot} - \bar{y}_{\cdot j} + \bar{y}_{\cdot\cdot})$ on the x deviations $(x_{ijk} - \bar{x}_{i\cdot k} - \bar{x}_{\cdot jk} + \bar{x}_{\cdot\cdot k})$. The residual sum of squares $\mathbf{g}'\mathbf{g}$ is determined from the residual vector $\mathbf{g}$ obtained from the residuals from the regression of $(y_{ij} - \bar{y}_{i\cdot})$ on $(x_{ijk} - \bar{x}_{\cdot jk})$. The residual sum of squares $\mathbf{m}'\mathbf{m}$ is determined from the residual vector $\mathbf{m}$ obtained from the residuals from the regression of $(y_{ij} - \bar{y}_{\cdot j})$ on $(x_{ijk} - \bar{x}_{\cdot jk})$. The formulae for s_v^2, s_λ^2 and s_μ^2 the estimators of σ_v^2, σ_λ^2 and σ_μ^2 respectively are complex and will not be presented here. The expressions can be obtained from Fuller and Battese (1974).

Autocorrelation

For the simple linear regression model the first order autocorrelation assumption for the error term was studied in Section 3.3.8. The generalization to the multiple regression model is given by

$$y_t = \beta_0 + \sum_{j=1}^{p} x_{tj}\beta_j + \mu_t \qquad t = 1, 2, \ldots, T$$

where $\mu_t = \rho\mu_{t-1} + \varepsilon_t$ and

$$E[\varepsilon_t] = 0, \quad E[\varepsilon_t^2] = \sigma_\varepsilon^2 \quad \text{and} \quad E[\varepsilon_t\varepsilon_s] = 0 \quad t \neq s, \quad s, t = 1, 2, \ldots, T.$$

The covariance matrix for this linear model is given by

$$\boldsymbol{\Omega} = \frac{\sigma_\varepsilon^2}{(1 - \rho^2)} \begin{bmatrix} 1 & \rho & \rho^2 & \cdots & \rho^{n-1} \\ \rho & 1 & \rho & \cdots & \rho^{n-2} \\ \rho^2 & \rho & 1 & \cdots & \vdots \\ \vdots & & & & \rho \\ \rho^{n-1} & \rho^{n-2} & \cdots & \rho & 1 \end{bmatrix}$$

and hence the generalized least squares estimator is given by

$$\hat{\mathbf{b}} = (\mathbf{X}'\mathbf{\Omega}^{-1}\mathbf{X})^{-1}(\mathbf{X}'\mathbf{\Omega}^{-1}\mathbf{y})$$

where

$$\mathbf{\Omega}^{-1} = \begin{bmatrix} 1 & -\rho & 0 & & \cdots & 0 \\ -\rho & 1+\rho^2 & -\rho & 0 & \cdots & 0 \\ 0 & -\rho & 1+\rho^2 & -\rho & 0 & \cdots & \vdots \\ \vdots & 0 & & & & \vdots \\ \vdots & \vdots & & & 1+\rho^2 & -\rho \\ 0 & 0 & \cdots & \cdots & 0 & -\rho & 1 \end{bmatrix}.$$

In this particular case the matrix $\mathbf{P}$ such that $\mathbf{P}'\mathbf{P} = \mathbf{\Omega}^{-1}$ can be written explicitly as

$$\mathbf{P} = \begin{bmatrix} \sqrt{1-\rho^2} & 0 & 0 & & \cdots & 0 \\ -\rho & 1 & 0 & & \cdots & 0 \\ 0 & -\rho & 1 & 0 & \cdots & 0 \\ \vdots & & & \ddots & & \vdots \\ 0 & \cdots & 0 & -\rho & 1 & 0 \\ 0 & \cdots & & 0 & -\rho & 1 \end{bmatrix}.$$

If ρ is known the transformed model can be determined and is given by

$$\mathbf{Py} = \mathbf{PX}\beta + \mathbf{Pu}.$$

The ordinary least squares estimator based on the transformed variables $\mathbf{y}^* = \mathbf{Py}$ and $\mathbf{X}^* = \mathbf{PX}$ is the Prais–Winsten estimator introduced in 3.3.8 generalized to more than one explanatory variable.

The techniques introduced in 3.3.8 are easily extended to the case of p explanatory variables and will not be discussed here. The Durbin–Watson statistic used to test for first order autocorrelation applies here as well except that the critical values depend on the number of explanatory variables.

The reader who is interested in a more extensive discussion of generalized least squares should consult an econometrics text such as Judge, Griffiths, Hill, Lütkepohl and Lee (1985), or Fomby, Hill and Johnson (1984).

Cited Literature and Additional References for Chapter 4

1. Andrews, D.F. and Pregibon, D. (1978). "Finding the Outliers that Matter." *Journal of the Royal Statistical Society, Series B* 40, 85–93.

2. Atkinson, A.C. (1985). *Plots, Transformations, and Regression.* Oxford: Clarendon Press.

3. Belsey, D.A. Kuh, E. and Welsch, R.E. (1980). *Regression Diagnostics: Identifying Influential Data and Sources of Collinearity.* New York: John Wiley and Sons, Inc.

4. Belsey, D.A. (1984). "Demeaning Conditioning Diagnostics Through Centering (with Discussion)." *The American Statistician* 38, 73–77.

5. Berk, K.N. (1977). "Tolerance and Condition in Regression Computations." *Journal of the American Statistical Association* 72, 46–53.

6. Box, G.E.P., Hunter, W.G. and Hunter, J.S. (1978). *Statistics For Experimenters: An Introduction to Design, Data Analysis and Model Building.* New York: John Wiley and Sons, Inc.

7. Breusch, T. and Pagan, A. (1979). "A Simple Test for Heteroscedasticity and Random Coefficient Variation." *Econometrica* 47, 1287–1294.

8. Chatterjee, S. and Hadi, A.S. (1988). *Sensitivity Analysis in Linear Regression.* New York: John Wiley and Sons, Inc.

9. Cook, R.D. and Weisberg, S. (1982). *Residuals and Influence in Regression.* New York: Chapman and Hall.

10. Draper, N.R. and John, J.A. (1981). "Influential Observations and Outliers in Regression." *Technometrics* 23, 21–26.

11. Fomby, T.B., Hill, R.C. and Johnson, S.R. (1984). *Advanced Econometric Methods.* New York: Springer–Verlag.

12. Freedman, D.A. (1983). "A Note on Screening Regression Equations." *American Statistician* 37, 152–155.

13. Fuller, W.A. and Battese, G.E. (1974). "Estimation of Linear Models with Crossed Error Structure." *Journal of Econometrics* 2, 67–78.

14. Furnival, G.M. and Wilson, R.B. (1974). "Regressions by Leaps and Bounds." *Technometrics* 16, 499–511.

15. Gallant, A.R. and Fuller, W.A. (1973). "Fitting Segmented Polynomial Regression Models Whose Join Points have to be Estimated." *Journal of the American Statistical Association* 68, 144–147.

16. Gibbons, D.I., McDonald, G.C. and Gunst, R.F. (1987). "The Complementary Use of Regression Diagnostics and Robust Estimators." *Naval Research Logistics* 34, 109–131.

17. Goldfeld, S.M. and Quandt, R.E. (1965). "Some Tests for Homoscedasticity." *Journal of the American Statistical Association* 60, 539–547.

18. Gray, J.B. and Ling, R.F. (1984). "K-Clustering as a Detection Tool for Influential Subsets in Regression." *Technometrics* 26, 305–318.

19. Graybill, F.A. (1976). *Theory and Application of the Linear Model*. Boston: Duxbury Press.

20. Gunst, R.F. and Mason, R.L. (1980). *Regression Analysis and its Applications*. New York: Marcel Dekker.

21. Hocking, R.R. (1976). "The Analysis and Selection of Variables in Linear Regression." *Biometrics* 32, 1–49.

22. Jobson, J.D. and Fuller, W.A. (1980). "Least Squares Estimation when the Covariance Matrix and Parameter Vector are Functionally Related." *Journal of the American Statistical Association* 75, 176–181.

23. Judge, G.G., Griffiths, W.E., Hill, R.C., Lütkepohl, H. and Lee, T. (1985). *The Theory and Practice of Econometrics*, Second Edition. New York: John Wiley and Sons, Inc.

24. Kerlinger, F.N. and Pedhauzer, E.J. (1973). *Multiple Regression in Behavioral Research*. New York: Holt, Rinehart and Winston, Inc.

25. Lovell, M.C. (1983). "Data Mining." *The Review of Economics and Statistics* 65, 1–12.

26. Mallows, C.L. (1986). "Augmented Partial Residuals." *Technometrics* 28, 313–319.

27. Mansfield, E.R. and Conerly, M.D. (1987). "Diagnostic Value of Residual and Partial Residual Plots." *American Statistician* 41, 107–116.

28. Marasingle, M.G. (1985). "A Multistage Procedure for Detecting Several Outliers in Linear Regression." *Technometrics* 27, 395–399.

29. Montgomery, D.C. (1984). *Design and Analysis of Experiments*, Second Edition. New York: John Wiley and Sons, Inc.

30. Montgomery, D.C. and Peck, E.A. (1982). *Introduction to Linear Regression Analysis*. New York: John Wiley and Sons, Inc.

31. Myers, R.H. (1976). *Response Surface Methodology*. Ann Arbor, Michigan: Edwards Brothers, Inc.

32. Myers, R.H. (1986). *Classical and Modern Regression With Applications*. Boston: Duxbury Press.

33. Neter, J., Wasserman, W. and Kutner, M.H. (1983). *Applied Linear Regression Models.* Homewood, Illinois: Richard D. Irwin, Inc.

34. Rawlings, J.O. (1988). *Applied Regression Analysis: A Research Tool.* Pacific Grove, California: Wadsworth & Brooks/Cole.

35. Smith, P.L. (1979). "Splines as a Convenient and Useful Statistical Tool." *American Statistician* 33, 57–62.

36. Theil, H. (1971). *Principles of Econometrics.* New York: John Wiley and Sons, Inc.

37. Thompson, M.L. (1978). "Selection of Variables in Multiple Regression: Part I. A Review and Evaluation and Part II. Chosen Procedures, Computations and Examples." *International Statistical Review* 46, 1–19, 129–146.

38. Transport Canada (1985). *Fuel Consumption Guide.* Ottawa, Canada.

39. Weisberg, S. (1980). *Applied Linear Regression*, Second Edition. New York: John Wiley and Sons, Inc.

40. Welsch, R.E. (1982). "Influence Functions and Regression Diagnostics," in *Modern Data Analysis*, eds. R. Laurier and A. Siegel. New York: Academic Press.

Exercises for Chapter 4

1. The exercises in this part are based on the data in Appendix Table D1. Multiple regression will be used to study the relationship between the dependent variable TMR and the explanatory variables SMIN, SMEAN, SMAX, PMIN, PMEAN, PMAX, PM2, PERWH, NONPOOR, GE65 and LPOP.

(a) Obtain the correlation matrix relating the twelve variables. What variables are most strongly related to TMR? What correlations among the eleven explanatory variables are relatively large?

(b) Fit the regression relating TMR to GE65 and then determine the residuals from this regression. Obtain a correlation matrix relating the residuals to the remaining ten explanatory variables. Comment on these correlations. What do they tell us about the importance of the remaining variables? How are these correlations related to multiple correlations?

(c) Using the correlations in (b) select a set of potential explanatory variables. Include only one of the **S** variables (SMIN, SMEAN, SMAX) and one of the **P** variables (PMIN, PMEAN, PMAX). Regress TMR on GE65 and your selection of variables. Comment on the regression results. Discuss the fitted relationship.

(d) For the multiple regression in (c) examine the residuals and influence diagnostics, and comment on the presence of outliers and on their influence on the fitted model. Include a comment on the individual DFBETA values for the individual explanatory variables. Obtain bivariate scatterplots relating TMR to any influential variables as suggested by the DFBETA values and comment on the results. Omit any influential outliers and fit a second multiple regression.

(e) Fit a multiple regression model with all eleven explanatory variables and comment on the results. Compare the model to the reduced model in (c) using an F-test. Is the model in (c) good enough in terms of R^2?

(f) Examine the residuals and influence diagnostics for the fitted model in (e) and comment on the presence of outliers and the influence that they have on the fit. Are the influential points the same as in (d)? Omit the influential points and refit the model with all eleven variables. Compare the two regressions.

(g) Use a variety of variable selection procedures to determine models relating TMR to the eleven explanatory variables. If prediction is the only purpose of the model which model would you choose? Compare the model you have chosen to the model obtained in (c).

(h) Divide the data set into two equal parts and fit a multiple regression model to each part separately using a model from question (g) or question (c). Use a Chow test to compare the two separate models to the overall model. Discuss your results. Use the estimated model for the first half of the data to predict TMR values for the second half of the data. Comment on the prediction error sum of squares and compare it to the error sum of squares obtained from the least squares fit for the second half of the data. You may wish to repeat the above without outliers and compare.

2. The exercises in this part are based on the data in Appendix Table D2. Multiple regression will be used to study the relationships between the dependent variable RETCAP and the explanatory variables GEARRAT, CAPINT, WCFTDT, WCFTCL, LOGSALE, LOGASST, CURRAT, INVTAST, NFATAST, PAYOUT, QUIKRAT and FATTOT.

(a) Obtain the correlation matrix relating the thirteen variables. What variables are most strongly related to RETCAP? What correlations among the twelve explanatory variables are relatively large?

(b) Fit the regression relating RETCAP to WCFTDT and then determine the residuals from this regression. Obtain a correlation matrix relating the residuals to the remaining eleven explanatory variables. Comment on these correlations. What do they tell us about the importance of the remaining variables? How are these correlations related to multiple correlations?

(c) Using the correlations in (b) select a set of potential explanatory variables. Regress RETCAP on WCFTDT and the variables you have selected. Comment on the regression results. Are there any variables that could be omitted? Suggest a modified model.

(d) Fit a model that you would recommend from (c). Obtain the residuals from this model. Obtain a correlation matrix relating the residuals to the variables not used in the model. Are there any variables which should be added to the model? If so fit a revised model and comment on the results. Discuss the fitted relationship.

(e) For your chosen model in (d) examine the residuals and influence diagnostics and comment on the presence of outliers and on their influence on the fitted model. Include a comment on the individual DFBETA values for the individual explanatory variables. Obtain bivariate scatterplots relating RETCAP to any influential variables as suggested by the DFBETA values and comment on the results.

Omit any influential outliers and fit a second multiple regression. Compare the two sets of regression results.

(f) Fit a multiple regression model with all twelve explanatory variables and discuss the results. Compare the model to the reduced model in (d) using an F-test. Is the model in (d) good enough in terms of R^2?

(g) Examine the residuals and influence diagnostics for the fitted model in (f) and comment on the presence of outliers and the influence they have on the fit. Are the influential points the same as in (e)? Omit the influential points and refit the model with all twelve explanatory variables. Compare the two regressions.

(h) Use a variety of variable selection procedures to determine models relating RETCAP to the twelve explanatory variables. If prediction is the only purpose of the model which model would you choose? Compare the model you have chosen to the model obtained in (d).

(i) Divide the data set into two equal parts and fit a multiple regression model to each part separately using a model from question (d) or question (h). Use a Chow test to compare the two separate models to the overall model. Discuss your results. Use the estimated model for the first half of the data to predict the RETCAP values for the second half of the data. Comment on the prediction error sum of squares and compare to the error sum of squares obtained from a least squares fit for the second half of the data. Do outliers have any impact on these results? If there are influential outliers, remove them and repeat the Chow test.

3. The exercises in this part are based on the data in Appendix Table D5. Multiple regression will be used to relate the dependent variable TRSBILL to the explanatory variables BANKCAN, USSPOT, USFORW and CPI.

(a) Examine the correlation matrix relating the five variables and discuss.

(b) Regress TRSBILL on the four explanatory variables. Comment on the results. Examine the standard errors of the regression coefficients. Carry out additional regressions which omit one of the variables and comment on the impact that this has on the regression coefficients and standard errors of these coefficients for the variables that remain.

(c) Obtain multicollinearity diagnostics as described in Chapter 4 and discuss these measures.

(d) Use ridge regression to estimate a relationship between TRSBILL and the four explanatory variables. Compare the results to those obtained in (b).

4. This exercise is based on the data in Appendix Table D4.

(a) Carry out a multiple regression of COMBRATE on the variable WEIGHT and discuss the results.

(b) Add the dummy variable FOR (FOR = 1 if foreign; FOR = 0 if domestic) to the model in (a) and compare the results to (a). Create a slope shifter variable FW = FOR*WEIGHT and regress COMBRATE on the variables WEIGHT, FOR and FW. Compare this estimated model to the one with FOR but without FW. Also compare this model to the one fitted in (a). For each comparison carry out an F-test and indicate what null hypothesis is being tested. Comment on the differences between the two groups.

(c) Repeat the exercise in (b) using effect coding by defining the variable FORE = 1 if foreign and FORE = -1 if domestic.

(d) Add dummy variables for CYLIND by creating dummy variables C4, C6 and C8 to represent CYLIND = 4, CYLIND = 6 and CYLIND = 8 respectively. Regress COMBRATE on WEIGHT and two of the three dummies (C4, C6 and C8) and discuss the results. Compare the results to the results in (a). Define slope shifters C4W = C4*WEIGHT, C6W = C6*WEIGHT, and C8W = C8*WEIGHT. Regress COMBRATE on WEIGHT also including two of the three dummies and two of three slope shifters. Discuss the results and compare the estimated model to the one with dummies without slope shifters and also to the estimated model in (a). For each comparison carry out an F-test and indicate what null hypothesis is being tested. Also carry out a different set of regressions with a new base case. Summarize the results by outlining how the three groups differ.

(e) Repeat the exercise in (d) using effect coding by defining the variables

$$
\begin{aligned}
C4E = \quad & 1 \text{ if CYLIND} = 4 \\
= \quad & 0 \text{ if CYLIND} = 6 \\
= \ & {-1} \text{ if CYLIND} = 8 \\
\\
C6E = \quad & 0 \text{ if CYLIND} = 4 \\
= \quad & 1 \text{ if CYLIND} = 6 \\
= \ & {-1} \text{ if CYLIND} = 8.
\end{aligned}
$$

Define corresponding slope shifters C4EW and C6EW. Also use an alternative set of effect coding variables which use a different base case.

(f) Combine the methodology in (b) and (d) to relate COMBRATE to WEIGHT with both the FOR variables and the CYLIND variables. Carry out a test for interaction between FOR and CYLIND and also carry out various other comparisons using F-tests. Provide summary tables showing the intercepts and slopes for the six groups. Compare the six groups.

(g) Combine the methodolgy in (c) and (e) to relate COMBRATE to WEIGHT with both the FOR variables and the CYLIND variables. Carry out a test for interaction between FOR and CYLIND and also carry out various other comparisons using F-tests. Provide summary tables showing the intercepts and slopes for the six groups. Compare the six groups.

(h) Create six cell mean dummies D1, D2, D3, D4, D5, D6, for the six cells derived from the cross classification of FOR with CYLIND. Also create the corresponding slope shifters D1W, D2W, D3W, D4W, D5W and D6W. Carry out analyses using transformation matrices which permit the testing for interaction, FOR effects and CYLIND effects. Discuss the results.

(i) Examine the influence diagnostics for the full model regression in (f). Discuss the results.

(j) Examine the influence diagnostics for the full model regression in (g). Discuss the results.

(k) Obtain the correlation matrix for the complete set of variables in (f) and comment on the correlations between the dummy variables and the slope shifters.

(l) Obtain the correlation matrix for the complete set of variables in (g) and comment on the correlations between the effect coding variables and the slope shifters.

5. The exercises in this part are based on the data in Appendix Table D3.

(a) Regress the variables LCURRENT on the variables EDUC, AGE, SENIOR and EXPER. Comment on the results. Add the variable LSTART to the explanatory variables and comment on the results. Compare the two fitted models. Be sure to interpret the regression coefficients in each case.

(b) Determine the variable LRATE = (LCURRENT – LSTART) and regress it on the variables EDUC, AGE, SENIOR and EXPER. Comment on the results. Interpret the regression coefficients.

(c) Regress the variable LCURRENT on EDUC including the dummy variable SEX and the slope shifter variable SEXED = SEX*EDUC. Comment on the results. Is there a difference between the two SEX groups? Is the slope shifter necessary?

(d) Regress the variable LCURRENT on EDUC including the dummy variable RACE and the slope shifter variable RACED = RACE* EDUC. Comment on the results. Is the relationship different for the two race groups? Is the slope shifter necessary?

(e) Combine the dummy variables and slope shifter variables from (c) and (d) to obtain a relationship between LCURRENT and EDUC including SEX, RACE, SEXED and RACED. Comment on the results. Give equations for the four SEX–RACE groups and comment on the differences. Is the interaction significant?

(f) Repeat exercises (c), (d) and (e) using effect coding.

(g) Regress LCURRENT on the variables EDUC, AGE and the interaction variable EDAGE = EDUC*AGE. Comment on the results. Interpret the coefficients. Describe the interaction effect.

(h) Regress the variable LCURRENT on EDUC including the categories of JOBCAT. Define cell mean indicator variables for the five JOB-CAT cells, say J1, J2, J3, J4 and J5. Define slope shifter variables J1ED, J2ED, J3ED, J4ED and J5ED by multiplying each of the dummy variables by EDUC. Obtain the cell means model regressions relating LCURRENT to EDUC. Use transformation matrices to test for equal slopes and equal intercepts. Comment on the results. Compare the five groups. Are the slope shifters necessary?

Questions for Chapter 4

1. (a) Use calculus to obtain the normal equations by differentiating the expression

$$\zeta^2 = (\mathbf{y} - \mathbf{X}\boldsymbol{\beta})'(\mathbf{y} - \mathbf{X}\boldsymbol{\beta})$$
$$= \mathbf{y}'\mathbf{y} + \boldsymbol{\beta}'\mathbf{X}'\mathbf{X}\boldsymbol{\beta} - 2\boldsymbol{\beta}'\mathbf{X}'\mathbf{y}$$

with respect to $\boldsymbol{\beta}$. Use the rules for vector differentiation given in the Appendix.

(b) Let $\mathbf{b}$ denote the solution for $\boldsymbol{\beta}$ in the normal equations and show that $E[\mathbf{b}] = \boldsymbol{\beta}$ and

$$E[(\mathbf{b} - \boldsymbol{\beta})(\mathbf{b} - \boldsymbol{\beta})'] = \sigma_u^2(\mathbf{X}'\mathbf{X})^{-1}$$

using the rules for expectation of vectors and matrices.

(c) Show that $\hat{\mathbf{y}} = \mathbf{H}\mathbf{y} = \mathbf{X}\boldsymbol{\beta} + \mathbf{H}\mathbf{u}$ where $\mathbf{H}$ is the hat matrix given by $\mathbf{H} = \mathbf{X}(\mathbf{X}'\mathbf{X})^{-1}\mathbf{X}'$, and

$$\mathbf{e} = \mathbf{y} - \hat{\mathbf{y}} = (\mathbf{I} - \mathbf{H})\mathbf{y} = (\mathbf{I} - \mathbf{H})\mathbf{u}.$$

The matrix $\mathbf{H}$ is a projection operator for the column space of $\mathbf{X}$ and hence $\mathbf{H}\mathbf{y}$ projects the vector $\mathbf{y}$ onto the column space of $\mathbf{X}$. Similarly the matrix $(\mathbf{I} - \mathbf{H})$ is a projection operator for the space which is orthogonal to the column space of $\mathbf{X}$ and hence $(\mathbf{I} - \mathbf{H})\mathbf{y}$ projects $\mathbf{y}$ onto the column space orthogonal to the columns of $\mathbf{X}$.

(d) Show that $\mathbf{H}' = \mathbf{H}$, $(\mathbf{I}-\mathbf{H})' = (\mathbf{I}-\mathbf{H})$, $\mathbf{H}'\mathbf{H} = \mathbf{H}$, $(\mathbf{I}-\mathbf{H})'(\mathbf{I}-\mathbf{H}) = (\mathbf{I} - \mathbf{H})$ and $\mathbf{H}'(\mathbf{I} - \mathbf{H}) = \mathbf{O}$.

(e) Show that $(\mathbf{y} - \mathbf{X}\mathbf{b})'(\mathbf{y} - \mathbf{X}\mathbf{b}) = \mathbf{y}'(\mathbf{I} - \mathbf{H})\mathbf{y}$ and $\hat{\mathbf{y}}'\hat{\mathbf{y}} = \mathbf{y}'\mathbf{H}\mathbf{y}$ and hence that

$$\mathbf{y}'\mathbf{y} = \hat{\mathbf{y}}'\hat{\mathbf{y}} + (\mathbf{y} - \mathbf{X}\mathbf{b})'(\mathbf{y} - \mathbf{X}\mathbf{b}).$$

(f) Define $\text{SST} = \mathbf{y}'\mathbf{y} - n\bar{y}^2$, $\text{SSR} = \hat{\mathbf{y}}'\hat{\mathbf{y}} - n\bar{y}^2$ and $\text{SSE} = (\mathbf{y} - \mathbf{X}\mathbf{b})'(\mathbf{y} - \mathbf{X}\mathbf{b})$. Show that $\text{SST} = \text{SSR} + \text{SSE}$.

(g) Show that $E[\mathbf{e}\mathbf{e}'] = \sigma_u^2(\mathbf{I} - \mathbf{H})$. What does this result tell us about the covariances among the residuals?

(h) Show that $E[\mathbf{e}'\mathbf{e}] = E[\mathbf{y}'(\mathbf{I} - \mathbf{H})\mathbf{y}] = E[tr\mathbf{y}'(\mathbf{I} - \mathbf{H})\mathbf{y}] = E[tr(\mathbf{I} - \mathbf{H})\mathbf{y}\mathbf{y}'] = tr(\mathbf{I} - \mathbf{H})E[\mathbf{y}\mathbf{y}'] = (n - p - 1)\sigma_u^2$.

(i) Let $\boldsymbol{\lambda}'\mathbf{b}$, $\boldsymbol{\lambda}(\rho + 1) \times 1$, denote a linear combination of the elements of $\hat{\boldsymbol{\beta}}$ and show that $V(\boldsymbol{\lambda}'\mathbf{b}) = \sigma_u^2\boldsymbol{\lambda}'\mathbf{c}^2\boldsymbol{\lambda}$, where $\mathbf{c}^2 = (\mathbf{X}'\mathbf{X})^{-1}$.

2. Let $\mathbf{T}\boldsymbol{\beta}$ denote a linear combination of the parameter vector $\boldsymbol{\beta}$.

(a) Assume that $\mathbf{T}$ is $(p+1) \times (p+1)$ and that $\mathbf{T}^{-1}$ exists. Show that the linear model can be written as

$$(\mathbf{y} - \mathbf{X}\mathbf{T}^{-1}\mathbf{a}) = \mathbf{X}\mathbf{T}^{-1}[\mathbf{T}\boldsymbol{\beta} - \mathbf{a}] + \mathbf{u}$$

and explain how you could use standard regression software to test $H_0 : \mathbf{T}\boldsymbol{\beta} = \mathbf{a}$. Show that the F-statistic for H_0 is equivalent to the F-statistic in (b).

(b) Assume that $\mathbf{T}$ is $k \times (p+1)$ where $k < p$ and assume that $\mathbf{T} = [\mathbf{T}_1\mathbf{T}_2]$ where $\mathbf{T}_1$ $(k \times k)$ has a unique inverse $\mathbf{T}_1^{-1}$. Let $\mathbf{T}\boldsymbol{\beta} = \mathbf{a}$ denote a set of k linear constraints on $\boldsymbol{\beta}$. Use the fact that $\boldsymbol{\beta}_1 = \mathbf{T}_1^{-1}(\mathbf{a} - \mathbf{T}_2\boldsymbol{\beta}_2)$ to write $\mathbf{y} - \mathbf{X}_1\mathbf{T}_1^{-1}\mathbf{a} = (\mathbf{X}_2 - \mathbf{X}_1\mathbf{T}_1^{-1}\mathbf{T}_2)\boldsymbol{\beta}_2 + \mathbf{U}$ and show that the least squares estimator of $\boldsymbol{\beta}$ subject to $\mathbf{T}\boldsymbol{\beta} = \mathbf{a}$ is given by $\mathbf{b}_1^* = \mathbf{T}_1^{-1}(\mathbf{b} - \mathbf{T}_2\mathbf{b}_2^*)$ and

$$\mathbf{b}_2^* = [(\mathbf{X}_2 - \mathbf{X}_1\mathbf{T}_1^{-1}\mathbf{T}_2)'(\mathbf{X}_2 - \mathbf{X}_1\mathbf{T}_1^{-1}\mathbf{T}_2)]^{-1}$$
$$\times (\mathbf{X}_2 - \mathbf{X}_1\mathbf{T}_1^{-1}\mathbf{T}_2)'(\mathbf{y} - \mathbf{X}_1\mathbf{T}_1^{-1}\mathbf{a})$$

and hence explain how ordinary regression software can be used to obtain $\mathbf{b}^*$ the estimator of $\boldsymbol{\beta}$ subject to $\mathbf{T}\boldsymbol{\beta} = \mathbf{a}$. Also show that if $(\mathbf{y} - \mathbf{X}_1\mathbf{T}_1^{-1}\mathbf{a})$ is regressed on $\mathbf{X}_1$ and $(\mathbf{X}_2 - \mathbf{X}_1\mathbf{T}_1^{-1}\mathbf{T}_2)$ this provides the full model for testing $H_0 : \mathbf{T}\boldsymbol{\beta} = \mathbf{a}$.

3. Assume that the matrix $\mathbf{X}$ contains n rows and $(p+1)$ columns with the first column being a column of unities denoted by $\mathbf{i}$.

(a) Show that the matrix $\mathbf{X}'\mathbf{X}$ can be written as

$$\mathbf{X}'\mathbf{X} = \begin{bmatrix} n & n\bar{\mathbf{x}}' \\ n\bar{\mathbf{x}} & \mathbf{X}_1'\mathbf{X}_1 \end{bmatrix}$$

where $\mathbf{X} = [\mathbf{i}\ \mathbf{X}_1]$ and $\bar{\mathbf{x}}$ is the vector of means $\bar{\mathbf{x}} = \mathbf{X}_1'\mathbf{i}/n$.

(b) Use the expression for the inverse of a partitioned matrix given in the Appendix to show that the components of

$$(\mathbf{X}'\mathbf{X})^{-1} = \begin{bmatrix} \alpha & \boldsymbol{\beta}' \\ \boldsymbol{\beta} & \boldsymbol{\Gamma} \end{bmatrix}$$

are given by

$$\alpha = \frac{1}{n} + \bar{\mathbf{x}}'[\mathbf{X}_1'\mathbf{X}_1 - n\bar{\mathbf{x}}\bar{\mathbf{x}}']^{-1}\bar{\mathbf{x}}$$

$$\boldsymbol{\beta}' = -\bar{\mathbf{x}}'[\mathbf{X}_1'\mathbf{X}_1 - n\bar{\mathbf{x}}\bar{\mathbf{x}}']^{-1}$$

$$\boldsymbol{\Gamma} = [\mathbf{X}_1'\mathbf{X}_1 - n\bar{\mathbf{x}}\bar{\mathbf{x}}']^{-1}.$$

(c) Show that the matrix $\mathbf{X}^{*\prime}\mathbf{X}^* = (\mathbf{X}_1'\mathbf{X}_1 - n\bar{\mathbf{x}}\bar{\mathbf{x}}')$ has elements of the form $\sum_{i=1}^{n}(x_{ij} - \bar{x}_{\cdot j})(x_{ik} - \bar{x}_{\cdot k})$ where $\mathbf{X}^* = (\mathbf{X}_1 - \mathbf{i}\bar{\mathbf{x}}')$ and $\mathbf{i}$ is an $(n \times 1)$ column of unities.

(d) Use the results in (b) and (c) to show that the hat matrix $\mathbf{H}$ is given by

$$\mathbf{H} = \frac{1}{n}\,\mathbf{i}\mathbf{i}' + \mathbf{X}^*(\mathbf{X}^{*\prime}\mathbf{X}^*)^{-1}\mathbf{X}^{*\prime}$$

and that each diagonal element of $\mathbf{H}$, say h_{ii}, is related to the Mahalanobis distance of a point $\mathbf{x}_i$ from the mean $\bar{\mathbf{x}}$.

(e) Use the results in (b) to show that if the sample correlations among the variables in $\mathbf{X}$ are zero then each of the elements of $\mathbf{b}$ excluding b_0 depends on only one of the $\mathbf{X}$ variables. Show also that the sample covariances among the elements of $\mathbf{b}$ are zero.

4. Partition the $\mathbf{X}$ $(n \times (p+1))$ matrix into two parts $\mathbf{X} = (\mathbf{X}_1\ \mathbf{x}_2)$ where $\mathbf{X}_1$ $(n \times p)$ contains the column of unities plus the first $(p-1)$ variables and $\mathbf{x}_2$ contains the last column of $\mathbf{X}$. Denote the ordinary least squares estimator of $\boldsymbol{\beta} = \begin{bmatrix} \boldsymbol{\beta}_1 \\ \beta_2 \end{bmatrix}$ by $\mathbf{b} = \begin{bmatrix} \mathbf{b}_1 \\ b_2 \end{bmatrix} = (\mathbf{X}'\mathbf{X})^{-1}\mathbf{X}'\mathbf{y}$ where $\boldsymbol{\beta}_1$ and $\mathbf{b}_1$ are the $(p \times 1)$ parameter vectors corresponding to $\mathbf{X}_1$ and β_2 and b_2 are the parameters corresponding to $\mathbf{x}_2$. Denote the ordinary least squares estimator of $\boldsymbol{\beta}_1$ based on $\mathbf{X}_1$ by $\mathbf{b}_1^* = (\mathbf{X}_1'\mathbf{X}_1)^{-1}\mathbf{X}_1'\mathbf{y}$.

(a) Show that if $\mathbf{x}_2$ is excluded from the model the least squares estimator of $\mathbf{y}$ in $\mathbf{y} = \mathbf{X}_1\boldsymbol{\beta}_1 + \mathbf{u}_1$ is given by $\hat{\mathbf{y}}^* = \mathbf{X}_1\mathbf{b}_1^* = \mathbf{H}_1\mathbf{y}$ and $\mathbf{e}_1 = (\mathbf{y} - \hat{\mathbf{y}}^*) = (\mathbf{I} - \mathbf{H}_1)\mathbf{y}$ where $\mathbf{H}_1 = \mathbf{X}_1(\mathbf{X}_1'\mathbf{X})^{-1}\mathbf{X}_1'$ and hence that $\hat{\mathbf{y}}^*$ and $\mathbf{e}_1$ are respectively in the column space of $\mathbf{X}_1$ and the column space orthogonal to $\mathbf{X}_1$.

(b) Consider the regression of $\mathbf{x}_2$ on $\mathbf{X}_1$ given by $\mathbf{x}_2 = \mathbf{X}_1\boldsymbol{\gamma} + \mathbf{v}$ and denote the ordinary least squares estimator by $\mathbf{g} = (\mathbf{X}_1'\mathbf{X}_1)^{-1}\mathbf{X}_1'\mathbf{x}_2$. Show that the residuals are given by $\mathbf{f} = \mathbf{x}_2 - \mathbf{X}_1\mathbf{g} = (\mathbf{I} - \mathbf{H}_1)\mathbf{x}_2$.

(c) Denote the regression model relating the $\mathbf{y}$ residuals, $\mathbf{e}_1$, to the $\mathbf{x}_2$ residuals, $\mathbf{f}$, by

$$\mathbf{e}_1 = \delta\mathbf{f} + \mathbf{w}.$$

(Note: δ is just the scalar regression parameter. This model has no intercept.) Show that the least squares estimator of δ is given by

$$d = [\mathbf{x}_2'(\mathbf{I} - \mathbf{H}_1)\mathbf{x}_2]^{-1}\mathbf{x}_2'(\mathbf{I} - \mathbf{H}_1)\mathbf{y}.$$

(Note: $(\mathbf{I} - \mathbf{H}_1)' = (\mathbf{I} - \mathbf{H}_1)$ and $(\mathbf{I} - \mathbf{H}_1)(\mathbf{I} - \mathbf{H}_1) = (\mathbf{I} - \mathbf{H}_1)$.)

(d) For $\mathbf{X} = (\mathbf{X}_1\ \mathbf{x}_2)$ show that the matrix $\mathbf{X}'\mathbf{X}$ is given by

$$\begin{bmatrix} \mathbf{X}_1'\mathbf{X}_1 & \mathbf{X}_1'\mathbf{x}_2 \\ \mathbf{x}_2'\mathbf{X}_1 & \mathbf{x}_2'\mathbf{x}_2 \end{bmatrix}.$$

Use the expression for the inverse of a partitioned matrix given in the Appendix to show that

$$[\mathbf{X'X}]^{-1} = \begin{bmatrix} \mathbf{A} & \mathbf{B} \\ \mathbf{B'} & c \end{bmatrix}$$

is given by

$$\mathbf{A} = (\mathbf{X_1'X_1})^{-1} + (\mathbf{X_1'X_1})^{-1}\mathbf{X_1'x_2}\,\mathbf{x_2'X_1}(\mathbf{X_1'X_1})^{-1}(\mathbf{x_2'}(\mathbf{I} - \mathbf{H_1})\mathbf{x_2})^{-1}$$

$$\mathbf{B} = -(\mathbf{X_1'X_1})^{-1}\mathbf{X_1'x_2}(\mathbf{x_2'}(\mathbf{I} - \mathbf{H_1})\mathbf{x_2})^{-1}$$

$$c = (\mathbf{x_2'}(\mathbf{I} - \mathbf{H_1})\mathbf{x_2})^{-1} \qquad \text{Note: } c \text{ is a scalar.}$$

(e) Use the result in (d) to show that

$$\mathbf{b_1} = \mathbf{b_1^*} - (\mathbf{X_1'X_1})^{-1}\mathbf{X_1'x_2}d \qquad \text{and} \qquad b_2 = d$$

and hence that the regression coefficient of $\mathbf{x_2}$ in the regression of $\mathbf{y}$ on $(\mathbf{X_1}\ \mathbf{x_2})$ is equivalent to the regression coefficient of the $\mathbf{y}$ residuals on the $\mathbf{x_2}$ residuals after removing or controlling for $\mathbf{X_1}$.

(f) For the three regression models

(i) $\mathbf{y} = \mathbf{X_1}\boldsymbol{\beta}_1 + \mathbf{v}$

(ii) $\mathbf{e_1} = \mathbf{f}\delta + \mathbf{w}$

(iii) $\mathbf{y} = \mathbf{X}\boldsymbol{\beta} + \mathbf{u}$.

Show that the regression sums of squares are given by

(i) $\text{SSR}_1 = \mathbf{b_1^{*'}X_1'X_1b_1^*} - n\bar{y}^2$

(ii) $\text{SSR}_2 = d\mathbf{x_2'}(\mathbf{I} - \mathbf{H_1})\mathbf{x_2}d$

(iii) $\text{SSR} = \text{SSR}_1 + \text{SSR}_2$

and hence the importance of the variable $\mathbf{x_2}$ to the regression after $\mathbf{X_1}$ depends on the strength of the relationship between $\mathbf{y}$ and $\mathbf{x_2}$ after removing $\mathbf{X_1}$, namely SSR_2.

5. Let $\mathbf{x_n'}$ $(1 \times (p+1))$ denote the last row of the matrix $\mathbf{X}$ and let $\mathbf{X}(n)$ $((n-1) \times (p+1))$ denote the $(n-1)$ rows of $\mathbf{X}$ excluding $\mathbf{x_n'}$.

(a) Show that $\mathbf{X'X}$ can be written as $\mathbf{X'}(n)\mathbf{X}(n) + \mathbf{x_n}\mathbf{x_n'}$.

(b) Given a full rank $[(p+1) \times (p+1)]$ matrix $\mathbf{A}$ and $(p+1) \times 1$ vectors $\mathbf{b}$ and $\mathbf{c}$ the inverse of the $(p+1) \times (p+1)$ matrix $[\mathbf{A} + \mathbf{bc'}]$ is given by $\mathbf{A}^{-1} - \dfrac{\mathbf{A}^{-1}\mathbf{bc'A}^{-1}}{(1 + \mathbf{b'A}^{-1}\mathbf{c})}$. Use this result and (a) to show that $(\mathbf{X'X})^{-1}$ can be written as

$$[\mathbf{X'}(n)\mathbf{X}(n)]^{-1} - [\mathbf{X'}(n)\mathbf{X}(n)]^{-1}\mathbf{x_n}\mathbf{x_n'}[\mathbf{X'}(n)\mathbf{X}(n)]^{-1}\left[\frac{1}{1 + h(n)}\right]$$

where $h(n) = \mathbf{x}'_n[\mathbf{X}'(n)\mathbf{X}(n)]\mathbf{x}_n$.

(c) Given the linear model

$$\begin{bmatrix} \mathbf{y}(n) \\ y_n \end{bmatrix} = \begin{bmatrix} \mathbf{X}(n) \\ \mathbf{x}'_n \end{bmatrix} \boldsymbol{\beta} + \mathbf{u}$$

where $\mathbf{y}(n)$ denotes the first $(n-1)$ observations of $\mathbf{y}$ and y_n denotes the last. Show that $\mathbf{X}'\mathbf{y} = \mathbf{X}'(n)\mathbf{y}(n) + \mathbf{x}_n y_n$.

(d) Show that the least squares estimator is given by

$$\mathbf{b} = \mathbf{b}(n) + [\mathbf{X}'(n)\mathbf{X}(n)]^{-1}\mathbf{x}_n[y_n - \mathbf{x}'_n\mathbf{b}(n)]\left[\frac{1}{1 + h(n)}\right],$$

where $\mathbf{b}(n) = [\mathbf{X}'(n)\mathbf{X}(n)]^{-1}\mathbf{X}'(n)\mathbf{y}(n)$.

(e) Use the result in (d) to discuss the impact on $\mathbf{b}$ of the addition of the observation $(\mathbf{x}'_n, y_n)$.

(f) Use the fact that $\mathbf{X}'(n)\mathbf{X}(n) = \mathbf{X}'\mathbf{X} - \mathbf{x}_n\mathbf{x}'_n$ and the matrix result in (b) to show that

$$[\mathbf{X}'(n)\mathbf{X}(n)]^{-1} = (\mathbf{X}'\mathbf{X})^{-1} + (\mathbf{X}'\mathbf{X})^{-1}\mathbf{x}_n\mathbf{x}'_n(\mathbf{X}'\mathbf{X})^{-1}\left[\frac{1}{1 - h_{nn}}\right]$$

where $h_{nn} = \mathbf{x}'_n(\mathbf{X}'\mathbf{X})^{-1}\mathbf{x}_n$.

(g) Use the result in (f) to show that

$$\mathbf{b}(n) = \mathbf{b} - (\mathbf{X}'\mathbf{X})^{-1}\mathbf{x}_n(y_n - \mathbf{x}'_n\mathbf{b})\left[\frac{1}{1 - h_{nn}}\right]$$

and use it to discuss the impact on $\mathbf{b}$ of the deletion of the observation $(\mathbf{x}'_n, y_n)$.

(h) The error sums of squares for the two models are given by

$$\text{SSE}(n) = \mathbf{y}'(n)[\mathbf{I} - \mathbf{X}(n)[\mathbf{X}'(n)\mathbf{X}(n)]^{-1}\mathbf{X}'(n)]\mathbf{y}(n)$$

and

$$\text{SSE} = \mathbf{y}'[\mathbf{I} - \mathbf{X}(\mathbf{X}'\mathbf{X})^{-1}\mathbf{X}']\mathbf{y}$$

for the models with $(n-1)$ observations and n observations respectively. Show that

$$\text{SSE} = \text{SSE}(n) + (y_n - \mathbf{x}'_n\mathbf{b}(n))^2\left[\frac{1}{1 + h(n)}\right].$$

What is the impact on SSE of adding the observation (x'_n, y_n)?

6. The mean shift outlier model is given by

$$y = X\beta + \phi z_n + U$$

where z_n is an $(n \times 1)$ vector with zeroes in all positions except the n-th position which is 1. ϕ is an unknown scalar parameter.

(a) Write the equation for the n-th observation and hence show that for the n-th observation the intercept is $(\beta_0 + \phi)$.

(b) Use the results of Question 4 to show that

$$d = [z_n'(I - H)z_n]^{-1}z_n'(I - H)y = e_n/(1 - hnn)$$

where $e_n = y_n - \hat{y}_n$, hnn is the n-th diagonal element of H, and $\hat{y}_n = x_n'(X'X)^{-1}X'y$.

(c) Show that the increase in regression sum of squares after fitting ϕ is given by

$$SSR_2 = d^2 z_n'(I - H)z_n = e_n^2/(1 - hnn).$$

(d) Show that the error sum of squares after fitting β and ϕ is given by

$$SSE = y'(I - H)y - d^2 z_n'(I - H)z_n = (n - p - 1)s^2 - e_n^2/(1 - hnn).$$

(e) Show that the F-test for testing $H_0 : \phi = 0$ is given by

$$F = \frac{SSR_2/1}{SSE/(n - p - 2)} = \frac{e_n^2(n - p - 2)}{[(n - p - 1)s^2(1 - hnn) - e_n^2]}$$

which is the square of the t-statistic for externally studentized residuals.

7. Use the results from Question 5 to show the following results.

(a) Cook's D statistic can be written as

$$D_n = (b(n) - b)'(X'X)(b(n) - b)/(p + 1)s^2$$

$$= \frac{hnn}{[1 - hnn]} \frac{e_n^2}{(p + 1)s^2(1 - hnn)}.$$

(b) The DFITS measure can be written as

$$(\hat{y}_n - \hat{y}_{(n)n})/s(n)\sqrt{hnn} = \left[\frac{hnn}{[1 - hnn]}\right]^{1/2} \frac{e_n}{s(n)(1 - hnn)^{1/2}}.$$

8. Suppose that the $\mathbf{X}$ matrix contains n rows and four columns. The first column is a column of unities for the intercept denoted by $\mathbf{i}$. The second column is a column of n observations on a variable $\mathbf{x}$. Assume that the n observations are from two groups with n_1 and n_2 observations respectively $n = (n_1 + n_2)$. Assume that the third column of $\mathbf{X}$ contains dummy variables for the first group and that column four contains the corresponding slope shifter for the first group. The $\mathbf{X}$ matrix is given by

$$
\mathbf{X} = \begin{bmatrix}
1 & x_{11} & 1 & x_{11} \\
1 & x_{21} & 1 & x_{21} \\
\vdots & \vdots & \vdots & \vdots \\
1 & x_{n_1 1} & 1 & x_{n_1 1} \\
1 & x_{12} & 0 & 0 \\
1 & x_{22} & 0 & 0 \\
\vdots & \vdots & \vdots & \vdots \\
1 & x_{n_2 2} & 0 & 0
\end{bmatrix}.
$$

(a) Show that the means of the columns are 1, $\bar{x}$, n_1/n and $n_1\bar{x}_1/n$, where $\bar{x} = \sum\limits_{j=1}^{2}\sum\limits_{i=1}^{n_j} x_{ij}/n$ and $\bar{x}_1 = \sum\limits_{i=1}^{n_1} x_{ij}/n_1$.

(b) Show that $\mathbf{X}'\mathbf{X}$ is given by

$$
\mathbf{X}'\mathbf{X} = \begin{bmatrix}
n & n\bar{x} & n_1 & n_1\bar{x}_1 \\
n\bar{x} & \sum\limits_{j=1}^{2}\sum\limits_{i=1}^{n_j} x_{ij}^2 & n_1\bar{x}_1 & \sum\limits_{i=1}^{n_1} x_{i1}^2 \\
n_1 & n_1\bar{x}_1 & n_1 & n_1\bar{x}_1 \\
n_1\bar{x}_1 & \sum\limits_{i=1}^{n_1} x_{i1}^2 & n_1\bar{x}_1 & \sum\limits_{i=1}^{n_1} x_{i1}^2
\end{bmatrix}.
$$

(c) Show that the matrix $\bar{\mathbf{x}}\bar{\mathbf{x}}'$ is given by

$$
\begin{bmatrix}
1 & \bar{x} & n_1/n & n_1\bar{x}_1/n \\
\bar{x} & \bar{x}^2 & n_1\bar{x}/n & n_1\bar{x}_1\bar{x}/n \\
n_1/n & n_1\bar{x}/n & n_1^2/n^2 & n_1^2\bar{x}_1/n^2 \\
n_1\bar{x}_1/n & n_1\bar{x}_1\bar{x}/n & n_1^2\bar{x}_1/n^2 & n_1^2\bar{x}_1^2/n^2
\end{bmatrix}
$$

where $\bar{\mathbf{x}}$ is the vector of means given in (a).

(d) Show that the matrix $[\mathbf{X}'\mathbf{X} - n\bar{\mathbf{x}}\bar{\mathbf{x}}']$ is given by

$$
\begin{bmatrix}
0 & 0 & 0 & 0 \\
0 & \left[\sum_{j=1}^{2}\sum_{i=1}^{n_j} x_{ij}^2 - n\bar{x}^2\right] & n_1(\bar{x}_1 - \bar{x}) & \left[\sum_{i=1}^{n_1} x_{i1}^2 - n_1\bar{x}_1\bar{x}\right] \\
0 & n_1(\bar{x}_1 - \bar{x}) & \left[n_1 - \frac{n_1^2}{n}\right] & \left[n_1\bar{x}_1 - n_1^2\bar{x}_1/n\right] \\
0 & \left[\sum_{i=1}^{n_1} x_{i1}^2 - n_1\bar{x}_1\bar{x}\right] & \left[n_1\bar{x}_1 - \frac{n_1^2\bar{x}_1}{n}\right] & \left[\sum_{i=1}^{n_1} x_{i1}^2 - \frac{n_1^2\bar{x}_1^2}{n}\right]
\end{bmatrix}.
$$

(e) Use the results in (d) to show that the correlation between the dummy for group 1 and the corresponding slope shifter is given by

$$
\frac{n_1\bar{x}_1(1 - n_1/n)}{\sqrt{\left[\sum x_{i1}^2 - \frac{n_1^2\bar{x}_1^2}{n}\right]\left[n_1(1 - n_1/n)\right]}}
$$

$$
= \frac{1}{\sqrt{\left[1 + \sum_{i=1}^{n_1}\left[\frac{x_{i1} - \bar{x}_1}{\bar{x}_1}\right]^2 \Big/ n_1\left(1 - \frac{n_1}{n}\right)\right]}}.
$$

Hint: Rewrite $\left[\sum_{i=1}^{n_1} x_{i1}^2 - \frac{n_1^2\bar{x}_1^2}{n}\right]$ as $\sum_{i=1}^{n_1}(x_{i1} - \bar{x}_1)^2 + n_1\bar{x}_1^2(1 - n_1/n)$.

(f) Use the result in (e) to discuss the factors which affect the correlation between a dummy and the corresponding slope shifter. What happens if $\bar{x}_1 = 0$?

5

Analysis of Variance and Experimental Design

Analysis of variance can be viewed as a special case of multiple regression where the explanatory variables are entirely qualitative. In the previous chapter it was demonstrated how indicator variables can be used in a multiple regression to represent qualitative explanatory variables. Most analysis of variance models can be conveniently represented using a multiple regression formulation. In this chapter the traditional analysis of variance models and the multiple regression model form will be introduced throughout. Indicator variables will be generated using dummy coding, effect coding and cell mean coding. Orthogonal coding will also be introduced.

Analysis of variance type models are sometimes employed to analyze data derived from planned experiments. In such circumstances, the data is derived from measurements made on a set of experimental units, which have been randomly assigned to one of a set of treatment combinations. Such data is said to be derived from randomization rather than random sampling. The techniques of *experimental design* are concerned with maximizing the information obtained from the experiment at minimum cost. The experimental design is the procedure employed to assign treatment combinations to experimental units. In some applications the potential number of influential factors is large while the number of homogeneous experimental units available is relatively small. In such applications experimental designs must be used to minimize the use of experimental units.

This chapter combines the techniques of analysis of variance and experimental design. The first part of the chapter provides a transition from the previous chapters by relating analysis of variance to the comparison of means and to the standard multiple linear regression model. The latter part of the chapter is concerned with providing an overview of some of the more complex experimental designs.

5.1 One-Way Analysis of Variance

One-way analysis of variance can be viewed as a special case of bivariate analysis. The two variables X and Y, whose relationship is to be studied, represent a quantitative variable Y and a qualitative variable X. The variable Y is assumed to be a continuous random variable with density $f(Y)$.

The variable X is assumed to be a classification variable with g categories. The interrelationship between the two variables therefore involves a comparison of the distribution of Y among the g categories of X. Such analyses may involve comparison of distribution parameters such as means or variances or may involve direct comparison of the density functions. The procedures that involve a comparison of the means have traditionally been referred to as analysis of variance. The term analysis of variance was first applied to these procedures because the statistical tests involve comparisons of various sums of squares or variation. The classification variable X is usually referred to as a *factor* and the categories of X are called *levels*. The one-way analysis of variance is also called a *single factor analysis of variance*.

Many scientific studies include the comparison of the properties of items which have been classified into types or brands or categories. Examples of such categories include varieties of an agricultural product such as corn, various brands of a consumer good such as soap and various methods of carrying out a task such as treating a patient or advertising a product.

This section outlines the techniques commonly used for one-way analysis of variance and introduces some terminology from experimental design. The section also includes a summary of multiple comparison procedures for comparing means and procedures for testing for homogeneity of variances. The use of the multiple regression model to carry out analysis of variance is also discussed.

Example

A marketing research organization monitors the food purchases of families in several cities. Each week each family submits a report detailing its purchases of food items including price paid, brand name, size and where purchased. These data are used by the research firm to monitor the impact of various advertising campaigns for various food items. A study was carried out to determine the effectiveness of four different ads designed to stimulate the consumption of milk products. The four ads were shown on local TV stations in different cities in five geographical areas. Four cities were chosen in each geographical area, one to receive each of the four ads. After two months of advertising, the consumption of milk products (measured in dollars) in the second month by each of six test families was determined. The data obtained from the study are given in Table 5.1. The four ads will be compared using one-way analysis of variance in this section.

Table 5.1. Expenditures on Milk Products by Test Families in Second Month of Advertising

Family Expenditures on Milk Products

Region	Ad #1	Ad #2	Ad #3	Ad #4	Family Size
1	12.35	21.86	14.43	21.44	1
1	20.52	42.17	22.26	31.21	2
1	30.85	49.61	23.99	40.09	3
1	39.35	63.65	36.98	55.68	4
1	48.87	73.75	42.13	65.81	5
1	58.01	85.95	54.19	76.61	6
2	28.26	13.76	14.44	30.78	1
2	37.67	24.59	29.63	45.75	2
2	44.70	37.30	38.27	56.37	3
2	57.54	49.53	51.59	70.19	4
2	67.57	59.25	59.09	79.81	5
2	77.70	67.68	71.69	94.23	6
3	10.97	0.00	2.90	6.46	1
3	26.70	2.41	17.28	18.61	2
3	36.81	16.10	19.62	30.14	3
3	51.34	22.71	29.53	39.12	4
3	62.69	30.19	38.57	51.15	5
3	72.68	41.64	48.20	59.11	6
4	0.00	11.90	4.48	27.62	1
4	4.52	27.75	18.01	42.63	2
4	13.71	42.22	21.96	59.20	3
4	27.91	56.06	34.42	74.92	4
4	38.57	66.16	40.14	92.37	5
4	42.71	78.71	57.06	98.02	6
5	13.11	8.00	10.90	14.36	1
5	16.89	18.27	28.22	26.37	2
5	27.99	27.72	38.62	34.15	3
5	36.35	42.04	48.31	54.02	4
5	48.85	48.50	60.23	59.90	5
5	61.97	59.92	71.39	74.79	6

5.1.1 The One-Way Analysis of Variance Model

The Sampling Model

One column of the data matrix contains observations on a continuous random variable Y. A second column of the data matrix contains observations on a classification variable X which classifies the n individuals or objects into g categories or groups. The X variable assumes one of the g integer values $1, 2, 3, \ldots, g$. We can view the n individuals as a random sample from a population which after classification according to X resulted in totals of $n_1, n_2, \ldots, n_g$ observations from the g groups respectively $\left(\sum_{j=1}^{g} n_j = n \right)$.

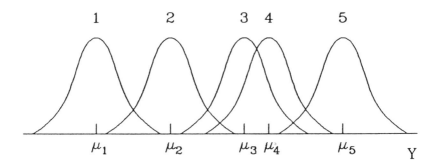

Figure 5.1. Comparison of Five Population Means

Equivalently, we could view the sample of size n as the result of selecting random samples of sizes $n_1, n_2, \ldots, n_g$ from the g groups respectively.

The means for the distribution of Y for the g groups are denoted by $\mu_1, \mu_2, \ldots, \mu_g$, while the variances are assumed to be homogeneous with magnitude σ^2. The distribution of Y is assumed to be normal for each group. Figure 5.1 represents a comparison of means for five population distributions in which the true means are all different. The objective in analysis of variance is to use the sample information to test the null hypothesis $H_0 : \mu_1 = \mu_2 = \cdots = \mu_g$; and if H_0 is rejected, to make comparisons among the g groups to characterize the differences. For the test marketing example, we wish to test the hypothesis that the mean family expenditure on milk products is the same for all four advertisement groups.

Given the g samples of observations, we denote by y_{ij} the i-th observation on Y taken in group j where $i = 1, 2, \ldots, n_j$ and $j = 1, 2, \ldots, g$. The sample means for the g groups are denoted by $\bar{y}_{.1}, \bar{y}_{.2}, \ldots, \bar{y}_{.g}$ where $\bar{y}_{.j} = \sum_{i=1}^{n_j} y_{ij}/n_j$.

The mean of all n observations is the grand mean $\bar{y}_{..} = \sum_{j=1}^{g} \sum_{i=1}^{n_j} y_{ij}/n$. The sample variances are denoted by s_j^2, $j = 1, 2, \ldots, g$ where $s_j^2 = \sum_{i=1}^{n_j} (y_{ij} - \bar{y}_{.j})^2/(n_j - 1)$. See Figure 5.2.

The g groups are sometimes referred to as cells. The population has been divided into g cells with cell means $\mu_1, \mu_2, \ldots, \mu_g$. The sample of size n is obtained by taking samples of $n_1, n_2, \ldots, n_g$ from the g cells respectively. A sampling model, describing the sampling process from the g cells, is required in order to derive a test procedure for testing the *equality of means*.

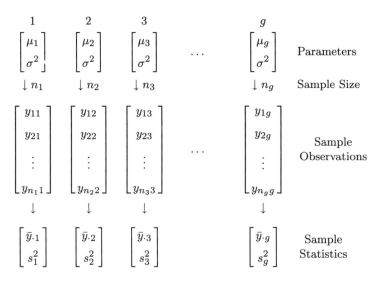

Figure 5.2. One-way Allocation of Sample to g Cells

Example

For the test marketing data displayed in Table 5.1 a sample of size 30 has been obtained for each of the four ads. The means and variances obtained from this sample for the four groups are given by

$$\bar{y}_{.1} = 37.239 \quad \bar{y}_{.2} = 39.647 \quad \bar{y}_{.3} = 34.951 \quad \bar{y}_{.4} = 51.030$$

$$s_1^2 = 430.32 \quad s_2^2 = 547.61 \quad s_3^2 = 354.24 \quad s_4^2 = 605.33. \tag{5.1}$$

Throughout this section the calculations for this example were performed using the SAS procedures PROC ANOVA or PROC REG.

The One-Way Anova Model

A model describing the behavior of the Y observations may be written as

$$y_{ij} = \mu_j + \varepsilon_{ij} \qquad i = 1, 2, \ldots, n_j; \quad j = 1, 2, \ldots, g; \tag{5.2}$$

where ε_{ij} represents a random variable with mean 0 and variance σ^2. By assumption the ε_{ij} are mutually independent. This model is sometimes called a *cell means model*. Under the assumption of *homogeneity of variance* the best unbiased estimator of μ_j is $\bar{y}_{.j}$. The model therefore can be written

$$y_{ij} = \bar{y}_{.j} + (y_{ij} - \bar{y}_{.j}) \qquad i = 1, 2, \ldots, n_j; \quad j = 1, 2, \ldots, g$$

with $\bar{y}_{.j}$ as an estimator of μ_j and $(y_{ij} - \bar{y}_{.j}) = e_{ij}$ as an estimator of ε_{ij}. The predicted values for y_{ij} from this model are $\hat{y}_{ij} = \bar{y}_{.j}$.

Sums of Squares

The *total variation in Y* over the sample, SST or the *total sum of squares*, is described by

$$\text{SST} = \sum_{j=1}^{g}\sum_{i=1}^{n_j}(y_{ij} - \bar{y}_{..})^2.$$

This sum of squares measures the variation in the Y values around the grand mean or overall mean given by $\bar{y}_{..} = \sum_{j=1}^{g}\sum_{i=1}^{n_j} y_{ij}/n$ where $n = \sum_{j=1}^{g} n_j$.

The variation explained by the fitted model is given by SSA, the *sum of squares among groups*. This sum of squares measures the *variation among the group means* and is determined as

$$\text{SSA} = \sum_{j=1}^{g}\sum_{i=1}^{n_j}(\hat{y}_{ij} - \bar{y}_{..})^2 = \sum_{j=1}^{g}\sum_{i=1}^{n_j}(\bar{y}_{\cdot j} - \bar{y}_{..})^2 = \sum_{j=1}^{g} n_j(\bar{y}_{\cdot j} - \bar{y}_{..})^2.$$

The variation in the sample within group j is given by $\sum_{i=1}^{n_j}(y_{ij} - \bar{y}_{\cdot j})^2$. This variation divided by $(n_j - 1)$ is denoted by s_j^2, and provides an unbiased estimator of σ^2 the true variance in group j. For all g groups the *total within-group sum of squares* is given by

$$\text{SSW} = \sum_{j=1}^{g}\sum_{i=1}^{n_j}(y_{ij} - \bar{y}_{\cdot j})^2.$$

It is easily demonstrated that $\text{SST} = \text{SSA} + \text{SSW}$.

Computation Formulae

The sums of squares can be computed more easily using the computing formulae

$$\text{SST} = \sum_{j=1}^{g}\sum_{i=1}^{n_j} y_{ij}^2 - n\bar{y}_{..}^2, \tag{5.3}$$

$$\text{SSW} = \sum_{j=1}^{g}\sum_{i=1}^{n_j} y_{ij}^2 - \sum_{j=1}^{g} n_j\bar{y}_{\cdot j}^2, \tag{5.4}$$

$$\text{SSA} = \sum_{j=1}^{g} n_j\bar{y}_{\cdot j}^2 - n\bar{y}_{..}^2. \tag{5.5}$$

The way in which SSA and SSW combine to form SST can be seen more clearly from these computational formulae.

Table 5.2. One-way Analysis of Variance Table (ANOVA Table)

Source	df	Sum of Squares	Mean Square	F
AMONG	$(g-1)$	SSA	$\text{MSA} = \text{SSA}/(g-1)$	MSA/MSW
WITHIN	$(n-g)$	SSW	$\text{MSW} = \text{SSW}/(n-g)$	
TOTAL	$(n-1)$	SST		

Estimation of σ^2

Under the assumption of homogeneity of variance the best unbiased estimator of σ^2 is given by $s^2 = \text{SSW}/(n-g)$. The estimator s^2 can be written as a weighted average of the individual group sample variances

$$s^2 = [(n_1 - 1)s_1^2 + (n_2 - 1)s_2^2 + \cdots + (n_g - 1)s_g^2]/(n - g).$$

The estimator for the common variance in the test marketing example is 484.38.

The ANOVA Table

Under H_0, $\text{SSA}/(g-1)$ and $\text{SSW}/(n-g)$ provide independent estimates of σ^2. Under the normality assumption the ratio $\dfrac{\text{SSA}}{\text{SSW}} \cdot \dfrac{(n-g)}{(g-1)}$ has an F distribution with $(g-1)$ and $(n-g)$ degrees of freedom. This test statistic can be used to test the hypothesis regarding equality of means for the g groups. The quantities which are used to compute F can be summarized in the ANOVA table, Table 5.2.

Expected Mean Squares

The expected value of the mean squares in the ANOVA table are given by

$$E[\text{MSA}] = \sigma^2 + \sum_{j=1}^{g} n_j(\mu_j - \mu)^2/(g-1),$$

where $\mu = \sum_{j=1}^{g} \mu_j/g$ and

$$E[\text{MSW}] = \sigma^2,$$

and hence under H_0 the expectations are equivalent.

Throughout this chapter the *expected mean squares* will be used to determine the proper ratio to use for F-tests for various hypotheses. Under the hypotheses being tested the ratio of the expected mean squares should have the value 1. In the case of MSA/MSW if H_0 is true $\mu_j = \mu$ and hence $E[\text{MSA}] = E[\text{MSW}]$.

Table 5.3. Analysis of Variance for Comparison of Mean Family Expenditures for Four Advertising Groups

Source	df	Sum of Squares	Mean Square	F
AMONG	3	4585.680	1528.560	3.16
WITHIN	116	56187.444	484.375	
TOTAL	119	60773.124		

Example

To apply the one-way analysis of variance model to the test marketing data given in Table 5.1, it is important that the assumptions required be clarified. For the purposes of this example it will be assumed that the advertising campaigns were randomly assigned to the various families without regard to region of the country or family size. Later in this chapter more restrictive assumptions will be used as more complex procedures are discussed. In addition, for any particular presentation of an ad, no more than one of the test families could have seen the ad. Thus each family receives a separate and distinct application of the treatment.

For the test marketing data the analysis of variance table is given in Table 5.3. The p-value for the F-statistic of 3.16 with 3 and 116 degrees of freedom is 0.0275. We conclude, therefore, that there are some differences among the means.

Upon examination of the four sample means in (5.1) it would appear that the average milk product consumption is larger for ad #4 than for the other three ads. A procedure is required which enables us to compare the means in pairs in such a way that we can make probability statements. A discussion of multiple comparison procedures will be provided below in 5.1.2.

Group Effects Model

An alternative model that is commonly used for one-way ANOVA is the *group effects model* given by

$$y_{ij} = \mu + \alpha_j + \varepsilon_{ij} \qquad i = 1, 2, \ldots, n; \quad j = 1, 2, \ldots, g$$

where

$$\sum_{j=1}^{g} \alpha_j = 0 \text{ and the } \varepsilon_{ij} \text{ are } N(0, \sigma^2). \tag{5.6}$$

Group Means

	1	2		g
cell means model	μ_1	μ_2	$\cdots$	μ_g
group effects model	$\mu + \alpha_1$	$\mu + \alpha_2$	$\cdots$	$\mu + \alpha_g$
base cell model	$\mu_g + \gamma_1$	$\mu_g + \gamma_2$	$\cdots$	μ_g

Figure 5.3. Group Mean Parameters

This particular parameterization uses $(g + 1)$ parameters to distinguish between the g means μ_j, $j = 1, 2, \ldots, g$ and hence the model is over-defined. For this reason the condition $\sum_{j=1}^{g} \alpha_j = 0$ must be added in order to determine the parameters uniquely. In this parameterization, the cell means μ_j are equivalent to $(\mu + \alpha_j)$, $j = 1, 2, \ldots, n$ where μ is the grand mean or overall mean, $\mu = \sum_{j=1}^{g} \mu_j / g$. A test of $H_0 : \mu_1 = \mu_2 = \cdots = \mu_g$ is therefore equivalent to a test of $H_0 : \alpha_1 = \alpha_2 = \cdots = \alpha_g = 0$. The minimum variance unbiased estimators of the parameters μ and α_j, $j = 1, 2, \ldots, g$ are given by $\bar{y}_{..}$ and $(\bar{y}_{.j} - \bar{y}_{..})$, $j = 1, 2, \ldots, g$ respectively. The advantage of this parameterization is that the g means are compared to an overall mean μ.

Given the model (5.6) the expected values of the mean squares are given by $E[\text{MSW}] = \sigma^2$ and $E[\text{MSE}] = \sigma^2 + \sum_{j=1}^{g} n_j \alpha_j^2 / (g - 1)$.

A Base Cell Model

A third alternative would be to write the model in the form of (5.6) but to leave out one of the g cells (say $j = g$)

$$y_{ij} = \mu + \gamma_j + \varepsilon_{ij} \quad i = 1, 2, \ldots, n_j; \quad j = 1, 2, \ldots, g - 1$$

$$y_{ig} = \mu + \varepsilon_{ig} \quad\quad i = 1, 2, \ldots, n_g.$$

In this case there are only g parameters being used. For this model the parameter $\mu = \mu_g$ and the values of γ_j denote deviations from μ_g, $\gamma_j = (\mu_j - \mu_g)$. The group g is therefore the base case for this model. The minimum variance unbiased estimators of μ_g and γ_j $j = 1, 2, \ldots, g - 1$ are given by $\bar{y}_{.g}$ and $(\bar{y}_{.j} - \bar{y}_{.g})$, $j = 1, 2, \ldots, g$ respectively. The group means are summarized in Figure 5.3 for the three alternative models.

Robustness

The analysis of variance procedure outlined above assumed that the observations were drawn at random from g normal distributions with common variance σ^2. This procedure is very robust to non-normality in the sense that the probability levels on the test statistics are robust to non-normality. The procedure may no longer be the most efficient, however. In the case of non-normality, transformations discussed in Chapter 2 may be useful. The ANOVA procedure is not generally robust in the presence of outliers. The sample distributions in each cell should be evaluated using the techniques of Chapter 2. Outliers can be detected at this stage.

If the group variances are unequal, the impact on the behavior of the F-test can be large, particularly if the n_j are unequal. If the large n_j are associated with the smaller σ_j^2 the F-statistic will tend to be too large. Transformations have been found to be useful when the magnitude of the variance is related to the magnitude of the mean. If the relationship is linear the $\log y$ transformation is recommended while if the relationship is curvilinear the $\sqrt{y}$ transformation is suggested.

Experimental Design and Randomization

A common source of data for the one-way ANOVA model is the *planned experiment*. The planned experiment is designed to compare the responses obtained from applying a variety of *treatments* to a population of available *experimental units*. In this case the experimental unit-treatment combinations are not randomly selected from a population of such combinations. Instead the set of available experimental units is fixed. The experimental units are assigned to receive one of the treatments being studied. To eliminate any systematic effects the experimental units should be randomly assigned to the various treatments. This experimental design is called a *completely randomized design*. The experimental units should not be assigned to treatments in any systematic or subjective manner. In the completely randomized design, all experimental units have the same chance of being assigned to any one of the treatments, or equivalently all possible treatment-experimental unit combinations are equally likely. Ideally the experimental units are reasonably homogeneous. By employing the principles of the completely randomized design, the one-way ANOVA model can be used to compare treatments even though the responses are not based on random samples from treatment populations. A discussion of the randomization principle is given in Kempthorne (1952).

The set of experimental units employed in an experiment may be viewed as a random sample of units selected from a population of experimental units. In this case the observed effects are called *random effects*, and the model is called a random effects model. Experiments involving human subjects are often viewed in this manner. It is also possible that the treatments employed in the experiment represent a random sample of treatments selected from a population of possible treatments. In this case, the observed effects would also be termed random effects. Random selections of treatments are rarely used. A discussion of random effects models is provided in Section 5.7.

As we shall see later, more complex designs may permit *repeated observations* on the same experimental unit-treatment combination, or different treatments assigned sequentially to the same experimental unit. Such designs are used to control for the presence of many influential factors while at the same time limiting the total number of experimental units. Note that in the test marketing data it was assumed that on any broadcast of an ad only one test family at the most would have seen the ad. While this may not be practical it is a necessary assumption for this design. Later in this chapter more appropriate designs will be discussed.

5.1.2 Multiple Comparisons

If the null hypothesis of equal means is rejected, a follow-up question would be: "which groups have different means?" Procedures for the comparison of cell means are called *multiple comparison procedures*. This section summarizes a variety of procedures commonly used to compare means. The procedures summarized in this section are all available in the SAS procedure PROC ANOVA.

LSD

For any two groups j and ℓ a $100(1 - \alpha)\%$ confidence interval for $(\mu_j - \mu_\ell)$ may be constructed using a t distribution. A $100(1 - \alpha)\%$ confidence interval for $(\mu_j - \mu_\ell)$ is given by

$$(\bar{y}_{\cdot j} - \bar{y}_{\cdot \ell}) \pm t_{\alpha/2;(n-g)} s \left[\frac{1}{n_j} + \frac{1}{n_\ell}\right]^{1/2}$$

where $s^2 = \text{SSW}/(n - g)$. Any pair of means may be compared using such an interval. If the interval includes 0 the means are said to be not significantly different at the α level. This procedure is commonly called the *least significant difference* or the LSD procedure.

Per Comparison Error and Experimentwise Error

The problem with multiple comparisons such as LSD is that if the number of groups, g, is large there are a total of $g(g-1)/2$ such comparisons. If a 5% type I error probability is used, we can expect $(.05)g(g-1)/2$ differences to appear significant even if there aren't any real differences. The t-test procedure therefore tends to find too many significant differences. The type I error α in this case is the *per comparison error* but not the *experimentwise error*. The experimentwise type I error rate refers to the probability of finding at least one significant difference over the complete set of all comparisons. Using the LSD procedure, the probability of finding at least one significant difference would be $1-(1-\alpha)^d$, where $d = g(g-1)/2$. For $g = 4$ and $\alpha = 0.05$, this would be $1 - (0.95)^6 = 0.265$. The probability of at least one significant difference is 0.265 if $\alpha = 0.05$ with four groups. The procedures outlined below are designed to reduce the experimentwise error rate.

Fisher's LSD

Use of multiple comparison procedures is usually preceeded by a simultaneous significance test for all means using the F-ratio as in Table 5.2. To protect the type I error level at α, R.A. Fisher recommended that the LSD procedure only be employed if the α level F-test for the equality of means hypothesis is rejected.

Bonferroni

An approximate procedure for controlling the experimentwise error rate at α can be obtained by using the *Bonferroni* method, introduced in Chapter 3 for a bivariate confidence ellipsoid. In the case of the g means there are a total of $g(g-1)/2 = p$ different pairs to be compared. Letting A_j, $j = 1, 2, \ldots, p$ denote the event that the j-th pair of means are declared equal we may write the probability that all p pairs are declared equal as $P[A_1 A_2 \ldots A_p]$. For an experimentwise rate of α we would like to have $(1 - \alpha) = P[A_1 A_2 \ldots A_p]$. By the Bonferroni inequality

$$P[A_1 A_2 \ldots A_p] \geq 1 - \sum_{j=1}^{p} P(\overline{A}_j)$$

where $\overline{A}_j$ denotes the complement of A_j. If we assume that $P(\overline{A}_j) = \alpha^\circ$ is constant for $j = 1, 2, \ldots, p$, we have $(1 - \alpha) \geq 1 - p\alpha^\circ$ and hence $\alpha^\circ \geq \alpha/p$ or $\alpha \leq p\alpha_0$. For a given α, the Bonferroni approach uses $\alpha^\circ = \alpha/g(g-1)/2$, and therefore uses $t_{\alpha^\circ/2;(n-g)}$ as the critical value of the test statistic rather than $t_{\alpha/2;(n-g)}$ as in the case of the LSD procedure.

The Bonferroni procedure is useful for making pairwise comparisons when the F-test of equality of means did not reject the equality hypothesis. In this case there may still be a large difference even though the F-test was not significant.

Sidak

An approach due to Sidak (1967) can also be used to obtain more conservative intervals. A $100(1 - \alpha)\%$ confidence interval for $(\mu_j - \mu_\ell)$ is given by

$$(\bar{y}_{\cdot j} - \bar{y}_{\cdot \ell}) \pm t_{\alpha^\circ/2;(n-g)} s \Big[\frac{1}{n_j} + \frac{1}{n_\ell}\Big]^{1/2}$$

where $\alpha^\circ = \big(1 - (1 - \alpha)^{2/g(g-1)}\big)$.

Scheffé

The most conservative approach to judging differences among means employs a result established by Scheffé (1959). For a paired comparison among means μ_j and μ_ℓ a $100(1 - \alpha)\%$ confidence interval is given by

$$(\bar{y}_{\cdot j} - \bar{y}_{\cdot \ell}) \pm s \Big[(g - 1)F_{\alpha;(g-1),(n-g)}\Big[\frac{1}{n_j} + \frac{1}{n_\ell}\Big]\Big]^{1/2}.$$

This method guarantees that the experimentwise error rate is controlled at level α for all possible paired comparisons among means.

To compare to the LSD procedure we can compare the interval length of

$$s\Big[\frac{1}{n_j} + \frac{1}{n_\ell}\Big]^{1/2}\big[(g - 1)F_{\alpha;(g-1)(n-g)}\big]^{1/2}$$

to the LSD interval length of

$$s\Big[\frac{1}{n_j} + \frac{1}{n_\ell}\Big]^{1/2} t_{\alpha/2;(n-g)} = s\Big[\frac{1}{n_j} + \frac{1}{n_\ell}\Big]^{1/2}[F_{\alpha;1,(n-g)}]^{1/2}.$$

Tukey

An alternative approach to the comparison of means developed by Tukey (1953) uses the studentized range distribution. Given g independent means $\bar{y}_{.j}$, $j = 1, 2, \ldots, g$ with common sample size $n_j = n_0$, $j = 1, 2, \ldots, g$ where $n = gn_0$ and $s^2 = \text{MSW}/n_0$, *the studentized range* is given by

$$q_{g,(n-g)} = \max_{\substack{1 \leq j \leq g \\ j \neq \ell}} |\bar{y}_{.j} - \bar{y}_{.\ell}|/s/\sqrt{n_0}.$$

The critical values of $q_{g,(n-g)}$ are denoted by $q_{\alpha;g,(n-g)}$ where $P[q_{g,(n-g)} \leq q_{\alpha;g,(n-g)}] = 1 - \alpha$. The critical values $q_{\alpha;g,(n-g)}$ are available in Table 5 of the Table Appendix.

For all $\binom{g}{2}$ paired comparisons among the g means a $100(1 - \alpha)\%$ confidence interval is given by

$$(\bar{y}_{.j} - \bar{y}_{.\ell}) - q_{\alpha;g,(n-g)}s/\sqrt{n_0} \leq (\mu_j - \mu_\ell) \leq (y_{.j} - y_{.\ell}) + q_{\alpha;g,(n-g)}s/\sqrt{n_0}.$$

If the interval does not include the value zero then we conclude that μ_j and μ_ℓ are significantly different. The confidence coefficient $(1 - \alpha)$ indicates that in $100(1 - \alpha)\%$ of such samples the above interval will correctly determine which of the $\binom{g}{2}$ pairwise comparisons are significant. In practice the g means are arranged in order from smallest to largest. If for any pair of means $\bar{y}_{.j}$, $\bar{y}_{.\ell}$ the difference $|\bar{y}_{.j} - \bar{y}_{.\ell}|$ exceeds $q_{\alpha;g,(n-g)}s/\sqrt{n_0}$ the two means are declared to be significantly different.

If the n_j are unequal an approximation suggested by Tukey is to replace $s/\sqrt{n_0}$ by $s\left[\frac{1}{2}\left(\frac{1}{n_j} + \frac{1}{n_\ell}\right)\right]^{1/2}$ for comparing $\bar{y}_{.j}$ and $\bar{y}_{.\ell}$. This method is commonly employed by computing software when the n_j are unequal. An alternative approach for unequal n_j suggested by Spjotvoll and Stoline (1973) replaces $\sqrt{n_0}$ in $s/\sqrt{n_0}$ by $\min(\sqrt{n_j}, \sqrt{n_\ell})$. A third alternative is to use the geometric mean of n_j and n_ℓ. All of these extensions are equivalent to Tukey's method when the n_j are equal.

GT2

An alternative group of extensions to the Tukey method when the n_j are unequal employs the *studentized maximum modulus* $m_{p,(n-g)}$, where

$$m_{p,(n-g)} = \max_{j \neq \ell} |y_{.j} - y_{.\ell}|/s\left(\frac{1}{n_j} + \frac{1}{n_\ell}\right)^{1/2}$$

where $p = g(g-1)/2$. The critical values of $m_{p,(n-g)}$ denoted by $m_{\alpha;p,(n-g)}$ are tabulated in Stoline and Ury (1979). Hochberg's (1974) GT2 method provides the $100(1 - \alpha)\%$ confidence interval

$$\bar{y}_{.j} - \bar{y}_{.\ell} \pm sm_{\alpha;p,(n-g)}\left[\frac{1}{n_j} + \frac{1}{n_\ell}\right]^{1/2}$$

for $(\mu_j - \mu_\ell)$.

Gabriel

A technique introduced by Gabriel (1978) also uses the studentized maximum modulus distribution. For each mean μ_j, the upper and lower $100(1 - \alpha)\%$ values L_j and U_j are determined using

$$L_j = \bar{y}_{\cdot j} - m_{\alpha;g^*,(n-g)}s/\sqrt{2n_j}$$

$$U_j = \bar{y}_{\cdot j} + m_{\alpha;g^*,(n-g)}s/\sqrt{2n_j},$$

and hence (L_j, U_j) provides a $100(1-\alpha)\%$ interval for μ_j. After calculating intervals for each mean, the intervals are compared. For pairs of means in which intervals do not overlap the means are declared significantly different. A confidence interval for $(\mu_j - \mu_\ell)$ is provided by

$$(L_j - U_\ell) \le (\mu_j - \mu_\ell) \le (U_j - L_\ell).$$

The advantage of Gabriel's method over other Tukey type methods for unequal n_j, is that only g intervals rather than $g(g - 1)/2$ need to be determined to make the $g(g - 1)/2$ comparisons. For equal sample sizes Gabriel's and Hochberg's methods are equivalent.

Duncan, S-N-K and REGW

An alternative approach to the comparison of means is known as a *range test*, which is a multiple stage approach. After rejecting the hypothesis that all g means are equal at level α_g each possible subset of $(g - 1)$ means is compared at level α_{g-1}. For every subset in which a rejection is obtained each possible subset of $(g - 2)$ means is compared at level α_{g-2}. This process continues as long as there is a rejection. For any subset in which a rejection is not obtained no further partitions are carried out on that subset. There are several alternative approaches to the multiple stage procedure. Three different tests are obtained by changing the formula for α_p where p is the number of means in the subset being compared. The three tests are Student-Newman-Keuls (S-N-K) with $\alpha_p = \alpha$, Duncan with $\alpha_p = 1 - (1-\alpha)^{p-1}$ and Ryan (1959, 1960), Einot and Gabriel (1975), Welsch (1977) (REGW) with $\alpha_p = 1 - (1-\alpha)^{p/g}$ for $p < (g-1)$ and $\alpha_p = \alpha$ for $p \ge g-1$.

The multiple stage approach begins by arranging the g means in order from smallest to largest. Adjacent means in the ordered set are compared using α_2. Means which are one mean apart are compared using α_3 and so on. Finally the smallest and largest means are compared using α_g.

The Duncan and S-N-K procedures employ the studentized range statistic defined above. For the REGW procedure either the studentized range statistic or the F-statistic can be used. The advantage of the REGW F-statistic is that it is compatible with the overall ANOVA F-statistic.

For the procedures using the studentized range statistic the comparison of a difference of two means is made with $(q_{\alpha_g;p,n-g})s\left[\frac{1}{2}\left(\frac{1}{n_j} + \frac{1}{n_\ell}\right)\right]^{1/2}$. For the REGW procedure, which uses an F distribution, the subset of means being tested are compared using a conventional F-test with $\alpha = \alpha_p$ and d.f. $(p - 1)$ and $(n - g)$.

Example

For the test market example introduced earlier a comparison of the four means was carried out using various paired comparisons procedures. An overall $\alpha = 0.05$ level was selected for the comparisons. The value of s is given by $\sqrt{\text{MSE}} = \sqrt{484.375} = 22.01$ and the value of $s\left(\frac{2}{n}\right)^{1/2} = 22.01\left(\frac{2}{30}\right)^{1/2} = 5.68$. For the LSD procedure $t_{\alpha/2;n-g} = t_{.025,116} = 1.98$. The critical difference for the LSD test is $(1.98)(5.68) = 11.25$. For the Bonferroni method $\alpha° = \alpha/g(g-1)/2 = .05/6 = .0083$ and the critical t-value is $t_{\alpha°/2;n-g} = t_{.0083/2;116} = 2.684$. The critical difference in this case is $(2.684)(5.68) = 15.245$. For the SIDAK approach $\alpha° = 1 - (1-\alpha)^{2/g(g-1)} = 1 - (.95)^{1/6} = .0085$ and the critical t-value is $t_{\alpha°/2;(n-g)} = t_{.0085/2;116} = 2.677$. The critical difference for the SIDAK test is $(2.677)(5.68) = 15.205$. For the observed means in the four groups 37.239, 39.647, 34.951 and 51.030 the LSD test concludes that group 4 is significantly different from the others. By the Sidak and Bonferroni procedures the only difference that is significant is the difference between groups 3 and 4 $(51.030 - 34.951)$.

To carry out Scheffé's procedure the critical F-statistic is given by $F_{.05,3,116} = 2.683$ and the critical difference is given by

$$s\left(\frac{1}{n}\right)^{1/2}\left[(g-1)F_{.05,(g-1),(n-g)}\right]^{1/2} = (5.682)\left(\frac{1}{30}\right)^{1/2}[3(2.683)]^{1/2} = 16.120.$$

In this case none of the means would be declared significantly different.

The critical value of the studentized range statistic $q_{\alpha,\,,n-g} = q_{.05,4,116}$ is equal to 3.686. For Tukey's test the critical difference is given by $q_{.05,4,116}\frac{s}{\sqrt{n}} = (3.686)(22.008)\left(\frac{1}{30}\right)^{1/2} = 14.810$. The critical value of the studentized maximum modulus statistic is $m_{\alpha;g(g-1)/2,(n-g)} = m_{.05;6,116} = 2.675$. The critical difference for Hochberg's GT2 method is given by $m_{.05;6,116}s\left(\frac{2}{n}\right)^{1/2} = (2.675)(5.682) = 15.199$. Both Tukey's method and the GT2 method find only the difference between groups 3 and 4 significant.

The critical values for the Duncan, S-N-K and REGW are based on the α values below

DUNCAN	$\alpha_2 = 1 - (1-\alpha)$	$\alpha_3 = 1 - (1-\alpha)^2$	$\alpha_4 = 1 - (1-\alpha)^3$
	$= .05$	$= .0975$	$= .143$
S-N-K	$\alpha_2 = \alpha$	$\alpha_3 = \alpha$	$\alpha_4 = \alpha$
	$= .05$	$= .05$	$= .05$
REGW	$\alpha_2 = 1 - (1-\alpha)^{2/4}$	$\alpha_3 = 1 - (1-\alpha)^{3/4}$	$\alpha_4 = 1 - (1-\alpha)$
	$= .025$	$= .038$	$= .05$

The resulting critical values of the studentized range multiplied by $s\left(\frac{1}{n}\right)^{1/2}$ are given by

	$p = 2$	$p = 3$	$p = 4$
DUNCAN	11.255	11.871	12.245
S-N-K	11.255	13.492	14.813
REGW	12.875	13.492	14.813

Both Duncan's and S-N-K's procedures conclude that the group 4 mean is significantly different from the other three means. For the REGW procedure the group 4 mean is significantly larger than the means for groups 3 and 1.

Contrasts

Paired comparisons among means outlined above are special cases of *contrasts*. A contrast is a linear combination of the means

$$L = \sum_{j=1}^{g} \lambda_j \mu_j \qquad \text{where} \quad \sum_{j=1}^{g} \lambda_j = 0.$$

Contrasts are useful for comparisons involving more than two means. If a reseacher believes a priori that $\mu_1 = \mu_3$ and $\mu_2 = \mu_4$ but that $(\mu_1, \mu_3) \neq (\mu_2, \mu_4)$ then a useful hypothesis to test is given by $H_0 : (\mu_1 + \mu_3)/2 - (\mu_2 + \mu_4)/2 = 0$. This hypothesis is a contrast with coefficients $1/2$, $-1/2$, $1/2$ and $-1/2$ for μ_1, μ_2, μ_3 and μ_4 respectively. The *Scheffé method* for comparing means can be extended to contrasts. A $100(1 - \alpha)\%$ confidence interval for the contrast $L = \sum_{j=1}^{g} \lambda_j \mu_j$ is given by

$$\sum_{j=1}^{g} \lambda_j \bar{y}_{\cdot j} \pm s \left[(g - 1) F_{\alpha;(g-1),(n-g)} \sum_{j=1}^{g} \lambda_j^2 / n_j \right]^{1/2}.$$

This method controls the experimentwise error rate at α, and hence for ALL possible contrast confidence intervals that could be constructed in this manner, the proportion of intervals that would not include zero is α if the g population means are equal. Since this method controls the experimentwise type I error it should be used for all contrasts which are not planned a priori.

On rare occasions it is possible for the F-test to reject $H_0 : \mu_1 = \mu_2 = \cdots = \mu_g$ at the α level and yet all possible paired comparisons among the means using the Scheffé procedure will not indicate any significant differences. Since in this case the F-test rejects H_0 there is at least one contrast that is significantly different from zero. The contrast weights, that provide the largest value of

$$\left[\sum_{j=1}^{g} \lambda_j \bar{y}_{\cdot j}\right]^2 \Big/ \left[s^2 \Sigma \lambda_j^2 / n_j\right],$$

are given by $\lambda_j = n_j(\bar{y}_{\cdot j} - \bar{y}_{\cdot\cdot})$, $j = 1, 2, \ldots, g$. This contrast may provide some insight into the nature of the differences among the means.

5.1.3 Multiple Regression Model Approach

The techniques outlined in Section 5.1.1 can be obtained as a special case of the multiple regression model by using dummy variables to represent qualitative variables. Dummy variables were first introduced in Section 4.5.

Dummy Variables

The one-way model can also be written in a multiple regression format by defining g dummy variables to represent the g groups. Define

$$
\begin{aligned}
D_1 &= 1 \quad \text{if the observation is from group 1} \\
&= 0 \quad \text{otherwise} \\[6pt]
D_2 &= 1 \quad \text{if the observation is from group 2} \\
&= 0 \quad \text{otherwise} \\[6pt]
&\vdots \\[6pt]
D_g &= 1 \quad \text{if the observation is from group } g \\
&= 0 \quad \text{otherwise.}
\end{aligned}
$$

While the g dummy variables can be used to characterize the differences among the means, it is sometimes more useful to delete one of the dummies (say the last one Dg), and instead retain a column of unities or intercept term as in the case of multiple regression. This approach permits a comparison of the means using the F-statistics used for testing no linear relationship in a multiple regression.

Linear Model Notation

Denoting the vector of Y observations by

$$\mathbf{y}' = [y_{11}, y_{21}, \ldots, y_{n_1 1}, y_{12}, y_{22}, \ldots, y_{n_2 2}, \ldots, y_{1g}, y_{2g}, \ldots, y_{n_g g}]$$

the resulting multiple regression model is given by

$$\mathbf{y} = \mathbf{X}\boldsymbol{\beta} + \mathbf{u},$$

where $\boldsymbol{\beta}' = [\beta_0, \beta_1, \ldots, \beta_{g-1}] = [\mu_g, \mu_1 - \mu_g, \mu_2 - \mu_g, \ldots, \mu_{g-1} - \mu_g]$ and

$$\mathbf{X} = \begin{bmatrix} 1 & 1 & 0 & \cdots & 0 \\ 1 & 1 & 0 & \cdots & 0 \\ \vdots & \vdots & \vdots & & \vdots \\ 1 & 1 & 0 & \cdots & 0 \\ 1 & 0 & 1 & \cdots & 0 \\ 1 & 0 & 1 & \cdots & 0 \\ \vdots & \vdots & \vdots & & \vdots \\ 1 & 0 & 1 & \cdots & 0 \\ \vdots & \vdots & \vdots & & \vdots \\ 1 & 0 & 0 & \cdots & 1 \\ \vdots & \vdots & \vdots & & \vdots \\ 1 & 0 & 0 & \cdots & 1 \\ 1 & 0 & 0 & \cdots & 0 \\ \vdots & \vdots & \vdots & & \vdots \\ 1 & 0 & 0 & \cdots & 0 \end{bmatrix} \qquad \mathbf{u} = \begin{bmatrix} \varepsilon_{11} \\ \varepsilon_{21} \\ \vdots \\ \varepsilon_{n_1 1} \\ \varepsilon_{21} \\ \varepsilon_{22} \\ \vdots \\ \varepsilon_{n_2 2} \\ \vdots \\ \varepsilon_{1g} \\ \varepsilon_{2g} \\ \vdots \\ \varepsilon_{n_g g} \end{bmatrix}.$$

The $\mathbf{X}$ matrix is usually called the *design matrix*. This model is equivalent to the base cell model introduced in the last section.

Least Squares Estimation

The least squares estimator of $\boldsymbol{\beta}$ is given by

$$\mathbf{b} = (\mathbf{X}'\mathbf{X})^{-1}\mathbf{X}'\mathbf{y}$$

and can be written as

$$\mathbf{b}' = [\bar{y}_{\cdot g}, (\bar{y}_{\cdot 1} - \bar{y}_{\cdot g}), \ldots, (\bar{y}_{\cdot g-1} - \bar{y}_{\cdot g})].$$

The ordinary least squares estimators for the parameters $\beta_0, \beta_1, \ldots, \beta_{g-1}$ are given by

$$b_j = (\bar{y}_{\cdot j} - \bar{y}_{\cdot g}), \quad j = 1, 2, \ldots, (g-1) \text{ and } b_0 = \bar{y}_{\cdot g}.$$

The predicted values of the y_{ij} are given by

$$\hat{y}_{ij} = b_0 + b_j \qquad j = 1, 2, \ldots, (g-1) \quad i = 1, 2, \ldots, n_j$$

$$\hat{y}_{ig} = b_0 \qquad i = 1, 2, \ldots, n_g$$

and hence $\hat{y}_{ij} = \bar{y}_{\cdot j}$, $i = 1, 2, \ldots, n_j$; $j = 1, 2, \ldots, g$.

As in multiple linear regression the sums of squares for this model are given by

$$\text{SST} = \sum_{j=1}^{g} \sum_{i=1}^{n_j} (y_{ij} - \bar{y}_{\cdot\cdot})^2 = \sum\sum y_{ij}^2 - n\bar{y}_{\cdot\cdot}^2 \qquad \text{as in (5.3)}$$

$$\text{SSR} = \sum_{j=1}^{g} \sum_{i=1}^{n_j} (\hat{y}_{ij} - \bar{y}_{\cdot\cdot})^2 = \sum_{j=1}^{g} n_j \bar{y}_{\cdot j}^2 - n\bar{y}_{\cdot\cdot}^2 = \text{SSA} \qquad \text{as in (5.5)}$$

$$\text{SSE} = \sum_{j=1}^{g} \sum_{i=1}^{n_j} (y_{ij} - \hat{y}_{ij})^2 = \sum_{j=1}^{g} \sum_{i=1}^{n_j} y_{ij}^2 - \sum_{j=1}^{g} n_j \bar{y}_{\cdot j}^2 = \text{SSW} \quad \text{as in (5.4)}$$

where $n = \sum_{j=1}^{g} n_j$.

The conventional F-test of $H_0 : \beta_1 = \beta_2 = \cdots = \beta_{g-1} = 0$ is therefore equivalent to $H_0 : \mu_1 = \mu_2 = \cdots = \mu_g$.

All multiple regression approach examples in this chapter have been carried out using the SAS procedure PROC REG.

Example

For the test marketing data summarized in Table 5.1, a comparison of the mean family expenditures for the four ads was carried out using one-way analysis of variance and is summarized in Table 5.3. To apply a dummy variable approach, the dummy variables AD1, AD2 and AD3 were defined to represent the first three ads; the fourth ad was chosen as the base case. For example purposes the **y** vector and **X** matrix for the observations in region 1 are illustrated in Table 5.4. The equation for the model is given by

$$\text{EXPEND} = \beta_0 + \beta_1 \text{ AD1} + \beta_2 \text{ AD2} + \beta_3 \text{ AD3} + U.$$

The mean expenditures for the four ad groups are therefore $(\beta_0 + \beta_1)$, $(\beta_0 + \beta_2)$, $(\beta_0 + \beta_3)$ and β_0 for the ads 1, 2, 3, and 4 respectively.

The estimated relationship based on the complete data set is given by

$$\text{EXPEND} = \underset{(.000)}{51.03} - \underset{(.017)}{13.79} \text{ AD1} - \underset{(.048)}{11.38} \text{ AD2} - \underset{(.006)}{16.08} \text{ AD3}$$

Table 5.4. **y** and **X** Matrix for Dummy Variable Coding for Region 1 in Test Marketing Example

$$
\mathbf{y} = \begin{bmatrix} 12.35 \\ 20.52 \\ 30.85 \\ 39.35 \\ 48.87 \\ 58.01 \\ 21.86 \\ 42.17 \\ 49.61 \\ 63.65 \\ 73.75 \\ 85.95 \\ 14.43 \\ 22.26 \\ 23.99 \\ 36.98 \\ 42.13 \\ 54.19 \\ 21.44 \\ 31.21 \\ 40.09 \\ 55.68 \\ 65.81 \\ 76.61 \end{bmatrix}
\qquad
\mathbf{X} = \begin{bmatrix}
1 & 1 & 0 & 0 \\
1 & 1 & 0 & 0 \\
1 & 1 & 0 & 0 \\
1 & 1 & 0 & 0 \\
1 & 1 & 0 & 0 \\
1 & 1 & 0 & 0 \\
1 & 0 & 1 & 0 \\
1 & 0 & 1 & 0 \\
1 & 0 & 1 & 0 \\
1 & 0 & 1 & 0 \\
1 & 0 & 1 & 0 \\
1 & 0 & 1 & 0 \\
1 & 0 & 0 & 1 \\
1 & 0 & 0 & 1 \\
1 & 0 & 0 & 1 \\
1 & 0 & 0 & 1 \\
1 & 0 & 0 & 1 \\
1 & 0 & 0 & 1 \\
1 & 0 & 0 & 0 \\
1 & 0 & 0 & 0 \\
1 & 0 & 0 & 0 \\
1 & 0 & 0 & 0 \\
1 & 0 & 0 & 0 \\
1 & 0 & 0 & 0
\end{bmatrix}
$$

with $R^2 = 0.075$. The regression sum of squares is 4585.7 and the error sum of squares is 56187.4. The means are therefore 37.24, 39.65, 34.95 and 51.03 respectively.

A conventional multiple regression F-test of the hypothesis $H_0 : \beta_1 = \beta_2 = \beta_3 = 0$ is equivalent to a test of equality of the four means. The F-statistic and p-value for this test are 3.16 and 0.028 respectively. The p-values for the t-statistics for the three coefficients are shown below the coefficients in the above equation. From the p-values we can conclude that the means for ads 1, 2 and 3 are all significantly different from AD4 at the 0.05 level. These t-statistic p-values are equivalent to the LSD test p-values given in Section 5.1.2.

Effect Coding

Effect coding introduced in Chapter 4 may also be used. The variables or columns of **X** are defined as

$$
\begin{aligned}
D_j &= \quad 1 \quad \text{for group } j \quad j = 1, 2, \ldots, (g-1) \\
&= \quad 0 \quad \text{for groups other than } j \text{ or } g \\
&= -1 \quad \text{for group } g.
\end{aligned}
$$

The **X** matrix contains a column of unities plus the $(g-1)$ variables D_1, D_2, D_{g-1}. The means for the groups are given by $(\beta_0+\alpha_j)$ $j = 1, 2, \ldots, (g-1)$ and $\left[\beta_0 - \sum_{j=1}^{g-1} \alpha_j\right]$ for group g. The parameter β_0 is the average of the means $\beta_0 = \sum_{j=1}^{g} \mu_j/g$. The parameters $\alpha_1, \alpha_2, \ldots, \alpha_{g-1}$, $\left(-\sum_{j=1}^{g-1} \alpha_j\right)$ measure the g group effects and hence the total of the group effects is zero. The effect coding approach is equivalent to the group effects model in (5.6).

Example

Using the effect coding approach for the above example the following variables are defined

$$
\begin{aligned}
ADV1 &= 1 \quad \text{for Ad 1} \\
&= 0 \quad \text{for Ads 2 and 3} \\
&= -1 \quad \text{for Ad 4}
\end{aligned}
$$

$$
\begin{aligned}
ADV2 &= 1 \quad \text{for Ad 2} \\
&= 0 \quad \text{for Ads 1 and 3} \\
&= -1 \quad \text{for Ad 4}
\end{aligned}
$$

$$
\begin{aligned}
ADV3 &= 1 \quad \text{for Ad 3} \\
&= 0 \quad \text{for Ads 1 and 2} \\
&= -1 \quad \text{for Ad 4.}
\end{aligned}
$$

The **X** matrix for the data of Table 5.4 using effect coding is given in Table 5.5.

The estimated model using the effect coding variables ADV1, ADV2, ADV3 defined above is given by

$$
EXPEND = \underset{(0.000)}{40.72} - \underset{(0.320)}{3.48} \; ADV1 - \underset{(0.759)}{1.07} \; ADV2 - \underset{(0.100)}{5.77} \; ADV3.
$$

The coefficient 40.72 represents the average of the means in the twenty cells. The coefficient –3.48 is obtained by subtracting 40.72 from the average of the five AD1 cell means (one for each of five regions). Similarly the coefficients –1.07 and –5.77 denote the deviations of the average of the AD2 cell and AD3 cell means from 40.72 respectively. The average of the cell means for AD4 is $40.72+5.77+3.48+1.07 = 51.04$. Since the design in this case has the same number of observations in each cell, the averages of cell means for a given ad are equivalent to the average of all the values for that ad. If the design was not balanced a weighted average of the cell means would be required to obtain the average of the values. The regression sum

Table 5.5. **X** Matrix for Effect Coding for Region 1 in Test Marketing Example

X

$$
\begin{bmatrix}
1 & 1 & 0 & 0 \\
1 & 1 & 0 & 0 \\
1 & 1 & 0 & 0 \\
1 & 1 & 0 & 0 \\
1 & 1 & 0 & 0 \\
1 & 1 & 0 & 0 \\
1 & 0 & 1 & 0 \\
1 & 0 & 1 & 0 \\
1 & 0 & 1 & 0 \\
1 & 0 & 1 & 0 \\
1 & 0 & 1 & 0 \\
1 & 0 & 1 & 0 \\
1 & 0 & 0 & 1 \\
1 & 0 & 0 & 1 \\
1 & 0 & 0 & 1 \\
1 & 0 & 0 & 1 \\
1 & 0 & 0 & 1 \\
1 & 0 & 0 & 1 \\
1 & -1 & -1 & -1 \\
1 & -1 & -1 & -1 \\
1 & -1 & -1 & -1 \\
1 & -1 & -1 & -1 \\
1 & -1 & -1 & -1 \\
1 & -1 & -1 & -1 \\
\end{bmatrix}
$$

of squares is 56187.4 as in the case of dummy variable coding. The p-values for the coefficients indicate that the means for the AD1, AD2 and AD3 are not significantly different from the overall average at the 0.10 level. In comparison to the results obtained from dummy coding, we can see that the choice of coding method determines the types of inferences produced by the t-statistics for the coefficients.

Orthogonal Coding

In the previous section we have seen that the use of either dummy coding or effect coding can facilitate the testing of certain hypotheses. For dummy coding, a base case is selected and all parameters measure deviations from this base. The overall goodness of fit measure R^2 reflects the variation of the various categories around a base category. If any of the dummy variables are deleted from the model, the base case changes and hence the values of the coefficients of the dummy variables will also change. In other words the dummy variables are not mutually orthogonal, and hence the parameter values depend on which dummy variables are included in the model.

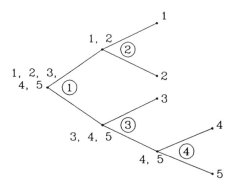

Figure 5.4. Nested Partitions for Groups

With effect coding, the parameters in the model include the overall mean plus the set of all paired comparisons between a base category and the remaining categories. The advantage of effect coding is that the explanatory variables are mutually orthogonal when the number of observations from each group n_j, $j = 1, 2, \ldots, g$, are all equal. If the columns are mutually orthogonal, the parameters remain the same regardless of how many of the variables are omitted. In this case, the set of all paired comparisons between a base category and the remaining categories are mutually orthogonal. It is possible to represent the variation by an alternative set of explanatory variables which are mutually orthogonal, and in addition provide a number of different comparisons among parameters. This form of coding is called *orthogonal coding*.

For a one-way analysis of variance, g means can be compared using a set of $(g - 1)$ mutually orthogonal columns in the design matrix. A useful way to determine a set of orthogonal columns is by a sequence of nested partitions of the g means. For a set of 5 means a set of 4 comparisons can be obtained from the nested partitions as shown in Figure 5.4.

In this case the four comparisons among the five means are

1. μ_1, μ_2 vs μ_3, μ_4, μ_5

2. μ_1 vs μ_2

3. μ_3 vs μ_4, μ_5

4. μ_4 vs μ_5.

For each of the comparisons a contrast $\sum_{j=1}^{g} c_j \mu_j$ is formed so that $\sum_{j=1}^{g} c_j =$

0. For the above four comparisons the contrasts are given by

1. $3(\mu_1 + \mu_2) - 2(\mu_3 + \mu_4 + \mu_5)$
2. $\mu_1 - \mu_2$
3. $2\mu_3 - (\mu_4 + \mu_5)$
4. $\mu_4 - \mu_5$.

The design matrix corresponding to the four contrasts is therefore

	1	2	3	4
μ_1	3	1	0	0
μ_2	3	-1	0	0
μ_3	-2	0	2	0
μ_4	-2	0	-1	1
μ_5	-2	0	-1	-1

It is easily seen that these four columns are mutually orthogonal. Thus if the design has an equal number of observations from each of the five groups the four columns will remain mutually orthogonal.

Examples

In the family expenditure study examined throughout this chapter four ads were compared using thirty observations for each ad. A possible set of three orthogonal comparisons is provided by defining the three variables A1, A2 and A3 as indicated by the following design matrix. The design matrix for one of the 30 repetitions is given by

Ad	A1	A2	A3
1	1	0	1
2	-1	0	1
3	0	1	-1
4	0	-1	-1

These three orthogonal variables permit the following three independent comparisons among the four means.

$H_{01} : (\mu_1 - \mu_2) = 0$, $H_{02} : (\mu_3 - \mu_4) = 0$ and $H_{03} : (\mu_1 + \mu_2) - (\mu_3 + \mu_4) = 0$

If the hypothesis $H_0 : \mu_1 = \mu_2 = \mu_3 = \mu_4$ is rejected the three contrasts illustrated above can also be tested at the same α. These contrasts are planned in that they have been built in to the design matrix. Any other comparisons among the means using multiple comparison procedures suffer from the usual problems associated with "post hoc" comparisons. Conservative procedures such as Tukey or Scheffé should be employed for such unplanned contrasts.

A Cell Means Model Approach

In the multiple regression model forms of the one-way analysis of variance model outlined above, the design matrix $\mathbf{X}$ always contained a column of unities. In such cases the effects measured by the remaining columns of $\mathbf{X}$ were measured relative to a base case or relative to the overall mean. Such approaches have been motivated by the classical form of the ANOVA model which defines parameters in terms of deviations from the mean, and also by the desire to employ early versions of multiple regression software which assumed that the first column of $\mathbf{X}$ contained a column of unities.

An alternative approach is to omit the column of unities in $\mathbf{X}$ and to let the columns of $\mathbf{X}$ contain indicator variables for the mean parameters μ_j, $j = 1, 2, \ldots, g$. In this case the linear model is given by

$$\mathbf{y} = \mathbf{X}\boldsymbol{\beta} + \mathbf{u}$$

where $\boldsymbol{\beta}(g \times 1) = \boldsymbol{\mu}(g \times 1)$ the column vector of mean parameter elements. Hypotheses regarding the elements of $\boldsymbol{\mu}$ can be tested using procedures designed to test $H_0 : \mathbf{T}\boldsymbol{\beta} = \boldsymbol{\beta}^*$ for a specified $\boldsymbol{\beta}^*$ in the multiple regression model.

For the usual test of equality for the elements of $\boldsymbol{\mu}$ the matrix $\mathbf{T}$ can be given by the $(g - 1) \times g$ matrix

$$\mathbf{T} = \begin{bmatrix} 1 & -1 & \cdots & & 0 \\ 1 & 0 & -1 & & \vdots \\ \vdots & & & -1 & 0 \\ 1 & 0 & \cdots & 0 & -1 \end{bmatrix}.$$

The hypothesis $H_0 : \mathbf{T}\boldsymbol{\mu} = 0$ is equivalent to the usual hypothesis of interest $H_0 : \mu_1 = \mu_2 = \cdots = \mu_g$. Many multiple regression software packages now permit this form of analysis. The F-test procedure for hypotheses of this type was outlined in Chapter 4. Individual t-tests can be used to test the individual components of $\mathbf{T}\boldsymbol{\mu} = 0$.

Example

For the test marketing data a model was fitted with all four dummy variables and no intercept. The model is given by

$$\text{EXPEND} = \beta_1 \text{ AD1} + \beta_2 \text{ AD2} + \beta_3 \text{ AD3} + \beta_4 \text{ AD4} + U.$$

The fitted model using the data in Table 5.1 is given by

$$\text{EXPEND} = 37.249 \text{ AD1} + 39.647 \text{ AD2} + 34.951 \text{ AD3} + 51.030 \text{ AD4}$$

and hence the regression coefficients are the four group or cell means. To test the equality of the means, the transformation matrix $\mathbf{T}$ is given by

$\mathbf{T} = \begin{bmatrix} 1 & -1 & 0 & 0 \\ 1 & 0 & -1 & 0 \\ 1 & 0 & 0 & -1 \end{bmatrix}$ and the restricted model must satisfy $\mathbf{T}\boldsymbol{\beta} = 0$,

where $\boldsymbol{\beta}$ is the vector containing the four regression coefficients defined above. The fitted model in this case is given by

$$\text{EXPEND} = 40.72(\text{AD1} + \text{AD2} + \text{AD3} + \text{AD4}).$$

The F-statistic for this restriction is given by 3.16 with 3 degrees of freedom and a p-value of 0.028. This test statistic is equivalent to the statistic obtained from dummy coding and effect coding earlier.

5.1.4 Nonparametric ANOVA

If the usual assumptions of normality and homogeneity of variance are not satisfied, there are nonparametric alternatives to the F-test for equality of group means.

Kruskal–Wallis

The most commonly used procedure is the *Kruskal-Wallis test* which uses the ranks of the observations. All of the observations y_{ij} are ranked jointly. When observations are tied, the tied observations are assigned the average of the rank positions they occupy. The rank scores R are then assigned to their respective groups and denoted by R_{ij}. The ranks in each group are summed to yield the rank sums $R_{\cdot j}$, $j = 1, 2, \ldots, g$. The test of equality of the g populations is carried out using the statistic

$$T = \left[\sum_{j=1}^{g} R_{\cdot j}^2 - \frac{n(n+1)^2}{4} \right] / s^2$$

where

$$s^2 = \frac{1}{n-1} \left[\sum_{j=1}^{g} \sum_{i=1}^{n_j} R_{ij}^2 - n \frac{(n+1)^2}{4} \right].$$

Under the null hypothesis the statistic T has a χ^2 distribution with $(g-1)$ degrees of freedom in large samples. Multiple comparisons among groups can be carried out by comparing $\left(\overline{R}_{\cdot j} - \overline{R}_{\cdot \ell} \right)$ to

$$t_{\alpha/2;n-g} \left(\frac{n-1-T}{n-g} \right)^{1/2} \left(\frac{1}{n_j} + \frac{1}{n_\ell} \right)^{1/2} s,$$

where $t_{\alpha/2;(n-g)}$ is the value of t such that $P[t > t_{\alpha/2;(n-g)}] = \alpha/2$.

Van Der Waerden Test

An alternative nonparametric procedure, which also begins with the ranks determined in the Kruskal-Wallis test, is called the *van der Waerden test*. Beginning with the ranks R determined above, the ranks are converted to normal scores by using the quantiles $R/(n+1)$ for the normal distribution. The normal scores are then reassigned to their respective groups and denoted by A_{ij}. The test statistic for the equality of distributions is given by

$$V = \sum_{j=1}^{g} n_j (\overline{A}_{.j})^2 \Big/ \Big(\frac{1}{n-1}\Big) \sum_{j=1}^{g} \sum_{i=1}^{n_j} A_{ij}^2.$$

Under the null hypothesis V has a χ^2 distribution with $(g-1)$ degrees of freedom. Mean differences $(\overline{A}_{.j} - \overline{A}_{.\ell})$ can be compared to

$$t_{\alpha/2;(n-g)} \Big(\frac{n-1-V}{n-g}\Big)^{1/2} \Big(\frac{1}{n_j} + \frac{1}{n_\ell}\Big)^{1/2} s.$$

Example

Using the test marketing example data, the Kruskal-Wallis and van der Waerden procedures were employed to compare the four advertisement groups. For the four ad groups 1, 2, 3, and 4 the rank sum scores were 1675, 1781.5, 1652.5 and 2241.0 respectively. The expected value of the rank sum scores under the equality hypothesis is 1815. The denominator s^2 of the T-statistic is given by 165. The computed value of T is 7.33 which has a χ^2 distribution with 3 d.f. if the hypothesis of equality holds. The p-value for this χ^2 is 0.06.

For the van der Waerden test the sums of the normal scores for ad groups 1, 2, 3 and 4 are given respectively by -4.57, -1.67, -6.95 and 13.19. The value of the statistic V is 8.72 which has a χ^2 distribution with 3 d.f. under the equality hypothesis. The p-value for χ^2 is 0.03. We can conclude from these two test procedures that there is some evidence that ad number 4 yields larger consumption values than the other three groups.

5.1.5 Homogeneity of Variance

In addition to comparing the means among several normal distributions it may also be of interest to compare the variances. One of the assumptions made in analysis of variance is that the variances are homogeneous. It may be of interest to test the hypothesis of equality of variances at the same time that the ANOVA is being carried out. Test procedures for the *homogeneity of variance* hypothesis are outlined below.

Bartlett's Test

Under the assumption of normality the homogeneity of variance null hypothesis $H_0 : \sigma_1^2 = \sigma_2^2 = \cdots = \sigma_g^2$ can be tested using *Bartlett's test*. Given the sample data of 5.1.1 the statistic $M/(c+1)$ can be used to test H_0 where s_j^2 and s^2 are defined in 5.1.1 and where

$$M = \sum_{j=1}^{g}(n_j - 1)\ln(s^2/s_j^2)$$

and

$$c = \left(\frac{1}{3(g-1)}\right)\left[\sum_{j=1}^{g}(\frac{1}{n_j-1}) - 1/\sum_{j=1}^{g}(n_j-1)\right].$$

Under H_0 the statistic $M/(c+1)$ is distributed as a χ^2 with $(g-1)$ d.f. in large samples. A statistic for testing H_0 which employs an F approximation is given by

$$F = \frac{d\,M}{(g-1)(d/f - M)},$$

where

$$f = (1-c) + 2/d$$

$$d = (g+1)/c^2.$$

Under H_0 in large samples this statistic has an F distribution with $(g-1)$ and d degrees of freedom.

Hartley's Test

For equal sample sizes $n_j = n_0$, $j = 1, 2, \ldots, g$, two additional test procedures can be used. *Hartley's test* uses the statistic

$$H = \frac{\max_j\{s_j^2\}}{\min_j\{s_j^2\}}.$$

Tables of this H-ratio in Pearson and Hartley (1970) can then be used to obtain a critical value.

The statistic H can also be used to obtain a confidence interval for σ_j^2/σ_ℓ^2. The interval $\left[\frac{1}{H_\alpha}\frac{s_j^2}{s_\ell^2}, H_\alpha \frac{s_j^2}{s_\ell^2}\right]$ is a $100(1-\alpha)\%$ confidence interval for σ_j^2/σ_ℓ^2 where H_α is the upper 100α percentile for the H distribution.

Cochran's Test

Cochran's test determines the ratio

$$C = \frac{\max_{j}\{s_j^2\}}{\sum_{j=1}^{g} s_j^2}.$$

Tables of the C-statistic in Pearson and Hartley (1970) can be used to test H_0.

Example

For the test marketing data introduced in the previous section, the sample variances were reported to be $s_1^2 = 430.32$, $s_2^2 = 547.61$, $s_3^2 = 354.24$ and $s_4^2 = 605.33$ and s^2, the estimator of the common variance, was 484.38. The value of the Bartlett-Box F-statistic is given by $F = 0.816$ which has a p-value of 0.485. For Cochran's C-statistic the value is 0.3124 which has a p-value of 0.549. On the basis of these two test statistics we cannot reject the hypothesis of equality of the variances for the four ad groups.

Some Nonparametric Alternatives

The above tests for homogeneity of variance are extremely sensitive to non-normality. Since the ANOVA test for equality of means is robust to non-normality several tests for homogeneity of variance in non-normal distributions make use of this robust property.

Levene's Method

A procedure suggested by Levene (1960) defines new variables

$$Z_{ij} = |y_{ij} - \bar{y}_{\cdot j}| \qquad i = 1, 2, \ldots, n_j; \quad j = 1, 2, \ldots, g.$$

An ANOVA procedure is then carried out using the Z_{ij} in place of the y_{ij} as outlined in 5.1.1. As in ANOVA the F-statistic with $(g - 1)$ and $(n - g)$ d.f. is used as a test statistic. Although the Z_{ij} are not mutually independent and identically distributed, and are not normally distributed, this test performs reasonably well.

In situations involving excessive skewness or excessive kurtosis Brown and Forsyth (1974) have proposed the use of the median in place of $\bar{y}_{\cdot j}$ when there is skewness, and a 10% trimmed mean in place of $\bar{y}_{\cdot j}$ when there is kurtosis.

A Jackknife Method

An alternative approach, which also uses the ANOVA test, employs a *jack-knife* procedure. A test developed by Miller (1968) computes variances within each group by leaving out one observation. Thus in each group j, n_j estimates of σ_j^2 are determined by leaving out each observation once.

$$s_{j(\ell)}^2 = \frac{1}{n_j - 2} \sum_{j \neq \ell}^{n_j} (y_{ij} - \bar{y}_{\cdot j(\ell)})^2$$

$$\bar{y}_{\cdot j(\ell)} = \frac{1}{n_j - 1} \sum_{j \neq \ell}^{n_j} y_{ij}.$$

Define

$$Z_{\ell j} = n_j \log s_j^2 - (n_j - 1) \log s_{j(\ell)}^2$$

and carry out a oneway ANOVA using the $Z_{\ell j}$ and the F distribution with $(g - 1)$ and $(n - g)$ d.f.

A Procedure Based on Ranks

A nonparametric procedure based on ranks given in Conover (1980) begins by determining the absolute differences for each group, as in Levene's test described above, $Z_{ij} = |y_{ij} - \bar{y}_{\cdot j}|$. The complete set of differences are ranked and denoted by R_{ij}. The sums of squares for the ranks are then determined for each group and denoted by $s_1, s_2, \ldots, s_g$. The test statistic for homogeneity of variance is given by

$$T = \left[\sum_{j=1}^{g} \frac{s_j^2}{n_j} - n\bar{s}^2 \right] / D^2$$

where

$$D^2 = \frac{1}{n - 1} \left(\sum_{j=1}^{g} \sum_{i=1}^{n_j} R_{ij}^4 - n\bar{s}^2 \right)$$

and

$$\bar{s} = \frac{1}{n} \sum_{j=1}^{g} s_j.$$

The statistic T is compared to a χ^2 distribution with $(g - 1)$ degrees of freedom. A pair of groups can be contrasted by comparing

$$\left| \frac{s_j}{n_j} - \frac{s_\ell}{n_\ell} \right| \quad \text{to} \quad t_{\alpha/2; n-g} \left[D^2 \left(\frac{n - 1 - T}{n - g} \right) \left(\frac{1}{n_j} + \frac{1}{n_\ell} \right) \right]^{1/2}.$$

5.2 Two-Way Analysis of Variance

In Section 5.1 the relationship between a quantitative variable Y and a qualitative variable X was studied using one-way analysis of variance. In this section a second qualitative variable say Z is introduced. The categories of the two qualitative variables jointly define a set of cells, and the variation in these cell means for the variable Y is to be studied. The model used in this case is called a *two-way analysis of variance model*. The second variable Z may define a second type of treatment or *factor*, or Z may represent some characteristic of the experimental unit which accounts for a lack of homogeneity among experimental units. In the first case the model is referred to as a *two factor model* while in the latter case the model is called a *randomized block design*. Although the analysis of variance procedures are the same for both models, each model will be discussed separately below.

In the discussion of one-way analysis of variance in 5.1 the objective was to compare the means for a particular random variable Y across g groups. The test for equality of group means employed the ratio of the mean square measuring variation among the means, MSA, to the mean square measuring variation within the groups, MSW. If this ratio was large the group means were declared to be significantly different. The ability of this procedure to detect differences among means therefore depends partly on the ability of the analyst to control other factors which can cause large variation among units within a group. Ideally the items in the group should be relatively homogeneous so that MSW will be relatively small. When there are other variables that have an impact on Y their effects should be eliminated so that this variation does not contribute to the variation among items in a group. If the variation due to the other variables cannot be eliminated it should be measured and removed from MSW before making a comparison to the mean squares among groups.

Suppose we wish to compare the fuel efficiency of various brands of gasoline for automobiles. While the variability of interest is the fuel efficiency across brands of gasoline, we must recognize that other factors will contribute to the variation among brands. One other important variable would be the type or make of automobile used. This variability could be minimized by restricting the study to only one make of automobile, however, our results would then only be valid for that make of automobile. Another possibility is to measure the variation contributed by automobile type and to include it in our analysis.

5.2.1 Randomized Block Design

In the randomized block design the experimental units are divided into b *blocks* in such a way that the units within a block are relatively homogeneous. Each block contains g experimental units. Each experimental unit within a block is then assigned randomly to one of the g treatments. The advantage of this design is that the variation among the means attributable to the b blocks can be removed from the within-group variation and hence reduce its size. The central focus of the experiment is still the comparison of the treatment effects. For the comparison of gasolines example, by blocking on 'automobile type' the variation in gasoline consumption due to automobile type can be taken into consideration. Thus each of the b automobile makes is assigned each of the g gasolines. An outline of the randomized block design analysis follows.

The Randomized Block Design Model

The most commonly used model for the randomized block design is given by

$$y_{ij} = \mu + \beta_i + \alpha_j + \varepsilon_{ij} \qquad i = 1, 2, \ldots, b; \quad j = 1, 2, \ldots, g$$

where

$$\sum_{i=1}^{b} \beta_i = 0, \qquad \sum_{j=1}^{g} \alpha_j = 0,$$

and the ε_{ij} are independent and identically distributed, $N(0, \sigma^2)$. The best linear unbiased estimators of μ_i, β_i and α_j are given by

$$\hat{\mu} = \bar{y}_{..}, \qquad \hat{\beta}_i = \bar{y}_{i.} - \bar{y}_{..} \qquad \text{and} \qquad \hat{\alpha}_j = \bar{y}_{.j} - \bar{y}_{..}$$

where

$$\bar{y}_{..} = \frac{1}{bg} \sum_{i=1}^{b} \sum_{j=1}^{g} y_{ij}; \qquad \bar{y}_{.j} = \frac{1}{b} \sum_{i=1}^{b} y_{ij} \qquad \text{and} \qquad \bar{y}_{i.} = \frac{1}{g} \sum_{j=1}^{g} y_{ij}$$

respectively. The model describes the cell means as shown in Figure 5.5.

For the gasoline brand comparison the variable y_{ij} may denote the volume of gasoline consumed by the automobile over a fixed distance on a particular roadway. The bg cells are formed by the b automobile types and the g gasoline brands. The parameter μ denotes the true average gasoline consumption over the bg cells. The parameters β_i, $i = 1, 2, \ldots, b$ and α_j, $j = 1, 2, \ldots, g$ denote the gasoline effects and auto make effects respectively. The parameter $(\mu + \beta_i + \alpha_j)$ for cell (i, j) denotes the true mean gasoline consumption for cell (i, j).

Groups

	1	2		g
1	$\mu + \alpha_1 + \beta_1$	$\mu + \alpha_2 + \beta_1$	$\ldots$	$\mu + \alpha_g + \beta_1$
2	$\mu + \alpha_1 + \beta_2$	$\mu + \alpha_2 + \beta_2$	$\ldots$	$\mu + \alpha_g + \beta_2$
	$\vdots$	$\vdots$	$\vdots$	$\vdots$
b	$\mu + \alpha_1 + \beta_b$	$\mu + \alpha_2 + \beta_b$	$\ldots$	$\mu + \alpha_g + \beta_b$

Blocks

Figure 5.5. Randomized Block Model Cell Parameters

Table 5.6. Observed Data for Ads by Region Randomized Block Design

Advertisement Type

Region	1	2	3	4	Means
1	12.35	21.86	14.43	21.44	17.52
2	28.26	13.76	14.44	30.78	21.81
3	10.97	0.00	2.90	6.46	5.08
4	0.00	11.90	4.48	27.62	11.00
5	13.11	8.00	10.90	14.36	11.59
					Grand Mean
Means	12.94	11.10	9.43	20.13	13.40

Example

A portion of the milk ad test marketing data was used to provide a randomized block design example with one observation per cell. The data corresponding to single person families for each advertisement group in each region provide a 4×5 design. The data are summarized in Table 5.6. The means are also given in the table. In order to use the randomized block design in this context it is assumed that the four families in each region were randomly assigned to receive one of the four advertisements. Anytime that the advertisement is broadcast by a particular station only one of the 20 test families can possibly receive it.

Table 5.7. Randomized Block ANOVA Table

Source	d.f.	S.S.	M.S.	F
Among	$(g-1)$	SSA	MSA=SSA/$(g-1)$	MSA/MSE
Blocks	$(b-1)$	SSB	MSB=SSB/$(b-1)$	MSB/MSE
Error	$(b-1)(g-1)$	SSE	MSE=SSE/$(b-1)(g-1)$	
Total	$bg-1$	SST		

The ANOVA Table

The analysis of variance tests for the equality of group means $H_{01} : \alpha_1 = \alpha_2 = \cdots = \alpha_g = 0$ and block means $H_{02} : \beta_1 = \beta_2 = \cdots = \beta_b = 0$ are carried out using the sums of squares

$$\text{SST} = \sum_{i=1}^{b}\sum_{j=1}^{g}(y_{ij} - \bar{y}_{..})^2 = \sum_{i=1}^{b}\sum_{j=1}^{g} y_{ij}^2 - gb\bar{y}_{..}^2,$$

$$\text{SSA} = b\sum_{j=1}^{g}(\bar{y}_{.j} - \bar{y}_{..})^2 = b\sum_{j=1}^{g} \bar{y}_{.j}^2 - gb\bar{y}_{..}^2,$$

$$\text{SSB} = g\sum_{i=1}^{b}(\bar{y}_{i.} - \bar{y}_{..})^2 = g\sum_{i=1}^{b} \bar{y}_{i.}^2 - gb\bar{y}_{..}^2,$$

$$\text{SSE} = \text{SST} - \text{SSA} - \text{SSB} = \sum_{i=1}^{b}\sum_{j=1}^{g}(y_{ij} - \bar{y}_{.j} - \bar{y}_{i.} + \bar{y}_{..})^2$$

$$= \sum_{i=1}^{b}\sum_{j=1}^{g} y_{ij}^2 - b\sum_{j=1}^{g} \bar{y}_{.j}^2 - g\sum_{i=1}^{b} \bar{y}_{i.}^2 + gb\bar{y}_{..}^2 .$$

The tests are summarized in the ANOVA table, Table 5.7.

Expected Mean Squares

The expected values of the mean squares in the ANOVA table are given by

$$E[\text{MSA}] = \sigma^2 + b\sum_{j=1}^{g}\alpha_j^2/(g-1)$$

$$E[\text{MSB}] = \sigma^2 + g\sum_{i=1}^{b}\beta_i^2/(b-1)$$

$$E[\text{MSE}] = \sigma^2,$$

and hence under the null hypotheses H_{01} and H_{02} the expected mean squares all have the value σ^2.

Test for Equality of Group Means

The major interest is in the test $H_{01} : \alpha_1 = \alpha_2 = \cdots = \alpha_g = 0$ that the group means are equal. If the hypothesis is true the statistic MSA/MSE has an F distribution with $(g-1)$ and $(b-1)(g-1)$ degrees of freedom. If H_{01} is rejected comparisons can be carried out using the multiple comparison techniques described in 5.1.2. The estimator of σ^2 is given by $s^2 = \text{MSE}$.

Test for Equality of Block Means

It is also important to test the hypothesis $H_{02} : \beta_1 = \beta_2 = \cdots = \beta_b = 0$ to determine if blocking was actually necessary and/or effective. If H_{02} is true the statistic MSB/MSE has an F distribution with $(b-1)$ and $(b-1)(g-1)$ degrees of freedom. If H_{02} is rejected multiple comparison procedures can be used to compare the block means using the methods of 5.1.2.

Efficiency of the Blocked Design

If H_{02} is accepted there may have been a loss in *efficiency* in testing H_{01} using a blocked design. The mean square in the denominator of F, MSE $=$ SSE/$(g-1)(b-1)$, may be larger than MSW $=$ SSW/$(bg-g)$, which would have been used if there was no blocking. In addition, the reduction in degrees of freedom in the denominator of F, results in a larger ratio being required for the same p-value. Since there is an equal number of blocks for each treatment the design is balanced, and hence estimates of the treatment effects will not change if the block effects are omitted from the model. The only change is in the mean square used in the denominator of the F-test for treatments.

Example

The ANOVA table for the data in Table 5.6 is presented in Table 5.8. From the p-values in the table we can conclude that the hypothesis of equality

Table 5.8. Analysis of Variance Table for Ads by Region Randomized Block Design

Source	d.f.	S.S.	M.S.	F	p-value
Ads	3	332.83	110.94	2.61	0.10
Regions	4	663.64	165.91	3.90	0.03
Error	12	509.88	42.49		
Total	19	1506.35			

for the 4 ad means can be rejected at the 0.10 level and that the hypothesis of equality of the 5 region means can be rejected at the 0.03 level.

A comparison of the 4 ad types using the LSD procedure at the 0.05 level yields a critical difference of 8.98. Therefore ad number 4 has a significantly larger mean than the other 3 ads. Using the Tukey procedure the 0.05 critical value is 12.24 and hence none of the four ad means could be declared significantly different from the others.

A comparison of the five regions using the same two multiple comparison procedures at the 0.05 level yields three groups of means for the LSD procedure and two groups of means for Tukey's comparison. Using the LSD procedure the critical difference is 10.04 which yields the following three groups of regions that are not significantly different: (2,1), (1,5,4) and (5,4,3). Using Tukey's procedure the critical value is 14.69 which produces the two groupings of non-significant differences, (2,1,5,4) and (1,5,4,3).

From this analysis we can conclude that for one-person families, only minor differences in consumption can be attributed to the four ads, although there is some evidence that ad number 4 may be more effective. The differences in the means by region are weakly significant with some evidence that region 3 has relatively low consumption and region 2 has relatively high consumption.

A Cell Means Model

A cell means model can also be used for the randomized block design. The model is given by

$$y_{ij} = \mu_{ij} + \varepsilon_{ij} \qquad i = 1, 2, \ldots, b; \quad j = 1, 2, \ldots, g.$$

The group means are given by the mean of the cell means in group j $\mu_{\cdot j} = \sum_{i=1}^{b} \mu_{ij}/b$, $j = 1, 2, \ldots, g$ and the block means are given by the cell means in block i, $\mu_{i\cdot} = \sum_{j=1}^{g} \mu_{ij}/g$, $i = 1, 2, \ldots, b$. The group means and block means are estimated as above by $\bar{y}_{\cdot j}$, $j = 1, 2, \ldots, g$ and $\bar{y}_{i\cdot}$, $i = 1, 2, \ldots, b$ respectively.

Groups

Blocks		1	2		$g-1$	g
	1	$\mu_{bg} + \alpha_1 + \beta_1$	$\mu_{bg} + \alpha_2 + \beta_1$	$\cdots$	$\mu_{bg} + \alpha_{g-1} + \beta_1$	$\mu_{bg} + \beta_1$
	2	$\mu_{bg} + \alpha_1 + \beta_2$	$\mu_{bg} + \alpha_2 + \beta_2$	$\cdots$	$\mu_{bg} + \alpha_{g-1} + \beta_2$	$\mu_{bg} + \beta_2$
		$\vdots$	$\vdots$	$\vdots$	$\vdots$	$\vdots$
	$b-1$	$\mu_{bg} + \alpha_1 + \beta_{b-1}$	$\mu_{bg} + \alpha_2 + \beta_{b-1}$	$\cdots$	$\mu_{bg} + \alpha_{g-1} + \beta_{b-1}$	$\mu_{bg} + \beta_{b-1}$
	b	$\mu_{bg} + \alpha_1$	$\mu_{bg} + \alpha_2$	$\cdots$	$\mu_{bg} + \alpha_{g-1}$	μ_{bg}

Figure 5.6. Cell Mean Parameters for Dummy Coding

A Base Cell Model

A third alternative, which omits from the model the parameters α_g and β_b corresponding to the last row and column in Figure 5.5, gives

$$y_{ij} = \mu_{bg} + \alpha_j + \beta_i + \varepsilon_{ij}, \qquad i = 1, 2, \ldots, b-1; \quad j = 1, 2, \ldots, g-1;$$

$$y_{ig} = \mu_{bg} + \beta_i + \varepsilon_{ig}, \qquad i = 1, 2, \ldots, b-1;$$

$$y_{bj} = \mu_{bg} + \alpha_j + \varepsilon_{bj}, \qquad j = 1, 2, \ldots, g-1;$$

$$y_{bg} = \mu_{bg} + \varepsilon_{bg}.$$

The parameter μ_{bg} is the mean for cell (b, g). In this model the bg cell is the base case to which all other cells are compared. The cell mean parameters are therefore given by Figure 5.6. In this formulation the parameters α_j now measure $(\mu_{\cdot j} - \mu_{\cdot g})$ $j = 1, 2, \ldots, (g-1)$ and the parameters β_i measure $(\mu_{i\cdot} - \mu_{b\cdot})$ $i = 1, 2, \ldots, (b-1)$. The parameters α_j and β_i are therefore estimated by $(\bar{y}_{\cdot j} - \bar{y}_{\cdot g})$ and $(\bar{y}_{i\cdot} - \bar{y}_{b\cdot})$ respectively $j = 1, 2, \ldots, k-1$; $i = 1, 2, \ldots, b-1$. The parameter μ_{bg} is estimated by $(\bar{y}_{b\cdot} + \bar{y}_{\cdot g} - \bar{y}_{\cdot\cdot})$.

These alternative formulations of the model are of interest when employing various computer programs for ANOVA. They will also be discussed in connection with the linear regression model approach later in this section.

A Nonparametric Approach

In Section 5.1.4 the Kruskal-Wallis test was introduced as an alternative approach to the classical one-way analysis of variance. The randomized block was introduced above as a method for taking into account the variation contributed by another variable. This randomized block method can also be employed with the Kruskal-Wallis test to obtain a nonparametric approach to the randomized block design. This approach is usually referred to as *Friedman's two-way analysis of variance*.

For the randomized block design introduced earlier in this section the y_{ij} $i = 1, 2, \ldots, b$; $j = 1, 2, \ldots, g$ are ranked within each block to obtain R_{ij}. The ranks are then summed in each group to obtain $R_{.j}$, $j = 1, 2, \ldots, g$. The test statistic

$$H = \frac{12}{bg(g+1)} \sum_{j=1}^{g} R_{.j}^2 - 3b(g+1)$$

has a χ^2 distribution with $(g-1)$ degrees of freedom in large samples if the populations within each block are identical.

A more recent statistic developed by Iman and Davenport (1980) uses an F approximation in place of the χ^2 approximation. The F-statistic is given by

$$T = (b-1)\left[\frac{1}{b}\sum_{j=1}^{g} R_{.j}^2 - bg(g+1)^2/4\right] \bigg/ \left[bg(g+1)(2g+1)/6 - \sum_{j=1}^{g} R_{.j}^2/b\right],$$

which has an F distribution with $(g-1)$ and $(b-1)(g-1)$ degrees of freedom in large samples, if the populations within each block are identical.

A Generalized Randomized Block Design

The *generalized randomized block design* is an extension of the randomized block design, where each block-treatment combination is assigned to c experimental units rather than only one experimental unit. In the example involving the study of gasoline brands the blocks used were automobile types. In the generalized randomized block design c automobiles of each type are used for each gasoline. Repetitions of the randomized block design in this fashion are called *replications*. The analysis for this design is similar to the analysis for the two-factor factorial design with interaction to be discussed next. The advantage of the design over the randomized block design is that it permits the measurement of interaction effects between treatments and blocks. Thus in the study of gasoline brands it may be of interest to determine whether some gasoline brands perform better for some automobile types than for others. The analysis for the *two-factor factorial* to be discussed next will illustrate the measurement of interaction.

5.2.2 The Two-Factor Factorial

If the relationship between a dependent variable Y and two qualitative variables is of interest then a *two factor model* is required. We assume that the first factor has g levels while the second factor has b levels. The cross classification of the two factors therefore has bg cells. If equal samples are drawn from each of the bg cells the design is said to be balanced and commonly called a *two-way or b.g factorial design*. It can also be referred to as a *balanced two-way layout* or a completely randomized design with two factors. We shall assume that c observations are randomly selected from each of the bg cells yielding a total of $n = bgc$ observations. Alternatively, a total of n experimental units are randomly assigned to the bg treatment combinations, in such a way that each combination is observed on c occasions. When each level of one factor occurs at least once with each level of the second factor the two factors are said to be *completely crossed*. If the number of observations obtained from each cell are not equal, the two-way model is said to be *unbalanced*. An unbalanced two-way model is outlined in 5.2.4.

In the example relating gasoline consumption to gasoline brands it was suggested that a randomized block design would permit the removal of variation due to automobile type. In the two factor model the automobile type is also a factor to be studied along with gasoline brand. If the number of observations per cell, c, is equal to one, the analysis for the two factor factorial design is the same as the randomized block design discussed above. The inference procedures for the blocks outlined above in this case become inference procedures for the second factor. In this section we outline the more general *two-way factorial* with $c > 1$. The analysis procedures for this balanced two-way layout could also be used for the generalized randomized block design which permits c observations for each block × treatment combination. As we shall see below, $c > 1$ observations per cell, permits the study of effects which measure interaction between the two factors. The balanced two-way layout or two-way factorial design is outlined next.

The Two-Way Factorial Model

The *balanced two-way model* is usually given by

$$y_{ijh} = \mu + \alpha_j + \beta_i + (\alpha\beta)_{ij} + \varepsilon_{ijh} \qquad \begin{array}{l} i = 1, 2, \ldots, b; \\ j = 1, 2, \ldots, g; \\ h = 1, 2, \ldots, c. \end{array} \qquad (5.7)$$

where

$$\sum_{i=1}^{b} \beta_i = \sum_{j=1}^{g} \alpha_j = \sum_{i=1}^{b} (\alpha\beta)_{ij} = \sum_{j=1}^{g} (\alpha\beta)_{ij} = 0$$

and the ε_{ijh} are mutually independent $N(0, \sigma^2)$. The parameters $(\alpha\beta)_{ij}$ are called *interaction parameters*.

In the balanced two-way model the population is divided into bg cells with cell means μ_{ij}, $i = 1, 2, \ldots, b$; $j = 1, 2, \ldots, g$. From each cell a total of c observations is obtained. The model can therefore be written in cell mean model form as

$$y_{ijh} = \mu_{ij} + \varepsilon_{ijh} \qquad i = 1, 2, \ldots, b; \quad j = 1, 2, \ldots, g;$$

$$h = 1, 2, \ldots, c.$$

Interaction

In the randomized block design with $c = 1$ discussed above, the model was expressed as $\mu_{ij} = \mu + \alpha_j + \beta_j$ where $\alpha_j = (\mu_{.j} - \mu)$ and $\beta_i = (\mu_{i.} - \mu)$ and hence $\mu_{ij} = \mu + (\mu_{.j} - \mu) + (\mu_{i.} - \mu)$. The cell mean was assumed to be composed of an overall mean, plus a row effect $(\mu_{i.} - \mu)$, and a column effect $(\mu_{.j} - \mu)$. Thus the difference between the means for any pair of rows is the same for all columns. Similarly the difference between the means for any pair of columns is the same for all rows. If these differences are not constant the means μ_{ij} are said to contain an interaction effect, and hence $\mu_{ij} - \{\mu + (\mu_{.j} - \mu) + (\mu_{i.} - \mu)\}$ is not necessarily zero. We denote the interaction by $(\alpha\beta)_{ij} = \mu_{ij} - \{\mu + (\mu_{.j} - \mu) + (\mu_{i.} - \mu)\} = (\mu_{ij} - \mu_{.j} - \mu_{i.} + \mu)$. In the randomized block design with $c = 1$ we obtain only one observation per cell and hence the parameter $(\alpha\beta)_{ij}$ cannot be distinguished from the error.

The two graphs in Figure 5.7 illustrate variation in cell means without and with interaction. For each row the column means are plotted as points and are joined by lines. For each row the plot is called a *profile*. A comparison of the various row profiles permits one to detect the presence of interaction. In panel (a) the parallel profiles for rows $i = 1$ and $i = 2$ compared over the 4 columns $j = 1, 2, 3$ and 4 illustrate NO INTERACTION. The differences between the cell means in each column are the same for both rows. In panel (b) the profiles are no longer parallel indicating an INTERACTION between rows and columns. In panel (b) the differences between the cell means in each column are not the same for the two rows. For $j = 1$ and $j = 2$ $\mu_{11} > \mu_{12}$ for $i = 1$ but $\mu_{21} < \mu_{22}$ for $i = 2$.

The interaction parameter $(\alpha\beta)_{ij}$ can also be viewed as

$$(\alpha\beta)_{ij} = \{\text{Column } (j) - \text{Average for All } j\}_{\text{Given Row } i}$$

$$- \{\text{Column } (j) - \text{Average for All } j\}_{\text{Averaged Over All } i}.$$

This is consistent with the interaction measure introduced with effect coding in Section 4.5. The interaction parameter for cell (i, j) is positive, if

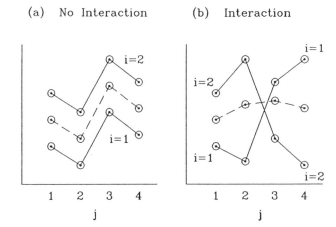

Figure 5.7. Graphical Display of Interaction Among Cell Means

the difference between the cell mean (i, j) and the average for all cells in column j, is larger than the difference between the average in column j for all rows and the overall average. In other words the j-th column effect in row i is larger than the j-th column effect averaged over all rows. In a similar fashion this interaction can also be characterized in terms of a comparison of the i-th row effect in column j and the average of the i-th row effect over all columns.

Interactions and Nonlinearity

A lack of *parallelism between two profiles* may be due to a nonlinear relationship between the dependent variable and one or more of the factors. Such interactions can sometimes be removed by a transformation applied to the dependent variable. Profiles which are nonparallel but still monotonic are sometimes indicative of such nonlinearities. In some applications the nonlinear relationship obtained may be an important finding in that the interaction effect is the effect of interest. In this case a transformation would not be appropriate.

Estimation

The best linear unbiased estimators of the parameters are given by

$$\hat{\alpha}_j = \bar{y}_{\cdot j \cdot} - \bar{y}_{\cdots}$$

$$\hat{\beta}_i = \bar{y}_{i \cdot \cdot} - \bar{y}_{\cdots}$$

Table 5.9. Balanced Two-Way Anova

Source	d.f.	S.S.	M.S.	F
Factor A	$(g-1)$	SSA	MSA=SSA/$(g-1)$	MSA/MSE
Factor B	$(b-1)$	SSB	MSB=SSB/$(b-1)$	MSB/MSE
Interaction	$(b-1)(g-1)$	SSI	MSI=SSI/$(b-1)(g-1)$	MSI/MSE
Error	$bg(c-1)$	SSE	MSE=SSE/$bg(c-1)$	
Total	$bgc-1$	SST		

$$(\widehat{\alpha\beta})_{ij} = [\bar{y}_{ij\cdot} - \bar{y}_{i\cdot\cdot} - \bar{y}_{\cdot j\cdot} + \bar{y}_{\cdots}]$$

$$= [\bar{y}_{ij\cdot} - (\bar{y}_{i\cdot\cdot} - \bar{y}_{\cdots}) - (\bar{y}_{\cdot j\cdot} - \bar{y}_{\cdots}) - \bar{y}_{\cdots}]$$

and $\hat{\mu}_{ij} = \bar{y}_{ij\cdot}$.

The ANOVA Table

For the balanced two-way design three independent hypothesis tests can be carried out in an ANOVA table using an F-statistic. The hypotheses are $H_{01} : \beta_1 = \beta_2 = \cdots = \beta_g = 0$; $H_{02} : \alpha_1 = \alpha_2 = \cdots = \alpha_b = 0$ and $H_{03} : (\alpha\beta)_{11} = (\alpha\beta)_{12} = \cdots = (\alpha\beta)_{bg} = 0$. The sums of squares are given by

$$\text{SST} = \sum_{i=1}^{b}\sum_{j=1}^{g}\sum_{h=1}^{c}(y_{ijh} - \bar{y}_{\cdots})^2 \qquad = \sum_{i=1}^{b}\sum_{j=1}^{g}\sum_{h=1}^{c}y_{ijh}^2 - bgc\bar{y}_{\cdots}^2$$

$$\text{SSA} = bc\sum_{j=1}^{g}(\bar{y}_{\cdot j\cdot} - \bar{y}_{\cdots})^2 \qquad = bc\sum_{j=1}^{g}\bar{y}_{\cdot j\cdot}^2 - bgc\bar{y}_{\cdots}^2$$

$$\text{SSB} = gc\sum_{i=1}^{b}(\bar{y}_{i\cdot\cdot} - \bar{y}_{\cdots})^2 \qquad = gc\sum_{i=1}^{b}\bar{y}_{\cdot j\cdot}^2 - bgc\bar{y}_{\cdots}^2$$

$$\text{SSI} = c\sum_{i=1}^{b}\sum_{j=1}^{g}(\bar{y}_{ij\cdot} - \bar{y}_{i\cdot\cdot} - \bar{y}_{\cdot j\cdot} + \bar{y}_{\cdots})^2 = c\sum_{i=1}^{b}\sum_{j=1}^{g}\bar{y}_{ij\cdot}^2 - bc\sum_{j=1}^{g}\bar{y}_{\cdot j\cdot}^2$$

$$- gc\sum_{i=1}^{b}\bar{y}_{i\cdot\cdot}^2 + bgc\bar{y}_{\cdots}^2$$

$$\text{SSE} = \sum_{i=1}^{b}\sum_{j=1}^{g}\sum_{h=1}^{c}(y_{ijh} - \bar{y}_{ij\cdot})^2 \qquad = \sum_{i=1}^{b}\sum_{j=1}^{g}\sum_{h=1}^{c}y_{ijh}^2 - c\sum_{i=1}^{b}\sum_{j=1}^{g}\bar{y}_{ij\cdot}^2$$

The corresponding ANOVA table is given in Table 5.9.

Expected Mean Squares

The expected values of the mean squares given in the above ANOVA table are

$$E[\text{MSA}] = \sigma^2 + cb \sum_{j=1}^{g} \alpha_j^2/(g-1)$$

$$E[\text{MSB}] = \sigma^2 + cg \sum_{i=1}^{b} \beta_i^2/(b-1)$$

$$E[\text{MSE}] = \sigma^2$$

$$E[\text{MSI}] = \sigma^2 + c \sum_{i=1}^{b} \sum_{j=1}^{g} (\alpha\beta)_{ij}/(b-1)(g-1)$$

all of which have the same expectation when the effects are zero.

Testing the Hypotheses

The test for H_{03} should be examined first. If H_{03} is accepted, then the interaction terms can be dropped from the model without changing the estimates of the main effects for the two factors. If the interaction terms are dropped the error sum of squares will increase to (SSE + SSI) and the error degrees of freedom increases to $(bgc - g - b + 1)$. In a similar fashion, if one of the main effects is not significant, it may also be dropped from the model without changing the estimates for the other main effect.

If H_{03} is rejected, but one or both of H_{01} and H_{02} is accepted, there is some difficulty in interpreting the outcome. In this case, the main effect parameters should not be removed from the model. A cell means model approach may be more useful in this case.

Example

The test marketing data given in Table 5.1 in Section 5.1.1 will be used here to provide an example of a balanced two-way design with more than one observation per cell. The cell means and row and column means are summarized in Table 5.10. Recall that for each ad-region combination, six families were observed. The assumption is made here that each family received independent broadcasts of a particular ad or that on any particular

Table 5.10. Cell Means for Test Marketing Data

| | Ad Group | | | | |
Region	1	2	3	4	Means
1	34.99	56.17	32.33	48.47	42.99
2	52.24	42.02	44.12	62.86	50.31
3	43.53	18.84	26.02	34.10	30.62
4	21.24	47.13	29.35	65.79	40.88
5	34.19	34.08	42.95	43.93	38.79
					Grand Mean
Means	37.24	39.65	34.95	51.03	40.72

Table 5.11. ANOVA Table for Ad by Region Two-Way Balanced Design

Source	d.f.	S.S.	M.S.	F	p-value
Ad	3	4585.68	1528.56	3.61	0.02
Region	4	4867.51	1216.88	2.87	0.02
Interaction	12	8937.92	744.83	1.76	0.07
Error	100	42382.02	423.82		
Total	119	60773.12			

broadcast of an ad by any station it is only possible for one of the test families to receive. This complexity will be removed later in this chapter.

The analysis of variance table for the two-way balanced design for the marketing data is given in Table 5.11. From the p-values for the three F-statistics we can conclude that there are significant differences among the ad means and among the region means at the 0.02 level. We can also conclude that there is a weak interaction between ad and region.

The nature of the interaction between the ad types and regions can be seen by looking at the estimates of the interaction parameters shown in Table 5.12. For ad number 1 the largest expenditure is in region 2 while for ad number 2 the largest expenditure is in region 1. A lack of parallelism among the four plots is an indication of interaction. The plots shown in Figure 5.8 can also be used for this purpose. Procedures for comparing the twenty cell means using a multiple regression model formulation will be introduced next.

5.2.3 A Multiple Regression Approach for the Balanced Two-Way Model

The two-way analysis of variance model was introduced for studying the relationship between a quantitative variable Y and two qualitative variables.

Table 5.12. Interactions for Test Marketing Data

| | | Ad Group | | |
Region	1	2	3	4
1	-4.52	14.25	-4.89	-4.84
2	5.41	-7.22	-0.42	2.23
3	16.39	-10.71	1.16	-6.84
4	-16.16	7.33	-5.77	14.60
5	-1.12	-3.65	9.92	-5.15

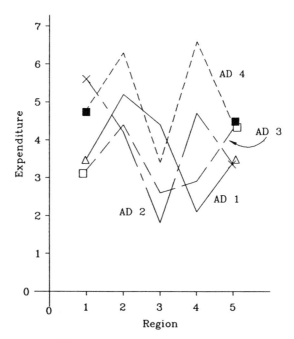

Figure 5.8. Variation in Mean Expenditure by Region and by Ad

The balanced two-way model was given by

$$y_{ij} = \mu + \alpha_i + \beta_j + (\alpha\beta)_{ij} + \varepsilon_{ijh}$$

$$i = 1, 2, \ldots, b; \quad j = 1, 2, \ldots, g; \quad h = 1, 2, \ldots, c.$$

In this case the bg cells defined by the categories of the two qualitative explanatory variables each contain c observations on Y. In this model the main effects α_i and β_j corresponding to the two variables can be tested independently of each other, but not of the interaction. If the interaction is ignored the resulting increase in the error mean square could result in the main effects being declared insignificant.

A more general two-way analysis of variance model is given by

$$y_{ijh} = \mu_{ij} + \varepsilon_{ijh} \qquad i = 1, 2, \ldots, b; \ j = 1, 2, \ldots, g;$$
$$h = 1, 2, \ldots, n_{ij}.$$

This model is not balanced in that the number of observations in each of the bg cells is not constant. Since the model is not balanced the two main effects cannot be tested independently. As in the case of multiple regression when the X columns are not orthogonal, the estimates of the coefficients depend on what variables are included. In an *unbalanced two-way model* the columns for one effect are not orthogonal to the columns for the other effect. The analysis for the *unbalanced* case will be outlined in 5.2.4.

Dummy Coding

As in the case of the multiple regression approach to the one-way model in Section 5.1.3, the two-way model can be expressed as a multiple regression model with the design matrix $\mathbf{X}$ containing bg columns representing the bg dummies for the bg cells. The least squares procedure yields the cell means $\bar{y}_{ij\cdot}$, $i = 1, 2, \ldots, b$; $j = 1, 2, \ldots, g$. To test the null hypothesis of equality of the cell means a reduced model with a single column of unities is fitted and compared to the previous model using an F-test. If the null hypothesis of equality of the cell means is rejected it is usually of interest to determine whether the differences among the cell means is due to one or both factors and possibly an interaction between them. The $\mathbf{X}$ matrix is usually more conveniently constructed by using dummy variables for the various categories of each of the explanatory variables. The interactions between the two factors are measured by the products of the elements in the corresponding columns of $\mathbf{X}$.

Salary-Sex-Occupation Example

The multiple regression approach to two-way ANOVA can be conveniently studied using the sex-occupation group example discussed in Chapter 4. The education variable covariate used in Chapter 4 will be omitted until the section on analysis of covariance. As illustrated in Table 5.13 the first column of $\mathbf{X}$ is a column of unities, the second column contains a sex dummy. The third column shows the occupation group dummy and the fourth column is the interaction column obtained from the products of

Table 5.13. **X** Matrix for Sex-Occupation Group Example

β_0	Sex α	Occupation γ	Interaction ρ		
1	0	0	0		
1	0	0	0		
$\vdots$	$\vdots$	$\vdots$	$\vdots$		Group A
1	0	0	0	Males	
1	0	1	0		
1	0	1	0		
$\vdots$	$\vdots$	$\vdots$	$\vdots$		Group B
1	0	1	0		
1	1	0	0		
1	1	0	0		
$\vdots$	$\vdots$	$\vdots$	$\vdots$		Group A
1	1	0	0	Females	
1	1	1	1		
1	1	1	1		
$\vdots$	$\vdots$	$\vdots$	$\vdots$		Group B
1	1	1	1		

columns 2 and 3. The means for the four cells are β_0 and $(\beta_0 + \alpha)$ for males and females in Group A, while for Group B the means are $(\beta_0 + \gamma)$ and $(\beta_0 + \alpha + \gamma + \rho)$ for males and females respectively. The sex effect is given by α, the group effect by γ and the interaction by ρ.

For the Sex-Occupation example the full model is given by

$$Y = \beta_0 + \gamma \text{ OCCUPATION} + \alpha \text{ SEX} + \rho \text{ SEX*OCCUPATION} + U$$

and is compared to each of the reduced models

$$Y = \beta_0 + \gamma \text{ OCCUPATION} + \rho \text{ SEX*OCCUPATION} + U$$

$$Y = \beta_0 + \alpha \text{ SEX} + \rho \text{ SEX*OCCUPATION} + U$$

$$Y = \beta_0 + \alpha \text{ SEX} + \gamma \text{ OCCUPATION} + U,$$

and hence each of the hypotheses $H_0 : \alpha = 0$, $H_0 : \gamma = 0$ and $H_0 : \rho = 0$ is tested. In general if the interaction is significant the main effect components are usually included for ease of interpretation. In other words the models fitted are usually hierarchical, in that the presence of any interaction term in the model requires the presence of the components that make up the interaction.

Test Marketing Example

The test marketing example introduced in Section 5.1 may be used here to illustrate dummy variable coding. In addition to the dummy variables AD1, AD2 and AD3 for the three ads, the dummy variables R1, R2, R3 and R4 are defined for the first four regions. The fifth region is the base case. The **X** matrix in Table 5.14 shows the first row of observations for each region from Table 5.1. The equation for EXPEND is given by

$$\text{EXPEND} = \beta_0 + \beta_1 \text{ AD1} + \beta_2 \text{ AD2} + \beta_3 \text{ AD3}$$

$$+ \gamma_1 \text{ R1} + \gamma_2 \text{ R2} + \gamma_3 \text{ R3} + \gamma_4 \text{ R4}$$

$$+ \rho_{11} \text{ AD1*R1} + \rho_{12} \text{ AD1*R2} + \rho_{13} \text{ AD1*R3} + \rho_{14} \text{ AD1*R4}$$

$$+ \rho_{21} \text{ AD2*R1} + \rho_{22} \text{ AD2*R2} + \rho_{23} \text{ AD2*R3} + \rho_{24} \text{ AD2*R4}$$

$$+ \rho_{31} \text{ AD3*R1} + \rho_{32} \text{ AD3*R2} + \rho_{33} \text{ AD3*R3} + \rho_{34} \text{ AD3*R4} + U.$$

The means for the twenty cells derived from the four ads and five regions are given in Figure 5.9.

For the complete test marketing data set the fitted model is given by

$$\text{EXPEND} = \underset{(0.000)}{43.93} - \underset{(0.415)}{9.74} \text{ AD1} - \underset{(0.409)}{9.86} \text{ AD2} - \underset{(0.934)}{0.99} \text{ AD3}$$

$$+ \underset{(0.703)}{4.54} \text{ R1} + \underset{(0.115)}{18.92} \text{ R2} - \underset{(0.410)}{9.83} \text{ R3} + \underset{(0.069)}{21.86} \text{ R4}$$

$$- \underset{(0.824)}{3.74} \text{ AD1*R1} - \underset{(0.959)}{0.88} \text{ AD1*R2} + \underset{(0.257)}{19.17} \text{ AD1*R3} - \underset{(0.041)}{34.82} \text{ AD1*R4}$$

$$+ \underset{(0.299)}{17.55} \text{ AD2*R1} - \underset{(0.515)}{10.98} \text{ AD2*R2} - \underset{(0.749)}{5.40} \text{ AD2*R3} - \underset{(0.602)}{8.80} \text{ AD2*R4}$$

$$- \underset{(0.749)}{15.16} \text{ AD3*R1} - \underset{(0.294)}{17.75} \text{ AD3*R2} - \underset{(0.674)}{7.10} \text{ AD3*R3} - \underset{(0.037)}{35.46} \text{ AD3*R4}.$$

The twenty cell means can be obtained by adding together the appropriate parameter estimates as shown above. The means first displayed in Table 5.10 are repeated in Table 5.15. The twenty cell means can be compared using the multiple comparison procedures introduced in Section 5.1.2. The F-statistic value for testing the equality of the twenty cell means is 2.28 with a p-value of 0.0045. The R^2 value is 0.303.

A useful method for judging the significance of the AD effects and REGION effects, as well as the interaction, is to test the hypothesis that the particular effect is not significant in the full model. The error sum of squares is given by 42382.02. The decreases in the regression sum of squares after dropping each of ADS, REGIONS and AD*REGIONS are respectively 4585.68, 4867.51 and 8937.92 respectively. The corresponding F-values are 3.61, 2.87 and 1.76 and p-values are 0.016, 0.027 and 0.066. These quantities are shown in Table 5.12 and in Table 5.11 in Section 5.2.2. The nature of the interaction was displayed in Figure 5.8 in Section 5.2.2.

Table 5.14. y and X Matrix for Dummy Variable Coding for First Four Observations in Each Region

| | | | | | | | | | | | | | | $\mathbf{X}$ | | | | | | |
y	β_0	β_1	β_2	β_3	γ_1	γ_2	γ_3	γ_4	ρ_{11}	ρ_{12}	ρ_{13}	ρ_{14}	ρ_{21}	ρ_{22}	ρ_{23}	ρ_{24}	ρ_{31}	ρ_{32}	ρ_{33}	ρ_{34}
12.35	1	1	0	0	1	0	0	0	1	0	0	0	0	0	0	0	0	0	0	0
21.86	1	0	1	0	1	0	0	0	0	0	0	0	1	0	0	0	0	0	0	0
14.43	1	0	0	1	1	0	0	0	0	0	0	0	0	0	0	0	1	0	0	0
21.44	1	0	0	0	1	0	0	0	0	0	0	0	0	0	0	0	0	0	0	0
28.26	1	1	0	0	0	1	0	0	0	1	0	0	0	0	0	0	0	0	0	0
13.76	1	0	1	0	0	1	0	0	0	0	0	0	0	1	0	0	0	0	0	0
14.44	1	0	0	1	0	1	0	0	0	0	0	0	0	0	0	0	0	1	0	0
30.78	1	0	0	0	0	1	0	0	0	0	0	0	0	0	0	0	0	0	0	0
10.97	1	1	0	0	0	0	1	0	0	0	1	0	0	0	0	0	0	0	0	0
0.00	1	0	1	0	0	0	1	0	0	0	0	0	0	0	1	0	0	0	0	0
2.90	1	0	0	1	0	0	1	0	0	0	0	0	0	0	0	0	0	0	1	0
6.46	1	0	0	0	0	0	1	0	0	0	0	0	0	0	0	0	0	0	0	0
0.00	1	1	0	0	0	0	0	1	0	0	0	1	0	0	0	0	0	0	0	0
11.90	1	0	1	0	0	0	0	1	0	0	0	0	0	0	0	1	0	0	0	0
4.48	1	0	0	1	0	0	0	1	0	0	0	0	0	0	0	0	0	0	0	1
27.62	1	0	0	0	0	0	0	1	0	0	0	0	0	0	0	0	0	0	0	0
13.11	1	1	0	0	0	0	0	0	0	0	0	0	0	0	0	0	0	0	0	0
8.00	1	0	1	0	0	0	0	0	0	0	0	0	0	0	0	0	0	0	0	0
10.90	1	0	0	1	0	0	0	0	0	0	0	0	0	0	0	0	0	0	0	0
14.36	1	0	0	0	0	0	0	0	0	0	0	0	0	0	0	0	0	0	0	0
	AD dummies				REGION dummies				AD*REGION Interactions											

	AD1	AD2	AD3	AD4
R1	$\beta_0 + \beta_1 + \gamma_1$ $+ \rho_{11}$	$\beta_0 + \beta_2 + \gamma_1$ $+ \rho_{21}$	$\beta_0 + \beta_3 + \gamma_1$ $+ \rho_{31}$	$\beta_0 + \gamma_1$
R2	$\beta_0 + \beta_1 + \gamma_2$ $+ \rho_{12}$	$\beta_0 + \beta_2 + \gamma_2$ $+ \rho_{22}$	$\beta_0 + \beta_3 + \gamma_2$ $+ \rho_{32}$	$\beta_0 + \gamma_2$
R3	$\beta_0 + \beta_1 + \gamma_3$ $+ \rho_{13}$	$\beta_0 + \beta_2 + \gamma_3$ $+ \rho_{23}$	$\beta_0 + \beta_3 + \gamma_3$ $+ \rho_{33}$	$\beta_0 + \gamma_3$
R4	$\beta_0 + \beta_1 + \gamma_4$ $+ \rho_{14}$	$\beta_0 + \beta_2 + \gamma_4$ $+ \rho_{24}$	$\beta_0 + \beta_3 + \gamma_4$ $+ \rho_{34}$	$\beta_0 + \gamma_4$
R5	$\beta_0 + \beta_1$	$\beta_0 + \beta_2$	$\beta_0 + \beta_3$	β_0

Figure 5.9. Cell Mean Parameters for Dummy Coding

Table 5.15. Means for AD*REGION Cells

	AD1	AD2	AD3	AD4
R1	34.99	56.17	32.33	48.47
R2	52.24	42.02	44.12	62.86
R3	43.53	18.84	26.02	34.10
R4	21.24	47.13	29.35	65.79
R5	34.19	34.08	42.95	43.93

Effect Coding

The effect coding method introduced above for the one-way model may also be employed in the two-way model. For the salary-sex-occupation example discussed above the effect coding variables are defined below.

$$\begin{aligned} \text{SEX} \quad &= \quad 1 \quad \text{for females} \\ &= -1 \quad \text{for males} \end{aligned}$$

$$\begin{aligned} \text{OCCUPATION} &= \quad 1 \quad \text{for group B} \\ &= -1 \quad \text{for group A.} \end{aligned}$$

The model is given by

$$Y = \beta_0 + \alpha \ \text{SEX} + \gamma \ \text{OCCUPATION} + \rho \ \text{SEX*OCCUPATION} + U.$$

Table 5.16. Cell Mean Parameters

	Males	Females
A	$\beta_0 - \alpha - \gamma + \rho$	$\beta_0 + \alpha - \gamma - \rho$
B	$\beta_0 - \alpha + \gamma - \rho$	$\beta_0 + \alpha + \gamma + \rho$

Table 5.17. **X** Matrix for Effect Coding

β_0	α	γ	ρ		
1	1	1	1		
1	1	1	1		Group B
$\vdots$	$\vdots$	$\vdots$	$\vdots$		
1	1	1	1	Females	
1	1	-1	-1		
1	1	-1	-1		Group A
$\vdots$	$\vdots$	$\vdots$	$\vdots$		
1	1	-1	-1		
1	-1	1	-1		
1	-1	1	-1		Group B
$\vdots$	$\vdots$	$\vdots$	$\vdots$		
1	-1	1	-1	Males	
1	-1	-1	1		
1	-1	-1	1		Group A
$\vdots$	$\vdots$	$\vdots$	$\vdots$		
1	-1	-1	1		

The parameters correspond to the cell means as shown in Table 5.16.

The parameter β_0 represents the overall average of the four cell means in Table 5.16. The average cell mean for males is given by $(\beta_0 - \alpha)$ while the average cell mean for females is $(\beta_0 + \alpha)$. The difference between these two averages is therefore 2α which is the SEX effect. The average cell mean for occupation group A is $(\beta_0 - \gamma)$ while the average cell mean for occupation group B is $(\beta_0 + \gamma)$. The difference between these two averages is the OCCUPATION effect 2γ. The difference between the male cell mean and the female cell mean for occupation group A is $(-2\alpha + 2\rho)$ while this difference for occupation group B is $(-2\alpha - 2\rho)$. The difference between these two sex effects $(A - B)$ is 4ρ which represents the INTERACTION effect. Comparing the differences between occupation groups for each sex would also yield the amount 4ρ.

The **X** matrix using effect coding is given in Table 5.17.

The columns of the **X** matrix are mutually orthogonal and hence the variation attributed to each effect does not depend on the other effects being present. In general, under effect coding in a balanced design the parameter estimates and sums of squares for each effect do not change when other effects are added or deleted. The contribution to the regression sum of squares for each effect, and also the parameter estimates remain fixed. This is an important advantage of effect coding in comparison to dummy coding.

Example

For the test marketing data, effect coding variables for the four ads and five regions are defined as follows

$$
\begin{aligned}
\text{ADV1} = \ &1 \quad \text{for Ad 1} \\
= \ &0 \quad \text{for Ads 2 and 3} \\
= -&1 \quad \text{for Ad 4} \\[6pt]
\text{ADV2} = \ &1 \quad \text{for Ad 2} \\
= \ &0 \quad \text{for Ads 1 and 3} \\
= -&1 \quad \text{for Ad 4} \\[6pt]
\text{ADV3} = \ &1 \quad \text{for Ad 3} \\
= \ &0 \quad \text{for Ads 1 and 2} \\
= -&1 \quad \text{for Ad 4} \\[6pt]
\text{REG1} = \ &1 \quad \text{for region 1} \\
= \ &0 \quad \text{for regions 2, 3 and 4} \\
= -&1 \quad \text{for region 5} \\[6pt]
\text{REG2} = \ &1 \quad \text{for region 2} \\
= \ &0 \quad \text{for regions 1, 3 and 4} \\
= -&1 \quad \text{for region 5} \\[6pt]
\text{REG3} = \ &1 \quad \text{for region 3} \\
= \ &0 \quad \text{for regions 1, 2 and 4} \\
= -&1 \quad \text{for region 5} \\[6pt]
\text{REG4} = \ &1 \quad \text{for region 4} \\
= \ &0 \quad \text{for regions 1, 2 and 3} \\
= -&1 \quad \text{for region 5.}
\end{aligned}
$$

The interaction variables are denoted by

$$\text{ADREG}_{ij} = \text{ADV}_i {}^*\text{REG}_j \quad \text{for } i = 1, 2, 3; \ j = 1, 2, 3, 4.$$

The estimated model is given by

$$Y = \underset{(0.000)}{40.72} - \underset{(0.288)}{3.48} \; ADV1 - \underset{(0.743)}{1.07} \; ADV2 - \underset{(0.080)}{5.77} \; ADV3$$

$$+ \underset{(0.547)}{2.27} \; REG1 + \underset{(0.012)}{9.59} \; REG2 - \underset{(0.009)}{10.09} \; REG3 + \underset{(0.966)}{0.16} \; REG4$$

$$- \underset{(0.489)}{4.52} \; ADREG11 + \underset{(0.408)}{5.41} \; ADREG12 + \underset{(0.013)}{16.39} \; ADREG13$$

$$- \underset{(0.015)}{16.16} \; ADREG14 + \underset{(0.031)}{14.25} \; ADREG21 - \underset{(0.270)}{7.22} \; ADREG22$$

$$- \underset{(0.103)}{10.71} \; ADREG23 + \underset{(0.263)}{7.33} \; ADREG24 - \underset{(0.454)}{4.89} \; ADREG31$$

$$- \underset{(0.948)}{0.42} \; ADREG32 + \underset{(0.859)}{1.16} \; ADREG33 - \underset{(0.378)}{5.77} \; ADREG34.$$

The regression sum of squares for this model is 18391.1, and the error sum of squares can be allocated to ADs, REGIONs and the INTERACTION as 4585.7, 4867.5, and 8937.9 respectively.

The estimated coefficients of ADV1, ADV2 and ADV3 for the above model are identical to the ones obtained using effect coding using only the variables ADV1, ADV2 and ADV3. Similarly the effect coding model which includes only the variables REG1, REG2, REG3 and REG4 yields the same coefficients as the above model. The model which includes only the interaction variables $ADREG_{ij}$, $i = 1, 2, 3$; $j = 1, 2, 3, 4$; also has the same estimated regression coefficients.

Cell Means Model Approach

Using a cell means model approach the column of unities is omitted from the **X** matrix. For each cell, dummy variables are defined and included as explanatory variables. For the test marketing data there are 20 cells derived from the four ads and five regions. The dummies will be denoted by D1 through D20 as shown in Table 5.18. The linear model is given by

$$EXPEND = \beta_1 \; D1 + \beta_2 \; D2 + \beta_3 \; D3 + \beta_4 \; D4 + \beta_5 \; D5$$

$$+ \beta_6 \; D6 + \beta_7 \; D7 + \beta_8 \; D8 + \beta_9 \; D9 + \beta_{10} \; D10$$

$$+ \beta_{11} \; D11 + \beta_{12} \; D12 + \beta_{13} \; D13 + \beta_{14} \; D14 + \beta_{15} \; D15$$

$$+ \beta_{16} \; D16 + \beta_{17} \; D17 + \beta_{18} \; D18 + \beta_{19} \; D19 + \beta_{20} \; D20 + U.$$

The conventional hypotheses of no INTERACTION, no AD effect and no REGION effect can be tested using various forms of the linear transformation matrix **T**. For the no INTERACTION hypothesis the matrix is

Table 5.18. Definition of Cell Dummies

	AD Type			
Region	1	2	3	4
1	D1	D2	D3	D4
2	D5	D6	D7	D8
3	D9	D10	D11	D12
4	D13	D14	D15	D16
5	D17	D18	D19	D20

Table 5.19. Transformation Matrix for Interaction Test

$$
\mathbf{T} =
\begin{array}{cccccccccccccccccccc}
1 & 2 & 3 & 4 & 5 & 6 & 7 & 8 & 9 & 10 & 11 & 12 & 13 & 14 & 15 & 16 & 17 & 18 & 19 & 20 \\
1 & -1 & 0 & 0 & -1 & 1 & 0 & 0 & 0 & 0 & 0 & 0 & 0 & 0 & 0 & 0 & 0 & 0 & 0 & 0 \\
1 & -1 & 0 & 0 & 0 & 0 & 0 & 0 & -1 & 1 & 0 & 0 & 0 & 0 & 0 & 0 & 0 & 0 & 0 & 0 \\
1 & -1 & 0 & 0 & 0 & 0 & 0 & 0 & 0 & 0 & 0 & 0 & -1 & 1 & 0 & 0 & 0 & 0 & 0 & 0 \\
1 & -1 & 0 & 0 & 0 & 0 & 0 & 0 & 0 & 0 & 0 & 0 & 0 & 0 & 0 & 0 & -1 & 1 & 0 & 0 \\
1 & 0 & -1 & 0 & -1 & 0 & 1 & 0 & 0 & 0 & 0 & 0 & 0 & 0 & 0 & 0 & 0 & 0 & 0 & 0 \\
1 & 0 & -1 & 0 & 0 & 0 & 0 & 0 & -1 & 0 & 1 & 0 & 0 & 0 & 0 & 0 & 0 & 0 & 0 & 0 \\
1 & 0 & -1 & 0 & 0 & 0 & 0 & 0 & 0 & 0 & 0 & 0 & -1 & 0 & 1 & 0 & 0 & 0 & 0 & 0 \\
1 & 0 & -1 & 0 & 0 & 0 & 0 & 0 & 0 & 0 & 0 & 0 & 0 & 0 & 0 & 0 & -1 & 0 & 1 & 0 \\
1 & 0 & 0 & -1 & -1 & 0 & 0 & 1 & 0 & 0 & 0 & 0 & 0 & 0 & 0 & 0 & 0 & 0 & 0 & 0 \\
1 & 0 & 0 & -1 & 0 & 0 & 0 & 0 & -1 & 0 & 0 & 1 & 0 & 0 & 0 & 0 & 0 & 0 & 0 & 0 \\
1 & 0 & 0 & -1 & 0 & 0 & 0 & 0 & 0 & 0 & 0 & 0 & -1 & 0 & 0 & 1 & 0 & 0 & 0 & 0 \\
1 & 0 & 0 & -1 & 0 & 0 & 0 & 0 & 0 & 0 & 0 & 0 & 0 & 0 & 0 & 0 & -1 & 0 & 0 & 1 \\
\end{array}
$$

given by the matrix **T** in Table 5.19. The first row compares the difference between ads 1 and 2 in regions 1 and 2. The next 3 rows compare the differences between ads 1 and 2 in region 1, to each of regions 3, 4 and 5. The next 4 rows compare the difference between ads 1 and 3 in region 1, to each of the remaining 4 regions. Finally the last set of 4 rows compare the differences between ads 1 and 4 in all five regions.

For the main effect hypotheses the **T** matrices for AD effects and RE-GION effects are given in Table 5.20 and 5.21 respectively. In Table 5.20 the first row compares ads 1 and 2 while the second and third rows compare ads 1 and 3 and ads 1 and 4 respectively. In Table 5.21 the first row compares regions 1 and 2 while the second row compares regions 1 and 3. The remaining two rows compare region 1 to each of regions 4 and 5 respectively. The equations for the various fitted models are summarized in Table 5.22.

Table 5.20.   Transformation Matrix for AD Effects Test

	1	2	3	4	5	6	7	8	9	10	11	12	13	14	15	16	17	18	19	20
	1	-1	0	0	1	-1	0	0	1	-1	0	0	1	-1	0	0	1	-1	0	0
$T=$	1	0	-1	0	1	0	-1	0	1	0	-1	0	1	0	-1	0	1	0	-1	0
	1	0	0	-1	1	0	0	-1	1	0	0	-1	1	0	0	-1	1	0	0	-1

Table 5.21.   Transformation Matrix For Region Effects Test

	1	2	3	4	5	6	7	8	9	10	11	12	13	14	15	16	17	18	19	20
	1	1	1	1	-1	-1	-1	-1	0	0	0	0	0	0	0	0	0	0	0	0
$T=$	1	1	1	1	0	0	0	0	-1	-1	-1	-1	0	0	0	0	0	0	0	0
	1	1	1	1	0	0	0	0	0	0	0	0	-1	-1	-1	-1	0	0	0	0
	1	1	1	1	0	0	0	0	0	0	0	0	0	0	0	0	-1	-1	-1	-1

Orthogonal Coding

In a two-way factorial design model, orthogonal coding is used to set up contrasts to compare the means over the levels of both factors. Interaction variables are then determined by multiplying corresponding elements of the design matrix main effects, one from each factor. All of the resulting design matrix columns are mutually orthogonal.

Example

In the two-way factorial model used to compare family expenditures across four ads and five regions, a set of mutually *orthogonal design* variables are defined below. For the five regions the design matrix to be used is given by

REGION	R1	R2	R3	R4
1	1	0	1	−1
2	−1	0	1	−1
3	0	0	0	4
4	0	1	−1	−1
5	0	−1	−1	−1 .

For the four ad types the design matrix chosen is

AD	A1	A2	A3
1	1	0	2
2	1	−1	−1
3	1	1	−1
4	−3	0	0 .

Table 5.22. Multiple Regression Results for Cell Means Coding

Variable	No Restrictions	No Interaction	No Ad Effect and No Interaction	No Region Effect and No Interaction
D1	34.99	39.51	42.99	37.24
D2	56.17	41.92	42.99	39.65
D3	32.33	37.22	42.99	34.95
D4	48.47	53.30	42.99	51.03
D5	52.24	46.83	50.31	37.24
D6	42.02	49.24	50.31	39.65
D7	44.12	44.54	50.31	34.95
D8	62.86	60.62	50.31	51.03
D9	43.53	27.14	30.62	37.24
D10	18.84	29.55	30.62	39.65
D11	26.02	24.86	30.62	34.95
D12	34.10	40.94	30.62	51.03
D13	21.24	37.40	40.88	37.24
D14	47.13	39.81	40.88	39.65
D15	29.35	35.11	40.88	34.95
D16	69.79	51.19	40.88	51.03
D17	34.19	35.31	38.79	37.24
D18	34.08	37.72	38.79	39.65
D19	42.95	33.02	38.79	34.95
D20	43.93	49.10	38.79	51.03
F-Statistic for Effect		1.7574	2.1273	2.0359
Degrees of Freedom		12	15	16
p-Value		0.0658	0.0142	0.0175

The seven design variables defined by the two design matrices are mutually orthogonal. Denoting the mean for the i-th region and the j-th ad by μ_{ij}, the five region means are denoted by $\mu_{i\cdot}$ and the four ad means by $\mu_{\cdot j}$. The coefficients of these design variables can be used to test the following seven independent hypotheses regarding the means.

$$H_{01} : \mu_{1\cdot} = \mu_{2\cdot}, \quad H_{02} : \mu_{4\cdot} = \mu_{5\cdot},$$

$$H_{03} : (\mu_{1\cdot} + \mu_{2\cdot}) = (\mu_{4\cdot} + \mu_{5\cdot}),$$

$$H_{04} : 4\mu_{3\cdot} = (\mu_{1\cdot} + \mu_{2\cdot} + \mu_{4\cdot} + \mu_{5\cdot}),$$

$$H_{05} : (\mu_{\cdot 1} + \mu_{\cdot 2} + \mu_{\cdot 3}) = 3\mu_{\cdot 4},$$

$$H_{06} : \mu_{\cdot 3} = \mu_{\cdot 2}, \quad H_{07} : 2\mu_{\cdot 1} = (\mu_{\cdot 2} + \mu_{\cdot 3}).$$

The seven design variables are summarized in Table 5.23.

Table 5.23. Orthogonal Design Matrix for Two Factor Factorial. – Test Marketing Study

Cell Mean	Reg-ion	Ad	R1	R2	R3	R4	A1	A2	A3	RA11	RA12	RA13	RA21	RA22	RA23	RA31	RA32	RA33	RA41	RA42	RA43
μ_{11}	1	1	1	0	1	-1	1	0	2	1	0	2	0	0	0	1	0	2	-1	0	-2
μ_{21}	2	1	-1	0	1	-1	1	0	2	-1	0	-2	0	0	0	1	0	2	-1	0	-2
μ_{31}	3	1	0	0	0	4	1	0	2	0	0	0	0	0	0	0	0	0	4	0	8
μ_{41}	4	1	0	1	-1	-1	1	0	2	0	0	0	1	0	2	-1	0	-2	-1	0	-2
μ_{51}	5	1	0	-1	-1	-1	1	0	2	0	0	0	-1	0	-2	-1	0	-2	-1	0	-2
μ_{12}	1	2	1	0	1	-1	1	-1	-1	1	-1	-1	0	0	0	1	-1	-1	-1	1	1
μ_{22}	2	2	-1	0	1	-1	1	-1	-1	-1	1	1	0	0	0	1	-1	-1	-1	1	1
μ_{32}	3	2	0	0	0	4	1	-1	-1	0	0	0	0	0	0	0	0	0	4	-4	-4
μ_{42}	4	2	0	1	-1	-1	1	-1	-1	0	0	0	1	-1	-1	-1	1	1	-1	1	1
μ_{52}	5	2	0	-1	-1	-1	1	-1	-1	0	0	0	-1	1	1	-1	1	1	-1	1	1
μ_{13}	1	3	1	0	1	-1	1	1	-1	1	1	-1	0	0	0	1	1	-1	-1	-1	1
μ_{23}	2	3	-1	0	1	-1	1	1	-1	-1	-1	1	0	0	0	1	1	-1	-1	-1	1
μ_{33}	3	3	0	0	0	4	1	1	-1	0	0	0	0	0	0	0	0	0	4	4	-4
μ_{43}	4	3	0	1	-1	-1	1	1	-1	0	0	0	1	1	-1	-1	-1	1	-1	-1	1
μ_{53}	5	3	0	-1	-1	-1	1	1	-1	0	0	0	-1	-1	1	-1	-1	1	-1	-1	1
μ_{14}	1	4	1	0	1	-1	-3	0	0	-3	0	0	0	0	0	-3	0	0	3	0	0
μ_{24}	2	4	-1	0	1	-1	-3	0	0	3	0	0	0	0	0	-3	0	0	3	0	0
μ_{34}	3	4	0	0	0	4	-3	0	0	0	0	0	0	0	0	0	0	0	-12	0	0
μ_{44}	4	4	0	1	-1	-1	-3	0	0	0	0	0	-3	0	0	3	0	0	3	0	0
μ_{54}	5	4	0	-1	-1	-1	-3	0	0	0	0	0	3	0	0	3	0	0	3	0	0

Interaction Effects

In addition to the seven main effect variables defined above, twelve interaction variables are defined by determining all possible combinations of one ad variable and one region variable. The interaction variables are obtained by multiplying corresponding elements of the two design variables that form the interaction. The resulting design matrix contains twelve orthogonal variables including the intercept. The design matrix for one replication of the 20 cells in the 4×5 factorial is summarized in Table 5.23. The actual data for the experiment summarized in Table 5.1 contains 6 observations for each cell. This design matrix is therefore repeated 6 times.

The twelve independent interaction hypotheses that would be tested using the interaction coefficients are summarized below.

RA11 $H_{11} : (\mu_{11} - \mu_{21}) + (\mu_{12} - \mu_{22}) + (\mu_{13} - \mu_{23}) = 3(\mu_{14} - \mu_{24})$
  - compares the difference between Region 1 and Region 2 for all 4 Ads. Ads 1, 2 and 3 are compared to 4.

RA12 $H_{12} : (\mu_{12} - \mu_{22}) = (\mu_{13} - \mu_{23})$
  - compares the difference between Region 1 and Region 2 for Ads 2 and 3.

RA13 $H_{13} : 2(\mu_{11} - \mu_{21}) = (\mu_{12} - \mu_{22}) + (\mu_{13} - \mu_{23})$
  - compares the difference between Region 1 and Region 2 for Ads 1, 2 and 3. Ad 1 is compared to Ads 2 and 3.

RA21 $H_{14} : (\mu_{41} - \mu_{51}) + (\mu_{42} - \mu_{52}) + (\mu_{43} - \mu_{53}) = 3(\mu_{44} - \mu_{54})$
  - compares the difference between Regions 4 and 5 for Ads 1, 2, 3, 4. Ads 1, 2, 3 are compared to 4.

RA22 $H_{15} : (\mu_{42} - \mu_{52}) = (\mu_{43} - \mu_{53})$
  - compares the difference between Regions 4 and 5 for Ads 2 and 3.

RA23 $H_{16} : 2(\mu_{41} - \mu_{51}) = (\mu_{42} - \mu_{52}) + (\mu_{43} - \mu_{53})$
  - compares the difference between Regions 4 and 5 for Ads 1, 2 and 3. Ads 2 and 3 are compared to 1.

RA31 $H_{17} : [(\mu_{11} + \mu_{21}) - (\mu_{41} + \mu_{51})] + [(\mu_{12} + \mu_{22}) - (\mu_{42} + \mu_{52})] + [(\mu_{13} + \mu_{23}) - (\mu_{43} + \mu_{53})] = 3[(\mu_{14} + \mu_{24}) - (\mu_{44} + \mu_{54})]$
  - compares the difference between Regions 1 and 2 and Regions 4 and 5; Ads 1, 2 and 3 are compared to 4.

RA32 $H_{18} : [(\mu_{12} + \mu_{22}) - (\mu_{42} + \mu_{52})] = [(\mu_{13} + \mu_{23}) - (\mu_{43} + \mu_{53})]$
  - compares the difference between Regions 1 and 2 and Regions 4 and 5 for Ad 2 versus Ad 3.

RA33 $H_{19} : 2[(\mu_{11} + \mu_{21}) - (\mu_{41} + \mu_{51})] = [(\mu_{12} + \mu_{22}) - (\mu_{42} + \mu_{52})] + [(\mu_{13} + \mu_{23}) - (\mu_{43} + \mu_{53})]$
  - compares the difference between Regions 1 and 2 and Regions 4 and 5 for Ad 1 versus Ads 2 and 3.

RA41 $H_{110} : [(\mu_{11} + \mu_{21} + \mu_{41} + \mu_{51}) - 4\mu_{31}] + [(\mu_{12} + \mu_{22} + \mu_{42} + \mu_{52}) - 4\mu_{32}] + [(\mu_{13} + \mu_{23} + \mu_{43} + \mu_{53}) - 4\mu_{33}] = 3[(\mu_{14} + \mu_{24} + \mu_{44} + \mu_{54}) - 4\mu_{34}]$
  - compares the difference between Region 3 and Regions 1, 2, 4 and 5 for Ad 4 versus Ads 1, 2 and 3.

RA42 $H_{111} : [(\mu_{12} + \mu_{22} + \mu_{42} + \mu_{52}) - 4\mu_{32}] = [(\mu_{13} + \mu_{23} + \mu_{43} + \mu_{53}) - 4\mu_{33}]$
  - compares the difference between Region 3 and Regions 1, 2, 4 and 5 for Ad 2 versus Ad 3.

RA43 $H_{112} : 2[(\mu_{11} + \mu_{21} + \mu_{41} + \mu_{51}) - 4\mu_{31}] = [(\mu_{12} + \mu_{22} + \mu_{42} + \mu_{52}) - 4\mu_{32}] + [(\mu_{13} + \mu_{23} + \mu_{33} + \mu_{53}) - 4\mu_{33}]$
  - compares the difference between Region 3 and Regions 1, 2, 4 and 5 for Ad 1 versus Ads 2 and 3.

The results of the regression of the variable EXPEN on the twenty design variables are summarized in Table 5.24. From the results in the table we can conclude that the coefficients of INTERCEPT, R3, R4, A1, RA21 and RA43 are significant at the approximate 0.10 level. We would therefore reject hypotheses H_{03}, H_{04} and H_{05} for the main effects and H_{14} and H_{112} for the interactions. It would appear that the means for Regions 1 and 2 exceed the means for Regions 4 and 5 and, also that the mean for Region 3 is less than the means for Regions 1, 2, 4 and 5. Ad type 4 results in a higher mean than Ad types 1, 2 and 3. We can also conclude that the difference between Regions 4 and 5 for Ads 1, 2 and 3 is less than for Ad 4. Equivalently, the differences between Ads 1, 2, 3 and Ad 4 are less for Region 4 than for Region 5. Finally when Ad 1 is compared to Ads 2 and 3, differences among ads for region 3 are much smaller than when Ad 1 is compared to Ads 2 and 3 for Regions 1, 2, 4 and 5.

Table 5.24 also contains the squared correlations between EXPEN and each of the twenty design variables. Since the design variables are mutually orthogonal, the sum of these squared correlations is the overall R^2 for the model. From the sums of squares given in Table 5.11 the proportion of the total variation can be allocated as 0.075 (Ads), 0.080 (Regions) and 0.147 (Interaction). From the above table we can conclude that the A1 contrast accounts for 0.070, the R3 and R4 contrasts 0.069 and the RA21 and RA43 contrasts 0.063. The total contribution to R^2 from the significant effects is 0.202 which represents 2/3 of the total R^2 of 0.3026.

5.2.4 An Unbalanced Two-Way Analysis

Dummy Variable Coding

In Table 4.41 of Chapter 4 a sample of observations relating COMBRATE to the variables WEIGHT, FOR and CYLIND was provided for a sample

Table 5.24. Relationship Between EXPEN and Twenty Design Variables

Variable	df	Parameter Estimate	t-value	p-value	Squared Correlation with EXPEN
Intercept	1	40.717	21.666	0.0001	0.0000
R1	1	−3.659	−1.231	0.2211	0.0106
R2	1	1.045	0.352	0.7257	0.0009
R3	1	3.409	1.622	0.1079	0.0184
R4	1	−2.524	−2.686	0.0085	0.0503
A1	1	−3.438	−3.168	0.0020	0.0700
A2	1	−2.348	−0.883	0.3791	0.0054
A3	1	−0.020	−0.013	0.9896	0.0000
RA11	1	1.177	0.686	0.4942	0.0033
RA12	1	−6.484	−1.543	0.1260	0.0166
RA13	1	−3.071	−1.266	0.2085	0.0112
RA21	1	−3.295	−1.921	0.0576	0.0257
RA22	1	−6.665	−1.586	0.1159	0.0175
RA23	1	−2.114	−0.871	0.3856	0.0053
RA31	1	1.003	0.826	0.4105	0.0048
RA32	1	−1.602	−0.539	0.5910	0.0020
RA33	1	1.770	1.031	0.3048	0.0074
RA41	1	0.570	1.050	0.2961	0.0077
RA42	1	1.484	1.117	0.2668	0.0087
RA43	1	1.764	2.299	0.0236	0.0368

$R^2 = 0.306$

SSR = 18391.109   19 d.f. $F = 2.284$ $p = 0.0045$
SSE = 42382.015   100 d.f.

of automobiles. In this section the relationship between COMBRATE and the two classification variables (FOR = 0 domestic, and = 1 foreign) and CYLIND (No. of cylinders = 4 or 6 or 8) will be studied. Table 5.25 shows the number of observations in the 6 categories defined by the cross classification of the two variables. The unequal frequencies in the six cells yields an unbalanced design.

Dummy variables C4, C6 and C8 will be used to represent the categories CYLIND = 4, CYLIND = 6 and CYLIND = 8 respectively. Interaction dummy variables were created using the products of FOR with each of C4, C6 and C8 respectively. These interaction dummies are denoted by FORC4, FORC6 and FORC8 respectively.

Table 5.27 contains the results of several regressions relating COM-BRATE to the various dummy variables. Using the results of the regressions in lines 1, 3, 4 and 6 of this table, the cell means shown in Table 5.26 can be determined. The regression in line 4 forces the INTERACTION to be zero while the regression in line 6 produces the actual cell means. With the no interaction model the FOR effect is constant at 3.96 for each level of CYLIND.

Table 5.25. Cell Sample Sizes

		FOR 0	FOR 1	Totals
	4	33	29	62
CYLIND	6	14	4	18
	8	16	2	18
	Totals	63	35	98

Table 5.26. Cell Means

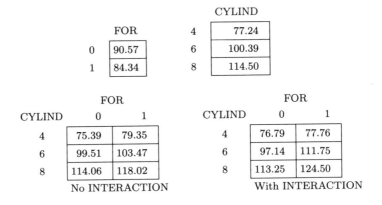

CYLIND	FOR 0	1
4	75.39	79.35
6	99.51	103.47
8	114.06	118.02

No INTERACTION

CYLIND	FOR 0	1
4	76.79	77.76
6	97.14	111.75
8	113.25	124.50

With INTERACTION

From the results of line 3 the two FOR means are significantly different at the 0.105 level. From the results of lines 1 and 2 the three CYLIND categories are significantly different at the 0.000 level. The results in lines 6 and 7 when compared to the results in lines 4 and 5 yield an F-value of

$$F = \frac{(0.723 - 0.705)/2}{(1 - 0.723)/92} = 2.99,$$

which has a p-value of 0.055, and hence the INTERACTION terms are marginally significant. An examination of the p-values for the INTERACTION terms in lines 6 and 7 indicates that the strongest INTERACTION occurs for 6-cylinder vehicles, where the difference between foreign and domestic vehicles is much larger than this difference is for 4-cylinder vehicles. This difference for 8-cylinder vehicles is also marginally significant in comparison to 4-cylinder vehicles.

Table 5.27. Multiple Regression Results Dummy Coding

	Intercept	C4	C6	C8	FOR	FORC4	FORC6	FORC8	R^2	F	p-value
1.	114.50 (0.000)	-37.26 (0.000)	-14.11 (0.000)	x	x	x	x	x	0.696	108.48	0.000
2.	77.24 (0.000)	x	23.15 (0.000)	37.26 (0.000)	x	x	x	x	0.696	108.48	0.000
3.	90.57 (0.000)	x	x	x	-6.26 (0.105)	x	x	x	0.705	74.980	0.000
4.	114.06 (0.000)	-38.67 (0.000)	-14.55 (0.000)	x	3.96 (0.080)	x	x	x	0.705	74.980	0.000
5.	75.39 (0.000)	x	24.12 (0.000)	38.67 (0.000)	3.96 (0.080)	x	x	x	0.705	74.980	0.000
6.	113.25 (0.000)	-36.46 (0.000)	-16.11 (0.000)	x	11.25 (0.132)	-10.28 (0.192)	3.36 (0.719)	x	0.723	48.099	0.000
7.	76.79 (0.000)	x	20.35 (0.000)	36.46 (0.000)	0.97 (0.700)	x	13.64 (0.029)	10.28 (0.192)	0.723	48.099	0.000

Effect Coding

Table 5.28 summarizes the results of several regressions using effect coding to differentiate between foreign and domestic vehicles, and to differentiate among the cylinder levels. The individual coefficients in the regressions allow comparisons to be made to the overall mean. From lines 1 and 2 we can conclude that 4-cylinder vehicles have a COMBRATE mean which is significantly lower than the overall mean, and that 8-cylinder vehicles have a COMBRATE mean significantly larger than the overall mean. Examining the INTERACTION coefficients we can conclude that the INTERACTION for 4-cylinder vehicles is significantly smaller than the average INTERACTION for all vehicles.

A Cell Means Approach

An alternative approach to comparing the means is provided by a cell means model. Dummy variables D1 to D6 are defined as follows:

D1 (FOR $= 0$, CYLIND $= 4$)   D2 (FOR $= 0$, CYLIND $= 6$)
D3 (FOR $= 0$, CYLIND $= 8$)   D4 (FOR $= 1$, CYLIND $= 4$)
D5 (FOR $= 1$, CYLIND $= 6$)   D6 (FOR $= 1$, CYLIND $= 8$)

Using a multiple linear regression model with no intercept the model becomes

$$\text{COMBRATE} = \beta_1 \text{ D1} + \beta_2 \text{ D2} + \beta_3 \text{ D3} + \beta_4 \text{ D4} + \beta_5 \text{ D5} + \beta_6 \text{ D6} + U$$

where each regression coefficient represents the mean of a particular cell. The estimated model is given by

$$\text{COMBRATE} = 76.79 \text{ D1} + 97.14 \text{ D2} + 113.25 \text{ D3}$$

$$+ 77.76 \text{ D4} + 111.75 \text{ D5} + 124.50 \text{ D6}.$$

Inference can be made using linear restrictions on the coefficients. To test the hypothesis of no INTERACTION the transformation matrix is given by

$$\mathbf{T} = \begin{bmatrix} 2 & -1 & -1 & -2 & 1 & 1 \\ -1 & 2 & -1 & 1 & -2 & 1 \end{bmatrix} \quad \text{for } \mathbf{T}\boldsymbol{\beta} = 0,$$

which uses the restrictions

$$(2\beta_1 - \beta_2 - \beta_3) - (2\beta_4 - \beta_5 - \beta_6) = 0$$
$$\text{and} \quad (2\beta_2 - \beta_1 - \beta_3) - (2\beta_5 - \beta_4 - \beta_6) = 0.$$

Table 5.28. Multiple Regression Results Effect Coding

	Intercept	C4E	C6E	C8E	FORE	FORC4E	FORC6E	FORC8E	R^2	F	p-value
1.	97.38 (0.000)	-20.14 (0.000)	3.01 (0.106)	x	x	x	x	x	0.696	108.48	0.000
2.	97.38	x	3.01 (0.106)	17.12 (0.000)	x	x	x	x	0.696	108.48	0.000
3.	87.44 (0.000)	x	x	x	-3.129 (0.105)	x	x	x	0.27	2.68	0.105
4.	98.30 (0.000)	-20.93 (0.000)	3.19 (0.084)	x	1.98 (0.080)	x	x	x	0.705	74.98	0.000
5.	98.30 (0.000)	x	3.19 (0.084)	17.74 (0.000)	1.98 (0.080)	x	x	x	0.705	74.98	0.000
6.	100.20 (0.000)	-22.92 (0.000)	4.25 (0.065)	x	4.47 (0.006)	-3.99 (0.026)	2.83 (0.217)	x	0.723	48.10	0.000
7.	100.20 (0.000)	x	4.25 (0.065)	18.67 (0.000)	4.47 (0.006)	x	2.832 (0.217)	1.154 (0.667)	0.723	48.10	0.000

These restrictions are equivalent (up to multiplication by a constant) to the more conventional forms

$$[\beta_1 - (\beta_1 + \beta_2 + \beta_3)/3] - [\beta_4 - (\beta_4 + \beta_5 + \beta_6)/3] = 0$$

$$[\beta_2 - (\beta_1 + \beta_2 + \beta_3)/3] - [\beta_5 - (\beta_4 + \beta_5 + \beta_6)/3] = 0.$$

The fitted model is given by

$$\text{COMBRATE} = 75.39 \text{ D1} + 99.51 \text{ D2} + 114.06 \text{ D3}$$

$$+ 79.35 \text{ D4} + 103.47 \text{ D5} + 118.02 \text{ D6}.$$

The F-statistic for testing the no INTERACTION hypothesis is given by $F = 2.998$ with 2 degrees of freedom and has a p-value of 0.055.

A fitted model with the no INTERACTION restriction and the no foreign effect is obtained by adding the row $\begin{bmatrix} 1 & 1 & 1 & -1 & -1 & -1 \end{bmatrix}$ to the $\mathbf{T}$ matrix. This restriction is equivalent to

$$[(\beta_1 + \beta_2 + \beta_3)/3 - (\beta_4 + \beta_5 + \beta_6)/3] = 0.$$

The fitted model obtained is given by

$$\text{COMBRATE} = 77.24 \text{ (D1} + \text{D4)} + 100.39 \text{ (D2} + \text{D5)} + 114.50 \text{ (D3} + \text{D6)}.$$

The F-statistic for testing this hypothesis has a value of 3.085 with 3 degrees of freedom and a p-value of 0.031.

Similarily a fitted model with no INTERACTIONS and no CYLIND effect is obtained by adding the two rows to $\mathbf{T}$ given by

$$\begin{bmatrix} 2 & -1 & -1 & 2 & -1 & -1 \\ -1 & 2 & -1 & -1 & 2 & -1 \end{bmatrix}$$

which is equivalent to the conditions

$$[(\beta_1 + \beta_4)/2 - (\beta_1 + \beta_2 + \beta_3 + \beta_4 + \beta_5 + \beta_6)/6] = 0$$

$$\text{and} \quad [(\beta_2 + \beta_5)/2 - (\beta_1 + \beta_2 + \beta_3 + \beta_4 + \beta_5 + \beta_6)/6] = 0.$$

The value of the F-statistic for this condition is 57.867 with 4 degrees of freedom and a p-value of 0.000.

Automobile Type (Block)	Gasoline Brand (Treatment) 1 2 3 4 5 6
A	* *
B	* *
C	* *

Figure 5.10. A Nested Treatment Design

5.2.5 A Two-Way Nested Treatment Design

In a randomized block design it may not be possible to assign all g treatment levels randomly to each block. It may be necessary to restrict certain treatment levels to particular blocks. In the study of the performance of various gasoline brands it may be necessary to restrict certain gasoline brands to certain automobile types so that gasoline brands are nested within automobile types. Such a design is called a *nested treatment design* or *hierarchical treatment design*. Recall that if each block contains all possible treatments, the treatment levels and blocks would be completely crossed; in this case they are nested not crossed.

Figure 5.10 illustrates a nested set of treatments for the gasoline brand comparison. In this experiment a total of 6 gasoline brands are compared using 3 types of automobiles with a total of 6 automobiles or experimental units. In comparison a randomized block design would require $3 \times 6 = 18$ automobiles or experimental units. In the test marketing study used throughout this chapter it may not be practical to use every ad type in every region. In that case the ads might be nested within the regions.

A model for the *two-way nested design* is given by

$$y_{ijh} = \mu + \beta_i + \alpha_{ij} + \varepsilon_{ijh} \qquad \begin{aligned} i &= 1, 2, \ldots, b; \\ j &= 1, 2, \ldots, g; \\ h &= 1, 2, \ldots, c; \end{aligned}$$

where the ε_{ijh} are mutually independent $N(0, \sigma^2)$. Each block contains g treatments and the total number of treatments is bg. There are c replications assumed for each treatment.

The cell mean for cell (i, j) is denoted by μ_{ij} and the block effect parameters by $(\mu_i. - \mu) = \beta_i$ where μ is the grand mean $\mu = \sum_{i=1}^{b} \sum_{j=1}^{g} \mu_{ij}/bg$. The treatment effects are given by $\alpha_{ij} = (\mu_{ij} - \mu_i.)$. The parameters therefore satisfy the conditions $\sum_{i=1}^{b} \beta_i = 0$ and $\sum_{i=1j=1}^{b\ \ g} \alpha_{ij} = 0$.

Table 5.29. Analysis of Variance for Nested Treatment Design

Source	d.f.	Sum of Squares	Mean Square	F
B	$(b-1)$	SSB	$\text{MSB} = \text{SSB}/(b-1)$	MSA/MSE
AB	$b(g-1)$	SSAB	$\text{MSAB} = \text{SSAB}/b(g-1)$	MSAB/MSE
ERROR	$bg(c-1)$	SSE	MSE	

The estimators of the effects are given by $\hat{\beta}_i = (\bar{y}_{i\cdot\cdot} - \bar{y}_{\cdots})$ and $\hat{\alpha}_{ij} = (\bar{y}_{ij\cdot} - \bar{y}_{i\cdot\cdot})$, $i = 1, 2, \ldots, b$; $j = 1, 2, \ldots, g$.

The corresponding sums of squares are given by $\text{SSB} = gc\sum_{i=1}^{b}\hat{\beta}_i^2$ and

$\text{SSAB} = c\sum_{i=1}^{b}\sum_{j=1}^{g}\hat{\alpha}_{ij}^2$. The total sum of squares and the error sum of squares

are given by $\text{SST} = \sum_{h=1}^{c}\sum_{j=1}^{g}\sum_{i=1}^{b}(y_{ijh} - \bar{y}_{\cdots})^2$ and $\text{SSE} = \sum_{h=1}^{c}\sum_{j=1}^{g}\sum_{i=1}^{b}(y_{ijh} - \bar{y}_{ij\cdot})^2$

respectively.

The analysis of variance table for this nested treatment design is given in Table 5.29.

If the block effects are ignored, and this design is analyzed as a completely randomized design, then if there are block effects the estimates of the treatment effects will be biased and inconsistent. The test will also be insensitive in that the error sum of squares will be inflated.

5.3 Analysis of Covariance

Up to this point our discussion of the measurement of other factors has assumed that the intervening variable is qualitative. It is also possible to measure and remove the effects of a quantitative variable using regression analysis. This technique is commonly referred to as the analysis of covariance. For the example involving the comparison of the fuel efficiency for various brands of gasoline, we may wish to take into account the weight of the automobiles used rather than the brand of the automobile. The quantitative variable weight in this case is called a *covariate*.

The Analysis of Covariance Model

In *analysis of covariance* each observation consists of a pair (y_{ij}, z_{ij}) where y_{ij}, $i = 1, 2, \ldots, n_j$; $j = 1, 2, \ldots, g$, denotes the i-th observation in group

j on the variable of interest and z_{ij} denotes the i-th observation in group j on the covariate. The relationship between y_{ij} and z_{ij} is assumed to be linear and is determined by a simple linear regression of the y_{ij} on the z_{ij}. The regression model can be expressed as

$$y_{ij} = \gamma_{0j} + \gamma_1 z_{ij} + v_{ij} \tag{5.8}$$

and hence the slope parameter γ_1 is assumed to be constant over the g groups. A more general model would allow the parameter γ_1 to vary over the g groups by assuming the model

$$y_{ij} = \gamma_{0j} + \gamma_{1j} z_{ij} + v_{ij}.$$

We discuss the more restricted version. The more general version will be discussed in the multiple regression approach. A special case of this model corresponding to $j = 2$ was discussed in Chapter 3 in connection with qualitative lurking variables.

Inference for the Model

The analysis of covariance begins by fitting the regression model (5.8) to obtain
$$\hat{y}_{ij} = \hat{\gamma}_{0j} + \hat{\gamma}_1 z_{ij}.$$
Removal of the effect of the covariate is then carried out by removing $\hat{\gamma}_1 z_{ij}$ from y_{ij} to obtain
$$\tilde{y}_{ij} = y_{ij} - \hat{\gamma}_1 z_{ij}.$$

A one-way analysis of variance is now carried out on the $\tilde{y}_{ij}$ in the same manner as discussed in 5.1.1. The degrees of freedom for SSW must now be reduced by 1 to account for the estimate of $\hat{\gamma}_1$. If desired, a test of the hypothesis that $\gamma_1 = 0$ can be carried out using the conventional F-test or t-test used in simple linear regression.

Example

Using the test marketing data given in Table 5.1 and the assumptions employed for the one-way model, the variable 'family size' can be treated as a covariate. These data therefore can be modelled using an analysis of covariance to control for family size. Using a one-way analysis of covariance the variation in expenditures was related to the ad type and family size using the data in Table 5.1. The results of the analysis of covariance are presented in Table 5.30. From the table we can conclude that family size has a significant impact on expenditure and that after controlling for family size the differences among the mean expenditures for the four ad groups are highly significant.

Table 5.30. Analysis of Covariance for Ad Types Controlling For Family Size

Source	d.f.	S.S.	M.S.	F	p-value
Ad	3	4585.68	1528.56	11.52	0.00
Family Size	1	40926.02	40926.02	308.39	0.00
Error	115	15261.43	132.71		

Multiple Regression Model Approach

The completely general form of the analysis of covariance model can be compared to the example used in Chapter 4 to relate supply S to price P allowing variation over quarters. The dummy variable/slope shifter approaches outlined in Chapter 4 can also be viewed as examples of the analysis of covariance. Analysis of covariance type analyses can be extended to the two-way analysis of variance model. The model outlined in Section 4.5 relating salary to education, sex and occupation group is an example of such a model. Other analysis of covariance type models can be obtained by adding additional quantitative explanatory variables. All such models may be viewed as special cases of the multiple regression model.

Example

For the test marketing data discussed throughout this chapter the variable family SIZE can be included in the analysis as a covariate. The model for EXPEN discussed in Section 5.4 contained the dummy variables AD1, AD2, AD3, REG1, REG2, REG3, REG4 and the interaction dummy variables $ADREG_{ij}$, $i = 1, 2, 3$; $j = 1, 2, 3, 4$. The variable SIZE can be added as a single variable or it can be included in slope shifter form with several of the variables AD1SIZE, AD2SIZE, AD3SIZE, REG1SIZE, REG2SIZE, REG3SIZE, REG4SIZE and $ADREG_{ij}SIZE$, $i = 1, 2, 3$; $j = 1, 2, 3, 4$.

In Section 5.2.2 the two-way model relating EXPEN to the AD types and REG types resulted in $R^2 = 0.303$ with partial sums of squares AD (4585.7), REG (4867.5) and AD*REG (8937.9) with 42382.0 as the error sum of squares. Adding the variable SIZE to the model increases R^2 to 0.976 and the partial sum of squares for SIZE is given by 40926.0. All the effects, AD, REG, AD*REG and SIZE have p values of 0.0001. Adding slope shifter variables to the model does not provide a significant increase in R^2.

The estimated model using the dummy variables defined above is given by

$$y = \underset{(0.000)}{6.08} - \underset{(0.000)}{9.74} \ \text{AD1} - \underset{(0.000)}{9.86} \ \text{AD2} - \underset{(0.657)}{0.99} \ \text{AD3}$$

$$+ \underset{(0.043)}{4.54} \ \text{REG1} + \underset{(0.000)}{18.92} \ \text{REG2} - \underset{(0.000)}{9.83} \ \text{REG3} + \underset{(0.000)}{21.86} \ \text{REG4}$$

$$- \underset{(0.235)}{3.74} \ \text{ADREG11} - \underset{(0.780)}{0.87} \ \text{ADREG12} + \underset{(0.000)}{19.17} \ \text{ADREG13}$$

$$- \underset{(0.000)}{34.82} \ \text{ADREG14} + \underset{(0.000)}{17.55} \ \text{ADREG21} - \underset{(0.001)}{10.998} \ \text{ADREG22}$$

$$- \underset{(0.088)}{5.40} \ \text{ADREG23} - \underset{(0.006)}{8.80} \ \text{ADREG24} - \underset{(0.000)}{15.16} \ \text{ADREG31}$$

$$- \underset{(0.000)}{17.75} \ \text{ADREG32} - \underset{(0.026)}{7.10} \ \text{ADREG33} - \underset{(0.000)}{35.46} \ \text{ADREG34}$$

$$+ \underset{(0.000)}{10.81} \ \text{SIZE.}$$

Comparing this equation to the model fitted in which excluded size reveals that the coefficients of the AD, REG and AD*REG variables have not changed. This reflects the orthogonality between the variable SIZE and these other effects. The fact that the design is balanced (one family of each size in each AD × REG cell) produces the orthogonal property. The p-values for the coefficients have been reduced considerably because of the large decrease in the error sum of squares brought about by the addition of the variable SIZE.

Example with an Unbalanced Design

While orthogonal designs are useful in obtaining simple analyses it is not always possible to obtain the data in this form. A second example of analysis of covariance without a balanced design is provided by the fuel efficiency data studied in Chapter 4. In Section 4.5 a complete model with all possible dummy variables and slope shifters was used to relate COMBRATE to WEIGHT controlling for CYLIND (number of cylinders) and FOR (foreign or domestic). It was found in 4.5 that almost all of the variables were significant at the 0.10 level.

Without the variable WEIGHT the dummy variables for FOR, CYLIND and FOR*CYLIND provide an $R^2 = 0.723$. The model is given by

$$\text{COMBRATE} = \underset{(0.000)}{113.25} - \underset{(0.000)}{36.46} \ \text{C4} - \underset{(0.000)}{16.11} \ \text{C6} + \underset{(0.132)}{11.25} \ \text{FOR}$$

$$- \underset{(0.192)}{10.28} \ \text{FORC4} + \underset{(0.719)}{3.36} \ \text{FORC6.}$$

The results indicate that in comparison to domestic 8-cylinder vehicles, domestic vehicles with 4 and 6 cylinders and foreign vehicles with 8 cylinders use much less fuel. Foreign vehicles with only 4 cylinders use more fuel than domestic 4-cylinder vehicles. Similarily for 6-cylinder vehicles.

When the covariate WEIGHT is added to the model the R^2 increases to 0.826. The fitted model is given by

$$\underset{(.000)}{\text{COMBRATE} = 42.870} - \underset{(.000)}{17.698}\ \text{C4} - \underset{(.006)}{8.588}\ \text{C6} - \underset{(.004)}{17.757}\ \text{FOR}$$

$$+ \underset{(.024)}{14.409}\ \text{FORC4} + \underset{(.810)}{1.797}\ \text{FORC6} + \underset{(.000)}{0.019}\ \text{WEIGHT}.$$

In comparison to the model without WEIGHT, the coefficients have changed because of the correlation between WEIGHT and the variables CYLIND and FOR. In addition to the increase in R^2 from 0.73 to 0.826 the main result of adding the variable WEIGHT is to change the coefficient of FOR from negative to positive. Thus if WEIGHT is held constant along with the number of cylinders, foreign automobiles tend to consume more fuel than domestic automobiles. If WEIGHT is not controlled, however, the fuel consumption of foreign cars after controlling for the number of cylinders is less than for domestic cars. This result would suggest that foreign automobiles probably weigh less, holding the number of cylinders fixed. The full model with all slope shifter variables yields an R^2 of 0.853. The equation for this model is given in Section 4.5.

If effect coding is used for the full model the fitted model is given by

$$\underset{(.000)}{\text{COMBRATE} = 74.656} - \underset{(.000)}{64.182}\ \text{C4} - \underset{(.036)}{53.512}\ \text{C6} - \underset{(.032)}{31.479}\ \text{FOR}$$

$$+ \underset{(.159)}{21.644}\ \text{FORC4} + \underset{(.189)}{33.353}\ \text{FORC6} + \underset{(.022)}{0.009}\ \text{WEIGHT}$$

$$- \underset{(.061)}{0.018}\ \text{FORW} + \underset{(.000)}{0.016}\ \text{C4W} - \underset{(.022)}{0.017}\ \text{C6W}$$

$$+ \underset{(.288)}{0.005}\ \text{FC4W} - \underset{(.169)}{0.010}\ \text{FC6W}.$$

From this equation we conclude that the average intercept for the six cells is 74.656. The intercept is significantly lower for 4-cylinder automobiles, 6-cylinder cars and for foreign made cars. For domestic 8-cylinder vehicles the intercept is therefore well above average. The fuel consumption of foreign automobiles and domestic automobiles with 6 cylinders is less sensitive to WEIGHT while for domestic cars with CYLIND = 4 the reverse is true. The interactions between number of cylinders and domestic vs foreign are not significant.

5.4 Some Three-way Analysis of Variance Models

In this section several *three-way analysis of variance* models are introduced to extend the concepts outlined in Sections 5.1 and 5.2. The first example combines a randomized block design with a two factor factorial arrangement of treatment combinations. This design is a simple extension of the two factor factorial allowing for a nuisance factor to be controlled through a randomized block design. A *Latin square design* is then introduced to extend the blocking concept to control for two nuisance factors. The Latin square is designed for economy in the use of experimental units but can only be used to estimate main effects. A *three factor factorial* is then outlined along with a discussion of the measurement of a three-way or second order interaction. Each treatment combination is assumed to be replicated a total of c times. While the above three examples are balanced designs the fourth example is an unbalanced design. The multiple regression model is used to illustrate the methodology using a particular example of an *unbalanced three-way model*.

5.4.1 A Two-Way Factorial in a Randomized Block Design

A common three-way analysis of variance model consists of the study of two treatment factors using a factorial arrangement along with a randomized block design to control for a nuisance factor. In the study of gasoline performance in automobiles both the gasoline brand and automobile type could be considered to be treatment factors, while the actual location of the test might be considered to be a blocking factor. The locations or test routes can vary with respect to driving conditions such as urban or rural, speed limits, weather etc.

Assume that the two treatment factors have g and ℓ levels respectively, and that the total number of treatment combinations is $g\ell$. Each treatment combination is randomly assigned to a total of b experimental units, one from each of b blocks. A total of $bg\ell$ experimental units are used. The experimental units are blocked so that there are b blocks each containing $g\ell$ relatively homogeneous units. The measured variable Y corresponding to block i, level j of factor 1 and level k of factor 2 is denoted by y_{ijk}, $i = 1, 2, \ldots, b$; $j = 1, 2, \ldots, g$; $k = 1, 2, \ldots, \ell$.

The analysis of variance model for this design is given by

$$y_{ijk} = \mu + \beta_i + \alpha_j + \gamma_k + (\alpha\gamma)_{jk} + \varepsilon_{ijk}$$

$i = 1, 2, \ldots, b$; $j = 1, 2, \ldots, g$; $k = 1, 2, \ldots, \ell$; where $\varepsilon_{ijk} \sim N(0, \sigma^2)$ and

$$\sum_{i=1}^{b} \beta_i = 0, \quad \sum_{j=1}^{g} \alpha_j = 0, \quad \sum_{k=1}^{\ell} \gamma_k = 0, \quad \sum_{j=1}^{g} (\alpha\gamma)_{jk} = 0, \quad \sum_{k=1}^{\ell} (\alpha\gamma)_{jk} = 0.$$ The treatment cell mean for block i is given by

$$\mu_{ijk} = \mu + \beta_i + \alpha_j + \gamma_k + (\alpha\gamma)_{jk}.$$

Table 5.31. ANOVA Table for Two-Way Factorial With Randomized Block Design

Source	d.f.	Sum of Squares	Mean Square	F
Blocks	$(b-1)$	SSB	$\text{MSB} = \text{SSB}/(b-1)$	MSB/MSE
Factor 1	$(g-1)$	SSA	$\text{MSA} = \text{SSA}/(g-1)$	MSA/MSE
Factor 2	$(\ell-1)$	SSL	$\text{MSL} = \text{SSL}/(\ell-1)$	MSL/MSE
Interaction	$(g-1)(\ell-1)$	SSAL	$\text{MSAL} = \text{SSAL}/(g-1)(\ell-1)$	MSAL/MSE
Error	$(b-1)(g\ell-1)$	SSE	$\text{MSE} = \text{SSE}/(b-1)(g\ell-1)$	

The effect parameters for the two treatments factors and the interaction parameter are given by $\alpha_j = (\mu_{\cdot j \cdot} - \mu)$, $\gamma_k = (\mu_{\cdot \cdot k} - \mu)$ and $(\alpha\gamma)_{jk} = (\mu_{\cdot jk} - \mu_{\cdot j \cdot} - \mu_{\cdot \cdot k} + \mu)$ respectively. The block parameter β_i is defined by $\beta_i = (\mu_{i\cdot\cdot} - \mu)$. The sample estimators of these parameters are determined by replacing each μ term by the corresponding $\bar{y}$ term. The estimators are given by $\hat{\alpha}_j = (\bar{y}_{\cdot j \cdot} - \bar{y}_{...})$, $\hat{\gamma}_k = (\bar{y}_{\cdot \cdot k} - \bar{y}_{...})$, $\hat{\beta}_i = (\bar{y}_{i\cdot\cdot} - \bar{y}_{...})$ and $(\widehat{\alpha\gamma})_{jk} = (\bar{y}_{\cdot ij} - \bar{y}_{\cdot j \cdot} - \bar{y}_{\cdot \cdot k} + \bar{y}_{...})$.

The total sum of squares $\text{SST} = \sum_{i=1}^{b} \sum_{j=1}^{g} \sum_{k=1}^{\ell} (y_{ijk} - \bar{y}_{...})^2$ is partitioned into sums of squares for each of the factors given by

$$\text{SSA} = b\ell \sum_{j=1}^{g} (\bar{y}_{\cdot j \cdot} - \bar{y}_{...})^2, \qquad \text{SSL} = bg \sum_{k=1}^{\ell} (\bar{y}_{\cdot \cdot \ell} - \bar{y}_{...})^2$$

and a sum of squares for interactions

$$\text{SSAL} = b \sum_{j=1}^{g} \sum_{k=1}^{\ell} (\bar{y}_{\cdot jk} - \bar{y}_{\cdot \cdot k} - \bar{y}_{\cdot j \cdot} + \bar{y}_{...})^2.$$

The block sum of squares is given by

$$\text{SSB} = g\ell \sum_{i=1}^{b} (\bar{y}_{i\cdot\cdot} - \bar{y}_{...})^2,$$

and the residual or error sum squares is given by the difference SST – SSA – SSL – SSB – SSAL. The analysis of variance table for the various effects is summarized in Table 5.31. An example of this design based on the family expenditure data is outlined next.

Table 5.32. Analysis of Variance for Family Expenditure Survey

Source	d.f.	Sum of Squares	F	p-value
REGIONS	4	4867.51	11.26	0.000
AD	3	4585.68	14.15	0.000
SIZE	5	40967.65	75.84	0.000
AD*SIZE	15	412.82	0.25	0.998
ERROR	92	9939.46		
Total	119	60773.12		

Example

For the family expenditure data assume that the ads and family size are to be viewed as treatment factors and that the five regions are to be considered blocks. Assume that the $4 \times 6 = 24$ ad-size combinations are randomly selected in each region. The experimental unit is the family unit. For each region therefore we are assuming that each of the 24 families are randomly selected so that there are 4 of each of the 6 sizes and that the television advertising has been administered separately to each of the 24 families. No two families will have seen the same sequence of television advertisements.

Table 5.32 shows the results of the analysis of the family expenditure data. From the table we can conclude that all main effects are significant but there is no interaction between the AD effect and the SIZE effect. The AD effects seem to be the same for all family sizes.

5.4.2 The Latin Square Design

An alternative three-way analysis of variance model involves a single treatment factor and two blocking or nuisance factors. A cross classification with respect to the levels of the two blocking factors yields a rectangular array of experimental units. If there are g treatment levels, then g rectangles of experimental units are required to compare the g treatments while controlling for the two blocking factors. The number of cells in each rectangle is equal to the number of combinations in the cross classification of the two blocking factors. If the number of cells in the rectangle is large then the total number of experimental units required for the experiment is also quite large. In situations where experimental units are scarce, a *Latin square* design can be used to substantially reduce the number of experimental units required. The cost of this reduction however is that only main effects can be studied.

A Latin square consists of a total of b^2 experimental units and is derived from the cross classification of b levels for each blocking factor. The number

Column Factor

	A	B	C	D
Row	B	A	D	C
Factor	C	D	A	B
	D	C	B	A

Figure 5.11. A 4×4 Latin Square

of levels of the treatment factor is also b. The Latin square design consists of randomly assigning the treatment levels to the Latin square cells, such that each row and each column of the square contain one of each of the b treatment levels. A 4×4 Latin square is shown in Figure 5.11. The four treatment levels are denoted by A, B, C and D. The levels of the blocking factors are the rows and columns respectively. Each replication of the experiment consists of one of these squares. In the example involving the study of gasoline performance, the automobile type and location factors can be used to construct the Latin square.

The analysis of variance model for the Latin square is given by

$$y_{ik(j)} = \mu + \beta_i + \gamma_k + \alpha_j + \varepsilon_{ik(j)}$$

where $i = 1, 2, \ldots, b; \ j = 1, 2, \ldots, b; \ k = 1, 2, \ldots, b;$ and $\varepsilon_{ik(j)} \sim N(0, \sigma^2)$. The parameters satisfy the conditions $\sum_{i=1}^{b} \beta_i = 0, \ \sum_{k=1}^{b} \gamma_k = 0, \ \sum_{j=1}^{b} \alpha_j = 0$. The two blocking factor effects are denoted by β_i and γ_k while α_j denotes the treatment effect.

To denote the estimators and sums of squares we use $y_{..j}$ to denote the sum over particular cells that correspond to treatment level j. The quantities $y_{i..}$ and $y_{.k.}$ denote sums over the i-th row and the k-th column respectively. The three main effects are estimated using

$$\hat{\beta}_i = (\bar{y}_{i..} - \bar{y}_{...}), \quad \hat{\gamma}_k = (\bar{y}_{.k.} - \bar{y}_{...}) \quad \text{and} \quad \hat{\alpha}_j = (\bar{y}_{..j} - \bar{y}_{...}).$$

The sums of squares are given by $\text{SST} = \sum_{i=1}^{b} \sum_{k=1}^{b} (y_{ik(j)} - \bar{y}_{...})^2$ for the total,

$\text{SSROW} = b \sum_{i=1}^{b} \hat{\beta}_i^2$ for rows, $\text{SSCOL} = b \sum_{k=1}^{b} \hat{\gamma}_k^2$ for columns and $\text{SSTR} =$

$b \sum_{j=1}^{b} \hat{\alpha}_j^2$ for treatments. The error sum of squares is given by

$$\text{SSE} = \text{SST} - \text{SSROW} - \text{SSCOL} - \text{SSTR}$$

$$= \sum_{i=1}^{b} \sum_{k=1}^{b} (y_{ik(j)} - \bar{y}_{i..} - \bar{y}_{.k.} - \bar{y}_{..j} + 2\bar{y}_{...})^2.$$

The analysis of variance table is shown in Table 5.33. Note that the subscript j is a function of the subscripts i and k, and hence j changes automatically as i and k cycle through the range 1 to b.

Table 5.33. Analysis of Variance Table for Latin Square

Source	d.f.	Sum of Squares	Mean Square	F
Rows	$(b-1)$	SSROW	MSROW $=$ SSROW$/(b-1)$	MSROW/MSE
Columns	$(b-1)$	SSCOL	MSCOL $=$ SSCOL$/(b-1)$	MSCOL/MSE
Treatments	$(b-1)$	SSTR	MSTR $=$ SSTR$/(b-1)$	MSTR/MSE
Error	$(b-1)(b-2)$	SSE	MSE $=$ SSE$/(b-1)(b-2)$	
Total	b^2-1	SST		

Multiple Regression Approach to Latin Squares

Using effect coding, $(b-1)$ indicator variables are defined for $(b-1)$ of the b levels for each of the row, column and treatment factors.

$$
\begin{aligned}
R_i &= \quad 1 \text{ if row } i \\
&= -1 \text{ if row } b \qquad i = 1, 2, \ldots, (b-1); \\
&= \quad 0 \text{ otherwise}
\end{aligned}
$$

$$
\begin{aligned}
C_k &= \quad 1 \text{ if column } k \\
&= -1 \text{ if column } b \qquad k = 1, 2, \ldots, (b-1); \\
&= \quad 0 \text{ otherwise}
\end{aligned}
$$

$$
\begin{aligned}
T_j &= \quad 1 \text{ if treatment } j \\
&= -1 \text{ if treatment } b \qquad j = 1, 2, \ldots, (b-1). \\
&= \quad 0 \text{ otherwise}
\end{aligned}
$$

The regression model is given by

$$
y_{ik(j)} = \mu + \sum_{i=1}^{(b-1)} \beta_i R_i + \sum_{k=1}^{(b-1)} \gamma_k C_k + \sum_{j=1}^{(b-1)} \alpha_j T_j + u_{ik(j)}.
$$

5.4.3 A Three-Way Factorial Model

Observations on a measured variable Y are obtained from a three factor experiment or *three-way factorial*. The three factors will be referred to as rows, columns and layers which together determine the three-dimensional cell (i, j, k). The number of observations per cell is assumed to be equal to c and the h-th observation in row i, column j and layer k is denoted by y_{ijkh}. The number of rows, columns and layers are denoted by b, g and ℓ respectively.

The linear model is given by

$$y_{ijkh} = \mu + \beta_i + \alpha_j + \gamma_k + (\beta\alpha)_{ij} + (\beta\gamma)_{ik} + (\alpha\gamma)_{jk} + (\beta\alpha\gamma)_{ijk} + \varepsilon_{ijkh}$$

$i = 1, 2, \ldots, b; \quad j = 1, 2, \ldots, g; \quad k = 1, 2, \ldots, \ell; \quad h = 1, 2, \ldots, c;$ where ε_{ijkh} are mutually independent, $N(0, \sigma^2)$. The cell means are denoted by μ_{ijk} and the parameters satisfy the conditions

$$\sum_{j=1}^{g} \alpha_j = \sum_{i=1}^{b} \beta_i = \sum_{k=1}^{\ell} \gamma_k = \sum_{j=1}^{g} (\beta\alpha)_{ij} = \sum_{k=1}^{\ell} (\beta\gamma)_{ik}$$

$$= \sum_{k=1}^{\ell} (\alpha\gamma)_{jk} = \sum_{i=1}^{b} (\beta\alpha)_{ij} = \sum_{i=1}^{b} (\beta\gamma)_{ik} = \sum_{j=1}^{g} (\alpha\gamma)_{jk}$$

$$= \sum_{k=1}^{\ell} (\beta\alpha\gamma)_{ijk} = \sum_{i=1}^{b} (\beta\alpha\gamma)_{ijk} = \sum_{j=1}^{g} (\beta\alpha\gamma)_{ijk} = 0.$$

The main effect and interaction parameters can be expressed as functions of the cell means μ_{ijk}. For the main effects the parameters are given by $\beta_i = (\mu_{i..} - \mu)$, $\alpha_j = (\mu_{.j.} - \mu)$ and $\gamma_k = (\mu_{..k} - \mu)$ and are estimated by the corresponding functions of the sample means given by

$$\hat{\beta}_i = (\bar{y}_{i...} - \bar{y}_{....}), \qquad \hat{\alpha}_j = (\bar{y}_{.j..} - \bar{y}_{....}) \qquad \text{and}$$
$$\hat{\gamma}_k = (\bar{y}_{..k.} - \bar{y}_{....}).$$

The three two-way interaction parameters can be expressed as $(\beta\alpha)_{ij} = (\mu_{ij.} - \mu_{i..} - \mu_{.j.} + \mu)$, $(\beta\gamma)_{ik} = (\mu_{i.k} - \mu_{i..} - \mu_{..k} + \mu)$ and $(\alpha\gamma)_{jk} = (\mu_{.jk} - \mu_{.j.} - \mu_{..k} + \mu)$ with the corresponding sample estimators given by $\widehat{(\beta\alpha)}_{ij} = (\bar{y}_{ij..} - \bar{y}_{i...} - \bar{y}_{.j..} + \bar{y}_{....})$, $\widehat{(\beta\gamma)}_{ik} = (\bar{y}_{i.k.} - \bar{y}_{i...} - \bar{y}_{..k.} + \bar{y}_{....})$ and $\widehat{(\alpha\gamma)}_{jk} = (\bar{y}_{.jk.} - \bar{y}_{.j..} - \bar{y}_{..k.} + \bar{y}_{....})$.

For the three-way interaction parameter $(\beta\alpha\gamma)_{ijk} = (\mu_{ijk} - \mu_{ij.} - \mu_{i.k} - \mu_{.jk} + \mu_{i..} + \mu_{.j.} + \mu_{..k} - \mu)$ and the estimator is given by

$$\widehat{(\beta\alpha\gamma)}_{ijk} = (\bar{y}_{ijk.} - \bar{y}_{ij..} - \bar{y}_{i.k.} - \bar{y}_{.jk.} + \bar{y}_{i...} + \bar{y}_{.j..} + \bar{y}_{..k.} - \bar{y}_{...}).$$

The total sum of squares SST $= \sum_{i=1}^{b} \sum_{j=1}^{g} \sum_{k=1}^{\ell} \sum_{h=1}^{c} (y_{ijkh} - \bar{y}_{....})^2$ can be partitioned into the components corresponding to the main effects and interactions defined above. These sums of squares are given by

$$\text{SSB} = cg\ell\sum_{i=1}^{b}\hat{\beta}_i^2, \qquad \text{SSA} = cb\ell\sum_{j=1}^{g}\hat{\alpha}_j^2, \qquad \text{SSL} = bgc\sum_{k=1}^{\ell}\hat{\gamma}_k^2,$$

$$\text{SSBA} = c\ell\sum_{i=1}^{b}\sum_{j=1}^{g}(\widehat{\beta\alpha})_{ij}^2, \qquad \text{SSBL} = cg\sum_{i=1}^{b}\sum_{k=1}^{\ell}(\widehat{\beta\gamma})_{ik}^2,$$

$$\text{SSAL} = cb\sum_{j=1}^{g}\sum_{k=1}^{\ell}(\widehat{\alpha\gamma})_{jk}^2 \quad \text{and} \quad \text{SSBAL} = c\sum_{i=1}^{b}\sum_{j=1}^{g}\sum_{k=1}^{\ell}(\widehat{\beta\alpha\gamma})_{ijk}^2,$$

where SST = SSB + SSA + SSL + SSBA + SSBL + SSAL + SSBAL + SSE, and SSE is the residual sum of squares given by $\text{SSE} = \sum_{i=1}^{b}\sum_{j=1}^{g}\sum_{k=1}^{\ell}\sum_{h=1}^{c}$ $(y_{ijk\ell} - \bar{y}_{ijk\cdot})^2$.

The analysis of variance tests for the various effects are summarized in Table 5.34.

Three-Way Interactions

The interpretation of two-way or first order interactions has already been discussed in Section 5.2. The *second order interaction* or *three-way inter-action* can be used to measure the variation in one of the two-way interactions over the categories of the third variable. The three-way interaction $(\beta\alpha\gamma)_{ijk}$ can be expressed as

$$(\beta\alpha\gamma)_{ijk} = (\mu_{ijk} - \mu_{ij\cdot} - \mu_{i\cdot k} + \mu_{i\cdot\cdot}) - (\mu_{\cdot jk} - \mu_{\cdot j\cdot} - \mu_{\cdot\cdot k} + \mu)$$

$$= \begin{pmatrix} \text{two-way interaction} \\ \text{between columns} \\ \text{and layers at row } i \end{pmatrix} - \begin{pmatrix} \text{overall interaction between} \\ \text{columns and layers averaged} \\ \text{over all rows} \end{pmatrix}$$

and hence measures the variation in the column × layer interactions over the various rows. In a similar fashion this second order interaction can be interpreted in terms of columns or layers.

Example

An experiment was designed to measure the impact of a fitness program for seniors on a sample of inactive individuals between the ages of 65 and 70. A fitness test was administered before and after a 6 months fitness training program. The differences in fitness scores are given in Table 5.35. The

Table 5.34. Analysis of Variance Table for a Three-way Factorial

Source	d.f.	Sum of Squares	Mean Square	F
Main Effects				
Rows	$(b-1)$	SSB	$SSB/(b-1) = MSB$	MSB/MSE
Columns	$(g-1)$	SSA	$SSA/(g-1) = MSA$	MSA/MSE
Layers	$(\ell-1)$	SSL	$SSL/(\ell-1) = MSL$	MSL/MSE
Two-Way Interactions				
Row/Column Interaction	$(b-1)(g-1)$	SSBA	$SSBA/(b-1)(g-1) = MSBA$	MSBA/MSE
Row/Layer Interaction	$(b-1)(\ell-1)$	SSBL	$SSBL/(b-1)(\ell-1) = MSBL$	MSBL/MSE
Column/Layer Interaction	$(g-1)(\ell-1)$	SSAL	$SSAL/(g-1)(\ell-1) = MSAL$	MSAL/MSE
Three-Way Interaction	$(b-1)(g-1)(\ell-1)$	SSBAL	$SSBAL/(g-1)(b-1)(\ell-1) = MSBAL$	MSBAL/MSE
Error	$bg\ell(c-1)$	SSE	$SSE/bg\ell(c-1) = MSE$	

Table 5.35. Change in Fitness Scores

		White		Non White	
Males	Non Smoker	54 54 58	55.33	52 52 48	50.67
	Smoker	44 40 44	42.67	18 22 18	19.33
Females	Non Smoker	44 40 40	41.33	6 2 2	3.33
	Smoker	22 18 22	20.6	24 20 24	22.67

Summary of Means

	White	Non White	Avg.	Smoker	Non Smoker
Male	49.0	35.0	42.0	31.0	53.0
Female	31.0	13.0	22.0	21.67	22.33
Avg.	40.0	24.0		26.33	37.67

	Smoker	Non Smoker
White	31.67	48.33
Non White	21.00	27.00

factors race, sex and smoking habit are also included. For each of the three factors there were two levels and hence the design has $2^3 = 8$ treatment levels. For each treatment level there were 3 observations. The means in each cell are also included in the table. The analysis of variance table is summarized in Table 5.36.

The analysis of variance table indicates that all main effects and interactions are highly significant with the exception of the RACE × SEX interaction which has a p-value of 0.05. From the tables of means in Table 5.35 we can see that average scores tend to be higher for males, non-smokers and whites in comparison to females, smokers and non-whites respectively. The two-way interactions can also be obtained from the tables. The difference between males and females tends to be larger for non-whites than whites and larger for non-smokers than smokers. The difference between whites and non-whites is larger for non-smokers than for smokers.

For the three-way interaction an examination of the cell means shows that for males the difference between smokers and non-smokers for whites $(55.33 - 42.67)$ is smaller than for non-whites $(50.67 - 19.33)$. The difference between these two is $(53.33 - 42.67) - (50.67 - 19.33) = -20.68$. For females the corresponding differences are $(41.33 - 20.67)$ for whites and $(3.33 - 22.67)$ for non-whites. The difference between these two is $(41.33$

Table 5.36. Analysis of Variance for Fitness Study

Source	d.f.	Sum of Square	Mean Square	F	p-value
SMOKING HABIT	1	770.67	770.67	114.50	0.000
RACE	1	1536.00	1536.00	288.00	0.000
SEX	1	2400.00	2400.00	450.00	0.000
SMOKE*SEX	1	682.67	682.67	128.00	0.000
SMOKE*RACE	1	170.67	170.67	32.00	0.000
RACE*SEX	1	24.00	24.00	4.50	0.050
SMOKE*RACE*SEX	1	1290.67	1290.67	242.00	0.000
ERROR	16	85.33	5.33		
Total	23	6960.00			

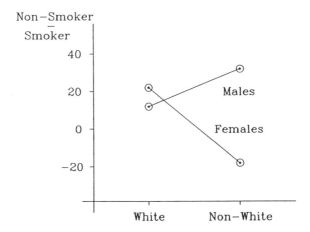

Figure 5.12. Second Order Interaction for Fitness Scores

$-20.67) - (3.33 - 22.67) = 40.00$. This interaction could be summarized by saying that for white males the score for non-smokers is slightly larger than for smokers while for non-white males the score for non-smokers is much larger than for smokers. For white females the score for non-smokers is much larger than for smokers but for non-white females the score for smokers exceeds that of non-smokers. The plot in Figure 5.12 characterizes this three-way or second order interaction. The vertical axis in the figure measures the difference in fitness score between non-smokers and smokers. The figure indicates that this difference is about the same for white males and white females. For non-white males the difference is larger, while for non-white females the difference is much smaller.

5.4.4 An Unbalanced Three-Way Analysis Example

An *unbalanced three-way analysis of variance* model will be introduced in this section using an example.

A sample of 103 observations was selected from a data file consisting of the RETURN (annual net income/annual net sales) for the years 1972, 1976 and 1982 for the top 800 companies in the United States. The sample was restricted to seven industries. For each observation it was also noted whether the company had undergone a change in CEO (chief executive officer) in that year. The variable CHANGE was denoted by 1 if there was no change, 2 if the change was routine with the CEO replaced from within, and 3 if the change was nonroutine with the CEO replaced from outside the company. The variable YR was defined to be 1, 2 or 3 corresponding to the years 1972, 1976 and 1982 respectively. The variable INDUSTRY was coded as 1 (Food & Kindred Products), 2 (Paper & Allied Products), 3 (Chemicals & Allied Products), 4 (Air Transportation), 5 (Retail Grocery Stores), 6 (State Banks) and 7 (National Banks). The data are displayed in Table 5.37. A perusal of the companies in Table 5.37 reveals some repetition. For the purposes of this example, the resultant impact on the ANOVA model error terms of repeated observations on the same company, is assumed to be negligible.

Examination of Cell Means

The purpose of analysis of variance is to study the variation in cell means. Table 5.38 shows the cell means for the data of Table 5.37. The variation in means by INDUSTRY, CHANGE and YR are shown first. The numbers in brackets under the means are the number of observations in the cell containing the mean. The variation in the cell means is also shown graphically in Figure 5.13, panels (a), (b) and (c). From panel (a) we observe a downward slope in mean RETURN by YR. In panel (b) it would appear that for nonroutine CHANGE the mean RETURN is much lower. From panel (c) we observe that there is considerable variation in mean RETURN by INDUSTRY.

Table 5.38 also shows the cell mean RETURN for the three two-way relationships among YR, CHANGE and INDUSTRY. The graphical representations for these cell means are shown in Figure 5.13 panels (d), (e) and (f). For the two-way relationship between YR and CHANGE it would appear from panel (f) that there is almost no interaction between YR and CHANGE. In each YR it would appear that for nonroutine CHANGE the mean RETURN is much lower than for both no CHANGE and routine CHANGE. In panel (e) it would appear that the variation in RETURN by INDUSTRY is similar for both no CHANGE and routine CHANGE. In the case of nonroutine CHANGE, however, the variation in mean RETURN is

Table 5.37. Company RETURN Data

Name	Return	Yr	Change	Industry
American Cyanimid Co	0.06495	2	2	3
American Fletcher Corp	0.04878	2	1	7
Anderson, Clayton & Co	0.04743	2	1	1
Arizona Bancwest Corp	0.06204	3	1	6
Atlantic Bancorp	0.03922	2	2	7
Bank of New York Co Inc	0.10092	2	1	6
Bankamerica Corp	0.06954	2	1	7
Bankamerica Corp	0.03016	3	3	7
Bankers Trust New York Corp	0.04083	2	1	6
Bankers Trust New York Corp	0.05183	3	1	6
Boise Cascade Corp	−0.14857	1	3	2
Campbell Soup Co	0.05525	1	2	1
Campbell Soup Co	0.06177	2	1	1
Carnation Co	0.05411	3	1	1
Chase Manhattan Corp	0.10157	1	1	7
Chase Manhattan Corp	0.03037	3	1	7
Chemical New York Corp	0.04426	3	1	6
Citizens & Southern GA Corp	0.13889	1	3	7
Commerce Bancshares Inc	0.18919	1	1	6
Consolidated Foods Corp	0.03666	1	1	1
Consolidated Foods Corp	0.03230	2	3	1
Continental Illinois Corp	0.01324	3	1	7
CPC International Inc	0.05613	1	3	1
Crocker National Corp	0.08216	1	1	7
Crown Zellerbach	0.04588	2	1	2
Delta Air Lines Inc	0.04578	2	1	4
Dominion Bankshares Corp	0.12963	1	1	6
Dominion Bankshares Corp	0.04959	3	2	6
Dow Chemical	0.10846	2	1	3
Du Pont (I.E.) De Nemours	0.02682	3	2	3
Eastern Air Lines	0.02521	2	2	4
Eastern Air Lines	−0.01990	3	1	4
Fidelcor	0.06306	3	1	6
First Chicago Corp	0.07648	2	2	7
First Empire State Corp	0.08333	1	1	6
First Empire State Corp	0.04819	3	1	6
First Hawaiian Inc	0.15385	1	1	6
First Tennessee Natl Corp	0.12658	1	1	7
FMC Corp	0.03271	1	3	3
FMC Corp	0.03730	2	1	3
Grace (W.R.) & Co	0.03651	2	1	3
Grace (W.R.) & Co	0.05222	3	1	3
Great Northern Nekoosa Corp	0.06864	2	1	2
Great Northern Nekoosa Corp	0.05524	3	1	2
Harris Bankcorp Inc	0.03960	3	1	6
Hercules Inc	0.07511	1	1	3
Hercules Inc	0.06704	2	1	3
Kellogg Co	0.08727	1	2	1
Kellogg Co	0.09632	3	1	1
Kimberly-Clark Corp	0.05539	1	2	2
Kimberly-Clark Corp	0.07634	2	1	2
Kroger Co	0.01210	3	1	5

Table 5.37. Company RETURN Data (continued)

Name	Return	Yr	Change	Industry
Lincoln First Banks	0.10000	1	1	7
Lincoln First Banks	0.06667	2	1	7
Lucky Stores Inc	0.01559	1	3	5
Lucky Stores Inc	0.01154	3	1	5
Marine Corp	0.08000	1	1	7
Marshall & Ilsley Corp	0.10417	2	1	7
Mead Corp	0.01594	1	1	2
Monsanto Co	0.05483	1	3	3
Morgan (J.P.) & Co	0.15984	1	2	6
National Distillers & Chemical	0.07785	2	1	3
NCNB Corp	0.07018	2	1	7
NCNB Corp	0.07096	3	1	7
Northern Trust Corp	0.13139	1	1	6
Northern Trust Corp	0.10887	2	1	6
Northwest Airlines Inc	0.04580	1	1	4
Northwest Airlines Inc	0.05394	2	1	4
Old Kent Financial Corp	0.08364	3	3	6
Pan American World Airways	−0.13052	3	3	4
Pillsbury Co	0.02451	1	1	1
Potlatch Corp	0.04509	1	3	2
PPG Industries Inc	0.06741	2	1	3
Quaker Oats Co	0.04410	1	1	1
Scott Paper Co	0.04791	1	2	2
Scott Paper Co	0.03227	3	2	2
Shawmut Corp	0.03593	2	1	6
Shawmut Corp	0.05592	3	1	6
Society Corp	0.08235	1	1	7
Society Corp	0.08824	2	2	7
Society Corp	0.05794	3	1	7
South Carolina Natl Corp	0.07391	3	1	7
Southland Corp	0.01639	1	1	5
Southland Corp	0.01633	3	1	5
St Regis Corp	0.05542	2	1	2
St Regis Corp	0.01761	3	1	2
Stauffer Chemical Co	0.07664	3	1	3
Stop & Shop Cos	0.00746	2	1	5
Stop & Shop Cos	0.01494	3	1	5
Texas Commerce Bancshares	0.18824	1	1	7
Texas Commerce Bancshares	0.13201	2	1	7
Trust Company of Georgia	0.12360	1	2	6
Trust Company of Georgia	0.06475	2	1	6
Trust Company of Georgia	0.10994	3	1	6
UAL Inc	0.01094	1	3	4
UAL Inc	0.00648	2	1	4
Union Camp Corp	0.06478	1	1	2
Union Carbide Corp	0.06286	1	2	3
Union Carbide Corp	0.06949	2	2	3
Union Carbide Corp	0.03421	3	3	3
Wachovia Corp	0.14625	2	1	7
Westvaco Corp	0.0618221	2	1	2
Winn-Dixie Stores Inc	0.0192897	2	1	5

Table 5.38. RETURN Means and Observation Frequencies

Industry

1	2	3	4	5	6	7
0.054	0.035	0.059	0.005	0.014	0.088	0.084
(11)	(14)	(16)	(8)	(8)	(22)	(24)

Change

YR	1	2	3
1	0.089	0.085	0.026
	(20)	(7)	(8)
2	0.064	0.061	0.032
	(30)	(6)	(1)
3	0.046	0.036	0.004
	(24)	(3)	(4)

Change

1	2	3
0.065	0.067	0.020
(74)	(16)	(13)

YR

1	2	3
0.073	0.063	0.040
(35)	(37)	(31)

Industry

Change	1	2	3	4	5	6	7
1	0.052	0.051	0.067	0.026	0.014	0.084	0.087
	(7)	(9)	(9)	(5)	(7)	(18)	(19)
2	0.071	0.045	0.056	0.025	–	0.064	0.068
	(2)	(3)	(4)	(1)		(3)	(3)
3	0.044	–0.052	0.041	–0.060	0.016	0.084	0.085
	(2)	(2)	(3)	(2)	(1)	(1)	(2)

Industry

YR	1	2	3	4	5	6	7
1	0.051	0.013	0.056	0.028	0.016	0.139	0.112
	(6)	(6)	(4)	(2)	(2)	(7)	(8)
2	0.047	0.062	0.066	0.033	0.013	0.070	0.084
	(3)	(5)	(8)	(4)	(2)	(5)	(10)
3	0.075	0.035	0.047	–0.075	0.014	0.061	0.046
	(2)	(3)	(4)	(2)	(4)	(10)	(6)

Table 5.38. RETURN Means and Observation Frequencies (continued)

Change	YR	Industry 1	2	3	4	5	6	7
	1	0.035	0.040	0.075	0.046	0.016	0.137	0.109
		(3)	(2)	(1)	(1)	(1)	(5)	(7)
1	2	0.055	0.062	0.066	0.035	0.013	0.070	0.091
		(2)	(5)	(6)	(3)	(2)	(5)	(7)
	3	0.075	0.036	0.064	−0.020	0.014	0.059	0.049
		(2)	(2)	(2)	(1)	(4)	(8)	(5)
	1	0.071	0.052	0.063	–	–	0.142	–
		(2)	(2)	(1)			(2)	
2	2	–	–	0.067	0.025	–	–	0.068
				(2)	(1)			(3)
	3	–	0.032	0.027	–	–	0.050	–
			(1)	(1)			(1)	
	1	0.056	−0.052	0.044	0.011	0.016	–	0.139
		(1)	(2)	(2)	(1)	(1)		(1)
3	2	0.032	–	–	–	–	–	–
		(1)						
	3	–	–	0.034	−0.131	–	0.084	0.030
				(1)	(1)		(1)	(1)

much greater. The difference between nonroutine CHANGE and both routine CHANGE and no CHANGE seems to be restricted to INDUSTRIES 2 and 4. In panel (d) the variation in mean RETURN by INDUSTRY by YR shows more complex interaction. It would appear that for YR = 2 the variation in mean RETURN by INDUSTRY is much less than the variation in YR's 1 and 3. In YR = 1, INDUSTRIES 6 and 7 are relatively high while for YR = 3 INDUSTRY 4 is relatively low. The final cell mean comparison in Table 5.38 shows the variation in mean RETURN by INDUSTRY by YR by CHANGE. From this table it can be seen that there are a number of empty cells and that for CHANGE = 2 or 3 the number of observations is generally quite small. Because of the empty cells we must assume that the three-way interaction terms are zero. In addition, because of the single observations for three cells in the CHANGE BY INDUSTRY table and one empty cell these interactions cannot be properly estimated. We assume this has no impact on the results although some omitted variable bias maybe present in some estimates.

(a)

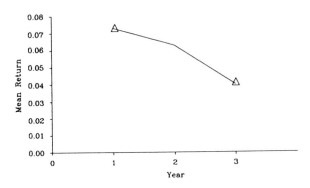

(b)

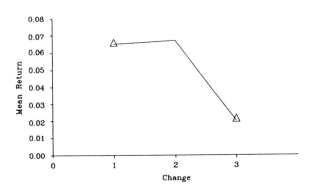

(c)

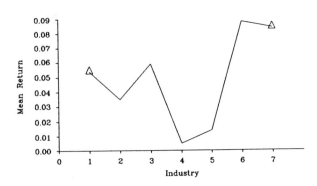

Figure 5.13. Return Cell Means

(d)

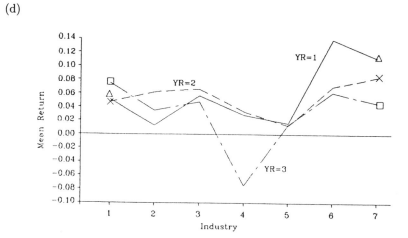

(e)

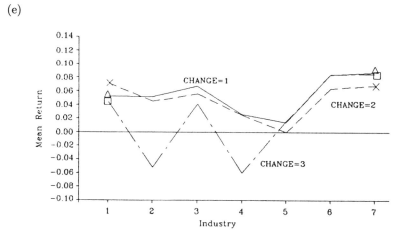

(f)

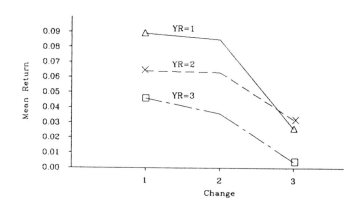

Figure 5.13.   Return Cell Means (continued)

Analysis of Variance Approach

The analysis of variance results for the variation in mean RETURN are shown in Table 5.39. Since the design is no longer balanced the effects are no longer orthogonal. The sums of squares due to each effect therefore cannot be determined independently of the other effects. In Table 5.39 the sum of squares shown for each effect is the marginal contribution to the total sum of squares arising from the addition to the model of the effect after the other effects had already been included. The resulting F-statistics are comparable to the F-statistics used in multiple regression analysis to test for significance of subsets of regression coefficients.

From one-way analyses numbered 1, 2 and 3 we can conclude that the variation in mean RETURN by each of CHANGE, YR and INDUSTRY is significant. For the two-way analyses numbered 5, 7 and 9 we can conclude that the YR by INDUSTRY interaction is significant but that the interactions between CHANGE and YR and between CHANGE and INDUSTRY are not. From each of the two-way analyses numbered 4, 6 and 8 the main effects ignoring interaction are always significant. All the main effects are also significant in the three-way analysis numbered 10.

The results of a three-way analysis with all two-way interactions present is shown in analysis 11 of Table 5.39. This analysis indicates that once again only the INDUSTRY by YR interaction is significant. Dropping the weakest interaction CHANGE by YR in analysis 12 the CHANGE by INDUSTRY interaction remains nonsignificant ($p = 0.135$). Analysis 13 includes only the one interaction effect for INDUSTRY by YR. Analysis 13 also indicates that the CHANGE effect is significant at $p = 0.083$. Since the prior belief of the researcher was that the category CHANGE = 3 would differ most from CHANGE = 1 and CHANGE = 2, these latter two categories were combined for analysis 14. For the categories CHANGE = 1 or 2 the new variable CH is coded 0 and for CHANGE = 3, CH is coded 1. The p-value for the CH effect in analysis 14 is now 0.025. The R^2 value for analysis 14 is 0.644. We can conclude from analysis 14 that after controlling for YR and INDUSTRY effects, there is a significant difference in mean RETURN between companies with a nonroutine CEO change and those with no change or a routine CEO change.

Multiple Regression Estimation

The multiple regression model with dummy variables for INDUSTRY, YR and CHANGE can be used to examine the variation in mean RETURN over the various cells. Using the last category of each of INDUSTRY (I7) and YR(YR3) along with CH = 0 as the base case, the dummy variables CH, YR1, YR2, I1, I2, ... , I6 were used. The fitted model corresponding to analysis 14 in Table 5.39 is given by

$$\text{RETURN} = \underset{(0.122)}{0.050} + \underset{(0.000)}{0.065} \text{ YR1} + \underset{(0.048)}{0.034} \text{ YR2} - \underset{(0.025)}{0.025} \text{ CH} + \underset{(0.350)}{0.025} \text{ I1}$$

Table 5.39. Analysis of Variance Results for Mean RETURN Variation by Change, YR and Industry

		df	Marginal Sum of Squares	F	p-value	R^2
1.	CHANGE	2	0.024	5.50	0.005	0.099
	Error	100	0.230			
2.	YR	2	0.019	4.37	0.015	0.080
	Error	100	0.220			
3.	INDUSTRY	6	0.081	8.19	0.000	0.202
	Error	96	0.158			
4.	CHANGE	2	0.029	7.48	0.001	0.202
	YR	2	0.025	6.33	0.003	
	Error	98	0.191			
5.	CHANGE	2	0.013	3.31	0.041	0.209
	YR	2	0.013	3.11	0.049	
	YR*CHANGE	4	0.002	0.20	0.938	
	Error	94	0.189			
6.	CHANGE	2	0.013	4.28	0.017	0.394
	INDUSTRY	6	0.070	7.61	0.000	
	Error	94	0.239			
7.	CHANGE	2	0.011	3.66	0.030	0.481
	INDUSTRY	6	0.050	5.55	0.000	
	CHANGE*INDUSTRY	11	0.021	1.26	0.261	
	Error	83	0.124			
8.	YR	2	0.023	7.90	0.001	0.434
	INDUSTRY	6	0.085	9.78	0.000	
	Error	94	0.239			
9.	YR	6	0.080	12.12	0.000	0.621
	INDUSTRY	2	0.013	5.77	0.005	
	YR*INDUSTRY	12	0.045	3.37	0.001	
	Error	82	0.091			
10.	CHANGE	2	0.016	6.02	0.004	0.499
	YR	2	0.025	9.69	0.000	
	INDUSTRY	6	0.071	9.10	0.000	
	Error	92	0.120			
11.	CHANGE	2	0.004	1.74	0.183	0.726
	YR	2	0.012	5.99	0.004	
	INDUSTRY	6	0.045	7.41	0.000	
	YR*INDUSTRY	12	0.028	2.29	0.017	
	YR*CHANGE	4	0.003	0.73	0.577	
	CHANGE*INDUSTRY	11	0.016	1.49	0.158	
	Error	65	0.065			
12.	CHANGE	2	0.013	2.22	0.116	0.714
	YR	2	0.004	6.53	0.003	
	INDUSTRY	6	0.049	8.30	0.000	
	YR*INDUSTRY	12	0.028	2.33	0.015	
	CHANGE*INDUSTRY	11	0.017	1.55	0.135	
	Error	69	0.068			

Table 5.39. Analysis of Variance Results for Mean RETURN Variation by Change, YR and Industry (continued)

		df	Marginal Sum of Squares	F	p-value	R^2
13.	CHANGE	2	0.005	2.57	0.083	0.644
	YR	2	0.015	6.81	0.002	
	INDUSTRY	6	0.069	10.79	0.000	
	YR*INDUSTRY	12	0.035	2.70	0.004	
	Error	80	0.239			
14.	CH	1	0.005	5.20	0.025	0.644
	YR	2	0.015	6.95	0.002	
	INDUSTRY	6	0.069	10.93	0.000	
	YR*INDUSTRY	12	0.035	2.75	0.004	
	Error	81	0.085			

$$- \underset{(0.510)}{0.015} \text{ I2} + \underset{(0.869)}{0.003} \text{ I3} - \underset{(0.000)}{0.113} \text{ I4} - \underset{(0.086)}{0.037} \text{ I5} + \underset{(0.439)}{0.013} \text{ I6}$$

$$- \underset{(0.009)}{0.086} \text{ YR1I1} - \underset{(0.009)}{0.079} \text{ YR1I2} - \underset{(0.088)}{0.050} \text{ YR1I3} + \underset{(0.303)}{0.038} \text{ YR1I4}$$

$$- \underset{(0.137)}{0.051} \text{ YR1I5} + \underset{(0.673)}{0.010} \text{ YR1I6} - \underset{(0.124)}{0.054} \text{ YR2I1} - \underset{(0.803)}{0.007} \text{ YR2I2}$$

$$- \underset{(0.411)}{0.021} \text{ YR2I3} + \underset{(0.065)}{0.062} \text{ YR2I4} - \underset{(0.299)}{0.034} \text{ YR2I5} - \underset{(0.274)}{0.027} \text{ YR2I6}$$

From the analysis we can conclude that after controlling for YR and INDUSTRY effects, the RETURN for companies that have experienced an outside CEO change experienced on average a return which was significantly below those with no change or an internal change. The difference in return of 0.025 had a p-value of 0.025. The YR effects and INDUSTRY effects are somewhat complex. In comparison to YR3, YR1 and YR2 on average showed significantly higher returns. In comparison to INDUSTRY $= 7$, average returns in INDUSTRY $= 4$ and INDUSTRY $= 5$ tended to be lower. Significant negative interactions appeared for average return in YR $= 1$ for INDUSTRY $= 1$ and 2 and a significant positive interaction occurred for YR $= 2$ for INDUSTRY $= 4$.

Using the fitted model the estimated cell means can be obtained. Table 5.40 shows the cell means determined from this fitted model and in brackets the observed cell means. The observed cell means for CHANGE $= 1$ and CHANGE $= 2$ have been combined to obtain the cell means for CH $= 0$. The cells which were empty have been assigned the estimated value of 0.0. A comparison of the estimated cell means to the observed cell means shows the goodness of fit of the model. The sum of squared differences between the estimated means and the RETURN observations is given by the R^2 for the model $= 0.644$.

Table 5.40. Estimated and (Actual) Cell Means for Return Model

INDUSTRY

CH	YR	1	2	3	4	5	6	7
	1	0.055	0.022	0.069	0.041	0.029	0.139	0.116
		(0.050)	(0.046)	(0.069)	(0.046)	(0.016)	(0.139)	(0.109)
0	2	0.056	0.062	0.066	0.033	0.013	0.070	0.084
		(0.055)	(0.062)	(0.066)	(0.033)	(0.013)	(0.070)	(0.084)
	3	0.075	0.035	0.054	-0.063	0.014	0.063	0.050
		(0.075)	(0.035)	(0.052)	(-0.020)	(0.014)	(0.058)	(0.049)
	1	0.030	-0.003	0.044	0.016	0.003	0.000	0.091
		(0.056)	(-0.052)	(0.044)	(0.011)	(0.016)	(−)	(0.139)
1	2	0.030	0.000	0.000	0.000	0.000	0.000	0.000
		(0.032)	(−)	(−)	(−)	(−)	(−)	(−)
	3	0.000	0.000	0.018	-0.088	0.000	0.038	0.025
		(−)	(−)	(0.034)	(-0.131)	(−)	(0.084)	(0.030)

Effect Coding

An alternative model for the data can be obtained using effect coding. Using the categories YR = 3, CH = 0 and INDUSTRY = 7 as the −1 values for the variables YR1 YR2, CH, I1, I2, I3, I4, I5, and I6 the estimated model is given below.

$$
\begin{aligned}
\text{RETURN} = \; & 0.039 + 0.016 \text{ YR1} + 0.003 \text{ YR2} - 0.013 \text{ CH} + 0.010 \text{ I1} \\
& (0.000) \quad (0.005) \qquad\quad (0.527) \qquad\quad (0.025) \qquad\quad (0.297)
\end{aligned}
$$

$$
\begin{aligned}
& - 0.012 \text{ I2} + 0.011 \text{ I3} - 0.048 \text{ I4} - 0.033 \text{ I5} + 0.039 \text{ I6} \\
& \;\; (0.161) \qquad (0.166) \qquad (0.000) \qquad (0.003) \qquad (0.000)
\end{aligned}
$$

$$
\begin{aligned}
& - 0.023 \text{ YR1I1} - 0.033 \text{ YR1I2} - 0.010 \text{ YR1I3} + 0.022 \text{ YR1I4} \\
& \;\; (0.073) \qquad\quad (0.004) \qquad\quad (0.428) \qquad\quad (0.180)
\end{aligned}
$$

$$
\begin{aligned}
& - 0.006 \text{ YR1I5} + 0.032 \text{ YR1I6} - 0.010 \text{ YR2I1} - 0.019 \text{ YR2I2} \\
& \;\; (0.733) \qquad\quad (0.002) \qquad\quad (0.500) \qquad\quad (0.109)
\end{aligned}
$$

$$
\begin{aligned}
& - 0.000 \text{ YR2I3} + 0.026 \text{ YR2I4} - 0.009 \text{ YR2I5} - 0.024 \text{ YR2I6}. \\
& \;\; (0.991) \qquad\quad (0.071) \qquad\quad (0.597) \qquad\quad (0.028)
\end{aligned}
$$

The interpretation of the coefficients is now in terms of a comparison to the overall mean RETURN calculated as the mean of the observed cell means in Table 5.40. For YR1, RETURNS tend to be higher while for YR = 3, RETURNS are lower than average by $[0.016 + 0.003] = 0.019$. INDUSTRY = 4 or 5 RETURNS tend to be lower than the overall average

while for INDUSTRY = 6 RETURNS tend to be higher. For INDUS-
TRY = 7 RETURNS tend to be higher than the average by $-[0.010 -
0.012 + 0.011 - 0.048 - 0.033 + 0.039] = 0.033$. Significant interactions oc-
cur for YR1I1, YR1I2, YR1I6, YR2I2, YR2I4 and YR2I6. The interaction
for YR = 1, INDUSTRY = 7 is given by $-[-0.023 - 0.033 - 0.010 + 0.022 -
0.006 + 0.032] = 0.018$ and for YR = 2, INDUSTRY = 7 the interaction is
given by $-[-0.010 + 0.019 - 0.000 + 0.026 - 0.009 - 0.024] = -0.002$.

The advantage of the effect coding approach in this case is that comparisons are made to the overall mean of the cell means rather than to a
base case. In the dummy variable method illustrated above the coefficients
depend on the choice of base case. If the objective of the analysis is to
study the variable CHANGE and its impact on RETURN after controlling
for INDUSTRY and YR effects then in this example either model is equally
useful. If however interactions between CHANGE and one or more of YR
and INDUSTRY were also fitted the interpretation would be different for
the two methods. In general the effect coding is preferable unless a particular base case yields a natural comparison. If the focus of the anlysis
is simply to compare the individual cells as defined by the $3 \times 3 \times 7 = 63$
categories, a one-way analysis which compares the cell means can be used.
This cell means comparison is outlined next.

A Cell Means Model Approach

In the previous two analyses both dummy coding and effect coding were
used to study the variation in RETURN by the factors YR, CHANGE and
INDUSTRY. In both these analyses attempts were made to measure main
effects and interactions between various pairs of factors. An alternative approach is provided by the *cell means model* approach which is a one factor
model with the levels of the factor defined by the totality of cells resulting
from the cross classification of the three original factors. In Table 5.38 the
42 non-empty cells represent the 42 levels of a single factor. A dummy coding or effect coding approach can then be used to define up to 41 variables
to characterize the variation among the means in the 42 cells. A useful
approach in this example is to use effect coding since for this method the
regression coefficients measure differences from the overall mean of the 42
cell means (note that this is not a simple mean of all 102 observations but
a mean of the 42 cell means). The results from this analysis are described
below.

The multiple regression model was estimated with 41 effect coding variables. The cell corresponding to YR = 3, CHANGE = 3 and INDUS-
TRY = 7 was coded –1 throughout. As a result the 41 regression coefficients represent deviations of the 41 cell means from the overall mean of the
cell means. The regression coefficients that were significant at the 0.05 level
corresponded to the following cells (1, 2, 7), (1, 3, 4), (1, 3, 5), (2, 1, 6), (1,

1, 6), (1, 1, 7), (3, 1, 2), (3, 1, 7) and (3, 3, 4) where the numerical codes refer to the categories of the variables CHANGE, YR, and INDUSTRY respectively. All of the coefficients were positive with the exception of (1, 3, 4), (1, 3, 5), (3, 1, 2) and (3, 3, 4). It would appear that all significant coefficients involving YR = 3 are negative. For CHANGE = 1 or 2 the significant coefficients were positive except for the two cases in YR = 3. For CHANGE = 3 two of the three significant coefficients are negative. The one CHANGE = 3 coefficient that was significant and positive was in INDUSTRY = 7 which has generally high values of RETURN. The overall mean of the 42 cell means was 0.047 which was significantly different from zero at the 0.0001 level. The value of R^2 for the model was 0.73. The cell means model is therefore useful for comparing the cell means in Table 5.38.

Because of so many empty cells in the three-way table portion of Table 5.38 it could be argued that the cell means model is the only proper approach particularly if the zero three-way interaction assumption is not valid. The cell means model approach should always be employed and compared to the effects model results when empty cells are present. An extensive discussion of the cell mean model for unbalanced data is provided in Searle (1987). Also, additional discussion of unbalanced data is provided in Milliken and Johnson (1984).

5.5 Some Basics of Experimental Design

In the first part of this Chapter the techniques of multiple linear regression and analysis of variance have been combined with concepts of experimental design. In this section an overview of some concepts of experimental design will be provided before a discussion of more complex designs appears in the latter sections of this chapter.

Experimental design is concerned with the planning of experiments in order to maximize the information obtained subject to resource constraints and physical limitations. Resource constraints may restrict the availability of experimental units, while physical limitations may have an impact on the homogeneity of experimental units. The two basic components of experimental design are the treatment combinations and the experimental units. In an ideal world there are infinitely many homogeneous experimental units available, so that a large number of replications of each treatment combination can be randomly assigned to the experimental units in a completely randomized design. Regardless of the number of factors the total number of possible treatment combinations, no matter how large, can be replicated several times to provide an estimate of error. In such a world there is little need to consider the more elaborate procedures available for experimental design. Given that such an ideal state does not usually exist experimental design techniques are required and are presented in the remainder of this chapter.

Terminology for Describing Experimental Designs

Using the terminology of Milliken and Johnson (1984) an experimental design consists of two independent structures, the *treatment structure* and the *design structure*. The experimental design combines the two structures. The treatment structure is based on the properties of the underlying factors or treatments that are to be compared, while the design structure is based on the properties of the experimental units. Common assumptions are that the blocks of homogeneous experimental units used are a random sample selected from a population of such blocks, and that there is no interaction between blocks and response to treatments.

Treatment Structure

The treatment structure can be a one-way treatment structure in which no relationship is assumed between the treatment combinations, or a *factorial arrangement* treatment structure in which a set of treatment combinations is derived from all possible levels of two or more treatment factors. Finally, a *fractional factorial arrangement* describes a treatment structure which consists of a set of treatment combinations which represent a subset of the set of all possible treatment combinations.

Design Structure

The design structure may consist of a completely randomized design in which all experimental units are assumed to be homogeneous, or a randomized block design in which experimental units are grouped into subsets or blocks of homogeneous units. In a completely randomized design the treatment combinations are randomly assigned to the experimental units, while in the randomized block design the treatment combinations are randomly assigned to experimental units within each block. The randomized block design assumes that each treatment combination appears in each block an equal number of times. If this is not true an *incomplete block design* occurs, which usually implies that the number of treatment combinations exceeds the number of experimental units in each block. A randomized block design which permits only one of each treatment combination in each block does not permit the measurement of interaction between block and treatment which as stated above is assumed to be zero.

A special type of incomplete block design with only one experimental unit per block is the Latin square. The blocks are derived from the cross classification of two blocking factors having the same number of classifications. Other special cases of incomplete block designs are the balanced incomplete block, the squared lattice design and the split plot design which will be defined later.

The replication of treatment combinations is used to measure experimental error. The greater the number of replications the greater the degrees of freedom and hence the greater the power and reliability. In the domain of scarce resources, techniques which involve the subsampling of experimental units, repeated measures on the same experimental unit, and the subdivision of experimental units, are used to provide economy and to reduce experimental error. These techniques will also be outlined in the latter sections of this chapter.

5.6 Multifactor Factorials, Fractional Replication, and Incomplete Blocks

In some applications there are potentially many factors which could influence the outcome of an experiment or process. As the number of factors increases the factorial design requires an increasing number of experimental units. If there are p factors with factor levels $b_1, b_2, \ldots, b_p$ then the product $(b_1, b_2, \ldots b_p)$ denotes the total number of experimental units that would be required for a single replication of the $(b_1 \times b_2 \times \ldots \times b_p)$ factorial. Since the number of experimental units required can be prohibitive, methods are required that will reduce the number of units needed. If there are potentially many factors and the nature of the influence of these factors and their interactions are initially unknown, a multistage process may be required to eliminate factors and combinations that are superfluous. Various methods will be discussed in this section.

A useful preliminary design is the 2^p *design* which is simple to use and to analyze and only requires 2^p experimental units. As a preliminary step each factor is represented at only 2 levels. Significant factors can then at a later stage be introduced at more levels. In addition to the 2^p design the 3^p design is also available as a simple design.

If the 2^p or 3^p designs still require too many experimental units the size of the experiment can be reduced further by using only a fraction of a complete factorial a procedure which is commonly referred to as *fractional replication*. The fractional replication sacrifices the higher level interactions and assumes that they are negligible.

In situations where the experimental units come in blocks it may not be possible to assume that a large block containing a large number of experimental units will yield experimental units which are relatively homogeneous within the block. In such circumstances the complete factorial design is split into blocks at the cost of being unable to detect higher level interactions. In such designs some effects will be *confounded* with other effects. The arrangements of treatments within blocks is carried out so that higher order interactions which are assumed to be negligible are confounded with lower order interactions and main effects. An alternative to reduction of block size with *confounding* is the balanced incomplete block design which eliminates confounding while maintaining small block size.

Table 5.41. Treatment Combinations for a 2^3 Factorial

$a_1b_1c_1$	$a_1b_1c_2$	$a_1b_2c_1$	$a_1b_2c_2$
$a_2b_1c_1$	$a_2b_1c_2$	$a_2b_2c_1$	$a_2b_2c_2$

5.6.1 The 2^p Design

A useful exploratory design when there are a large number of factors is the 2^p design. In this design there are p treatment factors each at 2 levels and hence a total of 2^p treatment combinations. For each replication or repetition of this design 2^p experimental units are required. If p is large the number of experimental units required for each replication is also large.

In factorial designs it is of interest to estimate and test the main effects as well as many of the interactions. In a 2^p design there are p main effects, $\binom{p}{2}$ first order (two-way) interactions, $\binom{p}{3}$ second order (three-way) interactions, ..., $\binom{p}{q+1}$ q-th order interactions, ... and finally one $(p-1)$-th (p-way) interaction. Each one of these interactions has one degree of freedom. To test these interactions requires an error mean-square with as many degrees of freedom as possible and hence the greater the number of replications (say b) the greater the error degrees of freedom. If the number of experimental units is large the replications can be based on b blocks each containing 2^p relatively homogeneous experimental units. In the case of blocks the block effects can also be estimated and tested. After estimating block effects the degrees of freedom available for error is $(2^p - 1)(b - 1)$.

Notation for Treatment Combinations

In a design which contains so many factors it is difficult to construct a system for expressing each interaction in terms of the underlying factors. It is customary to designate each treatment factor by an uppercase letter and the levels of the factor by lowercase letters and subscripts. Thus for a 2^3 design the factors are denoted by A, B and C and the levels by (a_1, a_2), (b_1, b_2) and (c_1, c_2) respectively. A single replication of a 2^3 design would therefore contain the 8 treatment combinations shown in Table 5.41.

Determining Interaction Effects Using Effect Coding

Effect coding introduced earlier in this chapter provides a useful approach to determining the levels of each factor that are present in any interaction. To illustrate this concept the effect coding design matrix for the above 2^3 design is shown in Table 5.42. The interactions are obtained by taking the

Table 5.42. Effect Coding for 2^3 Design

Cell	μ	A	B	C	AB	AC	BC	ABC
$a_1b_1c_1$	1	-1	-1	-1	1	1	1	-1
$a_1b_1c_2$	1	-1	-1	1	1	-1	-1	1
$a_1b_2c_1$	1	-1	1	-1	-1	1	-1	1
$a_1b_2c_2$	1	-1	1	1	-1	-1	1	-1
$a_2b_1c_1$	1	1	-1	-1	-1	-1	1	1
$a_2b_1c_2$	1	1	-1	1	-1	1	-1	-1
$a_2b_2c_1$	1	1	1	-1	1	-1	-1	-1
$a_2b_2c_2$	1	1	1	1	1	1	1	1

products of the elements in the corresponding main effect columns. From the table the contribution of each of the 8 cells to the particular interaction term can be determined from the particular column in the table.

The A effect is measured by determining the difference between the four cells containing a_2 and the four cells containing a_1, and is given by

$$[(a_2b_2c_2) + (a_2b_2c_1) + (a_2b_1c_2) + (a_2b_1c_1)]-$$
$$[(a_1b_1c_1) + (a_1b_1c_2) + (a_1b_2c_1) + (a_1b_2c_2)].$$

The ABC interaction is determined from

$$[(a_1b_1c_2) + (a_1b_2c_1) + (a_2b_1c_1) + (a_2b_2c_2)]-$$
$$[(a_1b_1c_1) + (a_1b_2c_2) + (a_2b_1c_2) + (a_2b_2c_1)]$$
$$= [(a_1b_1c_2) - (a_2b_1c_2)] + [(a_1b_2c_1) - (a_2b_2c_1)]$$
$$- [(a_1b_1c_1) - (a_2b_1c_1)] - [(a_1b_2c_2) - (a_2b_2c_2)].$$

Expressions for all of the other effects can be obtained in a similar manner.

Analysis of Variance Table for 2^p Design

Assuming that there are b *replications* or blocks the analysis of variance table is given by Table 5.43. The reader should note that the block effects are also examined in this table. If the replications are not blocks but are multiple samples from each of the eight cells then the block effect sum of squares is added to the error sum of squares as in the three-way factorial discussed in Section 5.4.

Table 5.43. Analysis of Variance Table for 2^p Design

Source	d.f.	Sum of Squares	Mean Squares	F
Blocks	$(b-1)$	SSBLOCKS	MSBLOCKS	MSBLOCKS/MSE
A	1	SSA	MSA	MSA/MSE
B	1	SSB	MSB	MSB/MSE
C	1	SSC	MSC	MSC/MSE
AB	1	SSAB	MSAB	MSAB/MSE
AC	1	SSAC	MSAC	MSAC/MSE
BC	1	SSBC	MSBC	MSBC/MSE
ABC	1	SSABC	MSABC	MSABC/MSE
ERROR	$7(b-1)$	SSE	MSE	
TOTAL	$8b-1$			

Table 5.44. Analysis of Variance for Fitness Survey with Blocks

Source	d.f.	Sum of Squares	F	p-value
Blocks	2	69.33	5.69	0.016
SMOKE	1	770.67	126.44	0.000
SEX	1	2400.00	393.75	0.000
RACE	1	1536.00	252.00	0.000
SEX*RACE	1	24.00	3.94	0.067
SEX*SMOKE	1	682.67	112.00	0.000
RACE*SMOKE	1	170.67	28.00	0.000
SEX*RACE*SMOKE	1	1290.67	211.75	0.000
ERROR	14	85.33		
TOTAL	23	7029.33		

Example

The 2^3 *design* with 3 replications introduced as an example in Section 5.4 will be used here to provide an example. We assume that the 3 replications are 3 blocks representing 3 different fitness coordinators. A block effect was added to the raw data used in Section 5.4. The ANOVA table with block effects is given in Table 5.44. Block effects were created artifically by adding 2 to the observations in block 1, 3 to the observations in block 2 and −1 to the observations in block 3. The other effects remain as in Section 5.4 except that the grand mean has been increased by 1 1/3.

Table 5.45. Analysis of Variance for a Single Replication of a 2^p Design with Example for $p = 6$

Source	d.f.	d.f. for $p = 6$
Main effects	$\binom{p}{1} = p$	6
First-order interactions	$\binom{p}{2} = p(p-1)/2$	15
Second-order interactions	$\binom{p}{3} = p(p-1)(p-2)/6$	20
Error	$2^p - \binom{p}{1} - \binom{p}{2} - \binom{p}{3} - 1$	22
Total	$(2^p - 1)$	63

5.6.2 Single and Fractional Replication

Single Replication

If experimental units are scarce it may not be possible to replicate the 2^p design. In such cases there are no degrees of freedom for error. If it can be assumed that the higher order interactions are truly zero then the sums of squares for these interactions can be combined to determine an estimate of the mean square error. In general higher order interactions are often zero. If this is not the case the error sum of squares will be too large and hence the F-test for the main effects and lower order interactions will be too conservative.

If one or more higher order interactions are known to be non-zero they can always be omitted from the error sum of squares *Provided This Decision is Made in Advance*. As a rule of thumb two-way interactions are not used for the error sum of squares while four-way interactions and higher usually are. Three-way interactions may or may not depending on the degrees of freedom for error and any a priori rationale for their omission.

Degrees of Freedom for Error in a 2^p Design

As an example the analysis of variance for a 2^p *factorial* would have the analysis of variance table shown in Table 5.45. Notice how the four-way and higher interactions have been used for error. The table also includes the degrees of freedom for $p = 6$. In the case of $p = 6$ the interactions from four-way and higher account for 22 degrees of freedom.

Effect Coding for a 2^4 Design

To illustrate a 2^4 *design* the design matrix using effect coding is shown in Table 5.46. As can be seen from the table the column corresponding to ABCD shows that this interaction would be measured by subtracting observations (2, 3, 5, 8, 9, 12, 14, 15) from observations (1, 4, 6, 7, 10, 11, 13, 16). If the three-way and *four-way interactions* are assumed negligible an estimate of error could be obtained which has 5 degrees of freedom.

Fractional Replication and Aliases

When the experimental units are scarce and there are a large number of treatment combinations it may be necessary to use a *fractional replication* which includes only a fraction of the total number of 2^p treatment combinations. As a result not all effects can be estimated and for the effects that can be estimated they will be *aliased* with one or more other effects. Two effects are *aliases* of each other if the contrast used to estimate one of them is identical to the contrast used to estimate the other. The procedures for fractional replication are concerned with designing the fraction so that the effects required are only aliased with effects that can be assumed a priori to be negligible.

Aliasing in a 2^4 Design

To illustrate the aliasing concept it will be useful to employ the 2^4 design laid out in Table 5.46. A one-half fractional replication of this 2^4 design involves removing half of the 16 cells. If all cells corresponding to ABCD = -1 (the 8 cells 2, 3, 5, 8, 9, 12, 14, 15) are removed then the ABCD interaction cannot be estimated. The ABCD interaction in this case is usually called the *defining contrast*. Table 5.47 shows the remaining 8 cells after elimination of these 8 cells.

An examination of the columns of the 8 cells table reveals that the main effect columns A, B, C and D are identical to the three-way interaction columns BCD, ACD, ABD and ABC respectively. Thus we conclude that the aliases are (A = BCD, B = ACD, C = ABD and D = ABC). The pattern for these aliases provides a useful way of determining the aliases. In addition the columns of the table also reveal the aliases AB = CD, AC = BD, and AD = BC.

This one-half fractional replication of the 2^4 factorial results in the four three-way interactions being aliased with the four main effects and three of the six two-way interactions being aliased with the remaining three two-way interactions. If we can assume that the four three-way interactions and the four-way interaction are negligible and if we can assume that all

Table 5.46. Effect Coding for 2^4 Factorial Design

Obs. No.	Cell	μ	A	B	C	D	AB	AC	AD	BC	BD	CD	ABC	ABD	BCD	ACD	ABCD
1	$a_1b_1c_1d_1$	1	-1	-1	-1	-1	1	1	1	1	1	1	-1	-1	-1	-1	1
2	$a_1b_1c_1d_2$	1	-1	-1	-1	1	1	1	-1	1	-1	-1	-1	1	1	1	-1
3	$a_1b_1c_2d_1$	1	-1	-1	1	-1	1	-1	1	-1	1	-1	1	-1	1	1	-1
4	$a_1b_1c_2d_2$	1	-1	-1	1	1	1	-1	-1	-1	-1	1	1	1	-1	-1	1
5	$a_1b_2c_1d_1$	1	-1	1	-1	-1	-1	1	1	-1	-1	1	1	1	1	-1	-1
6	$a_1b_2c_1d_2$	1	-1	1	-1	1	-1	1	-1	-1	1	-1	1	-1	-1	1	1
7	$a_1b_2c_2d_1$	1	-1	1	1	-1	-1	-1	1	1	-1	-1	-1	1	-1	1	1
8	$a_1b_2c_2d_2$	1	-1	1	1	1	-1	-1	-1	1	1	1	-1	-1	1	-1	-1
9	$a_2b_1c_1d_1$	1	1	-1	-1	-1	-1	-1	-1	1	1	1	1	1	-1	1	-1
10	$a_2b_1c_1d_2$	1	1	-1	-1	1	-1	-1	1	1	-1	-1	1	-1	1	-1	1
11	$a_2b_1c_2d_1$	1	1	-1	1	-1	-1	1	-1	-1	1	-1	-1	1	1	-1	1
12	$a_2b_1c_2d_2$	1	1	-1	1	1	-1	1	1	-1	-1	1	-1	-1	-1	1	-1
13	$a_2b_2c_1d_1$	1	1	1	-1	-1	1	-1	-1	-1	-1	1	-1	-1	1	1	1
14	$a_2b_2c_1d_2$	1	1	1	-1	1	1	-1	1	-1	1	-1	-1	1	-1	-1	-1
15	$a_2b_2c_2d_1$	1	1	1	1	-1	1	1	-1	1	-1	-1	1	-1	-1	-1	-1
16	$a_2b_2c_2d_2$	1	1	1	1	1	1	1	1	1	1	1	1	1	1	1	1

Table 5.47. A One-Half Fraction of a 2^4 Design with Four-Way Interaction as Defining Contrast

Obs. No.	Cell	A	B	C	D	AB	AC	AD	BC	BD	CD	ABC	ABD	BCD	ACD	ABCD
1	$a_1b_1c_1d_1$	-1	-1	-1	-1	1	1	1	1	1	1	-1	-1	-1	-1	1
4	$a_1b_1c_2d_2$	-1	-1	1	1	1	-1	-1	-1	-1	1	1	1	-1	-1	1
6	$a_1b_2c_1d_2$	-1	1	-1	1	-1	1	-1	-1	1	-1	1	-1	-1	1	1
7	$a_1b_2c_2d_1$	-1	1	1	-1	-1	-1	1	1	-1	-1	-1	1	-1	1	1
10	$a_2b_1c_1d_2$	1	-1	-1	1	-1	-1	1	1	-1	-1	1	-1	1	-1	1
11	$a_2b_1c_2d_1$	1	-1	1	-1	-1	1	-1	-1	1	-1	-1	1	1	-1	1
13	$a_2b_2c_1d_1$	1	1	-1	-1	1	-1	-1	-1	-1	1	-1	-1	1	1	1
16	$a_2b_2c_2d_2$	1	1	1	1	1	1	1	1	1	1	1	1	1	1	1

Table 5.48.   Fractional Replications of Some 2^p Designs in which Main Effects and Two-Way Interactions Can Be Estimated

No. of Factors p	No. of Experimental Units	No. of Replications Possible	Block Size
5	16	1	16
6	32	2	16
7	64	8	8
8	128	8	16

of the two-way interactions with one of the factors (say A) are negligible then the one-half fractional replication should be a useful design.

To get an estimate of error it is necessary that this one-half replication be repeated at least once. By using the same number of experimental units as in a single replication two separate one-half fractional replications can be generated. If not, an assumption of negligible two-way interactions would permit the sum of three sums of squares to be used to provide an estimate of error.

Experimental Units Requirements for Various Fractional Replications of 2^p Designs

In general *fractional factorial designs* are useful when 1) a complete factorial requires too many experimental units, 2) all main effects and two-way interactions are considered important and 3) experimental error is substantial. Table 5.48 illustrates some uses of fractional replications of 2^p designs in which the main effects and two-way interactions can be estimated. From the table we can see that for the 2^5 design there will be no estimate of error available other than the two-way interactions. For the remaining examples there are at least 2 replications of each fraction and hence an estimate of error is obtainable independent of the two-way interactions. Cochran and Cox (1957) provide a summary of a variety of fractional designs.

For a 2^p factorial fractional replications can be constructed using a $1/2^q$ fraction. The resulting fractional factorial is a 2^{p-q} *fractional factorial design*. The number of defining contrasts for a $1/2^q$ fraction is q.

A Fractional Replication for a 3^5 Factorial

A 3^5 factorial design requires 243 experimental units while a 1/3 replication of this design requires 81 units. The 81 experimental units can be arranged

Table 5.49. One Third Fraction of a 3^5 Factorial

Block No.	Experimental Units*		
1	11111	11312	11213
1	21233	21131	21332
1	31322	31223	31121
1	12221	12122	12323
1	22313	22211	22112
1	32132	32333	32231
1	13331	13232	13133
1	23123	23321	23222
1	33212	33113	33311
2	11231	11132	11333
2	21323	21221	21122
2	31112	31313	31211
2	12311	12212	12113
2	22133	22331	22232
2	32222	32123	32321
2	13121	13322	13223
2	23213	23111	23312
2	33332	33233	33131
3	11222	11123	11321
3	21311	21212	21113
3	31133	31331	31232
3	12332	12233	12131
3	22121	22322	22223
3	32213	32111	32312
3	13112	13313	13211
3	23231	23132	23333
3	33323	33221	33122

*Levels of the factors are given in order of A, B, C, D, and E and are denoted by the integers 1, 2, and 3.

as 3 replications of 27 units each with all main effects and two-way interactions estimable. Using this design the 80 degrees of freedom are allocated as 2 for blocks or replications, 10 for main effects, 40 for two-factor interactions and 28 for error. The plan for this design is illustrated in Table 5.49. This design along with many other examples are available in Cochran and Cox (1957).

Additional discussion of fractional factorial designs can be found in Cochran and Cox (1957) and Kempthorne (1952).

5.6.3 Confounding and Incomplete Block Designs

In some applications of large factorial experiments it is not possible to obtain a sufficient number of homogeneous experimental units. In such cases a block design must be used to control for one or more nuisance factors and hence the number of experimental units per block is less than the number of treatment combinations. In such cases one or more complete factorials are split into smaller blocks. The splitting of the complete factorial arrangements results in one or more of the effects being *confounded* with the block effect. The purpose of confounding is to control for a nuisance factor by reducing block size. The number of experimental units is not the major concern.

For example in the 2^4 design discussed above, the 16 units can be split into two blocks so that the ABCD column has $+1$ for one block and -1 for the second block. (See Table 5.46.) In this case the ABCD interaction will be confounded with the block effect.

If a large number of experimental units are available replications of the design can be constructed. If the complete factorial can be repeated with different effects being confounded each time, the confounded effects can also be estimated. In this case the confounding is called *partial confounding*. More extensive discussions of confounding are available in Cochran and Cox (1957) and Kempthorne (1952).

Incomplete Block Designs

In randomized block designs where the number of treatments combinations is large it may not be possible to obtain a sufficient number of homogeneous experimental units so that each block can accommodate all treatment combinations. Often such deficiencies occur because of shortages in equipment or laboratory facilities. If the design is such that only a portion of the set of treatment combinations occur in each block the design is said to be an *incomplete block design*. As discussed above the use of confounding to reduce block size results in the loss of information with respect to some interactions which are confounded with the blocking factor.

An alternative strategy to reduction of block size through confounding is the *balanced incomplete block design*. In this design every pair of treatment combinations appear together in a block the same number of times. This does not mean that the treatment pair appears together in every block; in fact, they may only appear together once. In a balanced incomplete block design each treatment difference is estimated with the same degree of precision. Ideally the experimental units within each block are homogeneous while the blocks themselves are quite different from one another.

If there are a total of g treatment combinations, r replications of each treatment combination, and c experimental units per block, then the number of times two treatments appear together in a block is given by

$$\lambda = r(c-1)/(g-1) = n(c-1)/g(g-1)$$
where $n = gr = $ number of experimental units   and
       $b = n/c = $ number of blocks.

A special case of a balanced incomplete block design is called a *balanced lattice design* in which the total number of treatments g is a perfect square, $g = k^2$, where the block size is k experimental units and there are $(k + 1)$ replications of each treatment. Using the above expression for λ in this case $\lambda = (k + 1)(k - 1)/(k^2 - 1) = 1$ and hence each treatment occurs with every other treatment only once.

An outline of the analysis for a balanced incomplete block design is given in Montgomery (1984) while lattice designs are discussed in Petersen (1985). More extensive discussion of these topics are given in Cochran and Cox (1957) and Kempthorne (1952).

5.7 Random Effects Models and Variance Components

For all of the analysis of variance models discussed in the previous sections of this chapter, the effects due to various factors have always been assumed to be fixed. In every case the observations were assumed to be based on random samples from population groups or upon measurements derived from experimental units which have been randomly assigned to various treatment combinations. In all cases the right hand side of the analysis of variance models studied contained the sum of one or more fixed effects plus a single random variable called the error term. The variance of the observation on the left hand side of the model therefore is equal to the variance of the error term on the right.

In this section the analysis of variance model is extended to include one or more *random effects* and hence the variance of the dependent variable is composed of more than one term on the right. The model therefore can be said to contain *variance components*. A random effect is generated if the treatments applied represent a random sample derived from a larger population of possible treatments, or if the experimental units employed represent a random sample derived from a population of experimental units. These random effects are considered to be important if their contribution to the variance of the observed measurements is relatively large. In the case of random effects models the inferences are made with respect to the variances of the effects rather than the means. In some cases the analysis of variance model can be composed of a mixture of fixed effects and random effects in which case the model is usually called a *mixed model*.

The section begins with a discussion of one-way and two-way random effects models followed by a discussion of a two-way mixed model. A more general discussion of variance components models is then provided. The section concludes with a discussion of nested designs. The random effects models, mixed models and nested design models are special cases of the general class of linear models called variance components models.

5.7.1 The One-Way Random Effects Model

The inferences derived from a one-way ANOVA are made with respect to the g populations from which the samples were selected. These g groups are fixed and the observed effects α_j, $j = 1, 2, \ldots, g$ in the model (5.6) of Section 5.1.1 are called fixed effects. There are situations in practice where the total number of groups to be compared is large, and hence a random sample of g groups must be selected from the population of groups before sampling from within each group begins. In such circumstances the group effects α_j, $j = 1, 2, \ldots, g$ represent a random sample of g group effects from the population of all such groups effects. In this case the model (5.6) is called a *random effects model*. In the context of the administration of treatments to experimental units the treatments may represent a random sample from a population of possible treatments.

Analysis of Variance

The mean μ in (5.6) now represents the overall mean of the population of group means while α_j denotes the deviation of a particular group mean from this overall mean. The effects α_j are therefore observations on a random variable α which has mean 0 and variance σ_α^2. The null hypothesis of equality of <u>ALL</u> the possible group means is equivalent to the null hypothesis $H_0 : \sigma_\alpha^2 = 0$. If α is assumed to be normally distributed then under H_0 the statistic

$$\frac{\text{SSA}/(g-1)}{\text{SSW}/(n-g)}$$

has an F distribution with $(g-1)$ and $(n-g)$ d.f. The test statistic for H_0 in the random effects model is therefore equivalent to the test statistic used to test equality of means in the fixed effects model in 5.1.1.

In general

$$E[\text{SSA}/(g-1)] = \sigma^2 + \frac{1}{(g-1)}\left[n - \sum_{j=1}^{g} n_j^2/n\right]\sigma_\alpha^2$$

and

$$E[\text{SSW}/(n-g)] = \sigma^2.$$

If H_0 is rejected $\bar{y}_{..}$ can be used to estimate μ and the quantity [MSA − MSW]/c can be used to estimate σ_α^2 where $c = \frac{1}{(g-1)}\left[n - \sum_{j=1}^{g} n_j^2/n\right]$. This estimator of σ_α^2 is usually referred to as a method of moments estimator since it is derived from expressions which are unbiased estimators of functions of the required variances.

Table 5.50. ANOVA Table for Random Effects Model for Regions

Source	d.f.	Sum of Squares	Mean Squares	F
AMONG	4	4790.531	1197.633	2.70
WITHIN	25	11090.125	443.605	
TOTAL	29	15880.656		

Example

For the family expenditure data introduced in Table 5.1 a one-way random effects model was fitted using the data for all regions for ad type 2. The five regions are assumed to be a random selection from a larger population of possible geographical regions. The ANOVA table for this data is given in Table 5.50. The p-value for the F-statistic is 0.054. We can conclude therefore that for significance levels above 0.054 the region effects for ad type 2 are not constant.

The means for Regions 1 through 5 are respectively 56.165, 42.018, 18.842, 47.133 and 34.075. An estimate of the overall mean μ is provided by the grand mean 33.039. The effects for the five regions are 23.126, 8.979, -15.197, 14.094 and 1.036 respectively. An estimate of the variance of the region effects σ_α^2 is provided by

$$[1197.633 - 443.605]/\frac{1}{4}\left(30 - \frac{180}{30}\right) = 754.028.$$

An estimate of the variance of a mean within regions is provided by $443.605/6 = 73.93$. The variance among the region effects is therefore ten times the variance within regions. The region effects variance is therefore too large to be considered negligible which is supported by the F-test p-value of 0.054.

Intraclass Correlation

Unlike the fixed effects model the observations y_{ij} and $y_{\ell j}$ are now correlated. Because both y_{ij} and $y_{\ell j}$ contain a common random variable α_j the covariance $\text{Cov}(y_{ij}, y_{\ell j}) = \sigma_\alpha^2$ while $V(y_{ij}) = \sigma_\alpha^2 + \sigma^2$, $i = 1, 2, \ldots, b$; $j = 1, 2, \ldots, g$. The correlation between observations *treated alike* is therefore given by $\sigma_\alpha^2/(\sigma_\alpha^2 + \sigma^2)$. This correlation is called the *intraclass correlation* coefficient in that it measures the correlation among observations in the same treatment class. This lack of independence among the y_{ij} does not affect the analysis of variance procedures.

5.7.2 Two-Way Models with Random Effects and Mixed Models

In the two-way factorial model discussed in Section 5.2.2 the population consisted of bg cells with means μ_{ij}, $i = 1, 2, \ldots, b$; $j = 1, 2, \ldots, g$. A total of c observations was randomly selected from each cell yielding a total of bgc observations y_{ijh}, $h = 1, 2, \ldots, c$; $i = 1, 2, \ldots, b$; $j = 1, 2, \ldots, g$. The model for y_{ijh} is given by

$$y_{ijh} = \mu + \alpha_j + \beta_i + (\alpha\beta)_{ij} + \varepsilon_{ijh}$$

where $V(\varepsilon_{ijh}) = \sigma^2$. The cell mean parameters are given by $\mu_{ij} = \mu + \alpha_j + \beta_i + (\alpha\beta)_{ij}$.

A Two-Way Random Effects Model

For a two-way random effects model the effects α_j, β_i and $(\alpha\beta)_{ij}$ are assumed to be observations on mutually independent normal random variables α, β and $(\alpha\beta)$ where $\alpha \sim N(0, \sigma_\alpha^2)$, $\beta \sim N(0, \sigma_\beta^2)$ and $(\alpha\beta) \sim N(0, \sigma_{\alpha\beta}^2)$ and hence $[\alpha + \beta + (\alpha\beta)] \sim N(0, \sigma_\alpha^2 + \sigma_\beta^2 + \sigma_{\alpha\beta}^2)$. Assuming $V(\varepsilon_{ijh}) = \sigma^2$, $V(y_{ijh}) = \sigma_\alpha^2 + \sigma_\beta^2 + \sigma_{\alpha\beta}^2 + \sigma^2$, a variance components model. The hypotheses of interest are $H_{01} : \sigma_\alpha^2 = 0$, $H_{02} : \sigma_\beta^2 = 0$ and $H_{03} : \sigma_{\alpha\beta}^2 = 0$.

Analysis of Variance

The sample error mean squares MSA, MSB, MSI and MSE defined in Table 5.9 in Section 5.2.2 are used to test the null hypotheses of interest. The expected values of the mean squares are given by

$$E[\text{MSA}] = \sigma^2 + c\sigma_{\alpha\beta}^2 + cb\sigma_\alpha^2$$
$$E[\text{MSB}] = \sigma^2 + c\sigma_{\alpha\beta}^2 + cg\sigma_\beta^2$$
$$E[\text{MSI}] = \sigma^2 + c\sigma_{\alpha\beta}^2$$
$$E[\text{MSE}] = \sigma^2.$$

Estimates of the variances using the method of moments are therefore given by

$$\hat{\sigma}^2 = \text{MSE}$$
$$\hat{\sigma}_\alpha^2 = (\text{MSA} - \text{MSI})/cb$$
$$\hat{\sigma}_\beta^2 = (\text{MSB} - \text{MSI})/cg$$
$$\hat{\sigma}_{\alpha\beta}^2 = (\text{MSI} - \text{MSE})/(c - 1).$$

Under H_{01} the ratio MSA/MSI has an F distribution with $(g - 1)$ and $(g - 1)(b - 1)$ d.f. Under H_{02} the ratio MSB/MSI has an F distribution with $(b - 1)$ and $(g - 1)(b - 1)$ d.f. Under H_{03} the ratio MSI/MSE has an F distribution with $(g - 1)(b - 1)$ and $bg(c - 1)$ d.f. In comparison to the fixed effects model of Table 5.9 the tests for the main effects in the random effects model use MSI in the denominator of F rather than MSE. The test for interaction however is the same in both cases.

Example

In a comparison of g gasoline brands controlling for b automobile types we may wish to assume that the b automobile types have been randomly selected from a larger population of automobile types. In addition it may be relevant to assume that the g gasoline brands are obtained from g different oil refineries which have been randomly selected from a larger population of oil refineries. For the random effects model described above a total of cg automobiles are required from each of the b automobile classes and hence a total of bcg experimental units.

A second example of a two-way random effects model is given by a study to measure price variation in confectionary products sold in a variety of types of retail outlets. Suppose that in a large geographical region the retail outlets are classified by type. In addition the confectionary products are classified by type and by manufacturer. A random sample of bgc retail outlets are selected so that there are gc outlets of each type. In addition a random sample of g products are selected from the population of confectionary products. The g products are then randomly assigned to the outlets so that each product-outlet type combination is repeated c times. If this design seems expensive, it is, and can be modified using repeated measures to be introduced in the next section.

Intraclass Correlation

Even though the effects are mutually independent there is still a correlation among responses. For observations with the same α treatment there is a correlation among responses given by

$$\rho_\alpha = \sigma_\alpha^2/(\sigma_\alpha^2 + \sigma_\beta^2 + \sigma_{\alpha\beta}^2 + \sigma^2).$$

This correlation is called an *intraclass correlation* coefficient reflecting the correlation among responses receiving the same α treatment.

In a similar fashion the intraclass correlation among responses receiving the same β treatment is given by

$$\rho_\beta = \sigma_\beta^2/(\sigma_\alpha^2 + \sigma_\beta^2 + \sigma_{\alpha\beta}^2 + \sigma^2).$$

Finally since there are c observations in each cell the intraclass correlation among responses from the same cell is given by

$$\rho_{\alpha\beta} = \sigma_{\alpha\beta}^2/(\sigma_\alpha^2 + \sigma_\beta^2 + \sigma_{\alpha\beta}^2 + \sigma^2).$$

Table 5.51. ANOVA Table for Two-Way Random Effects Model Ads by Regions

Source	d.f.	Sum of Squares	Mean Square	F	p-value
Ads	3	4585.68	1528.56	2.05	0.161
Regions	4	4867.51	1216.88	1.63	0.230
Interaction	12	8937.92	744.83	1.76	0.065
Error	100	42382.02	423.82		
Total	119	60773.12			

Example

Although the random effects model does not fit the design for the test marketing data it is used here to provide an example.

If we assume for the test marketing data in Table 5.1 that the region effects and ad effects are random then the analysis of variance table given by Table 5.11 must be revised as shown in Table 5.51. The F-statistics for the main effects are now obtained by dividing by MSI. The F-statistic for interaction remains unchanged. In comparison to Table 5.11 the F-values and corresponding p-values for the main effects have now been reduced since MSI exceeds MSE by the factor of 1.76.

Estimates of the variance components can be obtained from the error mean squares in Table 5.51. The estimates are given by

$$\hat{\sigma}^2 = 423.82, \quad \hat{\sigma}^2_{\text{ADS}} = (1528.56 - 744.83)/30 = 26.12$$
$$\hat{\sigma}^2_{\text{REGIONS}} = (1216.88 - 744.83)/24 = 19.67$$
$$\hat{\sigma}^2_{\text{ADS*REGIONS}} = (744.83 - 423.82)/5 = 64.20.$$

The total variance is therefore 533.81. The resulting intraclass correlation coefficients are given by $\hat{\rho}_{\text{ADS}} = 0.05$, $\hat{\rho}_{\text{REGIONS}} = 0.04$ and $\hat{\rho}_{\text{ADS*REGIONS}} = 0.12$.

Two-Way Mixed Model

If one of the main effects is assumed to be fixed while the other is considered to be random then the test statistics are a mixture of the test statistics for the fixed effects and random effects models. If in the two-way model discussed above the parameters α_j are fixed and the parameters β_i are random, the hypotheses of interest become $H_{01} : \alpha_1 = \alpha_2 = \cdots = \alpha_g = 0$, $H_{02} : \sigma^2_\beta = 0$ and $H_{03} : \sigma^2_{\alpha\beta} = 0$. It is not uncommon in practice to assume in a randomized block design that the treatment effects are fixed but that the block effects are random.

The F-statistics are MSA/MSI, MSB/MSE and MSI/MSE respectively. In this case the test for H_{01} is the same as in the fixed effects case while H_{02} and H_{03} are the same as in the random effects model. The assumption here is that the β_i effects and the $(\alpha\beta)_{ij}$ effects are mutually independent. A more general assumption is to assume the $(\alpha\beta)_{ij}$ and β_i are correlated with covariance matrix Σ for each j. This more complex model is useful for repeated measures experiments and will be discussed more extensively in Section 5.8.

Example

For the family expenditure data the two-way ANOVA was outlined in Table 5.11. Assuming that the ad effects are fixed and the region effects are random the F-statistics in Table 5.11 must be revised. The F-statistic for ads is now given by $1528.56/744.83 = 2.05$ which has 3 and 12 degrees of freedom and a p-value of 0.230. The ad effects are therefore no longer significant at conventional levels. The F-statistics for region effects and interaction effects remain unchanged as in Table 5.11. It would appear that when the five regions are viewed as a sample from a larger population of regions the variation among ads is not significant. There is still however a significant difference among the regions and there also appears to be some variation in the ad effects by region.

Intraclass Correlation

As in the case of the two-way random effects model discussed above, the mixed model version of a randomized block design also has correlation among responses receiving the same random effect (same block). This intrablock correlation is given by $\rho_\beta = \sigma_\beta^2/(\sigma^2 + \sigma_\beta^2)$.

In the case of the mixed model with interaction the intraclass correlation among units in the same block receiving the same treatment is given by

$$\rho_{\alpha\beta} = \sigma_{\alpha\beta}^2/(\sigma^2 + \sigma_\beta^2 + \sigma_{\alpha\beta}).$$

More extensive discussion of random effects within most design models can be found in Kirk (1982) and Neter, Wasserman and Kutner (1985).

5.7.3 The Variance Components Model

The random effects models discussed in this section are special cases of a more general class of models called *variance components models* or *random models*.

Using notation similar to the notation of Milliken and Johnson (1984) the general random model is given by

$$\mathbf{y} = \mu\mathbf{i} + \mathbf{X}_1\boldsymbol{\alpha}_1 + \mathbf{X}_2\boldsymbol{\alpha}_2 + \cdots + \mathbf{X}_r\boldsymbol{\alpha}_r + \mathbf{u}$$

where

(a) the $\boldsymbol{\alpha}_j$ $(s_j \times 1)$ are random treatment effects with $\boldsymbol{\alpha}_j \sim N(0, \sigma_j^2\mathbf{I}_{s_j})$, $j = 1, 2, \ldots, r$;

(b) $\mathbf{u} \sim N(0, \sigma_u^2\mathbf{I}_n)$, $\mathbf{u}(n \times 1)$ independent of $\boldsymbol{\alpha}_j$, $j = 1, 2, \ldots, r$;

(c) the matrices $\mathbf{X}_j$, $j = 1, 2, \ldots, r$ are $(s_j \times 1)$ design matrices;

(d) $\mathbf{y}(n \times 1) \sim N(\mu\mathbf{i}, \boldsymbol{\Omega}_y)$, $\boldsymbol{\Omega}_j = \sum_{j=1}^{r} \sigma_j^2\mathbf{X}_j\mathbf{X}_j' + \sigma_u^2\mathbf{I}_n$.

(e) $\mathbf{i}$ $(n \times 1)$ a column of unities.

Inference in the variance components model begins with the estimation of the variance parameters σ_j^2, $j = 1, 2, \ldots, r$; using analysis of variance type sums of squares. Analysis of variance sums of squares are *quadratic forms* of the form $Q_k = \mathbf{y}'\mathbf{A}_k\mathbf{y}$ where

$$E[Q_k] = tr\mathbf{A}_k\boldsymbol{\Omega} = \sum_{j=1}^{r} \sigma_j^2 tr\mathbf{A}_k\mathbf{X}_j\mathbf{X}_j' + \sigma_u^2 tr\mathbf{A}_k. \qquad (5.9)$$

Using a total of $(r + 1)$ quadratic forms of the form Q_k as in (5.9) a system of $(r + 1)$ equations is obtained in the $(r + 1)$ unknown variances given by $\mathbf{Q} = \mathbf{C}\boldsymbol{\sigma}$. The notation of (5.9) contains $\boldsymbol{\sigma}(r + 1) \times 1$, a vector of unknown variances σ_j^2, $j = 1, 2, \ldots, r$; $\mathbf{Q}(r+1) \times 1$, a vector of values Q_k, $k = 1, 2, \ldots, (r+1)$; and $\mathbf{C}$ an $(r+1) \times (r+1)$ matrix of coefficients obtained from $tr\mathbf{A}_k\boldsymbol{\Omega}$, $k = 1, 2, \ldots, (r + 1)$.

Replacing the expectations of the quadratic forms Q_k by their estimators $\widehat{Q}_k = \mathbf{y}'\mathbf{A}_k\mathbf{y}$ the estimators of the variances are given by $\hat{\boldsymbol{\sigma}} = \mathbf{C}^{-1}\widehat{\mathbf{Q}}$ where $\widehat{\mathbf{Q}}$ is the vector of estimators $\widehat{Q}_k$, $k = 1, 2, \ldots, (r + 1)$. This estimation method is commonly referred to as the *method of moments*.

For a balanced model the quadratic forms Q_k, $k = 1, 2, \ldots, (r + 1)$ are equivalent to the sums of squares used for the corresponding fixed effects model. In this case the quadratic forms are partitions of the sum of squares $\mathbf{y}'\mathbf{y}$ such that $\mathbf{y}'\mathbf{y} = n\bar{y}^2 + \sum_{k=1}^{r+1} Q_k$.

The component sums of squares Q_k, $k = 1, 2, \ldots, (r + 1)$ and $n\bar{y}^2$ are mutually independent. The expectations of the sums of squares normally used for fixed effects models provide the necessary equations to obtain the estimators of the variance components. Searle (1971) labels this approach the analysis of variance method. The estimation methods illustrated for the two random effects models discussed earlier in this section are special cases of this method.

A more extensive discussion of variance components models is given in Searle (1971) and Milliken and Johnson (1984).

5.7.4 Nested Designs and Variance Components

A useful example of a variance components model occurs in nested designs. In a *nested design* several treatment factors are arranged in a hierarchical fashion so that the levels of one factor are completely nested within the levels of a factor at a higher level. Alternatively we can view this design as an experiment in which large experimental units are randomly assigned a treatment level from a factor. The units are then subdivided into smaller experimental units. The smaller experimental units from a given larger unit are then randomly assigned to the levels of a second factor. This process of subdividing experimental units continues until all factors have been employed. The result is a hierarchical sequence of experimental units.

An example of such a design would be the study of a finished product. The raw material for the product is produced by one of several methods. The raw material is then used to make the finished product by one of a number of different methods. Within each of the methods there are several factories that exclusively use only one method. For each factory the finished product is produced in batches and the batches are then sampled to test the product.

A second example would be the study of the magnitude of clerical errors made by clerks in a large national retail chain. The retailer is divided into geographical regions for management purposes. Within each region retail stores are randomly selected and within each store departments are randomly selected. A random sample of invoices is then selected from the chosen departments.

The analysis of variance model for the nested design consists of a sum of nested random effects which are assumed to be mutually independent. For a design with four levels in the hierarchy the model is given by

$$y_{ijh\ell} = \mu + \beta_i + (\alpha\beta)_{ij} + (\gamma\alpha\beta)_{ijh} + \varepsilon_{ijh\ell}$$
$$i = 1, 2, \ldots, b;\ j = 1, 2, \ldots, g;$$
$$j = 1, 2, \ldots, c;\ \ell = 1, 2, \ldots, n;$$

where $\beta_i \sim N(0, \sigma_\beta^2)$, $(\alpha\beta)_{ij} \sim N(0, \sigma_{\alpha\beta}^2)$,

$(\gamma\alpha\beta)_{ijh} \sim N(0, \sigma_{\gamma\alpha\beta}^2)$ and $\varepsilon_{ijh\ell} \sim N(0, \sigma^2)$.

Table 5.52. Analysis of Variance for Nested Model with Four Random Factors

Source	d.f.	Sum of Squares	Mean Square
B	$(b-1)$	SSB	$\text{MSB} = \text{SSB}/(b-1)$
AB	$b(g-1)$	SSAB	$\text{MSAB} = \text{SSAB}/b(g-1)$
CAB	$bg(c-1)$	SSCAB	$\text{MSCAB} = \text{SSCAB}/bg(c-1)$
Error	$cbg(n-1)$	SSE	$\text{MSE} = \text{SSE}/cbg(n-1)$
Total	$ncbg-1$	SST	

The sums of squares for this model also form a nested sequence and are given by

$$\text{SSE} = \sum_{i=1}^{b}\sum_{j=1}^{g}\sum_{h=1}^{c}\sum_{\ell=1}^{n}(y_{ijh\ell} - \bar{y}_{ijh\cdot})^2$$

$$\text{SSCAB} = n\sum_{i=1}^{b}\sum_{j=1}^{g}\sum_{h=1}^{c}(\bar{y}_{ijh\cdot} - \bar{y}_{ij\cdot\cdot})^2$$

$$\text{SSAB} = cn\sum_{i=1}^{b}\sum_{j=1}^{g}(\bar{y}_{ij\cdot\cdot} - \bar{y}_{i\cdot\cdot\cdot})^2$$

$$\text{SSB} = gcn\sum_{i=1}^{b}(\bar{y}_{i\cdot\cdot\cdot} - \bar{y}_{\cdot\cdot\cdot\cdot})^2$$

$$\text{SST} = \sum_{i=1}^{b}\sum_{j=1}^{g}\sum_{h=1}^{c}\sum_{\ell=1}^{n}(y_{ijh\ell} - \bar{y}_{\cdot\cdot\cdot\cdot})^2.$$

The analysis of variance table for this model is given in Table 5.52.

The procedures for testing for the various effects and for estimating the variance components by the method of moments can be obtained from the expected mean squares summarized below. Each mean square can be compared to the one below it in the table to determine the required F-statistic. The expressions for the expected mean squares can be used to obtain method of moments equations for estimates of the variance components.

$$E[\text{MSB}] = \sigma^2 + n\sigma_{\gamma\alpha\beta} + nc\sigma_{\alpha\beta}^2 + ncg\sigma_{\beta}^2$$

$$E[\text{MSAB}] = \sigma^2 + n\sigma_{\gamma\alpha\beta}^2 + nc\sigma_{\alpha\beta}^2$$

$$E[\text{MSCAB}] = \sigma^2 + n\sigma_{\gamma\alpha\beta}^2$$

$$E[\text{MSE}] = \sigma^2.$$

5.8 Repeated Measures and Split Plots Designs

In a repeated measures experiment the same experimental unit is used repeatedly. The treatment combinations are applied sequentially so that each experimental unit receives all possible treatment combinations. The advantages of this design are that fewer experimental units are required and that by using the same experimental unit repeatedly, the variation among treatment responses can be measured without a contribution from the variation among experimental units.

In a *split plot design*, large experimental units called *whole plots* are exposed to one treatment, after which the experimental unit is subdivided into smaller experimental units called *subplots*, which are then randomly assigned to the levels of a second treatment factor. The number of subplots in each whole plot is equal to the number of levels of the second treatment factor. This split plot design is an example of a design which contains nested experimental units. The split plot design is also an incomplete block design in that each whole plot (block) only receives treatment combinations from one level of the whole plot treatment factor. The effect of the first treatment can be separated from block effects by replicating the experiment over many blocks.

A *repeated measures design* is similar to the split plot design in that the whole plot receives all possible treatments of one factor. The difference however is that in a split plot design the whole plot experimental unit is subdivided into subplots which are randomly assigned to the subplot treatments. In the repeated measures design the entire experimental unit is used repeatedly. In both repeated measures designs and split plot designs the model is a variance components type model in that there are at least two error terms. One term is due to the whole plot or subject and the second is due to the subplot or treatment.

5.8.1 Repeated Measures Designs

In randomized block designs a total of bg experimental units are used to compare g groups or treatments. The experimental units are divided into b blocks in such a way that within the blocks the variation among the experimental units is small. In some circumstances it is possible for the observations within a block to be repeated observations on the same subject or experimental unit. Thus there are only b experimental units required for the analysis since each experimental unit receives all g treatments. This special case of a randomized block design is called a *repeated measures* randomized block design. It is important in such a design that the order of application of the g treatments to the b experimental units be determined randomly to avoid any systematic effects resulting from the order of application of the treatments.

Applications

The repeated measures design is popular in the social sciences where the experimental units are usually people. Because of the natural high variability among responses from different people, there is a need to control this variability so that the factor of interest can be studied. In the repeated measures design, the subjects act as their own control in the sense that the variation due to treatments is measured only by the variation of each subject's responses around the subject mean. Each subject is a block. The variation between subjects in this case is the variation measured by the block sum of squares.

An alternative type of repeated measures design which occurs in medical science is the repeated administration of the same treatment to the same individual on several occasions. In a study to compare several drugs each subject is randomly assigned one of the drugs and then is treated with the drug on a number of occasions. In this case, the variable time records the number of times the individual has received the drug.

In the design to compare g gasoline brands the experimental units were b automobile types. The assumption in the randomized block design is that g different automobiles for each of the b automobiles types were required and hence a total of bg different automobiles are used. In a repeated measures randomized block design one automobile from each of b types is selected. Each of the selected automobiles is then used to test all g gasoline brands.

The advantage of this design is that variations in performance between automobiles of the same type is eliminated and hence the error sum of squares SSE should be smaller. Administration of this design, however, requires that much effort be devoted to insuring that the response to any treatment not be related to the order of receiving the treatments. Such carry-over effects could yield biased estimates of effects and thus should be avoided.

The Single Factor Repeated Measures Design

In the single factor repeated measures design a total of b subjects are randomly assigned to receive g treatments each. Each subject receives each treatment only once. This design is therefore a special case of a randomized block design where the subjects are the blocks. The most commonly used model for this design assumes that the treatment effects are fixed but that the subject effects are random. The subjects are assumed to be a random sample from a population of possible subjects. The model is therefore a special case of a mixed model introduced in Section 5.7. An additional complication of the repeated measures design is that with the same subject receiving all treatments, the assumption of independent errors across different treatments on the same subject, may not be valid. This complication will be addressed later in this section. Assuming for the moment that these errors are independent the analysis of variance model is outlined next.

Table 5.53. Analysis of Variance for Single Factor Repeated Measures Design

Source	d.f.	Sum of Squares	Mean Square	F
Treatments	$(g-1)$	SSA	$\text{MSA} = \text{SSA}/(g-1)$	MSA/MSE
Subjects	$(b-1)$	SSB	$\text{MSB} = \text{SSB}/(b-1)$	MSB/MSE
Error	$(b-1)(g-1)$	SSE	$\text{MSE} = \text{SSE}/(b-1)(g-1)$	
Total	$bg-1$	SST		

Analysis of Variance Model

The observation y_{ij} measures the response for subject i on treatment j, $i = 1, 2, \ldots, b$; $j = 1, 2, \ldots, g$. The model is given by

$$y_{ij} = \mu + \beta_i + \alpha_j + \varepsilon_{ij}$$

where $\varepsilon_{ij} \sim N(0, \sigma^2)$, $\beta_i \sim N(0, \sigma_\beta^2)$, $\sum_{j=1}^{g} \alpha_j = 0$ and the ε_{ij} are mutually independent, the β_i are mutually independent, and the β_i are independent of the ε_{ij}. It is assumed that the subject effects β_i and treatment effects α_j are additive, in that there is no interaction between treatment effect and subject effect. This model can be viewed as a special case of the mixed model with one observation per cell. The mixed model was introduced in the previous section.

The estimates of the parameters are given by $\hat{\mu} = \bar{y}_{..}$, $\hat{\beta}_i = (\bar{y}_{i.} - \bar{y}_{..})$ and $\hat{\alpha}_j = (\bar{y}_{.j} - \bar{y}_{..})$ and the sums of squares are given by

$$\text{SST} = \sum_{i=1}^{b}\sum_{j=1}^{g}(y_{ij} - \bar{y}_{..})^2, \quad \text{SSA} = \sum_{j=1}^{g}\hat{\alpha}_j^2, \quad \text{SSB} = \sum_{i=1}^{b}\hat{\beta}_i^2 \quad \text{and}$$

$$\text{SSE} = [\text{SST} - \text{SSA} - \text{SSB}] = \sum_{i=1}^{b}\sum_{j=1}^{g}(y_{ij} - y_{i.} - \bar{y}_{.j} + \bar{y}_{..})^2.$$

Under the assumption of no interaction the expected mean squares are given by

$$E[\text{SSA}/(g-1)] = \sigma^2 + b\sum_{j=1}^{g}\alpha_j^2/(g-1)$$

$$E[\text{SSB}/(b-1)] = \sigma^2 + g\sigma_\beta^2$$

$$E[\text{SSE}/(b-1)(g-1)] = \sigma^2.$$

The analysis of variance is summarized in Table 5.53. Even though one of the effects is random the analysis of variance is identical to the randomized block design discussed in Section 5.2.

The Correlation Structure

In the discussion of the random effects model and mixed model, a correlation occurred between observations containing the same random factor. Thus for the mixed model assumption used above for the single factor repeated measures design, the $(g \times g)$ covariance matrix among the observations is given by

$$\boldsymbol{\Sigma} = (\sigma_{jk}) \quad \text{where } \sigma_{jk} = \sigma_\beta^2 \qquad j \neq k; \ j, k = 1, 2, \ldots, g$$

$$= \sigma_\beta^2 + \sigma^2 \qquad j = k; \ j, k = 1, 2, \ldots, g.$$

This covariance structure is said to have the *equi-correlation structure* in that

$$\boldsymbol{\Sigma} = \begin{bmatrix} \sigma_y^2 & \rho\sigma_y^2 & \cdots & \rho\sigma_y^2 \\ \rho\sigma_y^2 & \sigma_y^2 & & \vdots \\ \vdots & & \ddots & \vdots \\ \rho\sigma_y^2 & \cdots & \rho\sigma_y^2 & \sigma_y^2 \end{bmatrix},$$

where $\sigma_y^2 = \sigma_\beta^2 + \sigma^2$ and $\rho = \sigma_\beta^2/(\sigma_\beta^2 + \sigma^2)$. This covariance structure is also referred to as the compound symmetry structure. If this structure holds then the F-test for treatments given by MSA/MSE is valid as was the case for the mixed model given in Section 5.7.

In repeated measures experiments, the same subject receives all possible treatments and hence the assumption of independence for errors associated with the same subject for different treatment may not be valid. In other words $\text{Cov}(\varepsilon_{ij}, \varepsilon_{ik})$ may not be zero. If this is the case then the compound symmetry assumption is no longer valid. For this reason it is common in repeated measures experiments to carry out a preliminary test for the covariance structure to determine if the F-test procedure for treatments is valid. As outlined below there is a more general covariance structure than compound symmetry that still results in the conventional F-test for treatments.

Compound Symmetry and Type H Covariance Matrices

The equi-correlation equal variance matrix given above is said to have *compound symmetry* because it has equal diagonal elements and equal off-diagonal elements. While this type of covariance structure is sufficient to assure that the F-test for treatments is valid it is not necessary.

A weaker condition that is necessary and sufficient is given by

$$V(y_{ij} - y_{ik}) = \sigma_{jj} + \sigma_{kk} - 2\sigma_{jk} = \text{a constant}$$
$$j, k = 1, 2, \ldots, g.$$

This condition is referred to as the Huynh and Feldt (1970) condition and is equivalent to the condition

$$\text{Cov}(y_{ij}, y_{ik}) = \lambda_j + \lambda_k + \alpha \quad j = k, \ j, k = 1, 2, \ldots, g;$$

$$= \lambda_j + \lambda_k \quad j \neq k, \ j, k = 1, 2, \ldots, g.$$

A covariance matrix that satisfies the *Huynh-Feldt condition* is often called a *type H covariance matrix*. The type H covariance matrix can also be written as

$$\Sigma = \alpha \mathbf{I} + \lambda \mathbf{i}' + \mathbf{i} \lambda'$$

where $\lambda(g \times 1)$, $\mathbf{i}(g \times 1)$ a vector of unities, $\mathbf{I}(g \times g)$ identity matrix.

The H covariance structure is equivalent to the condition that $\in = 1$ in the expression

$$\in = \frac{g^2(\bar{\sigma}_d - \bar{\sigma}_{..})^2}{(g-1)\left[\sum_{j=1}^{g}\sum_{k=1}^{g}\sigma_{jk}^2 - 2g\sum_{k=1}^{g}\bar{\sigma}_{.k}^2 + g^2\bar{\sigma}_{..}^2\right]},$$

where $\sigma_{jk} = \text{Cov}(y_{ij}, y_{ik})$, $\bar{\sigma}_d$ is the mean of the variances $\sum_{j=1}^{g}\sigma_{jj}/g$, $\bar{\sigma}_{.k}$ is the mean of the σ_{jk} for a given k, $\bar{\sigma}_{.k} = \sum_{j=1}^{g}\sigma_{jk}/g$, and $\bar{\sigma}_{..}$ is the mean of all σ_{jk}, $\bar{\sigma}_{..} = \sum_{j=1}^{g}\sum_{k=1}^{g}\sigma_{jk}/p^2$. This constant lies in the range $\frac{1}{(g-1)} < \in \leq 1$.

Some Modified Test Procedures

As outlined in Huynh and Feldt (1976) a sample estimate $\hat{\in}$ of $\in$ is computed using the sample variances and covariances. The degrees of freedom for the F-test for treatments are revised from $(g-1)$ and $(g-1)(b-1)$ to $\hat{\in}(g-1)$ and $\hat{\in}(g-1)(b-1)$ respectively. A more conservative procedure suggested by Geisser and Greenhouse (1958) uses the lower bound for $\in$ which is $1/(g-1)$ and revises the degrees of freedom for F to 1 and $(b-1)$ respectively.

Test for Compound Symmetry

A test statistic for the null hypothesis of compound symmetry is given by

$$-\left[(b-1) - \frac{g(g+1)^2(2g-3)}{6(g-1)(g^2+g-4)}\right]\ln L,$$

which has a χ^2 distribution with $[(1/2)g(g+1) - 2]$ degrees of freedom in large samples if the hypothesis of compound symmetry is true. L is the multivariate normal likelihood ratio statistic

$$L = |\mathbf{S}|/(\bar{s}^2)^g(1-\bar{r})^{g-1}(1 + (g-1)\bar{r}),$$

where $\bar{s}^2 = \sum_{j=1}^{g} s_{jj}/g$, $\bar{r} = \dfrac{1}{g(g-1)} \sum_{\substack{j=1 \\ j \neq k}}^{g} \sum_{k=1}^{g} s_{jk}/\bar{s}^2$, s_{jk} is the sample co-

variance between the y_{ij} and y_{ik} and $\mathbf{S}$ is the covariance matrix of these values; $i = 1, 2, \ldots, b$.

Test for Huynh-Feldt Pattern (Type H)

A test statistic for the type H covariance matrix is given by

$$-\left[(b-1) - \frac{2g^2 - 3g + 3}{6(g-1)}\right] \ln W,$$

where $W = (g-1)^{g-1}|\mathbf{CSC}'|/(tr\mathbf{CSC}')^{g-1}$ and where $\mathbf{C}((g-1) \times g)$ is a matrix whose rows are mutually orthogonal and also orthogonal to a row of unities. The elements in each row of $\mathbf{C}$ must also sum to one. The rows of $\mathbf{C}$ have the same form as the contrast coefficients used for orthogonal coding discussed in Section 5.2. In large samples the above test statistic has a χ^2 distribution with $[(1/2)g(g-1) - 1]$ degrees of freedom if the type H pattern is the true pattern.

Repeated measures designs are discussed more extensively in Kirk (1982), Edwards (1985), Milliken and Johnson (1984) and Winer (1971). Multivariate procedures for repeated measurements is discussed in Chapters 7 and 8 of Volume II of this text.

Example

A sample of 50 police officers responded to eight questions regarding the stress they felt in various work situations. The stress level was measured on a five point scale. The data is summarized in Table 5.54. The mean vector, covariance matrix and correlation matrix are summarized in Table 5.55. The variances appear in the main diagonal, the covariances appear to the right of the diagonal and the correlations to the left. The test for compound symmetry showed a χ^2 value of 35.6982 with a p-value of 0.3885 when compared to a χ^2 distribution with 34 degrees of freedom. The χ^2 test statistic for the type H covariance structure showed a value of 31.4994 and for 27 degrees of freedom the p-value is 0.2954. The use of the conventional F-test statistic to compare the eight stress means is therefore justified in this case. The SSA and SSE are given by 138.32 and 328.06 respectively with 7 and 343 degrees of freedom. The F-statistic value of 20.66 has a p-value of 0.000. We can conclude therefore that there are differences in the level of stress experienced in the eight different situations.

Table 5.54. Police Officer Stress Data

X_1	X_2	X_3	X_4	X_5	X_6	X_7	X_8	X_1	X_2	X_3	X_4	X_5	X_6	X_7	X_8
2	1	3	3	4	3	3	4	4	3	3	4	4	3	5	4
4	2	3	4	3	5	5	5	3	2	2	4	4	3	1	2
1	1	2	4	5	4	4	1	3	3	4	4	4	4	5	3
1	3	3	4	4	5	3	2	3	5	3	3	2	2	3	2
5	4	1	5	5	5	4	4	3	4	3	4	5	5	5	4
2	1	5	5	5	5	4	5	1	1	2	3	4	3	5	2
2	3	1	3	3	4	3	3	1	1	1	3	3	3	2	1
1	2	1	4	5	5	4	3	4	5	1	4	5	4	5	2
1	1	1	2	4	4	4	1	3	2	3	1	3	3	4	3
2	2	2	3	3	2	3	2	1	3	2	4	1	3	4	4
1	2	2	2	3	2	2	2	4	1	4	4	3	4	3	3
1	3	1	4	4	5	3	1	3	3	5	4	5	5	5	4
1	2	3	1	2	4	2	1	1	5	4	2	2	4	3	3
1	2	3	2	3	3	1	1	1	1	1	2	1	3	5	1
3	2	2	4	3	4	3	5	3	5	1	4	3	4	4	2
2	1	3	3	3	4	4	1	3	1	1	3	1	4	3	3
4	3	4	5	4	4	4	3	2	3	1	5	5	1	3	4
3	2	2	4	5	4	4	3	4	4	1	4	4	3	4	4
1	2	3	3	3	3	4	3	1	1	2	3	3	4	4	2
2	2	2	2	3	2	2	2	2	2	3	4	3	4	4	2
2	3	3	4	4	3	3	2	3	5	3	4	4	4	4	3
1	2	3	4	2	1	3	3	1	3	2	4	2	2	3	3
3	1	1	5	4	4	3	2	1	1	4	5	3	5	4	3
1	1	2	4	4	4	3	3	1	1	2	3	1	3	5	1
2	1	1	4	2	3	1	1	1	3	3	4	4	4	4	3

Variable Descriptions

X_1 Handling an investigation where there is serious injury or fatality.
X_2 Dealing with obnoxious or intoxicated people.
X_3 Tolerating verbal abuse in public.
X_4 Being unable to solve a continuing series of serious offenses.
X_5 Resources such as doctors, ambulances etc not being available when needed.
X_6 Poor presentation of a case by the prosecutor leading to dismissal of charge.
X_7 Unit members not getting along with unit commander.
X_8 Investigating domestic quarrels.

5.8.2 A Two-Way Split Plot Design

A *two-way split plot design* is a two-factor design in which the levels of a second factor are nested within the levels of a first factor. All the levels of the second factor must appear with each level of the first factor. The levels of the first factor consist of large experimental units which are usually called whole plots. This first factor is called the whole plot factor. Each of the whole plots or blocks is divided into subplots such that there is one subplot for each level of the second factor. The second factor which is called the subplot factor is nested within the whole plot factor. The whole plot factor levels are randomly assigned to the whole plots and the levels of

Table 5.55. Stress Data – Mean Vector, Covariance and Correlation Matrices (covariance matrix above diagonal, correlations below)

	Mean Vector	X_1	X_2	X_3	X_4	X_5	X_6	X_7	X_8
				Covariance and Correlation Matrices					
X_1	2.120	1.332	0.593	0.147	0.424	0.423	0.327	0.304	0.618
X_2	2.320	0.405	1.610	-0.098	0.191	0.284	-0.061	0.238	0.349
X_3	2.300	0.115	-0.069	1.235	0.161	0.190	0.316	0.249	0.402
X_4	3.540	0.369	0.151	0.146	0.988	0.546	0.418	0.244	0.516
X_5	3.380	0.316	0.193	0.147	0.474	1.342	0.582	0.288	0.372
X_6	3.500	0.255	-0.043	0.256	0.379	0.452	1.235	0.408	0.214
X_7	3.520	0.245	0.175	0.209	0.229	0.232	0.342	1.153	0.406
X_8	2.620	0.462	0.237	0.312	0.448	0.277	0.166	0.326	1.342

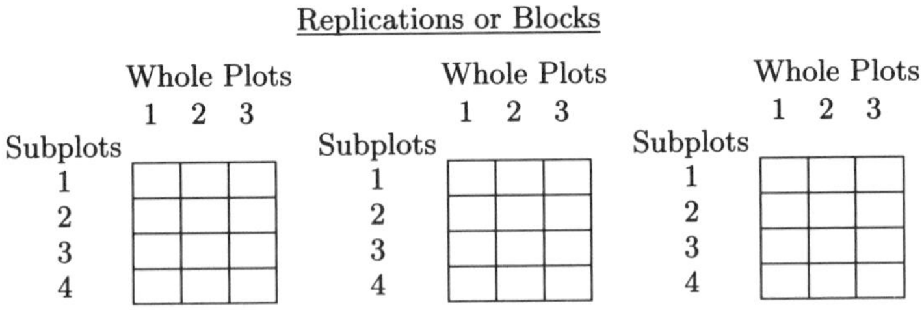

Figure 5.14. Example of a Split Plot Design

the subplot factor are randomly assigned to the subplots within each whole plot. The subplot factor is usually the factor of primary interest.

The terminology for the split plot design comes from agriculture where the whole plots and subplots represent plots of land used for crops. The subplot treatments may represent different types of fertilizer while the whole plots may represent different types of soil preparation by ploughing. The ploughing phase requires large plots of land while the fertilizer can be administered easily on smaller plots. Figure 5.14 illustrates a split plot design with 3 whole plots, 4 subplots and 3 replications. The 3 replications could represent 3 different blocks.

The split plot design is useful in a two factor experiment where the administration of one factor requires relatively large experimental units while the second factor requires only small experimental units. After the administration of the first factor to the large experimental units, these units are then subdivided before administration of the second factor. An example would be the comparison of several procedures for producing a finished product from a certain raw material. The raw material is produced in large batches and is then divided into small quantities for the production of the finished product. The raw material is produced in several different large vats with perhaps different conditions, properties or production methods. The primary interest, however, is the technique used to produce the finished product.

A second example would be the comparison of several different types of written examinations for students who have been taught a certain curriculum. The teaching is carried out in a classroom setting by a few teachers so that each student is only exposed to one teacher. Within each teacher's class however, all types of examinations can be tested by randomly assigning the different exams to the students within each class.

Model for the Split Plot Design

The observation taken from the h-th replication or block of the j-th subplot within the i-th whole plot is denoted by y_{ijh}, $i = 1, 2, \ldots, b$; $j = 1, 2, \ldots, g$; $h = 1, 2, \ldots, c$. For the j-th subplot within the i-th whole plot the mean response μ_{ij} is given by

$$\mu_{ij} = \mu + \beta_i + \alpha_j + (\alpha\beta)_{ij}$$

where $\sum_{i=1}^{b} \beta_i = 0$, $\sum_{j=1}^{g} \alpha_j = 0$, $\sum_{i=1}^{b} (\alpha\beta)_{ij} = \sum_{j=1}^{g} (\alpha\beta)_{ij} = 0$. The whole plot treatment effects are given by β_i while the subplot treatment effects are given by α_j. The interaction effects are given by $(\alpha\beta)_{ij}$. These fixed treatment effects can also be expressed as

$$\beta_i = (\mu_i. - \mu), \quad \alpha_j = (\mu_{.j} - \mu) \quad \text{and} \quad (\alpha\beta)_{ij} = (\mu_{ij} - \mu_i. - \mu_{.j} + \mu).$$

The model for y_{ijh} is given by

$$y_{ijh} = \mu + \beta_i + \alpha_j + (\alpha\beta)_{ij} + \gamma_h + \eta_{ih} + \varepsilon_{ijh}$$

where $\gamma_h \sim N(0, \sigma_\gamma^2)$, $\eta_{ih} \sim N(0, \sigma_\eta^2)$ and $\varepsilon_{ijh} \sim N(0, \sigma^2)$. The model contains a replication effect which is random, as well as an interaction effect between replications and whole plots, which is also random. The random error term is unique to each subplot. The three random components are all mutually independent.

The estimates of the treatment effects are given by $\hat{\beta}_i = (\bar{y}_{i..} - \bar{y}_{...})$, $\hat{\alpha}_j = (\bar{y}_{.j.} - \bar{y}_{...})$ and $\widehat{(\alpha\beta)}_{ij} = (\bar{y}_{ij.} - \bar{y}_{i..} - \bar{y}_{.j.} + \bar{y}_{...})$. The corresponding sums of squares are given by $\text{SSB} = \sum_{i=1}^{b} \hat{\beta}_i^2$, $\text{SSA} = \sum_{j=1}^{g} \hat{\alpha}_j^2$ and $\text{SSAB} = \sum_{i=1}^{b} \sum_{j=1}^{g} \widehat{(\alpha\beta)}_{ij}^2$.

The remaining sums of squares are given by

$$SST = \sum_{i=1}^{b}\sum_{j=1}^{g}\sum_{h=1}^{c}(y_{ijh} - \bar{y}...)^2 \text{ the total sum of squares,}$$

$$SSR = \sum_{h=1}^{c}(\bar{y}..h - \bar{y}...)^2 \text{ the sum of squares for replications,}$$

$$SSEW = \sum_{i=1}^{b}\sum_{h=1}^{c}(\bar{y}_{i\cdot h} - \bar{y}...)^2 - SSR - SSB = \text{whole plot error sum of squares}$$

and

$$SSES = SST - SSA - SSB - SSR - SSAB - SSEW$$
$$= \text{subplot error sum of squares.}$$

The expected mean squares are given by

$$E[SSR/(c - 1)] = \sigma^2 + b\sigma_\eta^2 + bg\sigma_\gamma^2$$

$$E[SSB/(b - 1)] = \sigma^2 + g\sigma_\eta^2 + gc\sum_{i=1}^{b}\beta_i^2/(b - 1)$$

$$E[SSEW/(c - 1)(b - 1)] = \sigma^2 + g\sigma_\eta^2$$

$$E[SSA/(g - 1)] = \sigma^2 + bc\sum_{j=1}^{g}\alpha_j^2/(g - 1)$$

$$E[SSAB/(b - 1)(g - 1)] = \sigma^2 + c\sum_{i=1}^{b}\sum_{j=1}^{g}(\alpha\beta)_{ij}^2/(b - 1)(g - 1)$$

$$E[SSES/b(g - 1)(c - 1)] = \sigma^2.$$

The analysis of variance table is given by Table 5.56. The F-statistics shown are the ones normally used to test for the two treatment effects and the interaction. Variations of this design are given by Anderson and McLean (1974), eg. an additional term can be added to reflect interaction between blocks and subplots. Further discussion of the split plot design is available in Milliken and Johnson (1984).

Table 5.56.

Source	d.f.	Sum of Squares	Mean Squares	F
Reps or Blocks	$(c-1)$	SSR	$\text{MSR}=\text{SSR}/(c-1)$	
Whole Plots	$(b-1)$	SSB	$\text{MSB}=\text{SSB}/(b-1)$	MSB/MSEW
Error (Whole Plots)	$(c-1)(b-1)$	SSEW	$\text{MSEW}=\text{SSEW}/(c-1)(b-1)$	
Subplots	$(g-1)$	SSA	$\text{MSA}=\text{SSA}/(g-1)$	MSA/MSES
Interaction	$(g-1)(b-1)$	SSAB	$\text{MSAB}=\text{SSAB}/(g-1)(b-1)$	MSAB/MSES
Error (Subplots)	$b(c-1)(g-1)$	SSES	$\text{MSES}=\text{SSES}/b(g-1)(c-1)$	

Table 5.57.

Source	d.f.	Sum of Squares	F-value	p-value
REGIONS	4	4867.51		
ADS	3	4585.68	2.05	0.160
REGION*AD	12	8937.92		
SIZE	5	40967.65	654.47	0.000
SIZE*AD	15	412.82	2.20	0.013
ERROR	80	1001.55		

Example

The family expenditure data summarized in Table 5.1 at the beginning of this chapter will be used to provide an example for the split plot design. For the purposes of this example it is assumed that each of the five regions represents a block or replication. Within each region there are four whole plot areas one for each of the four ads. Within each ad area or whole plot there are six family sizes. Thus the whole plot factor is the ad and the subplot factor is the family size. The analysis of variance for this split plot analysis is given in Table 5.57. From the table we conclude that the AD effects are not significant, while the SIZE effects are highly significant. The interaction AD*SIZE is significant at the 0.013 level.

From Table 5.57 it can be seen that the mean square for ERROR (REGION*AD) is relatively large when compared to the mean square for ERROR. As a result the mean square for ADs when compared to the REGION*AD mean square yielded a relatively small ratio.

5.8.3 A Repeated Measures Split Plot Design

In a split plot design the whole plots are large experimental units which are then subdivided to obtain smaller experimental units called subplots. In the repeated measures design the subjects are sampled repeatedly. These two concepts are combined in the repeated measures split plot design. The whole plot consists of a group of subjects. Each whole plot or group of subjects is exposed to one level of a treatment factor. Following this each subject is then sequentially treated by all levels of a second factor.

A model for this design is given by

$$y_{ijh} = \mu + \alpha_j + \beta_i + (\alpha\beta)_{ij} + \gamma_{ih} + \varepsilon_{ijh}$$
$$i = 1, 2, \ldots, b; \ j = 1, 2, \ldots, g; \ h = 1, 2, \ldots, c;$$

where $E[y_{ijh}] = \mu_{ijh}$, $\beta_i = (\mu_{i\cdot\cdot} - \mu)$, $\alpha_j = (\mu_{\cdot j\cdot} - \mu)$, $\gamma_{ih} = (\mu_{i\cdot h} - \mu_{i\cdot\cdot})$, $(\alpha\beta)_{ij} = (\mu_{ij\cdot} - \mu_{\cdot j\cdot} - \mu_{i\cdot\cdot} + \mu)$. The observation y_{ijh} denotes one observation made on the outcome of treatment j on the h-th subject in whole plot i. It is assumed that the two treatment factors, whole plot and subplot are both fixed but that the subject effects are random. As a result the assumptions for the model are given by

$$\varepsilon_{ijh} \sim N(0, \sigma^2), \quad \sum_{j=1}^{g} \alpha_j = 0, \quad \sum_{i=1}^{b} \beta_i = 0, \quad \sum_{i=1}^{b} (\alpha\beta)_{ij} = 0,$$

$$\sum_{j=1}^{g} (\alpha\beta)_{ij} = 0 \quad \text{and} \quad \gamma_h \sim N(0, \sigma_\gamma^2).$$

The model is therefore a mixed model with two fixed effects and one random effect.

The sums of squares for the various effects are given by

$$\mathrm{SST} = \sum_{i=1}^{b} \sum_{j=1}^{g} \sum_{h=1}^{c} (y_{ijh} - \bar{y}_{\ldots})^2 = \text{ total sum of squares,}$$

$$\mathrm{SSE} = \sum_{i=1}^{b} \sum_{j=1}^{g} \sum_{h=1}^{c} (y_{ijh} - \bar{y}_{i\cdot h} - \bar{y}_{ij\cdot} + \bar{y}_{i\cdot\cdot})^2 = \text{ error sum of squares,}$$

$$\mathrm{SSB} = cg \sum_{i=1}^{b} (\bar{y}_{i\cdot\cdot} - \bar{y}_{\ldots})^2 = \text{ first treatment or whole plot sum of squares,}$$

$$\mathrm{SSA} = cb \sum_{j=1}^{g} (\bar{y}_{\cdot j\cdot} - \bar{y}_{\ldots})^2 = \text{ second treatment or subplot sum of squares,}$$

Table 5.58. Analysis of Variance for Repeated Measures Split Plot Design

Source	d.f.	Sum of Squares	Mean Square	F
Whole Plot	$(b-1)$	SSB	MSB=SSB/$(b-1)$	MSB/MSC
Subjects	$b(c-1)$	SSC	MSC=SSC/$b(c-1)$	MSC/MSE
Subplot	$(g-1)$	SSA	MSA=SSA/$(g-1)$	MSA/MSE
Treatment Interaction	$(b-1)(g-1)$	SSAB	MSAB=SSAB/$(b-1)(g-1)$	MSAB/MSE
Error	$b(c-1)(g-1)$	SSE	MSE=SSE/$b(c-1)(g-1)$	
Total	$bcg-1$	SST		

$$\text{SSAB} = c\sum_{i=1}^{b}\sum_{j=1}^{g}(\bar{y}_{ij\cdot} - \bar{y}_{i\cdot\cdot} - \bar{y}_{\cdot j\cdot} + \bar{y}_{\cdots})^2 = \begin{array}{l}\text{treatment interaction}\\ \text{sum of squares,}\end{array}$$

$$\text{SSC} = g\sum_{i=1}^{b}\sum_{j=1}^{c}(\bar{y}_{i\cdot h} - \bar{y}_{i\cdot\cdot})^2 = \text{ subjects sum of squares.}$$

The analysis of variance table is given in Table 5.58.

The expected mean squares for this analysis of variance are given by

$$E[\text{MSE}] = \sigma^2$$

$$E[\text{MSB}] = \sigma^2 + g\sigma_\gamma^2 + cg\sum_{i=1}^{b}\beta_i^2/(b-1)$$

$$E[\text{MSA}] = \sigma^2 + cb\sum_{j=1}^{g}\alpha_j^2/(g-1)$$

$$E[\text{MSAB}] = \sigma^2 + c\sum_{i=1}^{b}\sum_{j=1}^{g}(\alpha\beta)_{ij}^2/(b-1)(g-1)$$

$$E[\text{MSC}] = \sigma^2 + g\sigma_\gamma^2$$

These expected mean squares can be used to justify the F-ratios shown in Table 5.58.

As in the case of the single factor repeated measures design, some attention should be paid to the validity of the necessary assumption of uncorrelated error terms ε_{ijh}. Since the same subject is being used for all treatment levels for the second factor (subplot) the compound symmetry or type H covariance matrix assumption may not be valid.

Further discussion of this design can be obtained from Kirk (1982), Edwards (1985) and Milliken and Johnson (1984).

Example

A sample of 25 subjects was selected from four different regions. Each region receives police services from a different administrative unit of the R.C.M.P. Each subject was asked to respond to six items regarding how safe the individual felt in the community. The individuals were asked to respond using one of six responses from 1 = very safe to 6 = very unsafe. The six items are:

X1: How safe do you feel in your town as a whole?
X2: How safe do you feel in your home?
X3: How safe do you feel walking alone in your neighborhood during the day?
X4: How safe do you feel walking in your neighborhood at night?
X5: How safe do you feel in downtown during the day?
X6: How safe do you feel in downtown at night?

The data and items are displayed in Table 5.59. The analysis of variance results for this data are shown in Table 5.60.

From the analysis of variance results we can conclude that the regions differ with respect to their opinions regarding safety in the community. In addition there is variation among the means on the six safety items as well as a significant interaction. A summary of the item means by region is provided in Table 5.61. Figure 5.15 shows a plot of the 24 means. From both the table and the figure it can be seen that the means in regions 1 and 3 are higher than the means in regions 2 and 4. In addition the means on items 4 and 6 tend to be higher while the means on items 2, 3 and 5 tend to be lower. The interaction is derived mainly from the relative differences between the means on items 4 and 6 and the remaining items. In regions 1 and 3 the means on 4 and 6 are much higher relative to the remaining items while in regions 2 and 4 the differences between items 4 and 6 and the remaining items is relatively small.

The chi-square test statistic for a type H covariance matrix in this case was 81.44 with 14 degrees of freedom yielding a p-value of 0.000. In this case we are not justified in assuming that the covariance structure permits the use of univariate analysis of variance procedures. Multivariate procedures which take into account the correlations among the six items are therefore required. These procedures will be outlined in the discussion of multivariate analysis of variance in Chapter 8 of Volume II.

Table 5.59. Observations from Public Safety Questionnaire

Region	X1	X2	X3	X4	X5	X6	Region	X1	X2	X3	X4	X5	X6
1	3	1	1	2	1	6	3	4	2	1	2	3	5
1	2	2	1	2	1	3	3	1	1	1	1	1	1
1	2	1	1	1	1	2	3	2	1	1	2	2	4
1	1	1	1	1	1	1	3	3	2	3	4	5	5
1	2	2	1	4	2	5	3	4	2	1	2	1	4
1	2	2	2	2	2	4	3	1	1	1	1	1	2
1	5	3	2	6	2	6	3	2	1	2	2	1	2
1	2	2	2	3	2	4	3	3	2	1	4	1	6
1	3	1	2	2	2	3	3	3	3	2	4	2	4
1	5	2	2	3	2	4	3	2	2	2	4	2	6
1	3	2	2	5	2	5	3	2	2	2	4	3	6
1	3	4	2	5	2	5	3	2	1	1	2	1	3
1	2	2	2	5	3	5	3	2	3	4	6	5	6
1	4	4	4	5	2	6	3	1	1	1	2	1	2
1	2	2	2	4	2	6	3	1	1	1	1	1	2
1	5	2	2	6	2	6	3	2	1	1	2	2	4
1	4	2	2	4	2	5	3	2	1	1	2	2	4
1	2	2	2	3	3	4	3	4	2	1	3	2	5
1	4	1	3	4	3	4	3	2	2	2	3	2	4
1	4	4	3	6	3	6	3	5	2	1	3	1	6
1	5	6	2	4	4	5	3	4	2	3	2	2	4
1	2	3	2	4	2	5	3	5	2	3	5	2	5
1	3	1	2	5	2	5	3	3	3	3	4	3	4
1	2	2	2	4	3	4	3	4	3	3	4	3	4
2	2	2	2	2	1	1	4	1	1	1	1	1	1
2	2	1	1	1	1	1	4	1	1	1	1	1	1
2	2	2	2	2	2	2	4	2	2	1	2	1	2
2	2	1	2	3	2	3	4	2	2	1	2	1	2
2	1	1	1	2	1	2	4	2	2	1	2	1	3
2	2	1	1	4	1	4	4	1	1	1	1	1	1
2	2	2	2	2	2	3	4	1	1	1	1	1	1
2	1	1	1	1	1	2	4	1	1	1	1	1	1
2	1	1	1	1	1	1	4	1	1	1	1	1	1
2	1	1	1	2	1	2	4	1	1	1	1	1	1
2	1	1	1	1	1	1	4	1	1	1	1	1	1
2	2	1	1	2	1	2	4	2	1	1	1	1	2
2	2	2	1	2	1	2	4	1	1	1	1	1	1
2	1	2	1	2	1	3	4	2	1	1	1	1	2
2	1	1	1	1	1	1	4	3	2	2	4	1	3
2	1	1	1	1	1	1	4	1	1	1	1	1	1
2	2	2	1	2	1	2	4	3	1	3	3	1	4
2	1	2	1	2	1	1	4	2	1	1	2	2	3
2	2	2	1	2	1	2	4	1	1	1	1	1	1
2	1	1	1	1	1	1	4	1	1	1	1	1	1
2	2	2	1	1	1	2	4	1	1	1	1	1	1
2	2	1	1	1	1	2	4	1	1	1	1	1	2
2	1	1	1	1	1	1	4	2	1	1	2	1	2
2	2	2	1	2	1	2	4	2	2	1	4	1	2

Table 5.60. Analysis of Variance Results for Police Service Survey

Source	d.f.	Sum of Squares	Mean Square	F	p-value
Region	3	299.340	99.78	31.39	0.000
Subjects	96	305.133	3.18	7.14	0.000
Items	5	180.260	36.05	80.96	0.000
Items*Region	15	68.660	4.58	10.28	0.000
Error	480	213.747	0.445		
Total	599	1067.140			

Table 5.61. Summary of Item Means by Region

Region	1	2	3	4	5	6
1	3.04	2.24	1.96	3.76	2.12	4.52
2	1.52	1.40	1.16	1.72	1.12	1.80
3	2.80	1.92	1.80	3.00	2.12	4.16
4	1.48	1.20	1.12	1.52	1.04	1.64

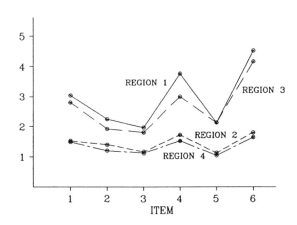

Figure 5.15. Comparison of Safety Item Means by Region

Cited Literature for Chapter 5

1. Anderson, V.L. and McLean, R.A. (1974). *Design of Experiments: A Realistic Approach.* New York: Marcel Dekker, Inc.

2. Brown, M.B. and Forsyth, A.B. (1974). "Robust Tests for the Equality of Variances," *Journal of the American Statistical Association* 69, 364–367.

3. Cochran, W.G. and Cox, G.M. (1957). *Experimental Design.* New York: John Wiley and Sons, Inc.

4. Conover, W.J. (1980). *Practical Nonparametric Statistics*, Second Edition. New York: John Wiley and Sons, Inc.

5. Edwards, A.L. (1985). *Experimental Design in Psychological Research*, Fifth Edition. New York: Harper and Row, Publishers.

6. Einot, I. and Gabriel, K.R. (1975). "A Study of the Powers of Several Methods of Multiple Comparisons," *Journal of American Statistical Association* 70, 351.

7. Gabriel, K.R. (1978). "A Simple Method of Multiple Comparisons of Means," *Journal of American Statistical Association,* 73, 364.

8. Geisser, S. and Greenhouse, S.W. (1958). "An Extension of Box's Results on the Use of the F-Distribution in Multivariate Analysis," *Annals of Mathematical Statistics* 29, 885–891.

9. Hochberg, Y. (1974). "Some Conservative Generalizations of the T-Method in Simultaneous Inference," *Journal of Multivariate Analysis* 4, 224–234.

10. Huynh, H. and Feldt, L.S. (1970), "Conditions under which Mean Square Ratios in Repeated Measurements Designs Have Exact F-Distributions," *Journal of the American Statistical Association* 65, 1582-1589.

11. Huynh, H. and Feldt, L.S. (1976). "Estimation of the Box Correction For Degrees of Freedom from Sample Data in Randomized Block and Split-Plot Designs," *Journal of Educational Statistics* 1, 69–82.

12. Iman, R.I. and Davenport, J.M. (1980). "Approximations of the Critical Region of the Friedman Statistics," *Communications in Statistics* A9, 571–595.

13. Kempthorne, O. (1952). *The Design and Analysis of Experiments.* New York: John Wiley and Sons, Inc.

14. Kirk, R.E. (1982). *Experimental Design: Procedures for the Behavioral Sciences*, Second Edition. Belmont, California: Brooks/Cole Publishing Co.

15. Levene, H. (1960). "Robust Tests for Equality of Variances" in I. Olkin, editor, *Contributions to Probability and Statistics*. Palo Alto, California: Stanford University Press, 278–292.

16. Miller, R.G. (1968). "Jackknifing Variances," *Annals of Mathematical Statistics* 39, 567–582.

17. Milliken, G.A. and Johnson, D.E. (1984). *The Analysis of Messy Data Volume I: Designed Experiments*. New York: Van Nostrand Reinhold Co.

18. Montgomery, D.C. (1984). *Design and Analysis of Experiments*, Second Edition. New York: John Wiley and Sons, Inc.

19. Neter, J., Wasserman, W. and Kutner, M.H. (1985). *Applied Linear Statistical Models*, Second Edition. Homewood, Illinois: Irwin.

20. Pearson, E.S. and Hartley, H.P. (1970). *Biometrika Tables for Statisticians*, Volumes I and II. Reprinted with corrections 1976. Cambridge: Cambridge Univ. Press.

21. Petersen, R. (1985). *Design and Analysis of Experiments*. New York: Marcel Dekker, Inc.

22. Ryan, T.A. (1959). "Multiple Comparisons in Psychological Research," *Psychological Bulletin* 56, 26–47.

23. Ryan, T.A. (1960). "Significance Tests for Multiple Comparisons of Proportions, Variances and Other Statistics," *Psychological Bulletin* 57, 318–328.

24. Scheffé, H. (1959). *The Analysis of Variance*, New York: John Wiley and Sons, Inc.

25. Searle, S.R. (1971). *Linear Models*. New York: John Wiley and Sons, Inc.

26. Searle, S.R. (1987). *Linear Models for Unbalanced Data*. New York: John Wiley and Sons, Inc.

27. Sidak, Z. (1967). "Rectangular Confidence Regions for the Means of Multivariate Normal Distributions," *Journal of the American Statistical Association*, 62, 626–633.

28. Spjotvoll, E. and Stoline, M.R. (1973). "An Extension of the T-method of Multiple Comparison to Include the Cases with Unequal Sample Sizes," *Journal of the American Statistical Association* 68, 975–978.

29. Stoline, M.R. and Ury, H.K. (1979). "Tables of the Studentized Maximum Modulus Distribution and an Application to Multiple Comparisons Among Means," *Technometrics* 21, 87–94.

30. Tukey, J.W. (1953). "The Problem of Multiple Comparisons," unpublished manuscript.

31. Welsch, R.E. (1977). "Stepwise Multiple Comparison Procedures," *Journal of American Statistical Association* 72, 359.

32. Winer, B.J. (1971). *Statistical Principles in Experimental Design*, Second Edition. New York: McGraw–Hill Book Company.

Exercises for Chapter 5

1. The following exercises are based on the data in Table D6 of the Data Appendix.

   (a) Carry out a one-way analysis of variance relating EXPEND separately to each of SIZE, REGION and AD. (Note: treat SIZE as categorical.) Use multiple comparison procedures to compare the means in each case. Discuss the results.

   (b) Carry out a two-way analysis of variance relating EXPEND to the variables REGION, AD and SIZE in pairs. (Note: treat SIZE as categorical.) Include tests for interaction. Discuss the results and compare to the results in (a).

   (c) Carry out an analysis of covariance relating EXPEND to AD controlling for SIZE. (Note: SIZE is to be treated as interval.) Repeat the analysis relating EXPEND to REGION controlling for SIZE. Now relate EXPEND to both AD and REGION simultaneously and control for the variable SIZE. Include an interaction term for AD by REGION. Discuss your results and compare to (a) and (b).

   (d) Define dummy variables for the variables AD and REGION. Relate EXPEND to AD and REGION including interaction using multiple regression. Repeat the process using effect coding. Discuss your results. Compare to the results obtained in (b). Also, add the variable SIZE as a covariate, and compare the multiple regression results to (c).

   (e) Set up a system of orthogonal variables for AD and REGION. Relate EXPEND to these variables including interaction. What hypotheses are you testing with the coefficients of the orthogonal variables? Discuss your results.

   (f) Define an indicator variable for each cell of the cross classification of AD with REGION. Regress EXPEND on these indicator variables without an intercept. Test appropriate hypotheses for main effects and interactions. Discuss your results.

   (g) Carry out a three-way analysis of variance relating EXPEND to the three variables AD, REGION and SIZE.

2. The following exercises are based on the data in Table D3 in the Data Appendix.

   (a) Relate the variables LCURRENT and LSTART to SEX and RACE using analysis of variance. Discuss your results. Use dummy coding and/or effect coding and multiple regression to study the relationships. Discuss your results.

(b) For the analysis in (a) use cell indicator variables and transformation matrices to test various hypotheses.

(c) Relate the variables LCURRENT and LSTART to SEX and JOBCAT. Determine first the number of observations in each cell of the cross classification of SEX with JOBCAT. For the nonempty cells define cell indicator variables and use multiple regression to estimate a relationship for each of LCURRENT and LSTART. Use transformation matrices to test appropriate hypotheses.

(d) Repeat the analysis in (c) for RACE and JOBCAT.

(e) Repeat the analysis in (c) for all three variables RACE, SEX and JOBCAT.

(f) Repeat the analysis in (b) using the variable EDUC as a covariate and discuss your results. Compare the results to (b).

3. The following exercises are based on the data in Table D4 in the Data Appendix.

(a) Determine the number of observations in the cells derived from the cross classification of FOR with CYLIND.

(b) Define dummy variables for the categories of FOR and CYLIND and use multiple regression to relate COMBRATE to these two variables including interaction. Discuss the results. Repeat the analysis using effect coding.

(c) Analyze the data using cell indicator variables and test appropriate hypotheses for main effects and interaction.

(d) Analyze the relationship between COMBRATE and FOR and CYLIND including WEIGHT as a covariate. Discuss your results and compare them to (b).

4. The following exercises are based on the data in Table D7 in the Data Appendix.

(a) Analyze the data as a three-way factorial as in Section 5.4.3. Discuss the results. Provide a discussion of the three-way interactions.

(b) Analyze the data as a 2^3 design with blocks as in Section 5.6.1. Discuss the results.

5. The following exercises are based on the data in Table D8 in the Data Appendix.

(a) Carry out a repeated measures one-way analysis of the eight stress variables as in Section 5.8.1. Discuss the results.

(b) Carry out a test for the type H covariance structure and for compound symmetry as outlined in Section 5.8.1.

6. The following exercises are based on the data in Table D9 in the Data Appendix.

(a) Analyze the data in this table as a repeated measures split plot design as outlined in Section 5.8.3. The four regions are the whole plots while the subplots are the six variables. Discuss the results.

(b) Carry out a test for type H covariance structure.

Questions for Chapter 5

1. In the one-way analysis of variance model $\text{SST} = \sum\limits_{j=1}^{g} \sum\limits_{i=1}^{n_j} (y_{ij} - \bar{y}_{..})^2$,

   $\text{SSA} = \sum\limits_{j=1}^{g} n_j (\bar{y}_{.j} - \bar{y}_{..})^2$ and $\text{SSW} = \sum\limits_{j=1}^{g} \sum\limits_{i=1}^{n_j} (y_{ij} - \bar{y}_{.j})^2$. Show that
   $\text{SST} = \text{SSW} + \text{SSA}$. Hint: In the expression for SST write $(y_{ij} - \bar{y}_{..}) = (y_{ij} - \bar{y}_{.j} + \bar{y}_{.j} - \bar{y}_{..})$ and expand.

2. For the one-way analysis of variance model assume that $y_{ij} \sim N(\mu_j, \sigma^2)$, $i = 1, 2, \ldots, n_j; \ j = 1, 2, \ldots, g$.

   (a) Show that $\bar{y}_{.j} \sim N(\mu_j, \sigma^2/n_j)$.

   (b) Show that $\text{Cov}(\bar{y}_{.j}, \bar{y}_{.\ell}) = 0$ if $j \neq \ell$.

   (c) Let $\boldsymbol{\ell}' \bar{\mathbf{y}}.$ denote a linear combination of the $(g \times 1)$ vector of means $\bar{\mathbf{y}}'. = (\bar{y}_{.1}, \bar{y}_{.2}, \ldots, \bar{y}_{.g})$. Show that $V(\boldsymbol{\ell}' \bar{\mathbf{y}}.) = \sigma^2 \sum\limits_{j=1}^{g} [\ell_j^2/n_j]$ and that
   $E[\boldsymbol{\ell}' \bar{\mathbf{y}}.] = \boldsymbol{\ell}' \boldsymbol{\mu}$ where $\boldsymbol{\mu}' = [\mu_1, \mu_2, \ldots, \mu_g]$.

   (d) Use the fact that $s^2 = \sum\limits_{j=1}^{g} s_j^2(n_j - 1)/(n - g)$ is an unbiased estimator
   of σ^2 with $(n - g)$ degrees of freedom and independent of the vector $\bar{\mathbf{y}}$.
   to show that $(\boldsymbol{\ell}' \bar{\mathbf{y}}. - \boldsymbol{\ell}' \boldsymbol{\mu}) / \sqrt{s^2 \sum\limits_{j=1}^{g} \ell_j^2/n_j}$ has a t distribution with $(n - g)$ degrees of freedom. Hint: The sum of normal random variables is also normal.

3. The one-way analysis of variance fixed effects model is given by
   $$y_{ij} = \mu_j + \varepsilon_{ij} \quad i = 1, 2, \ldots, n_j, \ j = 1, 2, \ldots, g;$$
   where $E[y_{ij}] = \mu_j$, the ε_{ij} are mutually independent and $\varepsilon_{ij} \sim N(0, \sigma^2)$.

   (a) Let $\bar{y}_{.j} = \sum\limits_{i=1}^{n_j} y_{ij}/n_j$ and $\bar{y}_{..} = \sum\limits_{j=1}^{g} \sum\limits_{i=1}^{n_j} y_{ij}/n = \sum\limits_{j=1}^{g} n_j \bar{y}_{.j}/n$ where $n = \sum\limits_{j=1}^{g} n_j$. Show that $E[\bar{y}_{.j}] = \mu_j$, $E[\bar{y}_{..}] = \bar{\mu}$ where $\bar{\mu} = \sum\limits_{j=1}^{g} n_j \mu_j/n$, $V[y_{ij}] = \sigma^2$, $V[\bar{y}_{.j}] = \sigma^2/n_j$, $V[\bar{y}_{..}] = \sigma^2/n$ and hence that $E[y_{ij}^2] = \sigma^2 + \mu_j^2$, $E[\bar{y}_{.j}^2] = (\frac{\sigma^2}{n_j} + \mu_j^2)$, $E[\bar{y}_{..}^2] = (\frac{\sigma^2}{n} + \bar{\mu}^2)$.

   (b) Show that
   $$\text{SST} = \sum\limits_{j=1}^{g} \sum\limits_{i=1}^{n_j} (y_{ij} - \bar{y}_{..})^2 = \sum\limits_{j=1}^{g} \sum\limits_{i=1}^{n_j} y_{ij}^2 - n\bar{y}_{..}^2.$$

$$\text{SSA} = \sum_{j=1}^{g} n_j (\bar{y}_{\cdot j} - \bar{y}_{\cdot\cdot})^2 = \sum_{j=1}^{g} n_j \bar{y}_{\cdot j}^2 - n\bar{y}_{\cdot\cdot}^2$$

$$\text{SSW} = \sum_{j=1}^{g}\sum_{i=1}^{n_j} (y_{ij} - \bar{y}_{\cdot j})^2 = \sum_{j=1}^{g}\sum_{i=1}^{n} y_{ij}^2 - \sum_{j=1}^{g} n_j \bar{y}_{\cdot j}^2.$$

(c) Use the results in (a) and (b) above to show that $E[\text{SST}] = \sigma^2(n - 1) + \sum_{j=1}^{g} n_j(\mu_j - \bar{\mu})^2$, $E[\text{SSA}] = \sigma^2(g-1) + \sum_{j=1}^{g} n_j(\mu_j - \bar{\mu})^2$, $E[\text{SSW}] = \sigma^2(n - g)$, and hence that $E[\text{MSA}] = \sigma^2 + \sum_{j=1}^{g} n_j(\mu_j - \bar{\mu})^2/(g - 1)$, $E[\text{MSW}] = \sigma^2$.

(d) Define $\mu_j = \mu + \alpha_j$, where $\sum_{j=1}^{g} \alpha_j = 0$ and show that $\mu = \sum_{j=1}^{g} \mu_j/g$, $\bar{\mu} = \sum_{j=1}^{g} n_j\mu_j/n = \mu + \sum_{j=1}^{g} n_j\alpha_j/n = \mu + \bar{\alpha}$, where $\bar{\alpha} = \sum_{j=1}^{g} n_j\alpha_j/n$. Use these results to show that $E[\text{MSA}]$ in (c) is given by $E[\text{MSA}] = \sigma^2 + \sum_{j=1}^{g} n_j(\alpha_j - \bar{\alpha})^2/(g - 1)$.

(e) Show that if $n_j = n_0$ a constant for all $j = 1, 2, \ldots, g$ then $\bar{\alpha} = 0$, $\mu = \bar{\mu}$ and $E[\text{MSA}] = \sigma^2 + \sum_{j=1}^{g} n_j\alpha_j^2/(g - 1)$.

(f) Show that if $H_0 : \mu_1 = \mu_2 = \cdots = \mu_g$ or equivalently $H_0 : \alpha_1 = \alpha_2 = \cdots = \alpha_g = 0$ then $E[\text{MSA}] = E[\text{MSW}] = \sigma^2$.

4. For the one-way analysis of variance model $y_{ij} = \mu_j + \varepsilon_{ij}$ define $(g-1)$ dummy variables

$$D_j = 1 \quad \text{if the observation is from group } j, \ j = 1, 2, \ldots, (g - 1)$$

$$= 0 \quad \text{otherwise.}$$

Let the **X** matrix contain a column of n unities plus $(g-1)$ columns corresponding to the observations on the $(g - 1)$ dummies. Each dummy column will contain n_j unities and $(n - n_j)$ zeroes, $j = 1, 2, \ldots, (g-1)$.

(a) Show that $\mathbf{X'X}$ has the form

$$\mathbf{X'X} = \begin{bmatrix} n & n_1 & n_2 & \cdots & n_{g-1} \\ n_1 & n_1 & 0 & \cdots & 0 \\ n_2 & 0 & n_2 & & \vdots \\ \vdots & \vdots & & \ddots & 0 \\ n_{g-1} & 0 & \cdots & 0 & n_{g-1} \end{bmatrix} = \begin{bmatrix} n & \mathbf{n'} \\ \mathbf{n} & \mathbf{N} \end{bmatrix}$$

where $\mathbf{n}' = (n_1, n_2, \ldots, n_{g-1})$ and $\mathbf{N}$ is the diagonal matrix of diagonal elements n_j, $j = 1, 2, \ldots, (g-1)$.

(b) Use the formula for the inverse of a partitioned matrix to show that $[\mathbf{X}'\mathbf{X}]^{-1}$ in (a) is given by

$$[\mathbf{X}'\mathbf{X}]^{-1} = \begin{bmatrix} a & \mathbf{b}' \\ \mathbf{b} & \mathbf{C} \end{bmatrix}$$

where $a = \dfrac{1}{n_g}$, $\mathbf{b}' = -\begin{bmatrix} \dfrac{1}{n_g} & \dfrac{1}{n_g} & \cdots & \dfrac{1}{n_g} \end{bmatrix}$

$$\mathbf{C} = \begin{bmatrix} 1/n_1 & 0 & \cdots & 0 \\ 0 & 1/n_2 & \cdots & 0 \\ \vdots & \vdots & \ddots & \vdots \\ 0 & \cdots & 0 & 1/n_{g-1} \end{bmatrix} + \begin{bmatrix} \frac{1}{n_g} & \frac{1}{n_g} & \cdots & \frac{1}{n_g} \\ \vdots & \vdots & & \vdots \\ \frac{1}{n_g} & \frac{1}{n_g} & \cdots & \frac{1}{n_g} \end{bmatrix}.$$

(c) Show that $\mathbf{b} = (\mathbf{X}'\mathbf{X})^{-1}\mathbf{X}'\mathbf{y}$ provides the estimators $b_j = (\bar{y}_{\cdot j} - \bar{y}_{\cdot g})$, $j = 1, 2, \ldots, (g-1)$ and $b_g = \bar{y}_{\cdot g}$.

5. (a) Let $\boldsymbol{\mu}' = [\mu_1, \mu_2, \ldots, \mu_g]$ and show that the one-way anova model can be written as

$$\mathbf{y} = \mathbf{Z}\boldsymbol{\mu} + \boldsymbol{\varepsilon}$$

where $\mathbf{Z} = [\mathbf{z}_1 \ \mathbf{z}_2 \ldots \mathbf{z}_g]$ and $\mathbf{z}_j$ is a dummy variable for group j, $j = 1, 2, \ldots, g$.

(b) The dummy variable coding method uses the matrix $\mathbf{X} = [\mathbf{i} \ \mathbf{z}_1 \ \mathbf{z}_2 \ldots \mathbf{z}_{(g-1)}]$ where $\mathbf{i}$ is a vector of unities, and the model is defined by $\mathbf{y} = \mathbf{X}\boldsymbol{\beta} + \boldsymbol{\varepsilon}$. Show that $\mathbf{Z}\boldsymbol{\mu} = \mathbf{X}\boldsymbol{\beta}$ implies that $\beta_0 = \mu_g$ and $\beta_j = (\mu_j - \mu_g)$, $j = 1, 2, \ldots, (g-1)$ and hence that $b_0 = \bar{y}_{\cdot g}$ and $b_j = (\bar{y}_{\cdot j} - \bar{y}_{\cdot g})$.

(c) The effect coding method uses the matrix

$$\mathbf{X} = \begin{bmatrix} \mathbf{i}_1 & \mathbf{z}_1 & \mathbf{z}_2 & \cdots & \mathbf{z}_{g-1} \\ \mathbf{i}_2 & -\mathbf{i}_2 & -\mathbf{i}_2 & \cdots & -\mathbf{i}_2 \end{bmatrix}$$

where $\mathbf{z}_j$ contains unities for group j and zeroes for all other groups except group g. $\mathbf{i}_1$ and $\mathbf{i}_2$ are vectors of unities. Show that $\mathbf{y} = \mathbf{X}\boldsymbol{\beta} + \boldsymbol{\varepsilon}$ and $\mathbf{Z}\boldsymbol{\mu} = \mathbf{X}\boldsymbol{\beta}$ implies that $\beta_0 = \bar{\mu}$, $\beta_j = \mu_j - \bar{\mu}$, $j = 1, \ldots, g-1$, and hence $b_0 = \sum\limits_{j=1}^{g} \bar{y}_{\cdot j}/g$, $b_j = (\bar{y}_{\cdot j} - b_0)$, $j = 1, 2, \ldots, (g-1)$.

6. In the randomized block design two-way model the additional sum of squares SSB $= g \sum\limits_{i=1}^{b} (\bar{y}_{i \cdot} - \bar{y}_{\cdot \cdot})^2$ is included as well as the sum of squares

SSA and SST with $n_j = b$. The error sum of squares is defined to be SSE = SST$-$SSA$-$SSB. Show that SSE $= \sum_{i=1}^{b} \sum_{j=1}^{g} (y_{ij} - \bar{y}_{i\cdot} - \bar{y}_{\cdot j} + \bar{y}_{\cdot\cdot})^2$.

7. Show that for the balanced two-way analysis of variance model with c observations per cell the relationship among the sums of squares is SST $=$ SSA $+$ SSB $+$ SSI $+$ SSE where

$$SSA = bc \sum_{j=1}^{g} (\bar{y}_{\cdot j\cdot} - \bar{y}_{\cdots})^2,$$

$$SSB = gc \sum_{i=1}^{b} (\bar{y}_{i\cdot\cdot} - \bar{y}_{\cdots})^2,$$

$$SST = \sum_{i=1}^{b} \sum_{j=1}^{g} \sum_{h=1}^{c} (y_{ijh} - \bar{y}_{\cdots})^2,$$

$$SSI = c \sum_{i=1}^{b} \sum_{j=1}^{g} (\bar{y}_{ij\cdot} - \bar{y}_{\cdot j\cdot} - \bar{y}_{i\cdot\cdot} + \bar{y}_{\cdots})^2,$$

$$SSE = \sum_{h=1}^{c} \sum_{i=1}^{b} \sum_{j=1}^{g} (y_{ijh} - \bar{y}_{ij\cdot})^2.$$

8. Assume the model

$$y_{ij} = \mu_{ij} + \varepsilon_{ij} \qquad i = 1, 2, \ldots, b; \ j = 1, 2, \ldots, g;$$

and the ε_{ij} are mutually independent, $\varepsilon_{ij} \sim N(0, \sigma^2)$.

(a) Let $\bar{y}_{\cdot j} = \sum_{i=1}^{b} y_{ij}/b$, $\bar{y}_{i\cdot} = \sum_{j=1}^{g} y_{ij}/g$, $\bar{y}_{\cdot\cdot} = \sum_{i=1}^{b} \sum_{j=1}^{g} y_{ij}/bg$, $\mu_{\cdot j} = \sum_{i=1}^{b} \mu_{ij}/b$, $\mu_{i\cdot} = \sum_{j=1}^{g} \mu_{ij}/g$, $\mu = \sum_{i=1}^{b} \sum_{j=1}^{g} \mu_{ij}/bg$. Show that $E[y_{ij}] = \mu_{ij}$, $E[\bar{y}_{\cdot j}] = \mu_{\cdot j}$, $E[\bar{y}_{i\cdot}] = \mu_{i\cdot}$, $E[\bar{y}_{\cdot\cdot}] = \mu$, $V[y_{ij}] = \sigma^2$, $V[\bar{y}_{\cdot j}] = \sigma^2/b$, $V[\bar{y}_{i\cdot}] = \sigma^2/g$, $V[\bar{y}_{\cdot\cdot}] = \sigma^2/bg$, and hence that $E[y_{ij}^2] = \sigma^2 + \mu_{ij}^2$, $E[\bar{y}_{\cdot j}^2] = \sigma^2/b + \mu_{\cdot j}^2$, $E[\bar{y}_{i\cdot}^2] = \sigma^2/g + \mu_{i\cdot}^2$ and $E[\bar{y}_{\cdot\cdot}^2] = \sigma^2/bg + \mu^2$.

(b) Show that

$$SST = \sum_{i=1}^{b} \sum_{j=1}^{g} (y_{ij} - \bar{y}_{\cdot\cdot})^2 = \sum_{i=1}^{b} \sum_{j=1}^{g} y_{ij}^2 - bg\bar{y}_{\cdot\cdot}^2$$

$$SSA = b \sum_{j=1}^{g} (\bar{y}_{\cdot j} - \bar{y}_{\cdot\cdot})^2 = b \sum_{j=1}^{g} \bar{y}_{\cdot j}^2 - bg\bar{y}_{\cdot\cdot}^2$$

$$SSB = g\sum_{i=1}^{b}(\bar{y}_{i\cdot} - \bar{y}_{\cdot\cdot})^2 = g\sum_{i=1}^{b}\bar{y}_{i\cdot}^2 - bg\bar{y}_{\cdot\cdot}^2$$

$$SSE = \sum_{i=1}^{b}\sum_{j=1}^{g}(y_{ij} - \bar{y}_{i\cdot} - \bar{y}_{\cdot j} + \bar{y}_{\cdot\cdot}^2)$$

$$= \sum_{i=1}^{b}\sum_{j=1}^{g}y_{ij}^2 - b\sum_{j=1}^{g}\bar{y}_{\cdot j}^2 - g\sum_{i=1}^{b}\bar{y}_{i\cdot}^2 + bg\bar{y}_{\cdot\cdot}^2 \ .$$

(c) Use the results in (a) and (b) to show that

$$E[SST] = (bg - 1)\sigma^2 + \sum_{j=1}^{g}\sum_{i=1}^{b}(\mu_{ij} - \mu)^2$$

$$E[SSA] = \sigma^2(g - 1) + b\sum_{j=1}^{g}(\mu_{\cdot j} - \mu)^2$$

$$E[SSB] = \sigma^2(b - 1) + g\sum_{i=1}^{b}(\mu_{i\cdot} - \mu)^2$$

$$E[SSE] = \sigma^2(b - 1)(g - 1) + \sum_{i=1}^{b}\sum_{j=1}^{g}(\mu_{ij} - \mu_{\cdot j} - \mu_{i\cdot} + \mu)^2 \ ,$$

and hence

$$E[MSA] = \sigma^2 + b\sum_{j=1}^{g}(\mu_{\cdot j} - \mu)^2/(g - 1)$$

$$E[MSB] = \sigma^2 + g\sum_{i=1}^{b}(\mu_{i\cdot} - \mu)^2/(b - 1)$$

$$E[MSE] = \sigma^2 + \sum_{i=1}^{b}\sum_{j=1}^{g}(\mu_{ij} - \mu_{\cdot j} - \mu_{i\cdot} + \mu)^2/(b - 1)(g - 1).$$

(d) What does the term $(\mu_{ij} - \mu_{\cdot j} - \mu_{i\cdot} + \mu)$ measure? Explain why this must be assumed to be zero in order to carry out the customary tests in this two-way model. Let $\mu_{ij} = \mu + \alpha_j + \beta_i$ where $\sum_{j=1}^{g}\alpha_j = \sum_{i=1}^{b}\beta_i = 0$ and show that this implies that $(\mu_{ij} - \mu_{i\cdot} - \mu_{\cdot j} + \mu) = 0$. Suppose that $(\mu_{ij} - \mu_{\cdot j} - \mu_{i\cdot} + \mu)$ was nonzero in a randomized block design model. What impact would this result have on the customary tests in analysis of variance?

9. Assume a two-factor model given by

$$y_{ijh} = \mu_{ij} + \varepsilon_{ijh} \quad i = 1, 2, \ldots, b; \; j = 1, 2, \ldots, g; \; h = 1, 2, \ldots, c;$$

where the ε_{ijh} are mutually independent and $\varepsilon_{ijh} \sim N(0, \sigma^2)$.

(a) Define $\bar{y}_{i..}$, $\bar{y}_{.j.}$, $\bar{y}_{ij.}$ and $\bar{y}_{...}$ and determine $E[\bar{y}_{i..}^2]$, $E[\bar{y}_{.j.}^2]$, $E[\bar{y}_{ij.}^2]$ and $E[\bar{y}_{...}^2]$ following the procedures in Question 8.

(b) Show that

$$E[\text{SSA}] = \sigma^2(g-1) + cb\sum_{j=1}^{g}(\mu_{.j} - \mu)^2$$

$$E[\text{SSB}] = \sigma^2(b-1) + cg\sum_{i=1}^{b}(\mu_{i.} - \mu)^2$$

$$E[\text{SSI}] = \sigma^2(b-1)(g-1) + c\sum_{j=1}^{g}\sum_{i=1}^{b}(\mu_{ij} - \mu_{i.} - \mu_{.j} + \mu)^2$$

$$E[\text{SSE}] = \sigma^2 bg(c-1)$$

using the methods in Question 8.

10. Assume a two factor analysis of variance model with each factor having three levels, and hence there are nine cells or treatment combinations.

(a) Using effect coding show that the design matrix for main effects with one observation per cell is given by

$$
\mathbf{X} =
\begin{array}{c|ccccc}
 & A & B & C & D & E \\
\hline
 & 1 & 1 & 0 & 1 & 0 \\
 & 1 & 1 & 0 & 0 & 1 \\
 & 1 & 1 & 0 & -1 & -1 \\
 & 1 & 0 & 1 & 1 & 0 \\
 & 1 & 0 & 1 & 0 & 1 \\
 & 1 & 0 & 1 & -1 & -1 \\
 & 1 & -1 & -1 & 1 & 0 \\
 & 1 & -1 & -1 & 0 & 1 \\
 & 1 & -1 & -1 & -1 & -1 \\
\end{array}
$$

where columns B and C correspond to the first factor and columns D and E correspond to the second factor.

(b) Show that columns B, C, D and E are orthogonal to column A and that columns B and C are both orthogonal to columns D and E.

(c) Give a design matrix which includes interaction by generating the four product columns between B and C and D and E. Show that the four interaction columns are orthogonal to the four main effect columns but that they are not orthogonal to each other.

(d) Let $\mathbf{X}$ denote the design matrix determined in (c) and compute $(\mathbf{X'X})$ and $(\mathbf{X'X})^{-1}$. What advantage does the orthogonality yield when determining the least squares estimator $(\mathbf{X'X})^{-1}\mathbf{X'y}$?

(e) Suppose that for the nine cells in this design the observations per cell are given by

		FACTOR 1		
		1	2	3
	1	2	1	1
FACTOR 2	2	1	1	2
	3	1	2	1

Show that the design matrix in (c) which includes these new observations is given by

$$
\begin{bmatrix}
1 & 1 & 0 & 1 & 0 & 1 & 0 & 0 & 0 \\
1 & 1 & 0 & 1 & 0 & 1 & 0 & 0 & 0 \\
1 & 1 & 0 & 0 & 1 & 0 & 1 & 0 & 0 \\
1 & 1 & 0 & -1 & -1 & -1 & -1 & 0 & 0 \\
1 & 0 & 1 & 1 & 0 & 0 & 0 & 1 & 0 \\
1 & 0 & 1 & 0 & 1 & 0 & 0 & 0 & 1 \\
1 & 0 & 1 & -1 & -1 & 0 & 0 & -1 & -1 \\
1 & 0 & 1 & -1 & -1 & 0 & 0 & -1 & -1 \\
1 & -1 & -1 & 1 & 0 & -1 & 0 & -1 & 0 \\
1 & -1 & -1 & 0 & 1 & 0 & -1 & 0 & -1 \\
1 & -1 & -1 & 0 & 1 & 0 & -1 & 0 & -1 \\
1 & -1 & -1 & -1 & -1 & 1 & 1 & 1 & 1 \\
\end{bmatrix}
$$

(f) Show that the orthogonality relations established in (b) and (c) no longer hold.

11. For a one-way random effects model

$$y_{ij} = \mu + \alpha_j + \varepsilon_{ij} \quad i = 1, 2, \ldots, n_j; \; j = 1, 2, \ldots, g;$$

α_j are mutually independent, $\alpha_j \sim N(0, \sigma_\alpha^2)$, ε_{ij} are mutually independent, $\varepsilon_{ij} \sim N(0, \sigma^2)$ and the α_j and ε_{ij} are mutually independent.

(a) Show that $V(y_{ij}) = (\sigma_\alpha^2 + \sigma^2)$, $V(\bar{y}_{\cdot j}) = (\sigma_\alpha^2 + \sigma^2/n_j)$, $V(\bar{y}_{\cdot\cdot}) = \sigma_\alpha^2/g + \sigma^2/n$ where $n = \sum_{j=1}^{g} n_j$, $E[y_{ij}] = \mu$, $E[\bar{y}_{\cdot j}] = \mu$, $E[\bar{y}_{\cdot\cdot}] = \mu$ and hence $E[y_{ij}^2] = (\sigma_\alpha^2 + \sigma^2) + \mu^2$, $E[\bar{y}_{\cdot j}^2] = \sigma_\alpha^2 + \sigma^2/n_j + \mu^2$, $E[\bar{y}_{\cdot\cdot}^2] = \sigma_\alpha^2/g + \sigma^2/n + \mu^2$.

(b) Use the results in (a) to show that

$$E[\text{SST}] = E\left[\sum_{j=1}^{g}\sum_{i=1}^{n_j} y_{ij}^2 - n\bar{y}_{\cdot\cdot}^2\right] = \sigma^2(n-1) + \sigma_\alpha^2\left[\frac{n(g-1)}{g}\right]$$

$$E[\text{SSA}] = E\left[\sum_{j=1}^{g} n_j\bar{y}_{\cdot j}^2 - n\bar{y}_{\cdot\cdot}^2\right] = \sigma^2(g-1) + \sigma_\alpha^2\left[\frac{n(g-1)}{g}\right]$$

$$E[\text{SSW}] = E\left[\sum_{j=1}^{g}\sum_{i=1}^{n_j} y_{ij}^2 - \sum_{j=1}^{g} n_j\bar{y}_{\cdot j}^2\right] = (n-g)\sigma^2$$

and hence that $E[\text{MSA}] = \sigma^2 + n\sigma_\alpha^2/g$ and $E[\text{MSW}] = \sigma^2$.

(c) Show that $H_0 : \sigma_\alpha^2 = 0$ is consistent with the hypothesis $E[\text{MSW}] = E[\text{MSA}] = \sigma^2$.

(d) Show that $\text{Cov}(y_{ij}, y_{\ell j}) = \sigma_\alpha^2$ and hence that the correlation between y_{ij} and $y_{\ell j}$ is $\rho_\alpha = \sigma_\alpha^2/[\sigma_\alpha^2 + \sigma^2]$. Why is this called an intraclass correlation?

12. (a) Use effect coding to give the design matrix for a 2^5 design.

(b) Denote the factors by A, B, C, D and E. Suppose that only 16 experimental units are available for the experiment. Using the ABCDE interaction as the defining contrast eliminate 16 of the rows of the X matrix in (a), and hence determine a one-half replication of the 2^5 design.

(c) Determine the aliases that have been created for all the effects in (b).

(d) Suppose that the 2^5 design in (a) is employed in two blocks of 16 experimental units each. What effect is confounded with the block effect?

Appendix

1. Matrix Algebra

1.1 Matrices

Matrix

A *matrix* of order $(n \times p)$ is a rectangular array of elements consisting of n rows and p columns. A matrix is denoted by a bold face letter say $\mathbf{A}$ where

$$\mathbf{A} = \begin{bmatrix} a_{11} & a_{12} & \cdots & a_{1p} \\ a_{21} & a_{22} & \cdots & a_{2p} \\ a_{31} & a_{32} & \cdots & a_{3p} \\ \vdots & & & \vdots \\ a_{n1} & a_{n2} & \cdots & a_{np} \end{bmatrix}.$$

The elements are denoted by a_{ij}, $i = 1, 2, \ldots, n$; $j = 1, 2, \ldots, p$, where the first subscript i refers to the row location, and the second subscript j refers to the column location of the element. The matrix is also sometimes denoted by $((a_{ij}))$.

Example of a Matrix

The matrix $\mathbf{B} = \begin{bmatrix} 1 & 9 & -2 \\ 4 & 6 & 3 \end{bmatrix}$ has 2 rows and 3 columns and hence is a (2×3) matrix, while the matrix $\mathbf{C} = \begin{bmatrix} -6 & 2 \\ 4 & -5 \\ -3 & 8 \end{bmatrix}$ has 3 rows and 2 columns and hence is a (3×2) matrix.

Transpose of a Matrix

The *transpose* of the matrix $\mathbf{A}$ $(n \times p)$ is the matrix $\mathbf{B}$ $(p \times n)$ obtained by interchanging rows and column so that

$$b_{ij} = a_{ji} \quad i = 1, 2, \ldots, p; \; j = 1, 2, \ldots, n.$$

The transpose of $\mathbf{A}$ is usually denoted by $\mathbf{B} = \mathbf{A}'$. Some additional properties of a matrix transpose are

1. $\mathbf{A}' = \mathbf{B}'$ if and only if $\mathbf{A} = \mathbf{B}$

2. $(\mathbf{A}')' = \mathbf{A}$.

Example of a Matrix Transpose

The transpose of the matrices $\mathbf{B}$ and $\mathbf{C}$ defined in the above example are given by

$$\mathbf{B}' = \begin{bmatrix} 1 & 4 \\ 9 & 6 \\ -2 & 3 \end{bmatrix} \quad \text{and} \quad \mathbf{C}' = \begin{bmatrix} -6 & 4 & -3 \\ 2 & -5 & 8 \end{bmatrix}.$$

Exercise — Matrix Transpose

Using the matrices $\mathbf{B}$ and $\mathbf{C}$ defined above verify that $(\mathbf{B}')' = \mathbf{B}$ and $(\mathbf{C}')' = \mathbf{C}$.

Row Vector and Column Vector

A *row vector* is a matrix with only one row and is denoted by a lower case letter with bold face type

$$\mathbf{b} = [b_1, b_2, \ldots, b_p].$$

A *column vector* is a matrix with only one column and is also denoted by a lower case bold face letter

$$\mathbf{a} = \begin{bmatrix} a_1 \\ a_2 \\ \vdots \\ a_n \end{bmatrix}.$$

Example of Row and Column Vectors

The (4×1) matrix $\mathbf{d} = \begin{bmatrix} 2 \\ -4 \\ 4 \\ -6 \end{bmatrix}$ is a column vector while the (1×5) matrix $\mathbf{f} = [3 \quad 5 \quad -7 \quad 4 \quad -2]$ is a row vector.

Square Matrix

A matrix is *square* if the number of rows n is equal to the number of columns p $(n = p)$. A *square matrix* with m rows and columns is said to have *order m*.

Symmetric Matrix

A square matrix $\mathbf{A}$ of order m is *symmetric*, if the transpose of $\mathbf{A}$ is equal to $\mathbf{A}$.

$$\mathbf{A} = \mathbf{A}' \quad \text{if } \mathbf{A} \text{ symmetric.}$$

Diagonal Elements

The elements a_{ii}, $i = 1, 2, \ldots, m$ are called the *diagonal* elements of the square matrix $\mathbf{A}$ with elements a_{ij}, $i = 1, 2, \ldots, m$; $j = 1, 2, \ldots, m$.

Trace of a Matrix

The sum of the diagonal elements of $\mathbf{A}$ is called the *trace* of $\mathbf{A}$ and is denoted by $tr(\mathbf{A})$. The trace of the $(m \times m)$ matrix is given by $tr(\mathbf{A}) = \sum_{i=1}^{m} a_{ii}$.

Example — Square, Symmetric, Diagonal Elements, Trace

The matrix $\mathbf{H} = \begin{bmatrix} 8 & 6 & -4 \\ 6 & 2 & -1 \\ -4 & -1 & 7 \end{bmatrix}$ is a (3×3) or square matrix of order 3. The matrix is symmetric since $\mathbf{H}' = \mathbf{H}$. The diagonal elements of $\mathbf{H}$ are the elements 8, 2 and 7. The trace of $\mathbf{H} = tr\mathbf{H} = 17$ which is the sum of the diagonal elements.

Exercise — Symmetric Matrix, Trace

Verify that the matrix $\mathbf{A} = \begin{bmatrix} a & b & c \\ b & d & e \\ c & e & f \end{bmatrix}$ is symmetric and that $tr(\mathbf{A}) = (a + d + f)$.

Null or Zero Matrix

The *null or zero matrix* denoted by **0** is the matrix whose elements are all zero.

Identity Matrix

The *identity matrix* of order m is the square matrix whose diagonal elements are all unity, and whose off-diagonal elements are all zero. The identity matrix of order m is usually denoted by $\mathbf{I}_m$ and is given by

$$\mathbf{I}_m = \begin{bmatrix} 1 & 0 & 0 & \cdots & 0 \\ 0 & 1 & 0 & \cdots & 0 \\ 0 & 0 & 1 & & \vdots \\ \vdots & \vdots & & \ddots & 0 \\ 0 & 0 & \cdots & 0 & 1 \end{bmatrix}.$$

Diagonal Matrix

A *diagonal matrix* is a matrix whose off-diagonal elements are all zero. The identity matrix is a special case of a diagonal matrix.

Submatrix

A *submatrix* of a matrix **A** is a matrix obtained from **A** by deleting some rows and columns of **A**.

Example of Submatrix

The (2×2) matrix $\mathbf{C}^* = \begin{bmatrix} 4 & -4 \\ 2 & 1 \end{bmatrix}$ is a submatrix of the matrix $\mathbf{C} = \begin{bmatrix} 9 & -6 & 3 \\ 4 & -4 & 5 \\ 2 & 1 & -5 \end{bmatrix}$. $\mathbf{C}^*$ is obtained from **C** by deleting the first row and the third column.

1.2 Matrix Operations

Equality of Matrices

Two matrices $\mathbf{A}$ and $\mathbf{B}$ are *equal*, if and only if each element of $\mathbf{A}$ is equal to the corresponding element of $\mathbf{B}$: $a_{ij} = b_{ij}, \; i = 1, 2, \ldots, n; \; j = 1, 2, \ldots, p.$

Addition of Matrices

The *addition* of two matrices $\mathbf{A}$ and $\mathbf{B}$ is carried out by adding together corresponding elements. The two matrices must have the same order. The sum is given by $\mathbf{C}$ where
$$\mathbf{C} = \mathbf{A} + \mathbf{B}$$
and where $c_{ij} = a_{ij} + b_{ij}, \; i = 1, 2, \ldots, n; \; j = 1, 2, \ldots, p.$

Additive Inverse

A matrix $\mathbf{B}$ is the *additive inverse* of a matrix $\mathbf{A}$, if the matrices sum to the null matrix $\mathbf{0}$
$$\mathbf{B} + \mathbf{A} = \mathbf{0},$$
where $b_{ij} + a_{ij} = 0$ or $b_{ij} = -a_{ij}, \; i = 1, 2, \ldots, n; \; j = 1, 2, \ldots, p.$ This additive inverse is denoted by $\mathbf{B} = -\mathbf{A}$. If $\mathbf{C} = \mathbf{A} + \mathbf{B}$ then $\mathbf{C}' = \mathbf{A}' + \mathbf{B}'$.

Example

The sum of the matrices $\mathbf{A}$ and $\mathbf{B}$ given below is denoted by $\mathbf{C}$.

$$\mathbf{A} = \begin{bmatrix} 4 & 3 \\ -6 & 9 \\ 2 & -4 \end{bmatrix} \qquad \mathbf{B} = \begin{bmatrix} 2 & -5 \\ -2 & -2 \\ 4 & 6 \end{bmatrix}$$

$$\mathbf{C} = \begin{bmatrix} 4+2 & 3-5 \\ -6-2 & 9-2 \\ 2+4 & -4+6 \end{bmatrix} = \begin{bmatrix} 6 & -2 \\ -8 & 7 \\ 6 & 2 \end{bmatrix}.$$

The matrix $\mathbf{D} = \begin{bmatrix} -4 & -3 \\ 6 & -9 \\ -2 & 4 \end{bmatrix}$ is the additive inverse of the matrix $\mathbf{A}$

since $\mathbf{A} + \mathbf{D} = \begin{bmatrix} 0 & 0 \\ 0 & 0 \\ 0 & 0 \end{bmatrix}.$

Exercise

Verify that $\mathbf{C}' = \mathbf{A}' + \mathbf{B}'$ using the matrices $\mathbf{A}$, $\mathbf{B}$ and $\mathbf{C}$ in the previous example.

Scalar Multiplication of a Matrix

The *scalar multiplication* of a matrix $\mathbf{A}$ by a scalar k is carried out by multiplying each element of $\mathbf{A}$ by k. This scalar product is denoted by $k\,\mathbf{A}$ and the elements by ka_{ij}, $i = 1, 2, \ldots, n$; $j = 1, 2, \ldots, p$.

Product of Two Matrices

The *product* of two matrices $\mathbf{A}$ $(n \times p) = (a_{ij})$ and $\mathbf{B}$ $(p \times m) = (b_{jk})$ is denoted by $\mathbf{C} = \mathbf{AB}$, if the number of columns (p) of $\mathbf{A}$ is equal to the number of rows (p) of $\mathbf{B}$, and if the elements of $\mathbf{C} = (c_{ij})$ are given by

$$c_{ik} = \sum_{j=1}^{p} a_{ij} b_{jk}.$$

The order of the product matrix $\mathbf{C}$ is $(n \times m)$.

Example

The matrix $\mathbf{A} = \begin{bmatrix} 4 & 3 \\ -6 & 9 \\ 2 & -4 \end{bmatrix}$ when multiplied by the scalar 3 yields the

matrix $3\mathbf{A} = \begin{bmatrix} 12 & 9 \\ -18 & 27 \\ 6 & -12 \end{bmatrix}$.

The product of the two matrices $\mathbf{A}$ and $\mathbf{B}$ where $\mathbf{A}$ is given above and $\mathbf{B} = \begin{bmatrix} 0 & 6 & 2 \\ 1 & -2 & 3 \end{bmatrix}$ is given by

$$\mathbf{C} = \mathbf{AB} = \begin{bmatrix} (4)(0) + (3)(1) & (4)(6) + (3)(-2) & (4)(2) + (3)(3) \\ (-6)(0) + (9)(1) & (-6)(6) + (9)(-2) & (-6)(2) + (9)(3) \\ (2)(0) + (-4)(1) & (2)(6) + (-4)(-2) & (2)(2) + (-4)(3) \end{bmatrix}$$

$$= \begin{bmatrix} 3 & 18 & 17 \\ 9 & -54 & 15 \\ -4 & 20 & -8 \end{bmatrix}.$$

Some additional properties are:

1. $\mathbf{A} + \mathbf{B} = \mathbf{B} + \mathbf{A}$

2. $(\mathbf{A} + \mathbf{B}) + \mathbf{C} = \mathbf{A} + (\mathbf{B} + \mathbf{C})$

3. $a(\mathbf{A} + \mathbf{B}) = a\mathbf{A} + a\mathbf{B}$

4. $(a + b)\mathbf{A} = a\mathbf{A} + b\mathbf{A}$

5. $(\mathbf{A}\mathbf{B})\mathbf{C} = \mathbf{A}(\mathbf{B}\mathbf{C})$

6. $(\mathbf{A} + \mathbf{B})\mathbf{C} = \mathbf{A}\mathbf{C} + \mathbf{B}\mathbf{C}$

7. In general $\mathbf{A}\mathbf{B} \neq \mathbf{B}\mathbf{A}$ even if the dimensions conform for multiplication.

8. For square matrices $\mathbf{A}$ and $\mathbf{B}$

$$tr(\mathbf{A} + \mathbf{B}) = tr\mathbf{A} + tr\mathbf{B} \quad \text{and} \quad tr(\mathbf{A}\mathbf{B}) = tr(\mathbf{B}\mathbf{A}).$$

Exercise

Given $\mathbf{A} = \begin{bmatrix} 1 & -1 \\ 2 & 1 \end{bmatrix}$, $\mathbf{B} = \begin{bmatrix} 2 & 0 \\ 1 & -1 \end{bmatrix}$ and $\mathbf{C} = \begin{bmatrix} -1 & 2 \\ 0 & 3 \end{bmatrix}$ verify the properties 1 through 8 given above.

Multiplicative Inverse

The *multiplicative inverse* of the square matrix $\mathbf{A}$ $(m \times m)$ is the matrix $\mathbf{B}$ $(m \times m)$ satisfying the equation $\mathbf{A}\mathbf{B} = \mathbf{B}\mathbf{A} = \mathbf{I}_m$, where $\mathbf{I}_m$ is the identity matrix of order m. The multiplicative inverse of $\mathbf{A}$ is usually denoted by $\mathbf{B} = \mathbf{A}^{-1}$. If $\mathbf{A}$ has an inverse it is unique.

Some additional properties are:

1. If $\mathbf{C} = \mathbf{A}\mathbf{B}$ then $\mathbf{C}^{-1} = \mathbf{B}^{-1}\mathbf{A}^{-1}$

2. $(\mathbf{A}')^{-1} = (\mathbf{A}^{-1})'$

3. If $\mathbf{C} = \mathbf{A}\mathbf{B}$ then $\mathbf{C}' = \mathbf{B}'\mathbf{A}'$.

4. $(k\mathbf{A})^{-1} = (1/k)\mathbf{A}^{-1}$.

Example

The inverse of the matrix $\mathbf{A} = \begin{bmatrix} 4 & -2 \\ -2 & 2 \end{bmatrix}$ is given by $\mathbf{A}^{-1} = \begin{bmatrix} 1/2 & 1/2 \\ 1/2 & 1 \end{bmatrix}$

since $\mathbf{A}\mathbf{A}^{-1} = \begin{bmatrix} 4 & -2 \\ -2 & 2 \end{bmatrix} \begin{bmatrix} 1/2 & 1/2 \\ 1/2 & 1 \end{bmatrix} = \begin{bmatrix} 1 & 0 \\ 0 & 1 \end{bmatrix}$.

A Useful Result

Given a symmetric nonsingular matrix $\mathbf{A}$ $(p \times p)$ and matrices $\mathbf{B}$ and $\mathbf{C}$ of order $(p \times q)$, the inverse of the matrix $[\mathbf{A} + \mathbf{BC}']$ is given by

$$[\mathbf{A} + \mathbf{BC}']^{-1} = \mathbf{A}^{-1} - \mathbf{A}^{-1}\mathbf{B}[\mathbf{I} + \mathbf{C}'\mathbf{A}^{-1}\mathbf{B}]^{-1}\mathbf{C}'\mathbf{A}^{-1}.$$

An important special case of this result is when $\mathbf{B}$ and $\mathbf{C}$ are $(p \times 1)$ vectors say $\mathbf{b}$ and $\mathbf{c}$. The inverse of the matrix $[\mathbf{A} + \mathbf{bc}']$ is given by

$$[\mathbf{A} + \mathbf{bc}']^{-1} = \mathbf{A}^{-1} - \frac{\mathbf{A}^{-1}\mathbf{bc}'\mathbf{A}^{-1}}{1 + \mathbf{c}'\mathbf{A}^{-1}\mathbf{b}}.$$

Exercise

(a) Verify the above two results by using matrix multiplication.

(b) The equi-correlation matrix has the form $\mathbf{B} = (1 - \rho)\mathbf{I} + \rho\mathbf{ii}'$ where ρ is a constant correlation coefficient and $\mathbf{i}$ $(n \times 1)$ is a vector of unities. Show by using the above result that the matrix $\mathbf{B}^{-1}$ is given by

$$\mathbf{B}^{-1} = \frac{1}{(1 - \rho)}\mathbf{I} - \frac{\rho\mathbf{ii}'}{(1 - \rho)^2 + \rho(1 - \rho)n}.$$

Idempotent Matrix

An *idempotent* matrix $\mathbf{A}$ is a matrix which has the property $\mathbf{AA} = \mathbf{A}$. If $\mathbf{A}$ is idempotent and has full rank then $\mathbf{A}$ is an identity matrix.

Example

The matrix $\mathbf{B} = \begin{bmatrix} 1 & -1 \\ 0 & 0 \end{bmatrix}$ is idempotent since $\begin{bmatrix} 1 & -1 \\ 0 & 0 \end{bmatrix}\begin{bmatrix} 1 & -1 \\ 0 & 0 \end{bmatrix} = \begin{bmatrix} 1 & -1 \\ 0 & 0 \end{bmatrix}.$

Exercise

Show that the mean centering matrix $[\mathbf{I} - \frac{1}{n}\mathbf{ii}']$ is idempotent where $\mathbf{I}$ is an $(n \times n)$ identity matrix and $\mathbf{i}$ $(n \times 1)$ is a vector of unities.

Kronecker Product

If $\mathbf{A}(n \times m)$ and $\mathbf{B}(p \times q)$, the *Kronecker product* of $\mathbf{A}$ and $\mathbf{B}$ is denoted by $\mathbf{A} \otimes \mathbf{B}$ and is given by the matrix

$$\begin{bmatrix} a_{11}\mathbf{B} & a_{12}\mathbf{B} & \cdots & a_{1m}\mathbf{B} \\ a_{21}\mathbf{B} & a_{22}\mathbf{B} & \cdots & a_{2m}\mathbf{B} \\ \vdots & & & \\ a_{n1}\mathbf{B} & a_{n2}\mathbf{B} & \cdots & a_{nm}\mathbf{B} \end{bmatrix}.$$

Example

Given $\mathbf{A} = \begin{bmatrix} 2 & -1 \\ 3 & 4 \end{bmatrix}$ $\mathbf{B} = \begin{bmatrix} 3 \\ 5 \end{bmatrix}$ then

$$\mathbf{A} \otimes \mathbf{B} = \begin{bmatrix} 2 \begin{bmatrix} 3 \\ 5 \end{bmatrix} & -1 \begin{bmatrix} 3 \\ 5 \end{bmatrix} \\ 3 \begin{bmatrix} 3 \\ 5 \end{bmatrix} & 4 \begin{bmatrix} 3 \\ 5 \end{bmatrix} \end{bmatrix} = \begin{bmatrix} 6 & -3 \\ 10 & -5 \\ 9 & 12 \\ 15 & 20 \end{bmatrix} \quad \text{and}$$

$$\mathbf{B} \otimes \mathbf{A} = \begin{bmatrix} 3 \begin{bmatrix} 2 & -1 \\ 3 & 4 \end{bmatrix} \\ 5 \begin{bmatrix} 2 & -1 \\ 3 & 4 \end{bmatrix} \end{bmatrix} = \begin{bmatrix} 6 & -3 \\ 9 & 12 \\ 10 & -5 \\ 15 & 20 \end{bmatrix}.$$

Some properties of Kronecker products are

1. $(\mathbf{A} \otimes \mathbf{B}) \otimes \mathbf{C} = \mathbf{A} \otimes (\mathbf{B} \otimes \mathbf{C})$

2. $(\mathbf{A} + \mathbf{B}) \otimes \mathbf{C} = (\mathbf{A} \otimes \mathbf{C}) + (\mathbf{B} \otimes \mathbf{C})$

3. $(\mathbf{A} \otimes \mathbf{B})(\mathbf{C} \otimes \mathbf{D}) = \mathbf{AC} \otimes \mathbf{BD}$

4. $(\mathbf{A} \otimes \mathbf{B})' = \mathbf{A}' \otimes \mathbf{B}'$

5. $tr(\mathbf{A} \otimes \mathbf{B}) = tr(\mathbf{A})tr(\mathbf{B})$

6. For vectors $\mathbf{a}$ and $\mathbf{b}$

$$\mathbf{a}' \otimes \mathbf{b} = \mathbf{b} \otimes \mathbf{a}' = \mathbf{ba}'$$

7. In general $\mathbf{A} \otimes \mathbf{B} \neq \mathbf{B} \otimes \mathbf{A}$ as demonstrated in the above example.

Exercise

(a) Use the matrices **A** and **B** defined in the example above to show that $(\mathbf{A} \otimes \mathbf{B})' = (\mathbf{A}' \otimes \mathbf{B}')$.

(b) Use the matrices $\mathbf{A} = \begin{bmatrix} 1 & -1 \\ 3 & 2 \end{bmatrix}$ and $\mathbf{B} = \begin{bmatrix} -1 & 2 \\ 5 & 3 \end{bmatrix}$ to verify that $tr(\mathbf{A} \otimes \mathbf{B}) = tr(\mathbf{A})tr(\mathbf{B})$.

1.3 Determinants and Rank

Determinant

The *determinant* of a square matrix **A** is a scalar quantity denoted by $|\mathbf{A}|$ and is given by

$$|\mathbf{A}| = \sum_{k=1}^{n!} (-1)^{j(k)} a_{11(k)} a_{22(k)} \cdots a_{nn(k)}.$$

The determinant represents the sum of $n!$ terms each term consisting of the product of n elements of **A**. For each term of the summation, the first subscripts are the integers 1 to n in their natural order, and the second subscripts represent a particular permutation of the integers 1 to n. The power $j(k)$ is 1 or 2 depending on whether the second subscripts represent an odd or an even number of interchanges with the integers in their natural order.

For the 2×2 matrix

$$\mathbf{A} = \begin{bmatrix} a_{11} & a_{12} \\ a_{21} & a_{22} \end{bmatrix}$$

the determinant is given by

$$|\mathbf{A}| = \underset{\text{(0 interchanges)}}{a_{11}a_{22}} - \underset{\text{(1 interchange)}}{a_{12}a_{21}}.$$

For the 3×3 matrix

$$\mathbf{B} = \begin{bmatrix} b_{11} & b_{12} & b_{13} \\ b_{21} & b_{22} & b_{23} \\ b_{31} & b_{32} & b_{33} \end{bmatrix}$$

the determinant is given by

$$|\mathbf{B}| = \underset{\text{(0 interchanges)}}{b_{11}b_{22}b_{33}} - \underset{\text{(1 interchange)}}{b_{11}b_{23}b_{32}} + \underset{\text{(2 interchanges)}}{b_{12}b_{23}b_{31}}$$

$$- \underset{\text{(1 interchange)}}{b_{12}b_{21}b_{33}} + \underset{\text{(2 interchanges)}}{b_{13}b_{21}b_{32}} - \underset{\text{(1 interchange)}}{b_{13}b_{22}b_{31}}.$$

NOTE: $n! = 3! = 6$ terms.

Nonsingular

If $|\mathbf{A}| \neq 0$ then $\mathbf{A}$ is said to be *nonsingular*.

The determinant of the matrix $\mathbf{A}$ $(n \times n)$ can be evaluated in terms of the determinants of submatrices of $\mathbf{A}$. The determinant of the submatrix of $\mathbf{A}$ obtained after deleting the j-th row and k-th column of $\mathbf{A}$ is called the *minor* of a_{jk} and is denoted by $|A^{jk}|$. The *cofactor* of a_{jk} is the quantity $(-1)^{j+k}|A^{jk}|$. The determinant of $\mathbf{A}$ can be expressed by

$$|\mathbf{A}| = \sum_{j=1}^{n} a_{jk}(-1)^{j+k}|A^{jk}| \text{ for any } k \text{ and is called the cofactor expansion}$$

of $\mathbf{A}$.

Example

The determinant of the matrix $\mathbf{B} = \begin{bmatrix} 1 & 3 & -2 \\ 4 & 5 & 1 \\ -3 & -4 & 7 \end{bmatrix}$ can be determined

by expanding about the first row obtaining

$$1(-1)^2 \begin{vmatrix} 5 & 1 \\ -4 & 7 \end{vmatrix} + 3(-1)^3 \begin{vmatrix} 4 & 1 \\ -3 & 7 \end{vmatrix} + (-2)(-1)^4 \begin{vmatrix} 4 & 5 \\ -3 & -4 \end{vmatrix} = -52.$$

Equivalently expanding about the third column

$$-2(-1)^4 \begin{vmatrix} 4 & 5 \\ -3 & -4 \end{vmatrix} + 1(-1)^5 \begin{vmatrix} 1 & 3 \\ -3 & -4 \end{vmatrix} + 7(-1)^6 \begin{vmatrix} 1 & 3 \\ 4 & 5 \end{vmatrix} = -52.$$

Exercise

Verify the value of the determinant in the example by expanding about the second row.

Some useful properties of the determinant are:

1. $|\mathbf{A}| = |\mathbf{A}|'$

2. If each element of a row (or column) of $\mathbf{A}$ is multiplied by the scalar k then the determinant of the new matrix is $k|\mathbf{A}|$.

3. $|k\mathbf{A}| = |\mathbf{A}|k^p$ if $\mathbf{A}$ is $(p \times p)$

4. If each element of a row (or column) of $\mathbf{A}$ is zero then $|\mathbf{A}| = 0$.

5. If two rows (or columns) of $\mathbf{A}$ are identical then $|\mathbf{A}| = 0$.

6. The determinant of a matrix remains unchanged if the elements of one row (or column) are multipled by a scalar k and the results added to a second row (or column).

7. The determinant of the product $\mathbf{AB}$ of the square matrices $\mathbf{A}$ and $\mathbf{B}$ is given by $|\mathbf{AB}| = |\mathbf{A}||\mathbf{B}|$.

8. If $\mathbf{A}^{-1}$ exists then $|\mathbf{A}^{-1}| = |\mathbf{A}|^{-1}$.

9. $\mathbf{A}^{-1}$ exists if and only if $|\mathbf{A}| \neq 0$.

10. $|\mathbf{A} \otimes \mathbf{B}| = |\mathbf{A}|^n |\mathbf{B}|^m$ where $\mathbf{A}(n \times n)$ and $\mathbf{B}(m \times m)$.

11. $(\mathbf{A} \otimes \mathbf{B})^{-1} = \mathbf{A}^{-1} \times \mathbf{B}^{-1}$

12. If $\mathbf{A}(p \times p)$ is non-singular, $\mathbf{B}(p \times m)$ and $\mathbf{C}(m \times p)$ then $|\mathbf{A} + \mathbf{BC}| = |\mathbf{A}|^{-1}|\mathbf{I}_p + \mathbf{A}^{-1}\mathbf{BC}| = |\mathbf{A}^{-1}||\mathbf{I}_m + \mathbf{CA}^{-1}\mathbf{B}|$ where $\mathbf{I}_p(p \times p)$ and $\mathbf{I}_m(m \times m)$ are identity matrices.

Relation Between Inverse and Determinant

The inverse of the matrix $\mathbf{A}$ is given by $\mathbf{A}^{-1} = \frac{1}{|\mathbf{A}|}\mathbf{A}^*$, where $\mathbf{A}^*$ is the transpose of the matrix of cofactors of $\mathbf{A}$. The matrix $\mathbf{A}^*$ is called the *adjoint* of $\mathbf{A}$.

Example

For the (2×2) matrix $\mathbf{A} = \begin{bmatrix} 9 & -2 \\ -4 & 6 \end{bmatrix}$ the determinant is given by $54 - 8 = 46$. The matrix of cofactors is given by $\begin{bmatrix} 6 & 4 \\ 2 & 9 \end{bmatrix}$ and hence the adjoint matrix is $\mathbf{A}^* = \begin{bmatrix} 6 & 2 \\ 4 & 9 \end{bmatrix}$. The inverse of $\mathbf{A}$ is therefore given by $\mathbf{A}^{-1} = \frac{1}{46}\begin{bmatrix} 6 & 2 \\ 4 & 9 \end{bmatrix}$.

Exercise

(a) Verify that $|\mathbf{A}^{-1}| = |\mathbf{A}|^{-1}$ using the matrix $\mathbf{A}$ in the previous example.

(b) Verify the inverse of the matrix $\mathbf{A}$ in the example above using the adjoint and determinant of $\mathbf{A}^{-1}$ to get $\mathbf{A}$.

(c) Use the properties above to show that the determinant of the equi-correlation matrix is given by $|\rho\mathbf{I} + (1-\rho)\mathbf{ii}'| = (1-\rho)^{n-1}[1 + \rho(n-1)]$, where $\mathbf{i}$ $(n \times 1)$ is a vector of unities.

Rank of a Matrix

The *rank* of a matrix $\mathbf{A}$, rank $(\mathbf{A})$, is the order of the largest nonsingular submatrix of $\mathbf{A}$. If $\mathbf{A}$ is nonsingular then $\mathbf{A}$ is said to have *full rank*.

Example

Given the matrix $\mathbf{B} = \begin{bmatrix} 6 & 4 \\ 3 & 2 \\ -1 & 4 \end{bmatrix}$ the rank is the order of the largest non-singular submatrix of $\mathbf{B}$. Since $\mathbf{B}$ is (3×2) the rank cannot exceed 2 since a (2×2) matrix is the largest square submatrix of $\mathbf{B}$. The three possible submatrices are

$$\begin{bmatrix} 6 & 4 \\ 3 & 2 \end{bmatrix}, \quad \begin{bmatrix} 3 & 2 \\ -1 & 4 \end{bmatrix} \quad \text{and} \quad \begin{bmatrix} 6 & 4 \\ -1 & 4 \end{bmatrix}.$$

The first of these submatrices has determinant zero and hence is singular. The remaining two submatrices are nonsingular. The rank of the matrix $\mathbf{B}$ is therefore 2.

Exercise

(a) Verify that the rank of the matrix $\mathbf{A}$ given by $\mathbf{A} = \begin{bmatrix} 2 & 3 & -4 \\ 1 & 4 & 2 \\ 6 & 11 & 0 \end{bmatrix}$ is 2 and determine all (2×2) matrices that have rank 2.

(b) Using the matrix $\mathbf{B}$ in the above example and $\mathbf{A}$ in (a) verify that rank $(\mathbf{BA}) = 2$.

The following properties are useful.
1. If $\mathbf{A}$ and $\mathbf{B}$ are nonsingular matrices with the appropriate dimensions and if $\mathbf{C}$ is arbitrary of appropriate dimension then:

   (a) rank $(\mathbf{AB}) \leq \min[\text{rank } (\mathbf{A}), \text{rank } (\mathbf{B})]$

   (b) rank $(\mathbf{AC}) = \text{rank } (\mathbf{C})$

   (c) rank $(\mathbf{CA}) = \text{rank } (\mathbf{C})$

   (d) rank $(\mathbf{ACB}) = \text{rank } (\mathbf{C})$.

2. The rank of an idempotent matrix is equal to its trace.

1.4 Quadratic Forms and Positive Definite Matrices

Quadratic Form

Given a symmetric matrix $\mathbf{A}$ $(n \times n)$ and an $(n \times 1)$ vector $\mathbf{x}$ a *quadratic form* is the scalar obtained from the product

$$\mathbf{x}'\mathbf{A}\mathbf{x} = \sum_{i=1}^{n}\sum_{j=1}^{n} a_{ij}x_i x_j.$$

The matrix $\mathbf{A}$ must be square, but need not be symmetric, since for any square matrix $\mathbf{B}$ the quadratic form $\mathbf{x}'\mathbf{B}\mathbf{x}$ can be written equivalently in terms of a symmetric matrix $\mathbf{A}$. Thus

$$\mathbf{x}'\mathbf{A}\mathbf{x} = \mathbf{x}'\mathbf{B}\mathbf{x}.$$

There is no loss of generality therefore, in assuming the matrix $\mathbf{A}$ in a quadratic form is symmetric.

Exercise

Given $\mathbf{x} = \begin{bmatrix} a \\ -2a \\ -a \end{bmatrix}$ and $\mathbf{A} = \begin{bmatrix} 1 & 3 & 0 \\ 3 & 2 & -1 \\ 0 & -1 & 4 \end{bmatrix}$ determine the value of $\mathbf{x}'\mathbf{A}\mathbf{x}$ and solve for a in the equation $\mathbf{x}'\mathbf{A}\mathbf{x} = 10$.

Congruent Matrix

A square matrix $\mathbf{B}$ is *congruent* to a square matrix $\mathbf{A}$, if there exists a nonsingular matrix $\mathbf{P}$ such that $\mathbf{A} = \mathbf{P}'\mathbf{B}\mathbf{P}$. By defining the linear transformation $\mathbf{y} = \mathbf{P}\mathbf{x}$ the quadratic form $\mathbf{y}'\mathbf{B}\mathbf{y}$ is equivalent to $\mathbf{x}'\mathbf{A}\mathbf{x}$. If $\mathbf{A}$ is of rank r, there exists a nonsingular matrix $\mathbf{P}$ such that the congruent matrix $\mathbf{B}$ is diagonal. In this case $\mathbf{B}$ has r nonzero diagonal elements and the remaining $(n - r)$ diagonal elements zero, therefore $\mathbf{y}'\mathbf{B}\mathbf{y} = \sum_{i=1}^{r} b_{ii}y_i^2$.

Positive Definite

A real symmetric matrix $\mathbf{A}$ $(n \times n)$ is *positive definite* if the quadratic form $\mathbf{x}'\mathbf{A}\mathbf{x}$ is positive for all $(n \times 1)$ vectors $\mathbf{x}$. If $\mathbf{A}$ is positive definite and nonsingular, then $\mathbf{A}$ is congruent to an identity matrix in that there exists a nonsingular matrix $\mathbf{P}$ such that

$$\mathbf{x}'\mathbf{A}\mathbf{x} = \mathbf{y}'\mathbf{y} \quad \text{where } \mathbf{y} = \mathbf{P}\mathbf{x}.$$

Positive Semidefinite, Negative Definite, Nonnegative Definite

The matrix $\mathbf{A}$ is *positive semidefinite* if $\mathbf{x}'\mathbf{A}\mathbf{x} \geq 0$ for all $\mathbf{x}$, and $\mathbf{x}'\mathbf{A}\mathbf{x} = 0$ for at least one $\mathbf{x}$. It is *negative definite* if $\mathbf{x}'\mathbf{A}\mathbf{x} < 0$ for all $\mathbf{x}$, and *negative semidefinite* if $\mathbf{x}'\mathbf{A}\mathbf{x} \leq 0$ for all $\mathbf{x}$ and $\mathbf{x}'\mathbf{A}\mathbf{x} = 0$ for at least one $\mathbf{x}$. The matrix $\mathbf{A}$ is *nonnegative definite* if it is not *negative definite*.

Exercise

Suppose that $\mathbf{A} = \mathbf{a}\mathbf{a}'$ where $\mathbf{a}(p \times 1)$ and show that $\mathbf{x}'\mathbf{A}\mathbf{x}$ is positive definite if $\mathbf{a} \neq \mathbf{0}$.

Some additional results are:

1. The determinant of a positive definite matrix $|\mathbf{A}|$ is positive.

2. If $\mathbf{A}$ is positive definite (semidefinite) then for any nonsingular matrix $\mathbf{P}$, $\mathbf{P}'\mathbf{A}\mathbf{P}$ is positive definite (semidefinite).

3. $\mathbf{A}$ is positive definite if and only if there exists a nonsingular matrix $\mathbf{V}$ such that $\mathbf{A} = \mathbf{V}'\mathbf{V}$.

4. If $\mathbf{A}$ is $(n \times p)$ of rank p then $\mathbf{A}'\mathbf{A}$ is positive definite and $\mathbf{A}\mathbf{A}'$ is positive semidefinite.

5. If $\mathbf{A}$ is positive definite then $\mathbf{A}^{-1}$ is positive definite.

1.5 Partitioned Matrices

A matrix $\mathbf{A}$ can be *partitioned* into submatrices by drawing horizontal and vertical lines between rows and columns of the matrix. Each element of the original matrix is contained in one and only one submatrix. A *partitioned matrix* is usually called a *block matrix*.

$$\mathbf{A} = \begin{bmatrix} a_{11} & a_{12} & a_{13} & a_{14} & a_{15} \\ a_{21} & a_{22} & a_{23} & a_{24} & a_{25} \\ a_{31} & a_{32} & a_{33} & a_{34} & a_{35} \\ a_{41} & a_{42} & a_{43} & a_{44} & a_{45} \end{bmatrix} = \begin{bmatrix} A_{11} & A_{12} & A_{13} \\ A_{21} & A_{22} & A_{23} \end{bmatrix}.$$

Product of Partitioned Matrices

The product of two block matrices can be determined if the column partitions of the first matrix correspond to the row partitions of the second matrix. As usual the total number of columns of the first matrix must be equal to the total number of rows of the second.

$$\mathbf{B} = \begin{bmatrix} b_{11} & b_{12} & b_{13} \\ b_{21} & b_{22} & b_{23} \\ b_{31} & b_{32} & b_{33} \\ b_{41} & b_{42} & b_{43} \\ b_{51} & b_{52} & b_{53} \end{bmatrix} = \begin{bmatrix} B_{11} & B_{12} \\ B_{21} & B_{22} \\ B_{31} & B_{32} \end{bmatrix} \quad \text{and using } \mathbf{A} \text{ above}$$

$$\mathbf{AB} = \begin{bmatrix} A_{11}B_{11} + A_{12}B_{21} + A_{13}B_{31} & A_{11}B_{12} + A_{12}B_{22} + A_{13}B_{32} \\ A_{21}B_{11} + A_{22}B_{21} + A_{23}B_{31} & A_{21}B_{12} + A_{22}B_{22} + A_{23}B_{32} \end{bmatrix}.$$

Inverse of a Partitioned Matrix

Let the $(n \times p)$ matrix $\mathbf{A}$ be partitioned into four submatrices $\mathbf{A}_{11}(n_1 \times p_1)$, $\mathbf{A}_{12}(n_1 \times (p - p_1))$, $\mathbf{A}_{21}((n - n_1) \times p_1)$ and $\mathbf{A}_{22}((n - n_1) \times (p - p_1))$ and hence $\mathbf{A} = \begin{bmatrix} \mathbf{A}_{11} & \mathbf{A}_{12} \\ \mathbf{A}_{21} & \mathbf{A}_{22} \end{bmatrix}$. If $\mathbf{B} = \mathbf{A}^{-1}$ exists then $\mathbf{B} = \begin{bmatrix} \mathbf{B}_{11} & \mathbf{B}_{12} \\ \mathbf{B}_{21} & \mathbf{B}_{22} \end{bmatrix}$ can be related to $\mathbf{A}$ by the following

1. $\mathbf{B}_{11} = [\mathbf{A}_{11} - \mathbf{A}_{12}\mathbf{A}_{22}^{-1}\mathbf{A}_{21}]^{-1} = [\mathbf{A}_{11}^{-1} + \mathbf{A}_{11}^{-1}\mathbf{A}_{12}\mathbf{B}_{22}\mathbf{A}_{21}\mathbf{A}_{11}^{-1}]$

2. $\mathbf{B}_{12} = -\mathbf{A}_{11}^{-1}\mathbf{A}_{12}[\mathbf{A}_{22} - \mathbf{A}_{21}\mathbf{A}_{11}^{-1}\mathbf{A}_{12}]^{-1} = -\mathbf{A}_{11}^{-1}\mathbf{A}_{12}\mathbf{B}_{22}$

3. $\mathbf{B}_{21} = -\mathbf{A}_{22}^{-1}\mathbf{A}_{21}[\mathbf{A}_{11} - \mathbf{A}_{12}\mathbf{A}_{22}^{-1}\mathbf{A}_{21}]^{-1} = -\mathbf{A}_{22}^{-1}\mathbf{A}_{21}\mathbf{B}_{11}$

4. $\mathbf{B}_{22} = [\mathbf{A}_{22} - \mathbf{A}_{21}\mathbf{A}_{11}^{-1}\mathbf{A}_{12}]^{-1} = [\mathbf{A}_{22}^{-1} + \mathbf{A}_{22}^{-1}\mathbf{A}_{21}\mathbf{B}_{11}\mathbf{A}_{12}\mathbf{A}_{22}^{-1}]$.

Exercise

1. (a) Verify the above expressions for $\mathbf{B} = \mathbf{A}^{-1}$ by multiplication to show that $\mathbf{AB} = \mathbf{I}$.
   (b) Verify the above expressions for $\mathbf{B} = \mathbf{A}^{-1}$ by solving for the four submatrices in $\mathbf{B}$ in the equation $\mathbf{AB} = \mathbf{I}$.
2. Use the formulae for the inverse of a partitioned matrix to invert the matrix $\mathbf{A} = \begin{bmatrix} 1 & 6 & 2 \\ 4 & 5 & -3 \\ -7 & 1 & 8 \end{bmatrix}$ by first determining the inverse of $\begin{bmatrix} 5 & -3 \\ 1 & 8 \end{bmatrix}$.

Determinant of a Partitioned Matrix

The determinant of $|\mathbf{A}|$ can be expressed as

1. $|\mathbf{A}| = |\mathbf{A}_{22}||\mathbf{A}_{11} - \mathbf{A}_{12}\mathbf{A}_{22}^{-1}\mathbf{A}_{22}| = |\mathbf{A}_{11}||\mathbf{A}_{22} - \mathbf{A}_{21}\mathbf{A}_{11}^{-1}\mathbf{A}_{12}|$

2. $|\mathbf{A}| = |\mathbf{A}_{22}|/|\mathbf{B}_{11}| = |\mathbf{A}_{11}|/|\mathbf{B}_{22}|$, where $\mathbf{B}_{11}$ and $\mathbf{B}_{22}$ are defined above with the inverse of a partitioned matrix.

If $\mathbf{A}_{21} = \mathbf{A}_{12} = \mathbf{0}$ then

1. $\mathbf{A}^{-1} = \begin{bmatrix} \mathbf{A}_{11}^{-1} & \mathbf{0} \\ \mathbf{0} & \mathbf{A}_{22}^{-1} \end{bmatrix}.$

2. $|\mathbf{A}| = |\mathbf{A}_{11}||\mathbf{A}_{22}|$.

Exercise

(a) Use the expression for $|\mathbf{A}|$ in (1) above to determine $|\mathbf{A}|$ in the previous exercise.

(b) Show that $|\mathbf{A}| = |\mathbf{A}_{11}||\mathbf{A}_{22}|$ using the matrix $\mathbf{A}$ in the previous exercise.

1.6 Expectations of Random Matrices

An $(n \times p)$ *random matrix* $\mathbf{X}$ is a matrix whose elements x_{ij} $i = 1, 2, \ldots, n$; $j = 1, 2, \ldots, p$ are random variables. The *expected value* of the random matrix $\mathbf{X}$ denoted by $E[\mathbf{X}]$ is the matrix of constants $E[x_{ij}]$ $i = 1, 2, \ldots, n$; $j = 1, 2, \ldots, p$ if these expectations exist.

Let $\mathbf{Y}$ $(n \times p)$ and $\mathbf{X}$ $(n \times p)$ denote random matrices whose expectations $E[\mathbf{Y}]$ and $E[\mathbf{X}]$ exist. Let $\mathbf{A}$ $(m \times n)$ and $\mathbf{B}$ $(p \times k)$ be matrices of constants. The following properties hold:

1. $E[\mathbf{AXB}] = \mathbf{A}E[\mathbf{X}]\mathbf{B}$-

2. $E[\mathbf{X} + \mathbf{Y}] = E[\mathbf{X}] + E[\mathbf{Y}]$.

Let $\mathbf{z}$ $(n \times 1)$ be a random vector and let $\mathbf{A}$ $(n \times n)$ be a matrix of constants. The expected value of the *quadratic form* $\mathbf{z}'\mathbf{A}\mathbf{z}$ is given by

$$E[\mathbf{z}'\mathbf{A}\mathbf{z}] = \sum_{i=1}^{n} \sum_{j=1}^{n} a_{ij} E[z_i z_j]$$ where $\mathbf{A}$ has elements a_{ij} and $\mathbf{z}$ has elements z_i.

Exercise

(a) Show that $E[(\mathbf{X} + \mathbf{A})'(\mathbf{X} + \mathbf{A})] = \mathbf{\Sigma} + \mathbf{A}'\mathbf{\Pi} + \mathbf{\Pi}'\mathbf{A} + \mathbf{A}'\mathbf{A}$ where $\mathbf{X}$ is a random matrix $E[\mathbf{X}] = \mathbf{\Pi}$ and $E[\mathbf{X}'\mathbf{X}] = \mathbf{\Sigma}$.

(b) Show that $E[\mathbf{x}'\mathbf{A}\mathbf{x}] = \sum\limits_{i=1}^{p} \sum\limits_{j=1}^{p} a_{ij}\sigma_{ij} + \sum\limits_{i=1}^{p} \sum\limits_{j=1}^{p} a_{ij}\mu_i\mu_j$ where $E[(x_i - \mu_i)(x_j - \mu_j)] = \sigma_{ij}$, $i, j = 1, 2, \ldots, p$.

1.7 Derivatives of Matrix Expressions

If the elements of the matrix $\mathbf{A}$ are functions of a random variable x then the derivative of $\mathbf{A}$ with respect to x is the matrix whose elements are the derivatives of the elements of $\mathbf{A}$. For $\mathbf{A} = (a_{ij})$, $\dfrac{d\mathbf{A}}{dx} = \left(\dfrac{da_{ij}}{dx}\right)$.

If $\mathbf{x}$ is a $(p \times 1)$ vector, the derivative of any function of $\mathbf{x}$, say $f(\mathbf{x})$, is the $(p \times 1)$ vector of elements $\left(\dfrac{\partial f}{\partial x_j}\right)$ $j = 1, 2, \ldots, p$.

Given a vector $\mathbf{x}$ $(p \times 1)$ and a vector $\mathbf{a}$ $(p \times 1)$, the vector derivative of the scalar $\mathbf{x}'\mathbf{a}$ with respect to the vector $\mathbf{x}$ is denoted by $\dfrac{\partial(\mathbf{x}'\mathbf{a})}{\partial \mathbf{x}} = \mathbf{a}$, which is equivalent to the vector

$$\begin{bmatrix} \dfrac{\partial(\mathbf{x}'\mathbf{a})}{\partial x_1} \\[2mm] \dfrac{\partial(\mathbf{x}'\mathbf{a})}{\partial x_2} \\[2mm] \vdots \\[2mm] \dfrac{\partial(\mathbf{x}'\mathbf{a})}{\partial x_p} \end{bmatrix}.$$

Given a vector $\mathbf{x}$ $(p \times 1)$ and a matrix $\mathbf{A}$ $(p \times p)$, the vector derivative of $\mathbf{A}\mathbf{x}$ is given by $\partial \mathbf{A}/\partial \mathbf{x} = \mathbf{A}'$ or $\partial \mathbf{A}/\partial \mathbf{x}' = \mathbf{A}$.

Given a vector $\mathbf{x}$ $(p \times 1)$ and a matrix $\mathbf{A}$ $(p \times p)$, the vector derivative of the quadratic form $\mathbf{x}'\mathbf{A}\mathbf{x}$ with respect to the vector $\mathbf{x}$ is denoted by $\dfrac{\partial(\mathbf{x}'\mathbf{A}\mathbf{x})}{\partial \mathbf{x}} = 2\mathbf{A}\mathbf{x}$, which is equivalent to the vector

$$\begin{bmatrix} \dfrac{\partial(\mathbf{x}'\mathbf{A}\mathbf{x})}{\partial x_1} \\[2mm] \dfrac{\partial(\mathbf{x}'\mathbf{A}\mathbf{x})}{\partial x_2} \\[2mm] \vdots \\[2mm] \dfrac{\partial(\mathbf{x}'\mathbf{A}\mathbf{x})}{\partial x_p} \end{bmatrix}.$$

Given a matrix $\mathbf{X}$ $(n \times m)$ then the derivative of a function of $\mathbf{X}$, say $f(\mathbf{X})$, with respect to the elements of $\mathbf{X}$ is the $(n \times m)$ matrix of partial derivatives $\left(\frac{\partial f}{\partial x_{ij}}\right)$. Some useful properties of the *matrix derivative* involving the trace operator are given by

1. $\frac{\partial}{\partial \mathbf{X}} tr(\mathbf{X}) = \mathbf{I}$

2. $\frac{\partial}{\partial \mathbf{X}} tr(\mathbf{AX}) = \mathbf{A}'$

3. $\frac{\partial}{\partial \mathbf{X}} tr(\mathbf{X'AX}) = (\mathbf{A} + \mathbf{A}')\mathbf{X}$.

The derivative of a determinant $|\mathbf{X}|$ of a matrix $\mathbf{X}$, with respect to the elements of the matrix, is given by $\frac{\partial |\mathbf{X}|}{\partial \mathbf{X}} = (\mathbf{X}')^* =$ adjoint of $\mathbf{X}'$.

Exercise

(a) Show that if $f = (\mathbf{a} - \mathbf{Bx})'(\mathbf{a} - \mathbf{Bx})$ where $\mathbf{a}$ $(n \times 1)$, $\mathbf{B}$ $(n \times p)$ and $\mathbf{x}$ $(p \times 1)$ then $\frac{\partial f}{\partial \mathbf{x}} = [2\mathbf{B'Bx} - 2\mathbf{B'a}]$.

(b) Show that if $\mathbf{f} = tr(\mathbf{A} - \mathbf{BX})'(\mathbf{A} - \mathbf{BX})$ where $\mathbf{A}$ $(n \times p)$, $\mathbf{B}$ $(n \times r)$ and $\mathbf{X}$ $(r \times p)$ then $\frac{\partial f}{\partial \mathbf{X}} = [2\mathbf{B'BX} - 2\mathbf{B'A}]$.

2. Linear Algebra

2.1 Geometric Representation for Vectors

A vector $\mathbf{x}' = (x_1, x_2, \ldots, x_n)$ can be represented geometrically in an *n-dimensional space*, as a *directed line segment* from the origin to the point with coordinates $(x_1, x_2, \ldots, x_n)$. The n-dimensional *vector space* is formed by n mutually perpendicular axes $X_1, X_2, \ldots, X_n$. The *coordinates* $(x_1, x_2, \ldots, x_n)$ are values of $X_1, X_2, \ldots, X_n$ respectively. Figure A.1 shows the vectors $\mathbf{x}_1' = (x_{11}, x_{21}, x_{31})$ and $\mathbf{x}_2' = (x_{12}, x_{22}, x_{32})$ in a three-dimensional space. The *addition of the two vectors* $\mathbf{x}_1' + \mathbf{x}_2' = (x_{11} + x_{12}, x_{21} + x_{22}, x_{31} + x_{32})$ is also shown in Figure A.1. Figure A.2 shows *scalar multiplication* by a scalar k for a two-dimensional vector $\mathbf{x}' = (x_{11}, x_{21})$, where $k\mathbf{x}' = (kx_{11}, kx_{21})$.

The *length of the vector* $\mathbf{x}' = (x_1, x_2, \ldots, x_n)$ is given by the *Euclidean distance* between the origin and the point $(x_1, x_2, \ldots, x_n)$ and is denoted by

$$\|\mathbf{x}\| = \sqrt{(x_1 - 0)^2 + (x_2 - 0)^2 + \cdots + (x_n - 0)^2} = \left[\sum_{i=1}^{n} x_i^2\right]^{1/2}.$$

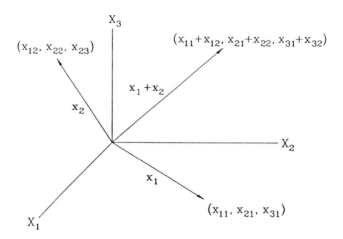

Figure A.1. Addition of Two Vectors

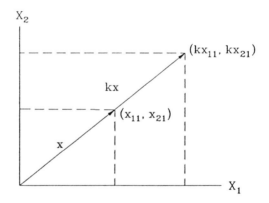

Figure A.2. Scalar Multiplication

The *angle θ between* the vector $\mathbf{x}_1 = (x_{11}, x_{21}, \ldots, x_{n1})$ and the vector $\mathbf{x}_2 = (x_{12}, x_{22}, \ldots, x_{n2})$ is given by

$$\text{Cos } \theta = \sum_{i=1}^{n} x_{i1}x_{i2} / \left[\sum_{i=1}^{n} x_{i1}^2 \right]^{1/2} \left[\sum_{i=1}^{n} x_{i2}^2 \right]^{1/2}.$$

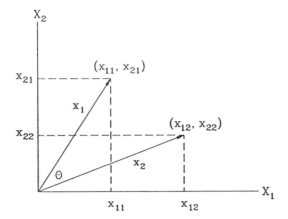

Figure A.3. Angle Between Two Vectors

The matrix product of the $(n \times 1)$ vectors $\mathbf{x}_1$ and $\mathbf{x}_2$ is given by $\mathbf{x}_1' \mathbf{x}_2 = \mathbf{x}_2' \mathbf{x}_1$

$$\mathbf{x}_1' \mathbf{x}_2 = \mathbf{x}_2' \mathbf{x}_1 = \sum_{i=1}^{n} x_{i1} x_{i2}$$

and hence

$$\text{Cos } \theta = \mathbf{x}_1' \mathbf{x}_2 / (\mathbf{x}_1' \mathbf{x}_1)^{1/2} (\mathbf{x}_2' \mathbf{x}_2)^{1/2}.$$

In Figure A.3 the angle θ between $\mathbf{x}_1' = (x_{11}, x_{21})$ and $\mathbf{x}_2' = (x_{12}, x_{22})$ is given by $\text{Cos } \theta = [x_{11} x_{12} + x_{21} x_{22}] / [x_{11}^2 + x_{21}^2]^{1/2} [x_{12}^2 + x_{22}^2]^{1/2}$. If $\text{Cos } \theta = \pm 1$ the vectors $\mathbf{x}_1$ and $\mathbf{x}_2$ are said to be *orthogonal*.

The distance from the vector point $\mathbf{x}_1' = (x_{11}, x_{21}, \ldots, x_{n1})$ to the vector point $\mathbf{x}_2' = (x_{12}, x_{22}, \ldots, x_{n2})$ is the Euclidean distance between the two points

$$d = \sqrt{(x_{11} - x_{12})^2 + (x_{21} - x_{22})^2 + \cdots + (x_{n1} - x_{n2})^2}.$$

The *projection* of the vector $\mathbf{x}_2$ on the vector $\mathbf{x}_1$ is a vector which is a scalar multiple of $\mathbf{x}_1$ given by

$$\mathbf{x}_2^* = \left[\frac{\|\mathbf{x}_2\|}{\|\mathbf{x}_1\|} \text{Cos } \theta \right] \mathbf{x}_1 = k \mathbf{x}_1,$$

where θ is the angle between $\mathbf{x}_2$ and $\mathbf{x}_1$, and k is the scalar given by $\frac{\|\mathbf{x}_2\|}{\|\mathbf{x}_1\|} \text{Cos } \theta$.

Exercise

(a) Given $\mathbf{x}_1 = \begin{bmatrix} 2 \\ 1 \end{bmatrix}$ and $\mathbf{x}_2 = \begin{bmatrix} 1 \\ 3 \end{bmatrix}$ plot the vectors $\mathbf{x}_1$ and $\mathbf{x}_2$ in the two-dimensional space formed by the axes X_1 and X_2.

(b) Plot the vector $(\mathbf{x}_1 + \mathbf{x}_2)$ and the vector $2\mathbf{x}_1$ in the two-dimensional space in (a).

(c) Determine the length of the vectors $\mathbf{x}_1$, $\mathbf{x}_2$ and $(\mathbf{x}_1 + \mathbf{x}_2)$.

(d) Determine the angle θ between the vectors $\mathbf{x}_1$ and $\mathbf{x}_2$.

(e) Determine the distance between the tips of the vectors $\mathbf{x}_1$ and $\mathbf{x}_2$.

(f) Determine the projection $\hat{\mathbf{x}}_2$ of the vector $\mathbf{x}_2$ on $\mathbf{x}_1$ and determine the vector $\mathbf{x}_2^* = \mathbf{x}_2 - \hat{\mathbf{x}}_2$. Show that $\hat{\mathbf{x}}_2$ and $\mathbf{x}_2^*$ are orthogonal.

2.2 Linear Dependence & Linear Transformations

Linearly Dependent Vectors

If one vector $\mathbf{x}$ in n-dimensional space can be written as a linear combination of other vectors in n-dimensional space, then $\mathbf{x}$ is said to be *linearly dependent* on the other vectors. If

$$\mathbf{x} = a_1\mathbf{v}_1 + a_2\mathbf{v}_2 + \cdots + a_p\mathbf{v}_p$$

then $\mathbf{x}$ is linearly dependent on $\mathbf{v}_1, \ldots, \mathbf{v}_p$.

The set of $(p+1)$ vectors $(\mathbf{x}, \mathbf{v}_1, \ldots, \mathbf{v}_p)$ is also a *linearly dependent set*, if any one of the vectors can be written as a linear combination of the remaining vectors in the set.

Linearly Independent Vectors

A set of vectors is *linearly independent*, if it is not possible to express one vector as a linear combination of the remaining vectors in the set. In an n-dimensional space the maximum number of linearly independent vectors is n. Thus given any set of n linearly independent vectors, all other vectors in the n-dimensional space can be expressed as a linear combination of the linearly independent set.

Basis for an n-Dimensional Space

The set of n linearly independent vectors is said to generate a *basis* for the n-dimensional space.

In two-dimensional space linearly dependent vectors are co-linear or lie along the same line. In three-dimensional space linearly dependent vectors are in the same two-dimensional plane and may be co-linear. If the set of linearly independent vectors are mutually orthogonal then the basis is said to be orthogonal.

Generation of a Vector Space and Rank of a Matrix

If the matrix $\mathbf{A}$ $(n \times p)$ has rank r, then the vectors formed by the p columns of $\mathbf{A}$ generate an r-dimensional vector space. Similarly the rows of $\mathbf{A}$ generate an r-dimensional vector space. In other words, the rank of a matrix is the maximum number of linearly independent columns, and equivalently the maximum number of linearly independent rows.

Exercise

(a) Given the vectors $\mathbf{x}_1 = \begin{bmatrix} 1 \\ -1 \\ 0 \end{bmatrix}$, $\mathbf{x}_2 = \begin{bmatrix} 0 \\ 1 \\ 2 \end{bmatrix}$ and $\mathbf{x}_3 = \begin{bmatrix} 2 \\ 0 \\ 1 \end{bmatrix}$ show that they are linearly independent by showing that each cannot be written as a linear combination of the remaining two.

(b) Let $\mathbf{A} = [\mathbf{x}_1, \mathbf{x}_2, \mathbf{x}_3]$ and determine $|\mathbf{A}|$ for the values of $\mathbf{x}_1, \mathbf{x}_2, \mathbf{x}_3$ given in (a).

Linear Transformation

Given a p-dimensional vector $\mathbf{x}$, a *linear transformation* of $\mathbf{x}$, is given by the matrix product

$$\mathbf{y} = \mathbf{A}\mathbf{x},$$

where $\mathbf{A}$ $(n \times p)$ is called a transformation matrix. The matrix $\mathbf{A}$ maps the n-dimensional vector $\mathbf{x}$ into a p-dimensional vector y. If $\mathbf{A}$ is a square nonsingular matrix, the transformation is *one to one*, in that $\mathbf{x} = \mathbf{A}^{-1}\mathbf{y}$ and hence each point in $\mathbf{x}$ corresponds to exactly one and only one point in $\mathbf{y}$.

Orthogonal Transformation, Rotation, Orthogonal Matrix

If $\mathbf{A}$ is a square matrix and $\mathbf{A}$ has the property that $\mathbf{A}' = \mathbf{A}^{-1}$, then the equation $\mathbf{y} = \mathbf{A}\mathbf{x}$ is an *orthogonal transformation* or *rotation*. The matrix $\mathbf{A}$ in this case is called an *orthogonal matrix*. If $\mathbf{A}$ is orthogonal, $\mathbf{A}$ also has the property that $|\mathbf{A}| = \pm 1$. The transformation is orthogonal because it can be viewed as a rotation of the coordinate axes through some angle θ. The angle θ between any pair of vectors remains the same after the transformation.

Example

In Figure A.4, the point M has coordinates (x_1, x_2) with respect to the $X_1 - X_2$ axes and has coordinates (z_1, z_2) with respect to the $Z_1 - Z_2$ axes. The angle ϕ between X_1 and Z_1 is the angle of rotation required to rotate the $X_1 - X_2$ axes into the $Z_1 - Z_2$ axes. The coordinates (z_1, z_2) of the point M in $Z_1 - Z_2$ space can be described in terms of the coordinates (x_1, x_2) of M in $X_1 - X_2$ space using the equations

$$z_1 = x_1 \cos \phi + x_2 \sin \phi$$
$$z_2 = -x_1 \sin \phi + x_2 \cos \phi.$$

In matrix notation the linear transformation to $\begin{bmatrix} z_1 \\ z_2 \end{bmatrix}$ from $\begin{bmatrix} x_1 \\ x_2 \end{bmatrix}$ can be

expressed as $\begin{bmatrix} z_1 \\ z_2 \end{bmatrix} = \begin{bmatrix} \cos \phi & \sin \phi \\ -\sin \phi & \cos \phi \end{bmatrix} \begin{bmatrix} x_1 \\ x_2 \end{bmatrix}$. The transformation matrix

$\mathbf{A} = \begin{bmatrix} \cos \phi & \sin \phi \\ -\sin \phi & \cos \phi \end{bmatrix}$ has the property that

$$\mathbf{A}^{-1} = \frac{1}{\cos^2 \phi + \sin^2 \phi} \begin{bmatrix} \cos \phi & -\sin \phi \\ \sin \phi & \cos \phi \end{bmatrix} = \begin{bmatrix} \cos \phi & -\sin \phi \\ \sin \phi & \cos \phi \end{bmatrix} = \mathbf{A}',$$

and hence the transformation is orthogonal as required.

Exercise

(a) Let $\mathbf{A} = \begin{bmatrix} \sqrt{2}/2 & -\sqrt{2}/2 \\ \sqrt{2}/2 & \sqrt{2}/2 \end{bmatrix}$ denote a transformation matrix. Show that $\mathbf{A}^{-1} = \mathbf{A}'$ and hence that $\mathbf{A}$ is an orthogonal transformation matrix.

(b) Use the results of the above example to show that the angle of rotation in (a) is $45°$.

(c) Give the transformation matrix corresponding to $\phi = 60°$.

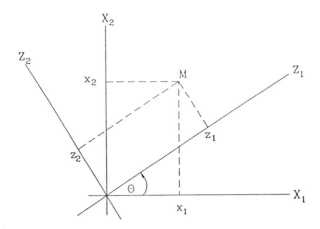

Figure A.4. Rotation of Axes

2.3 Systems of Equations

Let x_1, x_2, x_3 denote the unknowns in a system of 3 equations

$$a_{11}x_1 + a_{12}x_2 + a_{13}x_3 = b_1$$

$$a_{21}x_1 + a_{22}x_2 + a_{23}x_3 = b_2$$

$$a_{31}x_1 + a_{32}x_2 + a_{33}x_3 = b_3.$$

The *system of equations* can be represented by a matrix equation

$$\mathbf{Ax} = \mathbf{b}$$

where

$$\mathbf{A} = \begin{bmatrix} a_{11} & a_{12} & a_{12} \\ a_{21} & a_{22} & a_{23} \\ a_{31} & a_{32} & a_{33} \end{bmatrix}, \ \mathbf{x} = \begin{bmatrix} x_1 \\ x_2 \\ x_3 \end{bmatrix}, \ b = \begin{bmatrix} b_1 \\ b_2 \\ b_3 \end{bmatrix}.$$

Solution Vector for a System of Equations

The *solution vector* $\mathbf{x}$ is given by

$$\mathbf{x} = \mathbf{A}^{-1}\mathbf{b}.$$

Exercise

1. Given the three linearly independent vectors $\mathbf{x}_1 = \begin{bmatrix} 1 \\ -1 \\ 0 \end{bmatrix}$, $\mathbf{x}_2 = \begin{bmatrix} 0 \\ -1 \\ 2 \end{bmatrix}$

   and $\mathbf{x}_3 = \begin{bmatrix} 2 \\ 0 \\ 1 \end{bmatrix}$. If $\mathbf{y} = \begin{bmatrix} 5 \\ 3 \\ 2 \end{bmatrix}$ is contained in the space generated

   by $\mathbf{x}_1$, $\mathbf{x}_2$ and $\mathbf{x}_3$ then there exist coefficients a, b and c such that
   $\mathbf{y} = a\mathbf{x}_1 + b\mathbf{x}_2 + c\mathbf{x}_3$. Show that the equation for $\mathbf{y}$ can be written as
   $\mathbf{y} = \mathbf{A}\mathbf{f}$ where $\mathbf{f} = \begin{bmatrix} a \\ b \\ c \end{bmatrix}$ and solve for $\mathbf{f}$.

2. Solve the system of equations

$$
3x_1 + x_2 - 4x_3 = 3
$$
$$
2x_1 + 5x_2 + x_3 = 10
$$
$$
x_1 - 4x_2 + 3x_3 = -6
$$

   for x_1, x_2 and x_3 using the expression $\mathbf{x} = \mathbf{A}^{-1}\mathbf{y}$.

Homogeneous Equations — Trivial and Nontrivial Solutions

If the vector $\mathbf{b}$ in the system $\mathbf{A}\mathbf{x} = \mathbf{b}$ is the null vector $\mathbf{0}$, the system is
said to be a system of *homogeneous equations*. The obvious solution $\mathbf{x} = \mathbf{0}$
is said to be a *trivial solution*. If $\mathbf{A}$ is $(n \times p)$ of rank r, where $\mathbf{x}$ and $\mathbf{b}$ are
$(p \times 1)$, then the system of n homogeneous equations in p unknowns given
by $\mathbf{A}\mathbf{x} = \mathbf{b}$ has $(p - r)$ linearly independent solutions in addition to the
trivial solution $\mathbf{x} = \mathbf{0}$. In other words, a *nontrivial solution* exists if and
only if the rank of $\mathbf{A}$ is less than p. If $\mathbf{A}$ is a square matrix a nontrivial
solution exists if and only if $|\mathbf{A}| = 0$.

2.4 Column Spaces, Projection Operators and Least Squares

Column Space

Given an $(n \times p)$ matrix $\mathbf{X}$, the columns of $\mathbf{X}$ denoted by the p vectors
$\mathbf{x}_1, \mathbf{x}_2, \ldots, \mathbf{x}_p$ span a vector space called the *column space* of $\mathbf{X}$. The set of
all vectors $\hat{\mathbf{y}}$ $(n \times 1)$, defined by $\hat{\mathbf{y}} = \mathbf{X}\mathbf{b}$ for all vectors $\mathbf{b}$ $(p \times 1)$, $\mathbf{b} \neq \mathbf{0}$,
is the vector space generated by the columns of $\mathbf{X}$.

Orthogonal Complement

The set of all vectors $\mathbf{z}$ $(p \times 1)$ such that $\mathbf{Xz} = \mathbf{0}$ generates the *orthogonal complement* to the column space of $\mathbf{X}$.

Projection

Given a vector $\mathbf{y}$ $(n \times 1)$ and a p-dimensional column space defined by the matrix $\mathbf{X}$ $(n \times p)$, $p \leq n$, the vector $\hat{\mathbf{y}}$ $(n \times 1)$ is a *projection* of $\mathbf{y}$ onto the column space of $\mathbf{X}$, if there exists a $(p \times 1)$ vector $\mathbf{b}$ such that $\hat{\mathbf{y}} = \mathbf{Xb}$, and a vector $\mathbf{e}$ $(n \times 1)$ such that $\mathbf{e}$ is in the orthogonal complement to the column space of $\mathbf{X}$. If $\hat{\mathbf{y}}$ is the projection of $\mathbf{y}$ then $\mathbf{y} = \hat{\mathbf{y}} + \mathbf{e}$, and the vector $\mathbf{e}$ is the part of $\mathbf{y}$ that is orthogonal to the column space of $\mathbf{X}$.

Ordinary Least Squares Solution Vector

Given an $(n \times p)$ matrix $\mathbf{X}$ of rank p, and a vector $\mathbf{y}$ $(n \times 1)$, $n \geq p$, the projection of $\mathbf{y}$ onto the column space of $\mathbf{X}$ is given by the *ordinary least squares solution vector*

$$\hat{\mathbf{y}} = \mathbf{X}(\mathbf{X'X})^{-1}\mathbf{X'y} \quad \text{where } \mathbf{b} = (\mathbf{X'X})^{-1}\mathbf{X'y}.$$

Idempotent Matrix — Projection Operator

The matrix $\mathbf{X}(\mathbf{X'X})^{-1}\mathbf{X'}$ is an *idempotent matrix* and is called the *projection operator* for the column space of $\mathbf{X}$. The vector $\mathbf{e} = (\mathbf{y} - \hat{\mathbf{y}}) = (\mathbf{I} - \mathbf{X}(\mathbf{X'X})^{-1}\mathbf{X'})\mathbf{y}$ is orthogonal to the column space of $\mathbf{X}$. The idempotent matrix $(\mathbf{I} - \mathbf{X}(\mathbf{X'X})^{-1}\mathbf{X'})$ is called the projection operator for the vector space orthogonal to the column space of $\mathbf{X}$.

Exercise

1. Given $\mathbf{X}$ $(n \times p)$ and $\mathbf{y}$ $(n \times 1)$ demonstrate the following
   (a) $\mathbf{X}(\mathbf{X'X})^{-1}\mathbf{X'}$ is idempotent.

   (b) $(\mathbf{I} - \mathbf{X}(\mathbf{X'X})^{-1}\mathbf{X'})$ is idempotent.

   (c) $[\mathbf{I} - \mathbf{X}(\mathbf{X'X})^{-1}\mathbf{X'}][\mathbf{X}(\mathbf{X'X})^{-1}\mathbf{X'}] = \mathbf{0}$.

   (d) $\mathbf{y'}(\mathbf{y} - \hat{\mathbf{y}}) = \mathbf{0}$ where $\hat{\mathbf{y}} = \mathbf{X}(\mathbf{X'X})^{-1}\mathbf{X'y}$.

2. Given $\mathbf{X} = \begin{bmatrix} 1 & 1 \\ 1 & 2 \\ 1 & 3 \\ 1 & 4 \\ 1 & 5 \end{bmatrix}$ $\mathbf{y} = \begin{bmatrix} 3 \\ 5 \\ 4 \\ 7 \\ 6 \end{bmatrix}$ determine $\hat{\mathbf{b}} = (\mathbf{X}'\mathbf{X})^{-1}\mathbf{X}'\mathbf{y}$, $\hat{\mathbf{y}} = $
$\mathbf{X}(\mathbf{X}'\mathbf{X})^{-1}\mathbf{X}'\mathbf{y}$ and $(\mathbf{y} - \hat{\mathbf{y}})$.

3. Eigenvalue Structure and Singular Value Decomposition

3.1 Eigenvalue Structure for Square Matrices

Eigenvalues and Eigenvectors

Given a square matrix $\mathbf{A}$ of order n, the values of the scalars λ and $(n \times 1)$ vectors $\mathbf{v}$, $\mathbf{v} \neq \mathbf{0}$, which satisfy the equation

$$\mathbf{A}\mathbf{v} = \lambda\mathbf{v}, \tag{A.1}$$

are called the *eigenvalues* and *eigenvectors* of the matrix $\mathbf{A}$. The problem of finding λ and $\mathbf{v}$ in (A.1) is commonly referred to as the *eigenvalue problem*. From (A.1) it can be seen that if $\mathbf{v}$ is a solution then $k\mathbf{v}$ is also a solution where k is an arbitrary scalar. The eigenvectors are therefore unique up to a multiplicative constant. It is common to impose the additional constraint that $\mathbf{v}'\mathbf{v} = 1$, and hence that the eigenvector be normalized to have a length of 1.

Characteristic Polynomial, Characteristic Roots, Latent Roots, Eigenvalues

Rewriting the equation (A.1) as

$$(\mathbf{A} - \lambda\mathbf{I})\mathbf{v} = \mathbf{0}$$

we obtain a system of homogeneous equations. A nontrivial solution $\mathbf{v} \neq \mathbf{0}$ requires that

$$|\mathbf{A} - \lambda\mathbf{I}| = 0.$$

This equation yields a polynomial in λ of degree n and is commonly called the *characteristic polynomial*. The solutions of the equation are the roots of the polynomial, and are sometimes referred to as *characteristic roots* or *latent roots*, although the most common term used in statistics is *eigenvalues*.

The characteristic polynomial has n roots or eigenvalues some of which may be equal. For each eigenvalue λ there is a corresponding *eigenvector* $\mathbf{v}$ satisfying (A.1). The matrix $\mathbf{A}$ is singular if and only if at least one eigenvalue is zero.

Some additional properties of eigenvalues and eigenvectors are:

1. The eigenvalues of a diagonal matrix are the diagonal elements;

2. The matrices $\mathbf{A}$ and $\mathbf{A}'$ have the same eigenvalues but not necessarily the same eigenvectors;

3. If λ is an eigenvalue of $\mathbf{A}$ then $1/\lambda$ is an eigenvalue of $\mathbf{A}^{-1}$.

4. If λ is an eigenvalue of $\mathbf{A}$ and $\mathbf{v}$ the corresponding eigenvector then for the matrix $\mathbf{A}^k$, λ^k is an eigenvalue with corresponding eigenvector $\mathbf{v}$.

5. If the eigenvalues of $\mathbf{A}$ are denoted by $\lambda_1, \lambda_2, \ldots, \lambda_n$ then

$$tr\mathbf{A} = \sum_{j=1}^{n}\lambda_j \quad \text{and} \quad |\mathbf{A}| = \prod_{j=1}^{n}\lambda_j.$$

Eigenvalues and Eigenvectors for Real Symmetric Matrices and Some Properties

In statistics the matrix $\mathbf{A}$ will usually have real elements and be symmetric. In this case the eigenvalues and eigenvectors have additional properties:

1. If the rank of $\mathbf{A}$ is r then $(n - r)$ of the eigenvalues are zero;

2. If k of the eigenvalues are equal the eigenvalue is said to have multiplicity k. In this case there will be k orthogonal eigenvectors corresponding to the common eigenvalue;

3. If two eigenvalues are distinct then the corresponding eigenvectors are orthogonal;

4. An n-th order symmetric matrix produces a set of n orthogonal eigenvectors.

5. For idempotent matrices the eigenvalues are zero or one.

6. The maximum value of the quadratic form $\mathbf{v}'\mathbf{A}\mathbf{v}$ subject to $\mathbf{v}'\mathbf{v} = 1$ is given by $\lambda_1 = \mathbf{v}_1'\mathbf{A}\mathbf{v}_1$, where λ_1 is the largest eigenvalue of $\mathbf{A}$ and $\mathbf{v}_1$ is the corresponding eigenvector. Similarly the second largest eigenvalue $\lambda_2 = \mathbf{v}_2'\mathbf{A}\mathbf{v}_2$ of $\mathbf{A}$ is the maximum value of $\mathbf{v}'\mathbf{A}\mathbf{v}$ subject to $\mathbf{v}_2'\mathbf{v}_1 = 0$ and $\mathbf{v}_2'\mathbf{v}_2 = 1$, where $\mathbf{v}_2$ is the eigenvector corresponding to λ_2. The k-th largest eigenvalue $\lambda_k = \mathbf{v}_k'\mathbf{A}\mathbf{v}_k$ of $\mathbf{A}$ is the maximum value of $\mathbf{v}'\mathbf{A}\mathbf{v}$ subject to $\mathbf{v}_k'\mathbf{v}_1 = \mathbf{v}_k'\mathbf{v}_2 = \cdots = \mathbf{v}_k'\mathbf{v}_{(k-1)} = 0$ and $\mathbf{v}_k'\mathbf{v}_k = 1$, where $\mathbf{v}_k$ is the eigenvector corresponding to λ_k.

The above properties are essentially summarized by the following statement. If $\mathbf{A}$ is real symmetric of order n there exists an orthogonal matrix $\mathbf{V}$ such that

$$\mathbf{V'AV} = \mathbf{\Lambda} \quad \text{or} \quad \mathbf{A} = \mathbf{V\Lambda V'}$$

where $\mathbf{\Lambda}$ is a diagonal matrix of eigenvalues of $\mathbf{A}$ with diagonal elements $\lambda_1, \lambda_2, \ldots, \lambda_n$; and $\mathbf{V}$ is the matrix whose columns are the corresponding eigenvectors $\mathbf{v}_1, \mathbf{v}_2, \ldots, \mathbf{v}_n$. The orthogonal matrix $\mathbf{V}$ diagonalizes the matrix $\mathbf{A}$.

7. If $\mathbf{A}$ is non-singular and symmetric then $\mathbf{A}^n = \mathbf{V\Lambda^n V'}$ where $\mathbf{A} = \mathbf{V\Lambda V'}$.

8. If $\mathbf{A} = \mathbf{V\Lambda V'}$ then the quadratic form $\mathbf{x'Ax}$ can be written as $\mathbf{y'\Lambda y}$ where $\mathbf{y} = \mathbf{Vx}$.

Example

Given the matrix

$$\mathbf{A} = \begin{bmatrix} 6 & \sqrt{\frac{15}{2}} & 0 \\ \sqrt{\frac{15}{2}} & 4 & \sqrt{\frac{3}{2}} \\ 0 & \sqrt{\frac{3}{2}} & 2 \end{bmatrix}$$

the eigenvalues are obtained by solving the determinantal equation $|\mathbf{A} - \lambda\mathbf{I}| = 0$ which in this case is given by

$$\begin{vmatrix} (6-\lambda) & \sqrt{\frac{15}{2}} & 0 \\ \sqrt{\frac{15}{2}} & (4-\lambda) & \sqrt{\frac{3}{2}} \\ 0 & \sqrt{\frac{3}{2}} & (2-\lambda) \end{vmatrix} = 0.$$

The resulting characteristic polynomial is given by

$$(6-\lambda)(4-\lambda)(2-\lambda) - \frac{3}{2}(6-\lambda) - \frac{15}{2}(2-\lambda) = 0.$$

The characteristic roots or eigenvalues determined from this polynomial are $\lambda = 1$, 3 and 8.

The corresponding eigenvectors are obtained by solving the equation $(\mathbf{A} - \lambda\mathbf{I})\mathbf{v} = 0$ for each of the eigenvalues λ determined above. Corresponding to $\lambda = 3$ the equation becomes

$$\begin{bmatrix} 3 & \sqrt{\frac{15}{2}} & 0 \\ \sqrt{\frac{15}{2}} & 1 & \sqrt{\frac{3}{2}} \\ 0 & \sqrt{\frac{3}{2}} & -1 \end{bmatrix} \begin{bmatrix} v_1 \\ v_2 \\ v_3 \end{bmatrix} = 0.$$

Adding the condition that $v_1^2 + v_2^2 + v_3^2 = 1$ yields the eigenvector

$$\mathbf{v}' = \left[-\tfrac{1}{2} \quad \sqrt{\tfrac{3}{10}} \quad \sqrt{\tfrac{9}{20}} \right],$$

the remaining two eigenvectors are given by

$$\left[-\sqrt{\tfrac{3}{28}} \quad \sqrt{\tfrac{5}{14}} \quad -\sqrt{\tfrac{15}{28}} \right] \quad \text{and} \quad \left[\sqrt{\tfrac{9}{14}} \quad \sqrt{\tfrac{12}{35}} \quad \sqrt{\tfrac{1}{70}} \right]$$

corresponding to $\lambda = 1$ and $\lambda = 8$ respectively. The complete matrix of eigenvectors is given by

$$\mathbf{V} = \begin{bmatrix} -\sqrt{\tfrac{3}{28}} & -\tfrac{1}{2} & \sqrt{\tfrac{9}{14}} \\ \sqrt{\tfrac{5}{14}} & \sqrt{\tfrac{3}{10}} & \sqrt{\tfrac{12}{35}} \\ -\sqrt{\tfrac{15}{28}} & \sqrt{\tfrac{9}{20}} & \sqrt{\tfrac{1}{70}} \end{bmatrix}$$

corresponding to $\lambda = 1, 3$ and 8 respectively. The reader should verify that $\mathbf{V}'\mathbf{V} = \mathbf{I}$.

Exercise

(a) Show that the eigenvalues of the matrix $\mathbf{A} = \begin{bmatrix} 2 & 1 & 1 \\ 1 & 2 & 1 \\ 1 & 1 & 2 \end{bmatrix}$ are given by 1, 1 and 4.

(b) Show that $\mathbf{v}' = (\tfrac{1}{\sqrt{3}}, \tfrac{1}{\sqrt{3}}, \tfrac{1}{\sqrt{3}})$ is an eigenvector corresponding to $\lambda = 4$.

(c) Show that the remaining two eigenvectors corresponding to the double root $\lambda = 1$ are

$$\left(0, \ \tfrac{1}{\sqrt{2}}, \ -\tfrac{1}{\sqrt{2}} \right) \quad \text{and} \quad \left(-\tfrac{2}{\sqrt{6}}, \ \tfrac{1}{\sqrt{6}}, \ \tfrac{1}{\sqrt{6}} \right).$$

(d) Show that the 3×3 matrix of eigenvectors

$$\mathbf{V} = \begin{bmatrix} \tfrac{1}{\sqrt{3}} & 0 & -\tfrac{2}{\sqrt{6}} \\ \tfrac{1}{\sqrt{3}} & \tfrac{1}{\sqrt{2}} & \tfrac{1}{\sqrt{6}} \\ \tfrac{1}{\sqrt{3}} & -\tfrac{1}{\sqrt{2}} & \tfrac{1}{\sqrt{6}} \end{bmatrix}$$

satisfies $\mathbf{V}'\mathbf{V} = \mathbf{I}$.

Spectral Decomposition

The equation $\mathbf{A} = \mathbf{V\Lambda V}'$ can be written

$$\mathbf{A} = \lambda_1 \mathbf{v}_1 \mathbf{v}_1' + \lambda_2 \mathbf{v}_2 \mathbf{v}_2' + \cdots + \lambda_n \mathbf{v}_n \mathbf{v}_n'$$

where $\lambda_1, \lambda_2, \ldots, \lambda_n$ are the eigenvalues of $\mathbf{A}$ and $\mathbf{v}_1, \mathbf{v}_2, \ldots, \mathbf{v}_n$ are the corresponding eigenvectors. This equation gives the *spectral decomposition* of $\mathbf{A}$.

Matrix Approximation

Let $\mathbf{Z}$ $(n \times p)$ be a matrix such that $\mathbf{A} = \mathbf{Z}'\mathbf{Z}$ and let $\mathbf{A}(\ell)$ denote the first ℓ terms of the spectral decomposition $\mathbf{A}(\ell) = \sum_{j=1}^{\ell} \lambda_j \mathbf{v}_j \mathbf{v}_j'$, $\ell < n$. This expression minimizes $tr(\mathbf{Z} - \mathbf{X})(\mathbf{Z} - \mathbf{X})' = \sum_{i=1}^{n} \sum_{j=1}^{p} (a_{ij} - x_{ij})^2$ amongst all $(n \times p)$ matrices $\mathbf{X}$ of rank ℓ. Thus the first few terms of the spectral decomposition of $\mathbf{A} = \mathbf{Z}'\mathbf{Z}$ can be used to provide a matrix approximation to $\mathbf{A}$.

Example

For the matrix $\mathbf{A}$ of the previous example the matrix of eigenvalues can be approximated in decimal form by

$$\mathbf{V} = \begin{bmatrix} -0.327 & -0.500 & 0.802 \\ 0.598 & 0.547 & 0.586 \\ -0.732 & 0.671 & 0.120 \end{bmatrix}.$$

The spectral decomposition for $\mathbf{A}$ is given by

$$\mathbf{A} = 1 \begin{bmatrix} (0.327)^2 & -(0.327)(0.598) & (0.327)(0.732) \\ -(0.327)(0.598) & (0.598)^2 & -(0.598)(0.732) \\ (0.327)(0.732) & -(0.598)(0.732) & (0.732)^2 \end{bmatrix}$$

$$+ 3 \begin{bmatrix} (0.500)^2 & -(0.500)(0.547) & -(0.500)(0.671) \\ -(0.500)(0.547) & (0.547)^2 & (0.547)(0.671) \\ -(0.500)(0.671) & (0.547)(0.671) & (0.671)^2 \end{bmatrix}$$

$$+ 8 \begin{bmatrix} (0.802)^2 & (0.802)(0.586) & (0.802)(0.120) \\ (0.802)(0.586) & (0.586)^2 & (0.586)(0.120) \\ (0.802)(0.120) & (0.586)(0.120) & (0.120)^2 \end{bmatrix}.$$

This simplifies to

$$
\mathbf{A} = \begin{bmatrix} 0.107 & -0.196 & 0.239 \\ -0.196 & 0.358 & -0.438 \\ 0.239 & -0.438 & 0.536 \end{bmatrix} + \begin{bmatrix} 0.750 & -0.821 & -1.007 \\ -0.821 & 0.898 & 1.101 \\ -1.007 & 1.101 & 1.351 \end{bmatrix}
$$

$$
+ \begin{bmatrix} 5.146 & 3.760 & 0.770 \\ 3.760 & 2.747 & 0.563 \\ 0.770 & 0.563 & 0.115 \end{bmatrix}.
$$

Except for inaccuracies due to rounding this matrix should be equivalent to the decimal form of

$$
\mathbf{A} = \begin{bmatrix} 6.000 & 2.739 & 0.000 \\ 2.739 & 4.000 & 1.225 \\ 0.000 & 1.225 & 1.000 \end{bmatrix}.
$$

As a matrix approximation the term of the spectral decomposition corresponding to $\lambda = 8$, say $\widehat{\mathbf{A}}_1$, can be viewed as a first approximation to $\mathbf{A}$. The difference between the two matrices $\mathbf{A} - \widehat{\mathbf{A}}_1$ is given by

$$
\begin{bmatrix} 0.857 & 1.017 & -0.768 \\ 1.017 & 1.256 & 0.663 \\ -0.768 & 0.663 & 1.887 \end{bmatrix}.
$$

If two terms ($\lambda = 8$ and $\lambda = 3$) are used, we obtain $\widehat{\mathbf{A}}_2$ with the difference given by $(\mathbf{A} - \widehat{\mathbf{A}}_2)$ which is approximately the matrix term corresponding to $\lambda = 1$ in the spectral decomposition. It would appear that the variation in the magnitudes of the errors is quite small compared to the variation in the magnitudes in the original matrix. In other words the spectral decomposition approximation seems to be weighted towards the larger elements of $\mathbf{A}$.

Exercise

Determine the spectral decomposition for the matrix given in the eigenvalue exercise above and comment on the quality of the approximation based on the largest eigenvalue.

Eigenvalues for Nonnegative Definite Matrices

If the real symmetric matrix $\mathbf{A}$ is *positive definite* then the eigenvalues are all positive. If the matrix is *positive semidefinite* then the eigenvalues are nonnegative with the number of positive eigenvalues equal to the rank of $\mathbf{A}$.

3.2 Singular Value Decomposition

A real $(n \times p)$ matrix $\mathbf{A}$ of rank k can be expressed as the product of three matrices which have a useful interpretation. This decomposition of $\mathbf{A}$ is referred to as a *singular value decomposition* and is given by

$$\mathbf{A} = \mathbf{UDV}',$$

where

1. $\mathbf{D}$ $(k \times k)$ is a diagonal matrix with positive diagonal elements α_1, α_2, $\ldots, \alpha_k$ which are called the *singular values* of $\mathbf{A}$; (without loss of generality we assume that the α_j, $j = 1, 2, \ldots, k$ are arranged in descending order).

2. The k columns of $\mathbf{U}$ $(n \times k)$, $\mathbf{u}_1, \mathbf{u}_2, \ldots, \mathbf{u}_k$ are called the *left singular vectors* of $\mathbf{A}$ and the k columns of $\mathbf{V}$ $(p \times k)$, $\mathbf{v}_1, \mathbf{v}_2, \ldots, \mathbf{v}_k$ are called the *right singular vectors* of $\mathbf{A}$;

3. The matrix $\mathbf{A}$ can be written as the sum of k matrices, each with rank 1, $\mathbf{A} = \sum_{j=1}^{k} \alpha_j \mathbf{u}_j \mathbf{v}_j'$. The subtraction of any one of these terms from the sum results in a singular matrix for the remainder of the sum.

4. The matrices $\mathbf{U}$ $(n \times k)$ and $\mathbf{V}$ $(p \times k)$ have the property that $\mathbf{U}'\mathbf{U} = \mathbf{V}'\mathbf{V} = \mathbf{I}$; hence the columns of $\mathbf{U}$ form an orthonormal basis for the columns of $\mathbf{A}$ in n-dimensional space while the columns of $\mathbf{V}$ form an orthonormal basis for the rows of $\mathbf{A}$ in p-dimensional space.

5. Let $\mathbf{A}(\ell)$ denote the first ℓ terms of the singular value decomposition for $\mathbf{A}$; hence $\mathbf{A}(\ell) = \sum_{j=1}^{\ell} \alpha_j \mathbf{u}_j \mathbf{v}_j'$. This expression minimizes $tr[(\mathbf{A} - \mathbf{X})(\mathbf{A} - \mathbf{X})'] = \sum_{i=1}^{n} \sum_{j=1}^{p} (a_{ij} - x_{ij})^2$ amongst all $(n \times p)$ matrices $\mathbf{X}$ of rank ℓ. Thus the singular value decomposition can be used to provide a *matrix approximation* to $\mathbf{A}$.

Complete Singular Value Decomposition

A *complete singular value decomposition* can be obtained for $\mathbf{A}$ by adding $(n - k)$ orthonormal vectors $\mathbf{u}_{k+1}, \ldots, \mathbf{u}_n$ to the existing set $\mathbf{u}_1, \ldots, \mathbf{u}_k$. Similarly the orthonormal vectors $\mathbf{v}_{k+1}, \ldots, \mathbf{v}_p$ are added to the set $\mathbf{v}_1, \mathbf{v}_2$, $\ldots, \mathbf{v}_p$. Denoting the n column vectors $\mathbf{u}_j$, $j = 1, 2, \ldots, n$ by $\mathbf{U}^*$ and the p column vectors by $\mathbf{V}^*$ then $\mathbf{U}^{*\prime}\mathbf{U} = \mathbf{I}$ and $\mathbf{V}^{*\prime}\mathbf{V}^* = \mathbf{I}$ and $\mathbf{A} = \mathbf{U}^*\mathbf{D}^*\mathbf{V}^*$ is the complete singular value decomposition where $\mathbf{D}^* = \begin{bmatrix} \mathbf{D} & 0 \\ 0 & 0 \end{bmatrix}$.

Generalized Singular Value Decomposition

A *generalized singular value decomposition* permits the left and right singular vectors to be orthonormalized with respect to given diagonal matrices $\mathbf{\Omega}$ $(n \times n)$ and $\mathbf{\Phi}$ $(p \times p)$ where

$$\mathbf{N}'\mathbf{\Omega}\mathbf{N} = \mathbf{M}'\mathbf{\Phi}\mathbf{M} = \mathbf{I}.$$

The generalized singular value decomposition of $\mathbf{A}$ $(n \times p)$ is given by

$$\mathbf{A} = \mathbf{N}\mathbf{D}\mathbf{M}' = \sum_{j=1}^{k} \alpha_j \mathbf{n}_j \mathbf{m}_j'$$

where α_j, $j = 1, 2, \ldots, k$ are the diagonal elements of $\mathbf{D}$; $\mathbf{n}_j$, $j = 1, 2, \ldots, n$ are the columns of $\mathbf{N}$ and $\mathbf{m}_j$, $j = 1, 2, \ldots, p$ are the columns of $\mathbf{M}$. The columns of $\mathbf{N}$ and $\mathbf{M}$ are referred to as the *generalized left and right singular vectors* of $\mathbf{A}$ respectively. The diagonal elements of $\mathbf{D}$ are called the *generalized singular values*.

Let $\mathbf{A}(\ell)$ denote the first ℓ terms of the generalized singular value decomposition for $\mathbf{A}$ hence $\mathbf{A}(\ell) = \sum_{j=1}^{\ell} \alpha_j \mathbf{n}_j \mathbf{m}_j'$. This expression for $\mathbf{A}(\ell)$ minimizes $\sum_{j=1}^{p} \sum_{i=1}^{n} (a_{ij} - x_{ij})^2 \Phi_j \omega_i$ amongst all matrices $\mathbf{X}$ of rank at most ℓ where the Φ_j and ω_i are the diagonal elements of $\mathbf{\Phi}$ and $\mathbf{\Omega}$ respectively. The generalized singular value decomposition can be used to provide a *matrix approximation* to $\mathbf{A}$.

The generalized singular value decomposition of $\mathbf{A}$ given by $\mathbf{N}\mathbf{D}\mathbf{M}'$ where $\mathbf{N}'\mathbf{\Omega}\mathbf{N} = I$ and $\mathbf{M}'\mathbf{\Phi}\mathbf{M} = I$ can be related to the singular value decomposition by writing

$$\mathbf{\Omega}^{1/2}\mathbf{A}\mathbf{\Phi}^{1/2} = \mathbf{\Omega}^{1/2}\mathbf{N}\mathbf{D}\mathbf{M}'\mathbf{\Phi}^{1/2} = \mathbf{U}\mathbf{D}\mathbf{V}$$

where $\mathbf{U} = \mathbf{\Omega}^{1/2}\mathbf{N}$ and $\mathbf{V} = \mathbf{\Phi}^{1/2}\mathbf{M}$ and $\mathbf{U}'\mathbf{U} = \mathbf{V}'\mathbf{V} = \mathbf{I}$.

Relationship to Spectral Decomposition and Eigenvalues

If the matrix $\mathbf{A}$ is symmetric, the singular value decomposition is equivalent to the spectral decomposition

$$\mathbf{A} = \mathbf{U}\mathbf{D}\mathbf{V}' = \mathbf{V}\mathbf{\Lambda}\mathbf{V}'.$$

In this case the left and right singular vectors are equal to the eigenvectors and the singular values are equal to the eigenvalues.

The singular value decomposition of the $(n \times p)$ matrix $\mathbf{A}$ can also be related to the spectral decomposition for the symmetric matrices $\mathbf{AA}'$ and $\mathbf{A}'\mathbf{A}$. For the singular value decomposition

$$\mathbf{A} = \mathbf{UDV}'$$

the eigenvalues of $\mathbf{AA}'$ and $\mathbf{A}'\mathbf{A}$ are the squares of the singular values of $\mathbf{A}$. The eigenvectors of $\mathbf{AA}'$ are the columns of $\mathbf{U}$ and the eigenvectors of $\mathbf{A}'\mathbf{A}$ are the columns of $\mathbf{V}$. Since covariance matrices and correlation matrices can be written in the form $\mathbf{A}'\mathbf{A}$, the eigenvalues and eigenvectors are often determined using the singular value decomposition.

Data Appendix for Volume I

Introduction

This data appendix contains nine data tables which will be used in the chapter exercises throughout Volume I. The data sets summarized by these tables are similar to some of the data sets used to provide examples in the text. In all cases the data tables were derived from additional observations from the same data sets used for the examples. Outlines to the nine data sets followed by the data tables constitute the remainder of the Data Appendix.

Data Set D1   Air Pollution Data

This data set consists of an additional fifty observations from the population of cities used to generate the air pollution data given in Table 4.9 in Chapter 4. The twelve variables in this data set are also defined in Table 4.9 in Chapter 4.

Data Set D2   Financial Accounting Data

This data set consists of an additional forty observations from the population of financial accounting data introduced at the beginning of Chapter 4 in Tables 4.1 and 4.2. The thirteen variables in this data set are defined in Table 4.1.

Data Set D3   Bank Employee Salary Data

This data set consists of 100 observations selected from a larger data set consisting of salary data for bank employees. This data set has been used by SPSSX to provide examples for the SPSSX User's Guide. The variables in this data set are listed below.

LCURRENT -	ln (current salary)
LSTART -	ln (starting salary)
SEX -	male = 0, female = 1
JOBCAT -	1 = clerical, 2 = office trainee, 3 = security officer,
	4 = college trainee, 5 = MBA trainee

RACE -   white $= 0$, nonwhite $= 1$
SENIOR -   seniority with the bank in months
AGE -   age in years
EXPER -   relevant job experience in years

Additional observations from this data set were used to provide the data in Figure 3.10 and Table 4.54.

Data Set D4   Automobile Fuel Consumption Data

This data set consists of 99 observations taken from a larger data set of automobile fuel consumption data published by Transport Canada in 1985. Additional observations from this data set were introduced in Table 4.41 in Chapter 4.

Data Set D5   Canadian Financial Market Data

This data set consists of 60 observations from the Canadian Financial Market Data set introduced in Table 3.16 and Table 4.22. The five variables in the data set are defined in Chapter 4 immediately prior to Table 4.22.

Data Set D6   Test Marketing Data

This data set was derived by adding a normal random error term to the Test Marketing Data in Table 5.1 of Chapter 5. The means for various cells have also been changed by adding constants.

Data Set D7   Fitness Data

This data set was derived by adding a normal random term to the Fitness Data in Table 5.35 of Chapter 5.

Data Set D8   Police Officer Stress Data

This data set consists of an additional fifty observations from the same police officer stress data introduced in Table 5.54 of Chapter 5.

Data Set D9   Public Safety Data

This data set consists of an additional 100 observations (25 from each of four different regions) from the data set introduced in Chapter 5, Section 5.8.3, Table 5.59.

Table D1. Air Pollution Data

CITY	TMR	SMIN	SMEAN	SMAX	PMIN	PMEAN	PMAX	PM2	GE65	PERWH	NONPOOR	LPOP
Providence	1096	30	163	349	56	119	223	116.1	109	97.9	83.9	5.85
Jackson	789	29	70	161	27	74	124	21.3	64	60.0	69.1	5.27
Johnstown	1072	88	123	245	70	166	452	15.8	103	98.7	73.3	5.44
Jersey City	1199	155	229	340	63	147	253	1357.2	103	93.1	87.3	5.78
Huntington	967	60	70	137	56	122	219	18.1	93	97.0	73.2	5.40
Des Moines	950	31	88	188	61	183	329	44.8	97	95.9	87.1	5.42
Denver	841	2	61	188	54	126	229	25.4	82	95.8	86.9	5.96
Reading	1113	50	94	186	34	120	242	31.9	112	98.2	86.1	5.43
Toledo	1031	67	86	309	52	104	193	133.2	98	90.5	86.1	5.65
Fresno	845	18	34	198	45	119	304	6.1	81	92.5	78.5	5.56
Memphis	873	35	48	69	46	102	201	83.5	73	63.6	72.5	5.79
York	957	120	162	488	28	147	408	26.2	97	97.7	84.8	5.37
Milwaukee	921	65	134	236	49	150	299	150.2	88	94.4	90.4	6.07
Savannah	990	49	71	120	46	82	192	42.7	65	65.9	72.0	5.27
Omaha	922	20	74	148	39	107	198	29.9	90	94.0	86.4	5.66
Topeka	904	19	37	91	52	101	158	25.9	99	92.7	84.1	5.15
Columbus	877	94	161	276	74	119	190	127.2	79	88.1	86.3	5.83
Beaumont	728	27	71	144	32	76	190	23.5	58	79.3	79.9	5.48
Winston	802	28	58	128	72	147	306	44.7	62	75.8	79.9	5.27
Detroit	817	52	128	260	59	146	235	191.5	72	84.9	86.5	6.57
El Paso	618	47	87	207	49	150	373	29.8	45	96.7	77.9	5.49
Macon	869	18	27	128	22	122	754	28.6	62	69.0	73.7	5.25
Rockford	842	33	66	210	36	86	143	40.3	85	95.8	88.2	5.32
Jackson	928	41	52	138	39	77	124	18.7	90	94.3	86.5	5.12
Fall River	1157	62	79	136	18	102	254	71.7	116	98.7	82.9	5.60

Table D1. Air Pollution Data (continued)

CITY	TMR	SMIN	SMEAN	SMAX	PMIN	PMEAN	PMAX	PM2	GE65	PERWH	NONPOOR	LPOP
Boston	1112	42	163	337	55	141	252	174.5	109	97.3	88.5	6.49
Dayton	847	18	106	241	50	132	327	53.9	74	89.8	87.1	5.84
Charlotte	791	43	81	147	62	124	234	50.2	57	75.4	79.5	5.43
Miami	897	44	57	68	33	54	124	45.5	100	85.1	77.2	5.97
Bridgeport	938	137	205	308	32	91	182	103.3	94	94.7	90.7	5.81
Sioux Falls	795	18	55	121	25	108	358	10.6	92	99.3	82.4	4.93
Chicago	1000	75	166	328	88	182	296	167.5	86	85.2	89.4	6.79
South Bend	888	73	77	261	28	90	164	51.1	84	94.0	88.4	5.37
Norfolk	803	49	112	198	39	89	242	86.7	53	73.6	73.1	5.76
Cleveland	969	69	160	282	86	174	336	261.1	89	85.5	88.6	6.25
Austin	689	40	46	58	10	78	157	20.9	76	87.2	75.2	5.32
Knoxville	825	56	77	157	28	135	302	25.8	74	92.5	72.5	5.56
Indianapolis	969	50	139	269	92	178	275	173.5	85	85.6	87.2	5.84
Nashville	919	54	160	362	45	130	310	75.1	79	80.8	76.5	5.60
Seattle	938	1	47	179	32	69	141	26.2	96	95.2	88.8	6.04
Dallas	757	31	69	148	22	96	230	29.7	71	85.4	81.4	6.03
Mobile	823	47	67	248	29	129	284	25.3	57	67.7	74.6	5.49
Phoenix	758	15	86	266	98	247	573	7.2	72	94.5	80.9	5.82
Augusta	823	31	46	158	28	66	142	15.2	60	70.2	67.8	5.33
Youngstown	915	75	145	263	58	148	371	49.0	89	90.8	87.1	5.70
Chattanooga	940	10	105	191	69	186	361	27.7	77	82.4	74.0	5.45
Galveston	873	62	72	86	23	55	125	32.7	64	78.6	76.8	5.14
Fort Worth	789	9	32	73	28	79	152	35.8	73	89.3	80.7	5.75
Flint	747	64	80	229	49	124	468	58.2	62	90.1	87.8	5.57
Charleston	780	15	283	940	55	225	958	27.9	70	94.2	78.6	5.40

Table D2. Financial Accounting Data

OBS	RETCAP	WCFTCL	WCFTDT	LOGSALE	LOGASST	GEARRAT
1	0.19	0.16	0.16	5.22	4.83	0.15
2	0.22	0.26	0.16	4.14	4.34	0.54
3	0.17	0.26	0.20	5.38	4.88	0.49
4	0.12	0.08	0.08	4.12	3.93	0.39
5	0.21	0.34	0.34	4.77	4.58	0.11
6	0.12	0.25	0.25	4.15	3.90	0.19
7	0.15	0.25	0.16	5.69	5.55	0.35
8	0.10	0.12	0.09	4.41	4.21	0.39
9	0.08	0.04	0.04	4.71	4.51	0.50
10	0.31	0.12	0.11	4.46	4.19	0.41
11	0.21	0.36	0.33	4.38	4.23	0.08
12	0.22	0.37	0.37	4.02	3.83	0.16
13	0.20	0.48	0.48	3.85	3.87	0.13
14	0.11	0.18	0.15	3.90	3.86	0.23
15	0.38	0.25	0.20	5.16	4.66	0.27
16	0.23	0.24	0.24	5.71	4.97	0.00
17	0.32	0.09	0.09	4.71	4.31	0.11
18	0.13	0.06	0.05	4.67	4.49	0.55
19	0.29	0.60	0.60	4.52	4.87	0.00
20	0.09	0.10	0.09	4.98	4.40	0.28
21	-0.50	-1.28	-1.28	4.05	3.54	1.78
22	0.17	0.12	0.11	4.28	3.96	0.28
23	-0.04	-0.04	-0.04	4.76	4.31	0.46
24	0.26	0.23	0.23	4.24	3.87	0.00
25	0.21	0.40	0.30	4.41	4.38	0.20
26	0.15	0.30	0.21	4.39	4.36	0.66
27	0.23	0.07	0.07	4.83	4.43	0.11
28	0.20	0.33	0.28	4.20	4.03	0.33
29	0.19	0.16	0.14	4.31	4.17	0.30
30	0.08	0.18	0.10	4.95	4.86	0.35
31	0.19	0.15	0.14	5.57	5.44	0.19
32	0.20	0.63	0.35	4.77	4.86	0.21
33	0.14	0.27	0.20	4.99	4.82	0.30
34	0.04	0.07	0.07	4.17	3.91	0.18
35	0.10	0.15	0.12	5.76	5.78	0.13
36	-0.09	-0.46	-0.22	3.96	4.08	0.68
37	0.10	0.18	0.14	5.68	5.63	0.23
38	0.20	0.13	0.12	4.79	4.42	0.05
39	0.13	0.17	0.13	5.48	5.35	0.22
40	0.08	0.14	0.14	4.08	3.87	0.19

Table D2. Financial Accounting Data (continued)

OBS	CAPINT	NFATAST	FAT-TOT	INVTAST	PAY-OUT	QUIK-RAT	CURRAT
1	2.47	0.28	0.36	0.42	0.31	0.54	1.33
2	0.64	0.13	0.16	0.04	0.45	0.83	0.93
3	3.18	0.43	0.74	0.13	0.50	0.84	1.09
4	1.55	0.23	0.50	0.37	0.65	0.50	1.09
5	1.56	0.30	0.50	0.20	0.25	1.10	1.74
6	1.74	0.34	0.38	0.31	0.80	1.00	1.89
7	1.39	0.48	0.62	0.22	0.46	0.73	1.38
8	1.60	0.26	0.42	0.30	1.03	0.94	1.57
9	1.58	0.25	0.33	0.31	0.00	0.74	1.28
10	1.88	0.17	0.25	0.31	0.25	0.66	1.10
11	1.43	0.40	0.71	0.17	0.61	1.06	1.49
12	1.55	0.42	0.62	0.17	0.25	0.97	1.38
13	0.96	0.68	0.97	0.13	0.60	0.61	1.00
14	1.09	0.40	0.64	0.15	0.80	0.92	1.23
15	3.13	0.21	0.32	0.38	0.39	0.33	1.39
16	5.44	0.27	0.38	0.50	0.36	0.24	1.29
17	2.51	0.09	0.13	0.31	0.53	0.86	1.34
18	1.51	0.24	0.40	0.42	0.00	0.44	1.14
19	0.45	0.57	0.58	0.01	0.21	1.18	1.21
20	3.82	0.34	0.50	0.46	1.52	0.34	1.28
21	3.21	0.16	0.30	0.37	0.00	0.50	1.06
22	2.07	0.26	0.32	0.37	0.22	0.67	1.36
23	2.79	0.19	0.32	0.28	0.00	0.72	1.11
24	2.34	0.21	0.26	0.27	0.53	1.20	1.83
25	1.07	0.24	0.36	0.24	0.42	1.77	2.72
26	1.08	0.70	1.07	0.15	0.00	0.29	0.58
27	2.46	0.17	0.22	0.00	0.67	0.88	0.88
28	1.47	0.53	1.16	0.07	0.21	0.77	0.91
29	1.38	0.25	0.33	0.42	0.52	0.49	1.28
30	1.21	0.31	0.51	0.27	1.08	1.44	2.36
31	1.36	0.22	0.36	0.22	0.40	0.96	1.35
32	0.81	0.21	0.34	0.26	0.51	2.63	3.98
33	1.48	0.72	0.74	0.09	0.53	0.26	0.54
34	1.83	0.28	0.54	0.23	4.21	1.08	1.57
35	0.96	0.12	0.21	0.28	0.43	0.57	1.40
36	0.77	0.62	0.71	0.19	0.00	0.60	1.45
37	1.14	0.33	0.52	0.23	0.12	0.83	1.56
38	2.35	0.04	0.07	0.37	0.33	0.80	1.42
39	1.37	0.26	0.52	0.41	0.53	0.75	1.73
40	1.64	0.17	0.27	0.34	0.91	0.74	1.57

Table D3. Bank Employee Salary Data

LCURRENT	LSTART	SEX	JOB-CAT	RACE	EDUC	SENIOR	AGE	EXPER
10.63	10.08	0	5	0	16	73	40.33	12.50
9.99	9.23	0	5	0	15	83	31.08	4.08
9.86	9.07	0	4	0	16	93	31.17	1.83
10.21	9.47	0	4	0	18	80	29.50	2.42
9.41	8.74	0	3	0	12	77	52.92	26.42
9.99	9.25	0	4	0	17	93	32.33	2.67
9.85	9.48	0	5	0	19	64	31.92	2.25
10.00	9.51	0	4	0	19	81	30.75	5.17
10.22	9.30	0	4	0	17	89	34.17	3.17
10.24	9.07	0	4	0	16	65	28.00	1.58
9.99	9.54	0	5	0	19	65	39.75	10.75
10.20	9.54	0	5	0	17	83	30.17	0.75
10.35	9.30	0	4	0	18	91	30.17	3.92
9.99	9.47	0	5	0	18	75	41.17	10.42
10.28	9.48	0	4	0	19	78	32.92	3.75
9.41	8.74	0	3	0	8	78	63.75	35.75
9.96	9.03	0	4	0	16	93	30.67	4.00
9.41	8.69	0	3	0	8	84	63.42	31.67
10.10	9.48	0	5	0	19	68	29.50	0.75
10.18	9.85	0	5	0	16	86	42.42	12.50
10.34	9.39	0	4	0	19	93	31.67	0.58
10.07	9.39	0	5	0	20	89	35.58	0.50
10.06	9.07	0	4	0	16	84	30.25	1.08
10.16	9.41	0	4	0	17	78	29.75	2.17
9.41	8.74	0	3	0	12	80	61.67	38.33
9.87	9.08	0	4	0	16	76	32.67	5.08
9.82	9.04	0	4	0	16	93	29.75	2.92
10.16	9.45	0	4	0	19	69	28.83	6.17
9.94	9.30	0	5	0	19	80	45.67	18.42
9.93	9.58	0	5	0	18	78	39.42	12.42
10.56	9.54	0	5	0	19	91	34.33	5.67
9.51	8.69	0	3	0	12	83	50.25	23.67
9.53	8.48	1	2	0	12	77	24.33	0.33
9.54	8.66	1	1	0	16	93	31.50	0.67
9.02	8.47	1	1	0	8	79	50.17	5.83
9.16	8.59	1	1	0	12	98	47.33	20.33
9.19	8.29	1	1	0	12	92	44.00	3.67
9.33	8.53	1	2	0	15	81	27.17	1.58
9.07	8.51	1	1	0	8	74	59.83	26.50
9.37	8.71	1	2	0	12	72	25.75	2.50
9.16	8.41	1	2	0	12	83	25.83	1.33
9.37	8.64	1	2	0	12	63	25.08	0.75
9.19	8.59	1	2	0	12	78	27.17	3.92
9.65	8.69	1	1	0	16	97	30.58	1.42
9.26	8.59	1	2	0	15	94	29.50	0.25
9.18	8.57	1	1	0	12	88	54.42	8.92
8.92	8.34	1	1	0	15	90	58.00	4.50
9.60	8.88	1	2	0	15	75	28.75	0.42
9.61	8.74	1	1	0	12	73	54.08	11.00
9.18	8.47	1	2	0	12	79	24.33	0.67

Table D3. Bank Employee Salary Data (continued)

LCURRENT	LSTART	SEX	JOB-CAT	RACE	EDUC	SENIOR	AGE	EXPER
9.34	8.34	1	2	0	12	92	25.50	0.50
9.20	8.38	1	2	0	12	79	24.67	0.42
9.47	8.59	1	2	0	15	90	28.75	1.83
9.92	8.88	1	1	0	16	93	32.50	1.83
9.11	8.41	1	2	0	12	68	23.42	0.17
9.64	8.79	1	2	0	16	90	30.42	0.00
9.16	8.53	1	2	0	12	69	24.42	1.67
9.02	8.47	1	1	0	12	87	53.92	13.58
8.89	8.38	1	2	0	12	80	25.00	0.00
8.79	8.31	1	1	0	12	84	62.42	24.00
9.40	8.79	1	2	0	16	98	43.92	11.92
9.51	8.47	1	1	0	12	82	28.17	0.92
9.23	8.64	1	1	0	15	82	30.17	4.25
9.32	8.59	1	2	0	12	73	27.33	2.67
9.18	8.38	1	1	0	12	93	25.25	0.42
9.35	8.62	1	2	0	12	72	27.33	1.50
9.25	8.38	1	2	0	12	89	25.83	0.00
9.11	8.41	1	1	0	12	69	23.67	0.00
8.80	8.26	1	1	0	8	88	62.50	34.33
9.02	8.41	1	2	0	12	69	23.67	0.17
9.21	8.62	1	2	0	15	64	29.08	4.75
9.04	8.47	1	2	0	8	73	60.50	13.25
8.83	8.31	1	1	0	12	82	53.92	29.83
9.13	8.41	1	2	0	12	81	24.08	1.08
9.27	8.59	0	1	1	12	88	29.92	3.17
9.36	8.59	0	1	1	12	96	39.50	9.42
9.49	8.85	0	1	1	16	67	41.67	10.00
8.96	8.53	0	1	1	12	84	44.58	15.00
9.25	8.76	0	1	1	8	67	51.42	8.08
9.32	8.79	0	1	1	15	66	30.75	7.00
9.07	8.53	0	1	1	12	97	53.08	26.25
9.10	8.69	0	1	1	15	82	59.75	30.92
9.28	8.88	0	1	1	17	70	47.33	16.00
9.27	8.69	0	2	1	16	78	33.83	8.75
9.30	8.69	0	2	1	12	90	37.50	14.42
9.09	8.69	0	1	1	12	80	57.17	22.67
9.45	8.74	0	2	1	15	67	29.33	4.83
9.27	8.74	0	2	1	12	76	28.33	1.50
9.43	8.69	0	2	1	15	96	31.92	4.08
9.41	8.69	0	3	1	12	91	45.50	20.00
9.55	8.74	0	3	1	12	78	55.33	23.42
0.31	8.64	0	2	1	15	98	33.67	2.83
9.41	8.69	0	3	1	12	90	43.67	17.42
8.84	8.31	1	1	1	12	72	46.50	9.67
9.29	8.74	1	1	1	15	84	55.17	19.25
8.78	8.31	1	1	1	12	66	60.50	13.58
8.82	8.31	1	1	1	12	72	51.50	22.58
9.25	8.59	1	1	1	12	72	50.33	14.08
8.96	8.62	1	1	1	12	69	50.00	11.08
8.90	8.31	1	1	1	12	85	51.00	19.00

Table D4. Automobile Fuel Consumption Data

TYPE	COMB-RATE	WEIGHT	CYLIND	FOR	TYPE	COMB-RATE	WEIGHT	CYLIND	FOR
Mercedes 500SEL	138	4000	8	1	Subaru GLHB	80	2500	4	1
Chev Monte Carlo	131	3500	8	0	Chev Citation	76	2750	4	0
Subaru GL-10	82	2500	4	1	Volvo DLGL	90	3000	4	1
Chev Celebrity	79	3000	4	0	Subaru GLXT	81	2500	4	1
Pontiac Fiero	97	3000	6	0	Mercury Grand Marquis	128	4000	8	0
Buick Somerset	75	2750	4	0	Renault Alliance	80	2250	4	1
Buick Electra	109	3500	6	0	Dodge Omni	79	2500	4	0
Peugeot Turbo	99	3000	4	1	Nissan 300ZX	97	3500	6	1
Merkur XR4TI	101	3000	4	0	Honda Accord	78	2500	4	1
Chev Citation	73	2750	4	0	Dodge Omni	87	2500	4	0
Honda Civic	65	2000	4	1	Mercury Grand Marquis	132	4000	8	0
Olds Cutlass	79	3000	4	0	Plymouth Caravelle	89	3000	4	0
Plymouth Reliant	90	3000	4	0	Buick Skylark	91	3000	6	0
Honda Accord	68	2500	4	1	Chev Corvette	109	3500	8	0
Dodge Charger	83	2750	4	0	Pontiac Sunbird	85	2750	4	0
Buick Skylark	100	3000	6	0	Pontiac 6000	88	3000	6	0
Nissan 300ZX	116	3500	6	1	Honda Prelude	78	2500	4	1
Chev Camaro	97	3500	6	0	Ford Mustang	107	3500	8	0
Lada Signet	88	2500	4	1	Buick Skylark	85	2750	4	0
Dodge Colt	68	2250	4	0	Buick Skylark	90	3000	6	0
Dodge 600	89	3000	4	0	Olds Cutlass	90	3000	6	0
Pontiac Bonneville	109	3500	8	0	Olds Cutlass	97	3500	6	0
Nissan 300ZX	108	3500	6	1	Dodge Colt	68	2250	4	0
Volvo 760GLE	96	3000	6	1	Mazda 626	71	2500	4	1
Dodge Aries	81	2750	4	0	VW Jetta	83	2500	4	1

Table D4. Automobile Fuel Consumption Data (continued)

TYPE	COMB-RATE	WEIGHT	CYLIND	FOR	TYPE	COMB-RATE	WEIGHT	CYLIND	FOR
Olds Fierenza	102	2750	6	0	Subaru Turbo XT	74	2750	4	1
Ford Escort	68	2500	4	0	Cadillac Cimarron	100	3000	6	0
Nissan Micra	55	1750	4	1	Toyota Corolla Sport	68	2500	4	1
BMW 635CSIA	101	3500	6	1	Renault Alliance	54	2250	4	0
Chev Caprice	109	4000	8	0	Mercury Cougar	107	3500	8	0
Rolls-Royce Corniche	214	5500	8	1	BMW 325EA	87	3000	6	1
Toyota Corolla	61	2250	4	1	Maxda GLC	71	2250	4	1
Mazda GLC	62	2250	4	1	BMW 535I	110	3500	6	1
Chrysler Lebaron	89	3000	4	0	BMW 528EA	91	3000	6	1
Subaru GL-10	74	2500	4	1	Chrysler Laser	77	2750	4	0
Buick Lesabre	107	4000	8	0	Buick Skylark	72	2750	4	0
Pontiac 6000	90	3000	6	0	BMW 735IA	104	3500	6	1
Chrysler Daytona	84	3000	4	0	Mercury Marquis	107	3500	8	0
Lada Signet	79	2500	4	1	Peugeot 505 Turbo	99	3500	4	1
Chrysler Lebaron	76	3000	4	0	Dodge Charger	82	2750	4	0
Nissan Pulsar	73	2250	4	1	Pontiac 6000	79	3000	4	0
Plymouth Reliant	71	2750	4	0	Buick Somerset	90	3000	6	0
Olds Cutlass	99	3000	6	0	Ford Mustang	87	3000	4	0
Pontiac Grand Prix	113	3500	8	0	Plymouth Turismo	83	2750	4	0
Renault Fuego	78	2750	4	1	Pontiac Acadian	62	2250	4	0
Mercury Cougar	104	3500	4	0	Honda Prelude	71	2500	4	1
Nissan 200SX	85	2750	4	1	BMW 325E	84	3000	6	1
Hyundai Pony	76	2500	4	1	Pontiac Sunburst	60	2250	4	0
Pontiac Bonneville	113	3500	8	0	Mercury Marquis	87	3000	4	0
Volvo DLGL	80	3500	4	1					

Table D5. Canadian Financial Market Data

BANKCAN	TRSBILL	CPI	USSPOT	USFORW	BANKCAN	TRSBILL	CPI	USSPOT	USFORW
10400	8.59	61.2	1.01	1.01	13541	8.43	75.0	1.12	1.12
10317	8.70	41.5	0.99	1.00	13380	8.77	75.1	1.13	1.13
10580	9.04	61.7	0.98	0.99	14464	9.02	74.9	1.15	1.15
10559	8.97	62.0	0.98	0.98	16200	9.52	75.7	1.18	1.18
10504	8.94	62.5	0.98	0.98	13889	10.29	76.3	1.16	1.16
10946	8.99	62.9	0.97	0.98	15106	10.43	76.5	1.16	1.16
10995	9.02	63.0	0.96	0.97	14012	10.80	77.1	1.18	1.18
10806	9.12	63.3	0.97	0.98	14291	10.78	77.8	1.20	1.20
10830	8.97	63.6	0.97	0.97	14992	10.90	78.8	1.18	1.18
11002	9.07	64.0	0.97	0.97	14841	10.84	79.3	1.15	1.15
11759	8.88	64.2	0.97	0.97	14513	10.84	80.1	1.14	1.14
11843	8.41	64.5	1.03	1.03	15338	10.82	80.5	1.16	1.16
11362	8.08	65.0	1.00	1.01	14989	10.91	81.1	1.16	1.16
11176	7.67	65.6	1.02	1.02	15636	11.32	81.4	1.17	1.17
11687	7.61	66.3	1.04	1.05	16381	11.57	82.1	1.16	1.16
12049	7.55	66.7	1.05	1.05	15133	12.86	82.7	1.16	1.15
11830	7.26	67.2	1.04	1.04	15566	13.61	83.5	1.18	1.18
12132	7.07	67.7	1.05	1.05	15746	13.63	84.0	1.16	1.16
12348	7.12	68.3	1.05	1.05	14821	13.54	84.5	1.16	1.16
12024	7.16	68.6	1.07	1.07	15010	13.56	85.2	1.15	1.15
12107	7.09	69.0	1.07	1.07	17245	14.35	86.1	1.14	1.14
12503	7.19	69.6	1.07	1.07	15705	15.64	86.6	1.19	1.19
12303	7.25	70.1	1.10	1.10	16516	12.54	87.6	1.18	1.19
13416	7.18	70.6	1.10	1.10	15956	11.15	88.6	1.15	1.16
12471	7.14	70.8	1.09	1.09	16035	10.10	89.3	1.14	1.15
12948	7.24	71.3	1.10	1.10	17476	10.21	90.1	1.15	1.15
13439	7.62	72.1	1.11	1.11	16091	10.63	90.9	1.15	1.15
13783	8.18	72.3	1.13	1.13	17044	11.57	91.7	1.17	1.16
13119	8.13	73.3	1.12	1.12	17791	12.87	92.9	1.17	1.17
13804	8.24	73.9	1.11	1.11	17313	16.31	93.4	1.19	1.18

Table D6. Test Marketing Data

	AD#1			AD#2			AD#3			AD#4	
EXPEN	REGION	SIZE	EXPEN	REGION	SIZE	EXPEN	REGION	SIZE	EXPEN	REGION	SIZE
20.02	1	1	25.89	1	1	15.01	1	1	24.45	1	1
25.07	1	2	47.45	1	2	24.12	1	2	27.99	1	2
38.25	1	3	54.13	1	3	29.73	1	3	45.16	1	3
48.62	1	4	70.97	1	4	33.78	1	4	53.79	1	4
54.88	1	5	78.20	1	5	44.75	1	5	63.71	1	5
60.18	1	6	83.72	1	6	54.48	1	6	89.31	1	6
36.38	2	1	19.89	2	1	23.39	2	1	32.77	2	1
45.73	2	2	25.11	2	2	30.70	2	2	55.80	2	2
59.29	2	3	45.55	2	3	38.13	2	3	52.71	2	3
66.70	2	4	50.40	2	4	53.93	2	4	65.27	2	4
75.54	2	5	63.68	2	5	55.80	2	5	84.92	2	5
78.78	2	6	74.03	2	6	76.87	2	6	100.37	2	6
26.63	3	1	9.21	3	1	3.57	3	1	14.50	3	1
28.36	3	2	4.64	3	2	24.77	3	2	29.37	3	2
50.33	3	3	33.11	3	3	24.88	3	3	31.73	3	3
57.89	3	4	32.18	3	4	33.00	3	4	39.91	3	4
75.75	3	5	41.00	3	5	37.64	3	5	54.46	3	5
81.68	3	6	48.74	3	6	53.43	3	6	68.43	3	6
15.67	4	1	27.61	4	1	8.62	4	1	26.41	4	1
21.59	4	2	39.18	4	2	23.65	4	2	48.24	4	2
24.99	4	3	55.17	4	3	28.67	4	3	64.27	4	3
34.35	4	4	69.29	4	4	34.82	4	4	82.17	4	4
53.94	4	5	71.61	4	5	43.40	4	5	100.17	4	5
52.39	4	6	91.73	4	6	61.85	4	6	101.24	4	6
32.34	5	1	22.18	5	1	22.95	5	1	24.40	5	1
30.60	5	2	32.01	5	2	34.73	5	2	34.46	5	2
45.78	5	3	45.13	5	3	52.44	5	3	47.61	5	3
53.53	5	4	55.07	5	4	63.37	5	4	49.01	5	4
54.66	5	5	59.30	5	5	75.58	5	5	67.59	5	5
70.01	5	6	68.90	5	6	79.11	5	6	81.53	5	6

Table D7. Fitness Data

Score	Smoke	Race	Sex	Rep
54.95	1	1	1	1
46.05	0	1	1	1
45.95	1	1	0	1
22.16	0	1	0	1
46.37	1	0	1	1
17.85	0	0	1	1
4.37	1	0	0	1
24.68	0	0	0	1
51.64	1	1	1	2
39.71	0	1	1	2
38.41	1	1	0	2
20.70	0	1	0	2
51.59	1	0	1	2
23.24	0	0	1	2
6.48	1	0	0	2
19.09	0	0	0	2
56.00	1	1	1	3
47.05	0	1	1	3
42.17	1	1	0	3
16.79	0	1	0	3
55.89	1	0	1	3
16.94	0	0	1	3
2.11	1	0	0	3
26.12	0	0	0	3

Table D8. Police Officer Stress Data

X_1	X_2	X_3	X_4	X_5	X_6	X_7	X_8	X_1	X_2	X_3	X_4	X_5	X_6	X_7	X_8
2	3	2	2	2	2	3	2	1	3	3	3	3	2	4	2
1	1	2	3	1	1	3	2	4	3	2	3	4	4	5	4
1	2	3	5	2	4	5	1	2	1	2	3	2	1	3	2
2	2	1	3	3	4	3	3	2	2	2	3	3	3	3	2
1	3	2	3	4	4	3	3	4	2	3	3	4	2	3	2
1	2	2	3	4	3	3	3	5	1	2	3	4	3	4	2
2	3	2	4	4	3	4	3	4	3	2	3	4	5	3	4
3	1	1	3	2	3	3	1	1	2	3	2	2	3	2	2
2	1	1	4	4	4	3	4	2	2	2	3	1	2	1	2
2	1	2	3	2	5	4	5	3	2	1	2	2	1	2	1
2	1	3	2	2	2	2	1	4	3	2	3	2	4	4	1
2	1	2	3	5	5	4	2	2	3	3	4	3	3	4	2
4	3	4	3	4	3	3	3	3	3	2	4	4	3	3	3
3	3	4	3	4	4	3	2	1	3	2	3	2	3	3	1
2	3	3	3	2	3	2	2	3	3	4	3	2	1	4	3
2	3	5	4	4	4	5	2	1	1	5	3	3	3	3	1
1	3	3	2	4	4	1	4	5	3	3	3	4	4	5	4
2	2	1	3	3	3	2	4	3	3	2	3	1	2	1	2
1	1	2	3	1	1	1	1	1	1	1	5	5	4	5	4
3	4	4	2	3	5	3	4	3	2	2	3	2	2	4	2
3	2	1	3	2	2	2	3	3	2	3	3	3	3	3	3
1	1	3	3	1	5	1	2	3	3	3	4	4	3	4	5
1	2	4	3	3	3	3	4	1	1	1	2	2	4	1	3
3	3	2	4	4	3	4	4	2	2	1	3	3	4	3	1
3	4	5	4	4	5	5	4	1	1	3	4	4	2	3	4

Table D9. Public Safety Data

Region	X_1	X_2	X_3	X_4	X_5	X_6	Region	X_1	X_2	X_3	X_4	X_5	X_6
1	2	1	1	2	1	2	3	2	1	1	2	1	2
1	4	1	3	4	3	4	3	2	1	1	2	1	2
1	3	2	2	4	2	4	3	2	1	1	2	1	2
1	2	1	1	2	1	2	3	2	1	1	2	1	2
1	1	1	1	2	1	2	3	6	5	3	6	4	6
1	4	4	3	6	3	6	3	1	1	1	1	1	1
1	4	2	1	4	1	5	3	4	3	3	4	3	4
1	2	2	1	2	2	4	3	2	2	1	2	1	2
1	5	6	2	4	4	5	3	2	1	1	2	1	2
1	2	1	1	1	1	2	3	2	1	1	1	2	4
1	2	1	1	2	2	4	3	2	1	1	2	1	2
1	2	3	2	4	2	5	3	2	2	1	2	2	3
1	3	1	2	5	2	5	3	2	2	1	3	1	5
1	4	2	2	4	2	4	3	3	2	2	4	2	3
1	2	1	1	2	1	3	3	5	2	2	5	2	6
1	2	2	2	4	3	4	3	1	1	1	1	1	1
1	2	1	1	2	1	3	3	3	2	1	1	2	2
1	1	1	1	1	1	1	3	3	2	1	2	2	3
1	1	1	1	2	1	2	3	2	1	1	1	1	4
1	1	1	1	1	1	1	3	1	1	1	1	1	1
1	4	2	2	3	2	3	3	2	2	2	2	2	3
1	2	2	2	2	2	3	3	4	2	2	4	2	3
1	2	1	1	2	2	3	3	2	1	2	3	2	3
2	2	2	2	2	2	2	3	2	2	1	4	1	4
2	1	1	1	2	1	3	3	2	1	1	2	1	4
2	2	2	1	4	2	4	4	1	1	1	1	1	1
2	2	2	1	2	1	2	4	2	2	2	2	2	2
2	1	1	1	2	1	2	4	2	2	1	2	1	2
2	2	2	1	2	1	2	4	1	1	1	1	1	1
2	2	2	2	2	2	2	4	1	1	1	1	1	1
2	2	2	2	2	1	2	4	1	1	1	2	1	2
2	1	1	1	1	1	1	4	1	1	1	1	1	1
2	1	1	2	2	2	2	4	2	1	1	1	1	1
2	1	2	1	1	1	2	4	2	1	1	2	1	4
2	1	1	1	1	1	1	4	2	1	1	2	1	4
2	2	2	1	2	1	2	4	1	1	1	1	1	2
2	1	1	1	1	1	1	4	1	1	1	1	1	1
2	1	1	1	1	1	1	4	3	1	3	5	2	5
2	2	1	1	1	1	2	4	1	1	1	1	1	1
2	2	1	1	2	1	2	4	2	1	1	2	1	2
2	1	1	1	1	1	2	4	4	2	2	4	2	5
2	2	1	1	2	1	2	4	1	1	1	1	1	1
2	1	1	1	1	1	2	4	2	2	1	2	1	2
2	1	1	1	1	1	1	4	2	2	1	1	1	2
2	1	1	1	1	1	1	4	1	1	1	1	1	2
2	2	2	1	2	1	2	4	1	1	1	1	1	1
2	1	1	1	2	1	2	4	1	1	1	2	1	2
2	1	1	1	2	1	2	4	1	1	1	1	2	2
2							4	2	2	2	2	2	2
2							4	2	2	1	2	1	2

Table Appendix

Tables 1-4 in this appendix are reprinted from *Statistical Tables and Formulae* by Stephen Kokoska and Christopher Nevision, Springer-Verlag, 1989, with permission. Table 5 is reprinted from Harter, H. Leon (1960), "Tables of Range And Studentized Range," *Annals of Mathematical Statistics,* **31**, 1122-1147, with permission.

Table 1. Cumulative Distribution Function for the Standard Normal Random Variable

This table contains values of the cumulative distribution
function for the standard normal random variable
$\Phi(z) = P(Z \le z) = \int_{-\infty}^{z} \frac{1}{\sqrt{2\pi}} e^{-z^2/2} dz.$

z	.00	.01	.02	.03	.04	.05	.06	.07	.08	.09
-3.4	.0003	.0003	.0003	.0003	.0003	.0003	.0003	.0003	.0003	.0002
-3.3	.0005	.0005	.0005	.0004	.0004	.0004	.0004	.0004	.0004	.0003
-3.2	.0007	.0007	.0006	.0006	.0006	.0006	.0006	.0005	.0005	.0005
-3.1	.0010	.0009	.0009	.0009	.0008	.0008	.0008	.0008	.0007	.0007
-3.0	.0013	.0013	.0013	.0012	.0012	.0011	.0011	.0011	.0010	.0010
-2.9	.0019	.0018	.0018	.0017	.0016	.0016	.0015	.0015	.0014	.0014
-2.8	.0026	.0025	.0024	.0023	.0023	.0022	.0021	.0021	.0020	.0019
-2.7	.0035	.0034	.0033	.0032	.0031	.0030	.0029	.0028	.0027	.0026
-2.6	.0047	.0045	.0044	.0043	.0041	.0040	.0039	.0038	.0037	0036
-2.5	.0062	.0060	.0059	.0057	.0055	.0054	.0052	.0051	.0049	.0048
-2.4	.0082	.0080	.0078	.0075	.0073	.0071	.0069	.0068	.0066	.0064
-2.3	.0107	.0104	.0102	.0099	.0096	.0094	.0091	.0089	.0087	.0084
-2.2	.0139	.0136	.0132	.0129	.0125	.0122	.0119	.0116	.0113	.0110
-2.1	.0179	.0174	.0170	.0166	.0162	.0158	.0154	.0150	.0146	.0143
-2.0	.0228	.0222	.0217	.0212	.0207	.0202	.0197	.0192	.0188	.0183
-1.9	.0287	.0281	.0274	.0268	.0262	.0256	.0250	.0244	.0239	.0233
-1.8	.0359	.0351	.0344	.0336	.0329	.0322	.0314	.0307	.0301	.0294
-1.7	.0446	.0436	.0427	.0418	.0409	.0401	.0392	.0384	.0375	.0367
-1.6	.0548	.0537	.0526	.0516	.0505	.0495	.0485	.0475	.0465	.0455
-1.5	.0668	.0655	.0643	.0630	.0618	.0606	.0594	.0582	.0571	.0559
-1.4	.0808	.0793	.0778	.0764	.0749	.0735	.0721	.0708	.0694	.0681
-1.3	.0968	.0951	.0934	.0918	.0901	.0885	.0869	.0853	.0838	.0823
-1.2	.1151	.1131	.1112	.1093	.1075	.1056	.1038	.1020	.1003	.0985
-1.1	.1357	.1335	.1314	.1292	.1271	.1251	.1230	.1210	.1190	.1170
-1.0	.1587	.1562	.1539	.1515	.1492	.1469	.1446	.1423	.1401	.1379
-0.9	.1841	.1841	1788	.1762	.1736	.1711	.1685	.1660	.1635	.1611
-0.8	.2119	.2090	.2061	.2033	.2005	.1977	.1949	.1922	.1894	.1867
-0.7	.2420	.2389	.2358	.2327	.2296	.2266	.2236	.2206	.2177	.2148
-0.6	.2743	.2709	.2676	.2643	.2611	.2578	.2546	.2514	.2483	.2451
-0.5	.3805	.3050	.3015	.2981	.2946	.2912	2877	.2843	.2810	.2776
-0.4	.3446	.3409	.3372	.3336	.3300	.3264	.3228	3192	.3156	.3121
-0.3	.3821	.3783	.3745	.3707	.3669	.3632	.3594	.3557	.3520	.3483
-0.2	.4207	.4168	.4129	.4090	.4052	.4013	.3974	.3936	.3897	3859
-0.1	4602	.4562	.4522	.4483	.4443	.4404	.4364	.4325	.4286	.4247
-0.0	.5000	.4960	.4920	.4880	4840	.4801	.4761	.4721	.4681	.4641

Table 1. (cont.) Cumulative Distribution Function for the Standard Normal
Random Variable

z	.00	.01	.02	.03	.04	.05	.06	.07	.08	.09
0.0	.5000	.5040	.5080	.5120	.5160	.5199	.5239	.5279	.5319	.5359
0.1	.5398	.5438	.5478	.5517	.5557	.5596	.5636	.5675	.5714	.5753
0.2	.5793	.5832	.5871	.5910	.5948	.5987	.6026	.6064	.6103	.6141
0.3	.6179	.6217	.6255	.6293	.6331	.6368	.6406	.6443	.6480	.6517
0.4	.6554	.6591	.6628	.6664	.6700	.6736	.6772	.6808	.6844	.6879
0.5	.6915	.6950	.6985	.7019	.7054	.7088	.7123	.7157	.7190	.7224
0.6	.7257	.7291	.7324	.7357	.7389	.7422	.7454	.7486	.7517	.7549
0.7	.7580	.7611	.7642	.7673	.7704	.7734	.7764	.7794	.7823	.7852
0.8	.7881	.7910	.7339	7967	.7995	.8023	.8051	.8078	.8106	.8133
0.9	.8159	.8186	.8212	.8238	.8264	.8289	.8315	.8340	.8365	.8389
1.0	.8413	.8438	.8461	.8485	.8508	.8531	.8554	.8577	.8599	.8621
1.1	.8643	.8665	.8686	.8708	.8729	.8749	.8770	.8790	.8810	.8830
1.2	.8849	.8869	.8888	.8907	.8925	.8944	.8962	.8980	.8997	.9015
1.3	.9032	.9049	.9066	.9082	.9099	.9115	.9131	.9147	.9162	.9177
1.4	.9192	.9207	.9222	.9236	.9251	.9265	.9279	.9292	.9306	.9319
1.5	.9332	.9345	.9357	.9370	.9382	.9394	.9406	.9418	.9429	.9441
1.6	.9452	.9463	.9474	.9484	.9495	.9505	.9515	.9525	.9535	.9545
1.7	.9554	.9564	.9573	.9582	.9591	.9599	.9608	.9616	.9625	.9633
1.8	.9641	.9649	.9656	.9664	.9671	.9678	.9686	.9693	.9699	.9706
1.9	.9713	.9719	.9726	.9732	.9738	.9744	.9750	.9756	.9761	.9767
2.0	.9772	.9778	.9783	.9788	.9793	.9798	.9803	.9808	9812	.9817
2.1	.9821	.9826	.9830	.9834	.9838	.9842	.9846	.9850	.9854	.9857
2.2	.9861	.9864	.9868	.9871	.9875	.9878	.9881	.9884	.9887	.9890
2.3	.9893	.9896	.9898	.9901	.9904	.9906	.9909	.9911	.9913	.9916
2.4	.9918	.9920	9922	.9925	.9927	.9929	.9931	.9932	.9934	.9936
2.5	.9938	.9940	.9941	.9943	.9945	.9946	.9948	.9949	.9951	.9952
2.6	.9953	.9955	.9956	.9957	.9959	.9960	.9961	.9962	.9963	.9964
2.7	.9965	.9966	.9967	.9968	.9969	.9970	.9971	.9972	.9973	.9974
2.8	.9974	.9975	.9976	.9977	.9977	.9978	.9979	.9979	.9980	.9981
2.9	.9981	.9982	.9982	.9983	.9984	.9984	.9985	.9985	.9986	.9986
3.0	.9987	.9987	.9987	.9988	.9988	.9989	.9989	.9989	.9990	.9990
3.1	.9990	.9991	.9991	.9991	.9992	.9992	.9992	.9992	.9993	.9993
3.2	.9993	.9993	.9994	.9994	.9994	.9994	.9994	.9995	.9995	.9995
3.3	.9995	.9995	.9995	.9996	.9996	.9996	.9996	.9996	.9996	.9997
3.4	.9997	.9997	.9997	.9997	.9997	.9997	.9997	.9997	.9997	.9998

Critical Values, $P(Z \geq z_\alpha) = \alpha$

α	.10	.05	.025	.01	.005	.001	.0005	.0001
z_α	1.2816	1.6449	1.9600	2.3263	2.5758	3.0902	3.2905	3.7190

α	.00009	.00008	.00007	.00006	.00005	.00004	.0003	.00002	.00001
z_α	3.7455	3.7750	3.8082	3.8461	3.8906	3.9444	4.0128	4.1075	4.2649

Table 2. Critical Values for the t Distribution

This table contains critical values $t_{\alpha,\nu}$ for the
t distribution defined by $P(T \geq t_{\alpha,\nu}) = \alpha$.

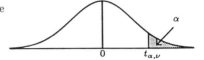

α

ν	.20	.10	.05	.025	.01	.005	.001	.0005	.0001
1	1.3764	3.0777	6.3138	12.7062	31.8205	63.6567	318.3088	636.6192	3183.9388
2	1.0607	1.8856	2.9200	4.3027	6.9646	9.9248	22.3271	31.5991	70.7001
3	.9785	1.6377	2.3534	3.1824	4.5407	5.8409	10.2145	12.9240	22.2037
4	.9410	1.5332	2.1318	2.7764	3.7469	4.6041	7.1732	8.6103	13.0337
5	.9195	1.4759	2.0150	2.5706	3.3649	4.0321	5.8934	6.8688	9.6776
6	.9057	1.4398	1.9432	2.4469	3.1427	3.7074	5.2076	5.9588	8.0248
7	.8960	1.4149	1.8946	2.3646	2.9980	3.4995	4.7853	5.4079	7.0634
8	.8889	1.3968	1.8595	2.3060	2.8965	3.3554	4.5008	5.0413	6.4420
9	.8834	1.3830	1.8331	2.2622	2.8214	3.2498	4.2968	4.7809	6.0101
10	.8791	1.3722	1.8125	2.2281	2.7638	3.1693	4.1437	4.5869	5.6938
11	.8755	1.3634	1.7959	2.2010	2.7181	3.1058	4.0247	4.4370	5.4528
12	.8726	1.3562	1.7823	2.1788	2.6810	3.0545	3.9296	4.3178	5.2633
13	.8702	1.3502	1.7709	2.1604	2.6503	3.0123	3.8520	4.2208	5.1106
14	.8681	1.3450	1.7613	2.1448	2.6245	2.9768	3.7874	4.1405	4.9850
15	.8662	1.3406	1.7531	2.1314	2.6025	2.9467	3.7328	4.0728	4.8800
16	.8647	1.3368	1.7459	2.1199	2.5835	2.9208	3.6862	4.0150	4.7909
17	.8633	1.3334	1.7396	2.1098	2.5669	2.8982	3.6458	3.9651	4.7144
18	.8620	1.3304	1.7341	2.1009	2.5524	2.8784	3.6105	3.9216	4.6480
19	.8610	1.3277	1.7291	2.0930	2.5395	2.8609	3.5794	3.8834	4.5899
20	.8600	1.3253	1.7247	2.0860	2.5280	2.8453	3.5518	3.8495	4.5385
21	.8591	1.3232	1.7207	2.0796	2.5176	2.8314	3.5271	3.8192	4.4929
22	.8583	1.3212	1.7171	2.0739	2.5083	2.8187	3.5050	3.7921	4.4520
23	.8575	1.3195	1.7139	2.0687	2.4999	2.8073	3.4850	3.7676	4.4152
24	.8569	1.3178	1.7109	2.0639	2.4922	2.7969	3.4668	3.7454	4.3819
25	.8562	1.3163	1.7081	2.0595	2.4851	2.7874	3.4502	3.7251	4.3517
26	.8557	1.3150	1.7056	2.0555	2.4786	2.7787	3.4350	3.7066	4.3240
27	.8551	1.3137	1.7033	2.0518	2.4727	2.7707	3.4210	3.6896	4.2987
28	.8546	1.3125	1.7011	2.0484	2.4671	2.7633	3.4081	3.6739	4.2754
29	.8542	1.3114	1.6991	2.0452	2.4620	2.7564	3.3962	3.6594	4.2539
30	.8538	1.3104	1.6973	2.0423	2.4573	2.7500	3.3852	3.6460	4.2340
40	.8507	1.3031	1.6839	2.0211	2.4233	2.7045	3.3069	3.5510	4.0942
50	.8489	1.2987	1.6759	2.0086	2.4033	2.6778	3.2614	3.4960	4.0140
60	.8477	1.2958	1.6706	2.0003	2.3901	2.6603	3.2317	3.4602	3.9621
120	.8446	1.2886	1.6577	1.9799	2.3578	2.6174	3.1595	3.3735	3.8372
∞	.8416	1.2816	1.6449	1.9600	2.3263	2.5758	3.0902	3.2905	3.7190

Table 3. Critical Values for the Chi-Square Distribution

This table contains critical values $\chi^2_{\alpha,\nu}$ for the Chi-Square distribution defined by $P(\chi^2 \geq \chi^2_{\alpha,\nu}) = \alpha$.

α

ν	.9999	.9995	.999	.995	.99	.975	.95	.90
1	$.0^7157$	$.0^6393$	$.0^5175$	$.0^4393$	.0002	.0010	.0039	.0158
2	.0002	.0010	.0020	.0100	.0201	.0506	.1026	.2107
3	.0052	.0153	.0243	.0717	.1148	.2158	.3518	.5844
4	.0284	.0639	.0908	.2070	.2971	.4844	.7107	1.0636
5	.0822	.1581	.2102	.4117	.5543	.8312	1.1455	1.6103
6	.1724	.2994	.3811	.6757	.8721	1.2373	1.6354	2.2041
7	.3000	.4849	.5985	.9893	1.2390	1.6899	2.1673	2.8331
8	.4636	.7104	.8571	1.3444	1.6465	2.1797	2.7326	3.4895
9	.6608	.9717	1.1519	1.7349	2.0879	2.7004	3.3251	4.1682
10	.8889	1.2650	1.4787	2.1559	2.5582	3.2470	3.9403	4.8652
11	1.1453	1.5868	1.8339	2.6032	3.0535	3.8157	4.5748	5.5778
12	1.4275	1.9344	2.2142	3.0738	3.5706	4.4038	5.2260	6.3038
13	1.7333	2.3051	2.6172	3.5650	4.1069	5.0088	5.8919	7.0415
14	2.0608	2.6967	3.0407	4.0747	4.6604	5.6287	6.5706	7.7895
15	2.4082	3.1075	3.4827	4.6009	5.2293	6.2621	7.2609	8.5468
16	2.7739	3.5358	3.9416	5.1422	5.8122	6.9077	7.9616	9.3122
17	3.1567	3.9802	4.4161	5.6972	6.4078	7.5642	8.6718	10.0852
18	3.5552	4.4394	4.9048	6.2648	7.0149	8.2307	9.3905	10.8649
19	3.9683	4.9123	5.4068	6.8440	7.6327	8.9065	10.1170	11.6509
20	4.3952	5.3981	5.9210	7.4338	8.2604	9.5908	10.8508	12.4426
21	4.8348	5.8957	6.4467	8.0337	8.8972	10.2829	11.5913	13.2396
22	5.2865	6.4045	6.9830	8.6427	9.5425	10.9823	12.3380	14.0415
23	5.7494	6.9237	7.5292	9.2604	10.1957	11.6886	13.0905	14.8480
24	6.2230	7.4527	8.0849	9.8862	10.8564	12.4012	13.8484	15.6587
25	6.7066	7.9910	8.6493	10.5197	11.5240	13.1197	14.6114	16.4734
26	7.1998	8.5379	9.2221	11.1602	12.1981	13.8439	15.3792	17.2919
27	7.7019	9.0932	9.8028	11.8076	12.8785	14.5734	16.1514	18.1139
28	8.2126	9.6563	10.3909	12.4613	13.5647	15.3079	16.9279	18.9392
29	8.7315	10.2268	10.9861	13.1211	14.2565	16.0471	17.7084	19.7677
30	9.2581	10.8044	11.5880	13.7867	14.9535	16.7908	18.4927	20.5992
31	9.7921	11.3887	12.1963	14.4578	15.6555	17.5387	19.2806	21.4336
32	10.3331	11.9794	12.8107	15.1340	16.3622	18.2908	20.0719	22.2706
33	10.8810	12.5763	13.4309	15.8153	17.0735	19.0467	20.8665	23.1102
34	11.4352	13.1791	14.0567	16.5013	17.7891	19.8063	21.6643	23.9253
35	11.9957	13.7875	14.6878	17.1918	18.5089	20.5694	22.4650	24.7967
36	12.5622	14.4012	15.3241	17.8867	19.2327	21.3359	23.2686	25.6433
37	13.1343	15.0202	15.9653	18.5858	19.9602	22.1056	24.0749	26.4921
38	13.7120	15.6441	16.6112	19.2889	20.6914	22.8785	24.8839	27.3430
39	14.2950	16.2729	17.2616	19.9959	21.4262	23.6543	25.6954	281958
40	14.8831	16.9062	17.9164	20.7065	22.1643	24.4330	26.5093	29.0505
50	21.0093	23.4610	24.6739	27.9907	29.7067	32.3574	34.7643	37.6886
60	27.4969	30.3405	31.7383	35.5345	37.4849	40.4817	43.1880	46.4589
70	34.2607	37.4674	39.0364	43.2752	45.4417	48.7576	51.7393	55.3289
80	41.2445	44.7910	46.5199	51.1719	53.5401	57.1532	60.3915	64.2778
90	48.4087	52.2758	54.1552	59.1963	61.7541	65.6466	69.1260	73.2911
100	55.7246	59.8957	61.9179	67.3276	70.0649	74.2219	77.9295	82.3581

Table 3. (cont.) Critical Values for the Chi-Square Distribution

$$\alpha$$

ν	.10	.05	.025	.01	.005	.001	.0005	.0001
1	.27055	3.8415	5.0239	6.6349	7.8794	10.2876	12.1157	15.1367
2	4.6052	5.9915	7.3778	9.2103	10.5966	13.8155	15.2018	18.4207
3	6.2514	7.8147	9.3484	11.3449	12.8382	16.2662	17.7300	21.1075
4	7.7794	9.4877	11.1433	13.2767	14.8603	18.4668	19.9974	23.5127
5	9.2364	11.0705	12.8325	15.0863	16.7496	20.5150	22.1053	25.7448
6	10.6446	12.5916	14.4494	16.8119	18.5476	22.4577	24.1028	27.8563
7	12.0170	14.0671	16.0128	18.4753	20.2777	24.3219	26.0178	29.8775
8	13.3616	15.5073	17.5345	20.0902	21.9550	26.1245	27.8680	31.8276
9	14.6837	16.9190	19.0228	21.6660	23.5894	27.8772	29.6658	33.7199
10	15.9872	18.3070	20.4832	23.2093	25.1882	29.5883	31.4198	35.5640
11	17.2750	19.6751	21.9200	24.7250	26.7568	31.2641	33.1366	37.3670
12	18.5493	21.0261	23.3367	26.2170	28.2995	32.9095	34.8213	39.1344
13	19.8119	22.3620	24.7356	27.6882	29.8195	34.5282	36.4778	40.8707
14	21.0641	23.6848	26.1189	29.1412	31.3193	36.1233	38.1094	42.5793
15	22.3071	24.9958	27.4884	30.5779	32.8013	37.6973	39.7188	44.2632
16	23.5418	26.2962	28.8454	31.9999	34.2672	39.2524	41.3081	45.9249
17	24.7690	27.5871	30.1910	33.4087	35.7185	40.7902	42.8792	47.5664
18	25.9894	28.8693	31.5264	34.8053	37.1565	42.3124	44.4338	49.1894
19	27.2036	30.1435	32.8523	36.1909	38.5823	43.8202	45.9731	50.7955
20	28.4120	31.4104	34.1696	37.5662	39.9968	45.3147	47.4985	52.3860
21	29.6151	32.6706	35.4789	38.9322	41.4011	46.7970	49.0108	53.9620
22	30.8133	33.9244	36.7807	40.2894	42.7957	48.2679	50.5111	55.5246
23	32.0069	35.1725	38.0756	41.6384	44.1813	49.7282	52.0002	57.0746
24	33.1962	36.4150	39.3641	42.9798	45.5585	51.1786	53.4788	58.6130
25	34.3816	37.6525	40.6465	44.3141	46.9279	52.6197	54.9475	60.1403
26	35.5632	38.8851	41.9232	45.6417	48.2899	54.0520	56.4069	61.6573
27	36.7412	40.1133	43.1945	46.9629	49.6449	55.4760	57.8576	63.1645
28	37.9159	41.3371	44.4608	48.2782	50.9934	56.8923	59.3000	64.6624
29	39.0875	42.5570	45.7223	49.5879	52.3356	58.3012	60.7346	66.1517
30	40.2560	43.7730	46.9792	50.8922	53.6720	59.7031	62.1619	67.6326
31	41.4217	44.9853	48.2319	52.1914	55.0027	61.0983	63.5820	69.1057
32	42.5847	46.1943	49.4804	53.4858	56.3281	62.4872	64.9955	70.5712
33	43.7452	47.3999	50.7251	54.7755	57.6484	63.8701	66.4025	72.0296
34	44.9032	48.6024	51.9660	56.0609	58.9636	65.2472	67.8035	73.4812
35	46.0588	49.8018	53.2033	57.3421	60.2748	66.6188	69.1986	74.9262
36	47.2122	50.9985	54.4373	58.6192	61.5812	67.9858	70.5881	76.3650
37	48.3634	52.1923	55.6680	59.8925	62.8833	69.3465	71.9722	77.7977
38	49.5126	53.3835	56.8955	61.1621	64.1814	70.7029	73.3512	79.2247
39	50.6598	54.5722	58.1201	62.4281	65.4756	72.0547	74.7253	80.6462
40	51.8051	55.7585	59.3417	63.6907	66.7660	73.4020	76.0946	82.0623
50	63.1671	67.5048	71.4202	76.1539	79.4900	86.6608	89.5605	95.9687
60	74.3970	79.0819	83.2977	83.3794	91.9517	99.6072	102.6948	109.5029
70	85.5270	90.5312	95.0232	100.4252	104.2149	112.3169	115.5776	122.7547
80	96.5782	101.8795	106.6286	112.3288	116.3211	124.8392	128.2613	135.7825
90	107.5650	113.1453	118.1359	124.1163	128.2989	137.2084	140.7823	148.6273
100	118.4980	124.3421	129.5612	135.8067	140.1695	149.4493	153.1670	161.3187

This table contains critical vlaues F_{α,ν_1,ν_2} for the F distribution defined by

$$P(F \geq F_{\alpha,\nu_1,\nu_2}) = \alpha.$$

Table 4. Critical Values for the F Distribution

$\alpha = .05$

$\nu_2 \backslash \nu_1$	1	2	3	4	5	6	7	8	9	10	15	20	30	40	60	120	∞
1	161.45	199.50	215.71	224.58	230.16	233.99	236.77	238.88	240.54	241.88	245.95	248.01	250.10	251.14	252.20	253.25	254.25
2	18.51	19.00	19.16	19.25	19.30	19.33	19.35	19.37	19.38	19.40	19.43	19.45	19.46	19.47	19.48	19.49	19.50
3	10.13	9.55	9.28	9.12	9.01	8.94	8.89	8.85	8.81	8.79	8.70	8.66	8.62	8.59	8.57	8.55	8.53
4	7.71	6.94	6.59	6.39	6.26	6.16	6.09	6.04	6.00	5.96	5.86	5.80	5.75	5.72	5.69	5.66	5.63
5	6.61	5.79	5.41	5.19	5.05	4.95	4.88	4.82	4.77	4.74	4.62	4.56	4.50	4.46	4.43	4.40	4.37
6	5.99	5.14	4.76	4.53	4.39	4.28	4.21	4.15	4.10	4.06	3.94	3.87	3.81	3.77	3.74	3.70	3.67
7	5.59	4.74	4.35	4.12	3.97	3.87	3.79	3.73	3.68	3.64	3.51	3.44	3.38	3.34	3.30	3.27	3.23
8	5.32	4.46	4.07	3.84	3.69	3.58	3.50	3.44	3.39	3.35	3.22	3.15	3.08	3.04	3.01	2.97	2.93
9	5.12	4.26	3.86	3.63	3.48	3.37	3.29	3.23	3.18	3.14	3.01	2.94	2.86	2.83	2.79	2.75	2.71
10	4.96	4.10	3.71	3.48	3.33	3.22	3.14	3.07	3.02	2.98	2.85	2.77	2.70	2.66	2.62	2.58	2.54
11	4.84	3.98	3.59	3.36	3.20	3.09	3.01	2.95	2.90	2.85	2.72	2.65	2.57	2.53	2.49	2.45	2.41
12	4.75	3.89	3.49	3.26	3.11	3.00	2.91	2.85	2.80	2.75	2.62	2.54	2.47	2.43	2.38	2.34	2.30
13	4.67	3.81	3.41	3.18	3.03	2.92	2.83	2.77	2.71	2.67	2.53	2.46	2.38	2.34	2.30	2.25	2.21
14	4.60	3.74	3.34	3.11	2.96	2.85	2.76	2.70	2.65	2.60	2.46	2.39	2.31	2.27	2.22	2.18	2.13
15	4.54	3.68	3.29	3.06	2.90	2.79	2.71	2.64	2.59	2.54	2.40	2.33	2.25	2.20	2.16	2.11	2.07
16	4.49	3.63	3.24	3.01	2.85	2.74	2.66	2.59	2.54	2.49	2.35	2.28	2.19	2.15	2.11	2.06	2.01
17	4.45	3.59	3.20	2.96	2.81	2.70	2.61	2.55	2.49	2.45	2.31	2.23	2.15	2.10	2.06	2.01	1.96
18	4.41	3.55	3.16	2.93	2.77	2.66	2.58	2.51	2.46	2.41	2.27	2.19	2.11	2.06	2.02	1.97	1.92
19	4.38	3.52	3.13	2.90	2.74	2.63	2.54	2.48	2.42	2.38	2.23	2.16	2.07	2.03	1.98	1.93	1.88
20	4.35	3.49	3.10	2.87	2.71	2.60	2.51	2.45	2.39	2.35	2.20	2.12	2.04	1.99	1.95	1.90	1.85
21	4.32	3.47	3.07	2.84	2.68	2.57	2.49	2.42	2.37	2.32	2.18	2.10	2.01	1.96	1.92	1.87	1.82
22	4.30	3.44	3.05	2.82	2.66	2.55	2.46	2.40	2.34	2.30	2.15	2.07	1.98	1.94	1.89	1.84	1.79
23	4.28	3.42	3.03	2.80	2.64	2.53	2.44	2.37	2.32	2.27	2.13	2.05	1.96	1.91	1.86	1.81	1.76
24	4.26	3.40	3.01	2.78	2.62	2.51	2.42	2.36	2.30	2.25	2.11	2.03	1.94	1.89	1.84	1.79	1.74
25	4.24	3.39	2.99	2.76	2.60	2.49	2.40	2.34	2.28	2.24	2.09	2.01	1.92	1.87	1.82	1.77	1.71
30	4.17	3.32	2.92	2.69	2.53	2.42	2.33	2.27	2.21	2.16	2.01	1.93	1.84	1.79	1.74	1.68	1.63
40	4.08	3.23	2.84	2.61	2.45	2.34	2.25	2.18	2.12	2.08	1.92	1.84	1.74	1.69	1.64	1.58	1.51
50	4.03	3.18	2.79	2.56	2.40	2.29	2.20	2.13	2.07	2.03	1.87	1.78	1.69	1.63	1.58	1.51	1.44
60	4.00	3.15	2.76	2.53	2.37	2.25	2.17	2.10	2.04	1.99	1.84	1.75	1.65	1.59	1.53	1.47	1.39
120	3.92	3.07	2.68	2.45	2.29	2.18	2.09	2.02	1.96	1.91	1.75	1.66	1.55	1.50	1.43	1.35	1.26
∞	3.85	3.00	2.61	2.38	2.22	2.10	2.01	1.94	1.88	1.84	1.67	1.58	1.46	1.40	1.32	1.23	1.00

Table 4. (cont.) Critical Values for the F Distribution

$\alpha = .01$

ν_1

ν_2	1	2	3	4	5	6	7	8	9	10	15	20	30	40	60	120	∞
2	98.50	99.00	99.17	99.25	99.30	99.33	99.36	99.37	99.39	99.40	99.43	99.45	99.47	99.47	99.48	99.49	99.50
3	34.12	30.82	29.46	28.71	28.24	27.91	27.67	27.49	27.35	27.23	26.87	26.69	26.50	26.41	26.32	26.22	26.13
4	21.20	18.00	16.69	15.98	15.52	15.21	14.98	14.80	14.66	14.55	14.20	14.02	13.84	13.75	13.65	13.56	13.47
5	16.26	13.27	12.06	11.39	10.97	10.67	10.46	10.29	10.16	10.05	9.72	9.55	9.38	9.29	9.20	9.11	9.03
6	13.75	10.92	9.78	9.15	8.75	8.47	8.26	8.10	7.98	7.87	7.56	7.40	7.23	7.14	7.06	6.97	6.89
7	12.25	9.55	8.45	7.85	7.46	7.19	6.99	6.84	6.72	6.62	6.31	6.16	5.99	5.91	5.82	5.74	5.65
8	11.26	8.65	7.59	7.01	6.63	6.37	6.18	6.03	5.91	5.81	5.52	5.36	5.20	5.12	5.03	4.95	4.86
9	10.56	8.02	6.99	6.42	6.06	5.80	5.61	5.47	5.35	5.26	4.96	4.81	4.65	4.57	4.48	4.40	4.32
10	10.04	7.56	6.55	5.99	5.64	5.39	5.20	5.06	4.94	4.85	4.56	4.41	4.25	4.17	4.08	4.00	3.91
11	9.65	7.21	6.22	5.67	5.32	5.07	4.89	4.74	4.63	4.54	4.25	4.10	3.94	3.86	3.78	3.69	3.61
12	9.33	6.93	5.95	5.41	5.06	4.82	4.64	4.50	4.39	4.30	4.01	3.86	3.70	3.62	3.54	3.45	3.37
13	9.07	6.70	5.74	5.21	4.86	4.62	4.44	4.30	4.19	4.10	3.82	3.66	3.51	3.43	3.34	3.25	3.17
14	8.86	6.51	5.56	5.04	4.69	4.46	4.28	4.14	4.03	3.94	3.66	3.51	3.35	3.27	3.18	3.09	3.01
15	8.68	6.36	5.42	4.89	4.56	4.32	4.14	4.00	3.89	3.80	3.52	3.37	3.21	3.13	3.05	2.96	2.87
16	8.53	6.23	5.29	4.77	4.44	4.20	4.03	3.89	3.78	3.69	3.41	3.26	3.10	3.02	2.93	2.84	2.76
17	8.40	6.11	5.19	4.67	4.34	4.10	3.93	3.79	3.68	3.59	3.31	3.16	3.00	2.92	2.83	2.75	2.66
18	8.29	6.01	5.09	4.58	4.25	4.01	3.84	3.71	3.60	3.51	3.23	3.08	2.92	2.84	2.75	2.66	2.57
19	8.18	5.93	5.01	4.50	4.17	3.94	3.77	3.63	3.52	3.43	3.15	3.00	2.84	2.76	2.67	2.58	2.50
20	8.10	5.85	4.94	4.43	4.10	3.87	3.70	3.56	3.46	3.37	3.09	2.94	2.78	2.69	2.61	2.52	2.43
21	8.02	5.78	4.87	4.37	4.04	3.81	3.64	3.51	3.40	3.31	3.03	2.88	2.72	2.64	2.55	2.46	2.37
22	7.95	5.72	4.82	4.31	3.99	3.76	3.59	3.45	3.35	3.26	2.98	2.83	2.67	2.58	2.50	2.40	2.31
23	7.88	5.66	4.76	4.26	3.94	3.71	3.54	3.41	3.30	3.21	2.93	2.78	2.62	2.54	2.45	2.35	2.26
24	7.82	5.61	4.72	4.22	3.90	3.67	3.50	3.36	3.26	3.17	2.89	2.74	2.58	2.49	2.40	2.31	2.22
25	7.77	5.57	4.68	4.18	3.85	3.63	3.46	3.32	3.22	3.13	2.85	2.70	2.54	2.45	2.36	2.27	2.18
30	7.56	5.39	4.51	4.02	3.70	3.47	3.30	3.17	3.07	2.98	2.70	2.55	2.39	2.30	2.21	2.11	2.01
40	7.31	5.18	4.31	3.83	3.51	3.29	3.12	2.99	2.89	2.80	2.52	2.37	2.20	2.11	2.02	1.92	1.81
50	7.17	5.06	4.20	3.72	3.41	3.19	3.02	2.89	2.78	2.70	2.42	2.27	2.10	2.01	1.91	1.80	1.69
60	7.08	4.98	4.13	3.65	3.34	3.12	2.95	2.82	2.72	2.63	2.35	2.20	2.03	1.94	1.84	1.73	1.61
120	6.85	4.79	3.95	3.48	3.17	2.96	2.79	2.66	2.56	2.47	2.19	2.03	1.86	1.76	1.66	1.53	1.39
∞	6.65	4.62	3.79	3.33	3.03	2.81	2.65	2.52	2.42	2.33	2.05	1.89	1.71	1.60	1.48	1.34	1.00

Table 4. (cont.) Critical Values for the F Distribution

$\alpha = .001$

ν_2									ν_1								
	1	2	3	4	5	6	7	8	9	10	15	20	30	40	60	120	∞
2	998.50	999.00	999.17	999.25	999.30	999.33	999.36	999.37	999.39	999.40	999.42	999.45	999.47	999.47	999.48	999.49	999.50
3	167.03	148.50	141.11	137.10	134.58	132.85	131.58	130.62	129.86	129.25	127.37	126.42	125.45	124.96	124.47	123.97	123.50
4	74.14	61.25	56.18	53.44	51.71	50.53	49.66	49.00	48.47	48.05	46.76	46.10	45.43	45.09	44.75	44.40	44.07
5	47.18	37.12	33.20	31.09	29.75	28.83	28.16	27.65	27.24	26.92	25.91	25.39	24.87	24.60	24.33	24.06	23.80
6	35.51	27.00	23.70	21.92	20.80	20.03	19.46	19.03	18.69	18.41	17.56	17.12	16.67	16.44	16.21	15.98	15.76
7	29.25	21.69	18.77	17.20	16.21	15.52	15.02	14.63	14.33	14.08	13.32	12.93	12.53	12.33	12.12	11.91	11.71
8	25.41	18.49	15.83	14.39	13.48	12.86	12.40	12.05	11.77	11.54	10.84	10.48	10.11	9.92	9.73	9.53	9.35
9	22.86	16.39	13.90	12.56	11.71	11.13	10.70	10.37	10.11	9.89	9.24	8.90	8.55	8.37	8.19	8.00	7.82
10	21.04	14.91	12.55	11.28	10.48	9.93	9.52	9.20	8.96	8.75	8.13	7.80	7.47	7.30	7.12	6.94	6.77
11	19.69	13.81	11.56	10.35	9.58	9.05	8.66	8.35	8.12	7.92	7.32	7.01	6.68	6.52	6.35	6.18	6.01
12	18.64	12.97	10.80	9.63	8.89	8.38	8.00	7.71	7.48	7.29	6.71	6.40	6.09	5.93	5.76	5.59	5.43
13	17.82	12.31	10.21	9.07	8.35	7.86	7.49	7.21	6.98	6.80	6.23	5.93	5.63	5.47	5.30	5.14	4.98
14	17.14	11.78	9.73	8.62	7.92	7.44	7.08	6.80	6.58	6.40	5.85	5.56	5.25	5.10	4.94	4.77	4.61
15	16.59	11.34	9.34	8.25	7.57	7.09	6.74	6.47	6.26	6.08	5.54	5.25	4.95	4.80	4.64	4.47	4.32
16	16.12	10.97	9.01	7.94	7.27	6.80	6.46	6.19	5.98	5.81	5.27	4.99	4.70	4.54	4.39	4.23	4.07
17	15.72	10.66	8.73	7.68	7.02	6.56	6.22	5.96	5.75	5.58	5.05	4.78	4.48	4.33	4.18	4.02	3.86
18	15.38	10.39	8.49	7.46	6.81	6.35	6.02	5.76	5.56	5.39	4.87	4.59	4.30	4.15	4.00	3.84	3.68
19	15.08	10.16	8.28	7.27	6.62	6.18	5.85	5.59	5.39	5.22	4.70	4.43	4.14	3.99	3.84	3.68	3.52
20	14.82	9.95	8.10	7.10	6.46	6.02	5.69	5.44	5.24	5.08	4.56	4.29	4.00	3.86	3.70	3.54	3.39
21	14.59	9.77	7.94	6.95	6.32	5.88	5.56	5.31	5.11	4.95	4.44	4.17	3.88	3.74	3.58	3.42	3.27
22	14.38	9.61	7.80	6.81	6.19	5.76	5.44	5.19	4.99	4.83	4.33	4.06	3.78	3.63	3.48	3.32	3.16
23	14.20	9.47	7.67	6.70	6.08	5.65	5.33	5.09	4.89	4.73	4.23	3.96	3.68	3.53	3.38	3.22	3.07
24	14.03	9.34	7.55	6.59	5.98	5.55	5.23	4.99	4.80	4.64	4.14	3.87	3.59	3.45	3.29	3.14	2.98
25	13.88	9.22	7.45	6.49	5.89	5.46	5.15	4.91	4.71	4.56	4.06	3.79	3.52	3.37	3.22	3.06	2.90
30	13.29	8.77	7.05	6.12	5.53	5.12	4.82	4.58	4.39	4.24	3.75	3.49	3.22	3.07	2.92	2.76	2.60
40	12.61	8.25	6.59	5.70	5.13	4.73	4.44	4.21	4.02	3.87	3.40	3.14	2.87	2.73	2.57	2.41	2.24
50	12.22	7.96	6.34	5.46	4.90	4.51	4.22	4.00	3.82	3.67	3.20	2.95	2.68	2.53	2.38	2.21	2.04
60	11.97	7.77	6.17	5.31	4.76	4.37	4.09	3.86	3.69	3.54	3.08	2.83	2.55	2.41	2.25	2.08	1.90
120	11.38	7.32	5.78	4.95	4.42	4.04	3.77	3.55	3.38	3.24	2.78	2.53	2.26	2.11	1.95	1.77	1.56
∞	10.86	6.93	5.44	4.64	4.12	3.76	3.49	3.28	3.11	2.97	2.53	2.28	2.01	1.85	1.68	1.47	1.00

This table contains critical values $Q_{\alpha,k,\nu}$ for the Studentized Range distribution defined by $P(Q \geq Q_{\alpha,k,\nu}) = \alpha$, k is the number of degrees of freedom in the numerator (the number of treatment groups) and ν is the number in the denominator (s^2).

Table 5. Critical Values for the Studentized Range Distribution

$\alpha = .05$

ν \ k	2	3	4	5	6	7	8	9	10	11	12	13	14	15	16	17	18	19	20
1	17.97	26.98	38.82	37.08	40.41	43.12	45.40	47.36	49.07	50.59	51.96	53.20	54.33	55.36	56.32	57.22	58.04	58.83	59.56
2	6.085	8.331	9.798	10.88	11.74	12.44	13.03	13.54	13.99	14.39	14.75	15.08	15.38	15.65	15.91	16.14	16.37	16.57	16.77
3	4.501	5.910	6.825	7.502	8.037	8.478	8.853	9.177	9.462	9.717	9.946	10.15	10.35	10.53	10.69	10.84	10.98	11.11	11.24
4	3.927	5.040	5.757	6.287	6.707	7.053	7.347	7.602	7.826	8.027	8.208	8.373	8.525	8.664	8.794	8.914	9.028	9.134	9.233
5	3.635	4.602	5.218	5.673	6.033	6.330	6.582	6.802	6.995	7.168	7.324	7.466	7.596	7.717	7.828	7.932	8.030	8.122	8.208
6	3.461	4.339	4.896	5.305	5.628	5.895	6.122	6.319	6.493	6.649	6.789	6.917	7.034	7.143	7.244	7.338	7.426	7.508	7.587
7	3.344	4.165	4.681	5.060	5.359	5.606	5.815	5.998	6.158	6.302	6.431	6.550	6.658	6.759	6.852	6.939	7.020	7.097	7.170
8	3.261	4.041	4.529	4.886	5.167	5.399	5.597	5.767	5.918	6.054	6.175	6.287	6.389	6.483	6.571	6.653	6.729	6.802	6.870
9	3.199	3.949	4.415	4.756	5.024	5.244	5.432	5.595	5.739	5.867	5.983	6.089	6.186	6.276	6.359	6.437	6.510	6.579	6.644
10	3.151	3.877	4.327	4.654	4.912	5.124	5.305	5.461	5.599	5.722	5.833	5.935	6.028	6.114	6.194	6.269	6.339	6.405	6.467
11	3.113	3.820	4.256	4.574	4.823	5.028	5.202	5.353	5.487	5.605	5.713	5.811	5.901	5.984	6.062	6.134	6.202	6.265	6.326
12	3.082	3.773	4.199	4.508	4.751	4.950	5.119	5.265	5.395	5.511	5.615	5.710	5.798	5.878	5.953	6.023	6.089	6.151	6.209
13	3.055	3.735	4.151	4.453	4.690	4.885	5.049	5.192	5.318	5.431	5.533	5.625	5.711	5.789	5.862	5.931	5.995	6.055	6.112
14	3.033	3.702	4.111	4.407	4.639	4.829	4.990	5.131	5.254	5.364	5.463	5.554	5.637	5.714	5.786	5.852	5.915	5.974	6.029
15	3.014	3.674	4.076	4.367	4.595	4.782	4.940	5.077	5.198	5.306	5.404	5.493	5.574	5.649	5.720	5.785	5.846	5.904	5.958
16	2.998	3.649	4.046	4.333	4.557	4.741	4.897	5.031	5.150	5.256	5.352	5.439	5.520	5.593	5.662	5.727	5.786	5.843	5.897
17	2.984	3.628	4.020	4.303	4.524	4.705	4.858	4.991	5.108	5.212	5.307	5.392	5.471	5.544	5.612	5.675	5.734	5.790	5.842
18	2.971	3.609	3.997	4.277	4.495	4.673	4.824	4.956	5.071	5.174	5.267	5.352	5.429	5.501	5.568	5.630	5.688	5.743	5.794
19	2.960	3.593	3.977	4.253	4.469	4.645	4.794	4.924	5.038	5.140	5.231	5.315	5.391	5.462	5.528	5.589	5.647	5.701	5.752
20	2.950	3.578	3.958	4.232	4.445	4.620	4.768	4.896	5.008	5.108	5.199	5.282	5.357	5.427	5.493	5.553	5.610	5.663	5.714
24	2.919	3.532	3.901	4.166	4.373	4.541	4.684	4.807	4.915	5.012	5.099	5.179	5.251	5.319	5.381	5.439	5.494	5.545	5.594
30	2.888	3.486	3.845	4.102	4.302	4.464	4.602	4.720	4.824	4.917	5.001	5.077	5.147	5.211	5.271	5.327	5.379	5.429	5.475
40	2.858	3.442	3.791	4.039	4.232	4.389	4.521	4.635	4.735	4.824	4.904	4.977	5.044	5.106	5.163	5.216	5.266	5.313	5.358
60	2.829	3.399	3.737	3.977	4.163	4.314	4.441	4.550	4.646	4.732	4.808	4.878	4.942	5.001	5.056	5.107	5.154	5.199	5.241
120	2.800	3.356	3.685	3.917	4.096	4.241	4.363	4.468	4.560	4.641	4.714	4.781	4.842	4.898	4.950	4.998	5.044	5.086	5.126
∞	2.772	3.314	3.633	3.858	4.030	4.170	4.286	4.387	4.474	4.552	4.622	4.685	4.743	4.796	4.845	4.891	4.934	4.974	5.012

Table 5. (cont.) Critical Values for the Studentized Range Distribution

$\alpha = .01$

								k												
ν	2	3	4	5	6	7	8	9	10	11	12	13	14	15	16	17	18	19	20
1	90.03	135.0	164.3	185.6	202.2	215.8	227.2	237.0	245.6	253.2	260.0	266.2	271.8	277.0	281.8	286.3	290.4	294.3	298.0
2	14.04	19.02	22.29	24.72	26.63	28.20	29.53	30.68	31.69	32.59	33.40	34.13	34.81	35.43	36.00	36.53	37.03	37.50	37.95
3	8.261	10.62	12.17	13.33	14.24	15.00	15.64	16.20	16.69	17.13	17.53	17.89	18.22	18.52	18.81	19.07	19.32	19.55	19.77
4	6.512	8.120	9.173	9.958	10.58	11.10	11.55	11.93	12.27	12.57	12.84	13.09	13.32	13.53	13.73	13.91	14.08	14.24	14.40
5	5.702	6.976	7.804	8.421	8.913	9.321	9.669	9.972	10.24	10.48	10.70	10.89	11.08	11.24	11.40	11.55	11.68	11.81	11.93
6	5.243	6.331	7.033	7.556	7.973	8.318	8.613	8.869	9.097	9.301	9.485	9.653	9.808	9.951	10.08	10.21	10.32	10.43	10.54
7	4.949	5.919	6.543	7.005	7.373	7.679	7.939	8.166	8.368	8.548	8.711	8.860	8.997	9.124	9.242	9.353	9.456	9.554	9.646
8	4.746	5.635	6.204	6.625	6.960	7.237	7.474	7.681	7.863	8.027	8.176	8.312	8.436	8.552	8.659	8.760	8.854	8.943	9.027
9	4.596	5.428	5.957	6.348	6.658	6.915	7.134	7.325	7.495	7.647	7.784	7.910	8.025	8.132	8.232	8.325	8.412	8.495	8.573
10	4.482	5.270	5.769	6.136	6.428	6.669	6.875	7.055	7.213	7.356	7.485	7.603	7.712	7.812	7.906	7.993	8.076	8.153	8.226
11	4.392	5.146	5.621	5.970	6.247	6.476	6.672	6.842	6.992	7.128	7.250	7.362	7.465	7.560	7.649	7.732	7.809	7.883	7.952
12	4.320	5.046	5.502	5.836	6.101	6.321	6.507	6.670	6.814	6.943	7.060	7.167	7.265	7.356	7.441	7.520	7.594	7.665	7.731
13	4.260	4.964	5.404	5.727	5.981	6.192	6.372	6.528	6.667	6.791	6.903	7.006	7.101	7.188	7.269	7.345	7.417	7.485	7.548
14	4.210	4.895	5.322	5.634	5.881	6.085	6.258	6.409	6.543	6.664	6.772	6.871	6.962	7.047	7.126	7.199	7.268	7.333	7.395
15	4.168	4.836	5.252	5.556	5.796	5.994	6.162	6.309	6.439	6.555	6.660	6.757	6.845	6.927	7.003	7.074	7.142	7.204	7.264
16	4.131	4.786	5.192	5.489	5.722	5.915	6.079	6.222	6.349	6.462	6.564	6.658	6.744	6.823	6.898	6.967	7.032	7.093	7.152
17	4.099	4.742	5.140	5.430	5.659	5.847	6.007	6.147	6.270	6.381	6.480	6.572	6.656	6.734	6.806	6.873	6.937	6.997	7.053
18	4.071	4.703	5.094	5.379	5.603	5.788	5.944	6.081	6.201	6.310	6.407	6.497	6.579	6.655	6.725	6.792	6.854	6.912	6.968
19	4.046	4.670	5.054	5.334	5.554	5.735	5.889	6.022	6.141	6.247	6.342	6.430	6.510	6.585	6.654	6.719	6.780	6.837	6.891
20	4.024	4.639	5.018	5.294	5.510	5.688	5.839	5.970	6.087	6.191	6.285	6.371	6.450	6.523	6.591	6.654	6.714	6.771	6.823
24	3.956	4.546	4.907	5.168	5.374	5.542	5.685	5.809	5.919	6.017	6.106	6.186	6.261	6.330	6.394	6.453	6.510	6.563	6.612
30	3.889	4.455	4.799	5.048	5.242	5.401	5.536	5.653	5.756	5.849	5.932	6.008	6.078	6.143	6.203	6.259	6.311	6.361	6.407
40	3.825	4.367	4.696	4.931	5.114	5.265	5.392	5.502	5.599	5.686	5.764	5.835	5.900	5.961	6.017	6.069	6.119	6.165	6.209
60	3.762	4.282	4.595	4.818	4.991	5.133	5.253	5.356	5.447	5.528	5.601	5.667	5.728	5.785	5.837	5.886	5.931	5.974	6.015
120	3.702	4.200	4.497	4.709	4.872	5.005	5.118	5.214	5.299	5.375	5.443	5.505	5.562	5.614	5.662	5.708	5.750	5.790	5.827
∞	3.643	4.120	4.403	4.603	4.757	4.882	4.987	5.078	5.157	5.227	5.290	5.348	5.400	5.448	5.493	5.535	5.574	5.611	5.645

Table 5. (cont.) Critical Values for the Studentized Range Distribution

$\alpha = .001$

k

ν	2	3	4	5	6	7	8	9	10	11	12	13	14	15	16	17	18	19	20
1	900.3	1351.	1643.	1856.	2022.	2158.	2272.	2730.	2455.	2532.	2600.	2662.	2718.	2770.	2818.	2863.	2904.	2943.	2980.
2	44.69	60.42	70.77	78.43	84.49	89.46	93.67	97.30	100.5	103.3	105.9	108.2	110.4	112.3	114.2	115.9	117.4	118.9	120.3
3	18.28	23.32	26.65	29.13	31.11	32.74	34.12	35.33	36.39	37.34	38.20	38.98	39.69	40.35	40.97	41.54	42.07	42.58	43.05
4	12.18	14.99	16.84	18.23	19.34	20.26	21.04	21.73	22.33	22.87	23.36	23.81	24.21	24.59	24.94	25.27	25.58	25.87	26.14
5	9.714	11.67	12.96	13.93	14.71	15.35	15.90	16.38	16.81	17.18	17.53	17.85	18.13	18.41	18.66	18.89	19.10	19.31	19.51
6	8.427	9.960	10.97	11.72	12.32	12.83	13.26	13.63	13.97	14.27	14.54	14.79	15.01	15.22	15.42	15.60	15.78	15.94	16.09
7	7.648	8.930	9.768	10.40	10.90	11.32	11.68	11.99	12.27	12.52	12.74	12.95	13.14	13.32	13.48	13.64	13.78	13.92	14.04
8	7.130	8.250	8.978	9.522	9.958	10.32	10.64	10.91	11.15	11.36	11.56	11.74	11.91	12.06	12.21	12.34	12.47	12.59	12.70
9	6.762	7.768	8.419	8.906	9.295	9.619	9.897	10.14	10.36	10.55	10.73	10.89	11.03	11.18	11.30	11.42	11.54	11.64	11.75
10	6.487	7.411	8.006	8.450	8.804	9.099	9.352	9.573	9.769	9.946	10.11	10.25	10.39	10.52	10.64	10.75	10.85	10.95	11.03
11	6.275	7.136	7.687	8.098	8.426	8.699	8.933	9.138	9.319	9.482	9.630	9.766	9.892	10.01	10.12	10.22	10.31	10.41	10.49
12	6.106	6.917	7.436	7.821	8.127	8.383	8.601	8.793	8.962	9.115	9.254	9.381	9.498	9.606	9.707	9.802	9.891	9.975	10.06
13	5.970	6.740	7.231	7.595	7.885	8.126	8.333	8.513	8.673	8.817	8.948	9.068	9.178	9.281	9.376	9.466	9.550	9.629	9.704
14	5.856	6.594	7.062	7.409	7.685	7.915	8.110	8.282	8.434	8.571	8.696	8.809	8.914	9.012	9.103	9.188	9.267	9.343	9.414
15	5.760	6.470	6.920	7.252	7.517	7.736	7.925	8.088	8.234	8.365	8.483	8.592	8.693	8.786	8.872	8.954	9.030	9.102	9.170
16	5.678	6.365	6.799	7.119	7.374	7.585	7.766	7.923	8.063	8.189	8.303	8.407	8.504	8.593	8.676	8.755	8.828	8.897	8.963
17	5.608	6.275	6.695	7.005	7.250	7.454	7.629	7.781	7.916	8.037	8.148	8.248	8.342	8.427	8.508	8.583	8.654	8.720	8.784
18	5.546	6.196	6.604	6.905	7.143	7.341	7.510	7.657	7.788	7.906	8.012	8.110	8.199	8.283	8.361	8.434	8.502	8.567	8.628
19	5.492	6.127	6.525	6.817	7.049	7.242	7.405	7.549	7.676	7.790	7.893	7.988	8.075	8.156	8.232	8.303	8.369	8.432	8.491
20	5.444	6.065	6.454	6.740	6.966	7.154	7.313	7.453	7.577	7.688	7.778	7.880	7.966	8.044	8.118	8.186	8.251	8.312	8.370
24	5.297	5.877	6.238	6.503	6.712	6.884	7.031	7.159	7.272	7.374	7.467	7.551	7.629	7.701	7.768	7.831	7.890	7.946	7.999
30	5.156	5.698	6.033	6.278	6.470	6.628	6.763	6.880	6.984	7.077	7.162	7.239	7.310	7.375	7.437	7.494	7.548	7.599	7.647
40	5.022	5.528	5.838	6.063	6.240	6.386	6.509	6.616	6.711	6.796	6.872	6.942	7.007	7.067	7.122	7.174	7.223	7.269	7.312
60	4.894	5.365	5.653	5.860	6.022	6.155	6.268	6.366	6.451	6.528	6.598	6.661	6.720	6.774	6.824	6.871	6.914	6.956	6.995
120	4.771	5.211	5.476	5.667	5.815	5.937	6.039	6.128	6.206	6.276	6.339	6.396	6.448	6.496	6.542	6.583	6.623	6.600	6.695
∞	4.654	5.063	5.309	5.484	5.619	5.730	5.823	5.903	5.973	6.036	6.092	6.144	6.191	6.234	6.274	6.312	6.347	6.380	6.411

Author Index

Subject Index

Springer Texts in Statistics *(continued from page ii)*